Array Beamforming Enabled Wireless Communications

This book investigates the most advanced theories and methodologies of array beamforming, with a focus on antenna array enabled wireless communication technology.

Combining with the current development needs and trends of wireless communication technology around the world, the authors explore the potentials and challenges of large-scale antenna array beamforming technology in next-generation mobile communication and some important emerging application scenarios. The book first introduces the basic structure of antenna array hierarchical codebook and channel estimation with high dimensionality, with which the time cost of searching the channel information can be effectively reduced. It then explicates high-efficiency beamforming transmission methods for point-to-point transmission, full-duplex point-to-point transmission, and point-to-multipoint transmission where array beamforming enabled non-orthogonal multiple access (NOMA) technologies for typical two-user systems and general multi-user systems are emphasized. The book also discusses array beamforming enabled unmanned aerial vehicle (UAV) communications and array beamforming enabled space/air/ground communications, with the uniqueness and relative solutions for single UAV systems and multi-UAV networks being analyzed.

This will be a vital reference for researchers, students, and professionals interested in wireless communications, array beamforming, and millimeter-wave communications.

Zhenyu Xiao is a full professor at the Department of Electronic and Information Engineering, Beihang University, China. His main research directions are millimeter-wave communications, array signal processing, unmanned aerial vehicle (UAV) communications, satellite communications, and SAGIN.

Lipeng Zhu is a Research Fellow at the Department of Electrical and Computer Engineering, National University of Singapore. His research interest is millimeter-wave communications, non-orthogonal multiple access, and UAV communications.

Lin Bai is a full professor at the School of Cyber Science and Technology, Beihang University, China. His research interests include multiple-input multiple-output (MIMO), the Internet of Things (IoT), and UAV communications.

Xiang-Gen Xia is the Charles Black Evans Professor at the Department of Electrical and Computer Engineering, University of Delaware, USA. His current research interests include space-time coding, MIMO and OFDM systems, digital signal processing, and SAR and ISAR imaging.

Array Beamforming Enabled Wireless Communications

Zhenyu Xiao
Lipeng Zhu
Lin Bai
Xiang-Gen Xia

CRC Press
Taylor & Francis Group
Boca Raton London New York

CRC Press is an imprint of the
Taylor & Francis Group, an **informa** business

This work was supported in part by the National Natural Science Foundation of China (NSFC) under Grant 61827901 and Grant 62171010 and in part by the Beijing Natural Science Foundation under Grant L212003.

MATLAB® is a trademark of The MathWorks, Inc. and is used with permission. The MathWorks does not warrant the accuracy of the text or exercises in this book. This book's use or discussion of MATLAB® software or related products does not constitute endorsement or sponsorship by The MathWorks of a particular pedagogical approach or particular use of the MATLAB® software.

First edition published 2023
by CRC Press
6000 Broken Sound Parkway NW, Suite 300, Boca Raton, FL 33487-2742

and by CRC Press
4 Park Square, Milton Park, Abingdon, Oxon, OX14 4RN

CRC Press is an imprint of Taylor & Francis Group, LLC

© 2023 Zhenyu Xiao, Lipeng Zhu, Lin Bai, Xiang-Gen Xia

ISBN: 978-1-032-43088-1 (hbk)
ISBN: 978-1-032-43242-7 (pbk)
ISBN: 978-1-003-36636-2 (ebk)

DOI: 10.1201/9781003366362

Typeset in Latin Modern font
by KnowledgeWorks Global Ltd.

Publisher's note: This book has been prepared from camera-ready copy provided by the authors.

Contents

Preface

Antenna arrays have a history of more than 100 years and have been evolving closely with the development of electronic and information technologies, playing an indispensable role in wireless communications. As the communication requirement explosively increases, conventional single-antenna transmission faces challenges to meet the insistent demands of high capacity, huge data rate, long distance, low latency, energy efficiency, and strong robustness. To meet the ever-increasing requirements of the future sixth generation (6G) wireless communications, it is promising to leverage different types of antennas with various beamforming technologies in wireless communication systems, bringing in advantages, such as considerable antenna gains, multiplexing gains, and diversity gains.

In order to pursue broadband communication, the exploitation of high-frequency bands, such as millimeter-wave frequencies with rich spectrum resources, has become a prevailing trend. However, the high-frequency bands also cause more severe propagation losses. Antenna arrays are a powerful option to achieve high directional gains by employing multiple connected antenna elements (AEs) to work cooperatively. By steering the radiation energy only to the desired directions, antenna arrays provide considerable antenna gains to compensate for propagation loss, supporting high-frequency broadband communications. At the same time, the improved signal-to-noise ratio (SNR) at the receiver is also beneficial for supporting long-distance transmission. The antenna arrays will play an important role in wireless communications with millimeter-wave frequency band, since the short wavelength of millimeter-wave signals makes it possible to pack a large number of AEs in a small area. However, large-scale antenna arrays require compact circuit implementation, expensive radio frequency (RF) chains, and high power consumption. The antenna layout, system integration, and power control should be considered in particular.

With large-scale antenna array, beamforming technology can be implemented to meet the requirements of quality of service and compensate for the propagation loss of the signals. With proper beamforming, the beams can be steered into the desired directions, which not only improves the received signal power at the target users but also reduces the interference to undesired users. Compared to the conventional directional millimeter-wave antennas and integrated antennas, antenna arrays have higher beam gains and more flexible beamforming capabilities. According to the hardware structure of an antenna array, beamforming architectures can be roughly divided into three categories, namely digital beamforming, analog beamforming, and hybrid beamforming. Fully digital beamforming is one of the signal processing approaches in baseband, where each antenna is driven by an independent RF chain, and multiple data streams can be transmitted simultaneously. However, the digital beamforming architecture results in unaffordable hardware cost and energy consumption in the millimeter-wave-band with large antenna arrays. In contrast, analog beamforming, where the antennas share only one RF chain, is an energy-efficient alternative. Nevertheless, one RF chain can support only one data stream in general, which limits the spectrum efficiency. In consideration of the compromise between energy efficiency and spectrum efficiency, hybrid beamforming was proposed and preferred. With a small number of RF chains connected to a large number of antennas, beam gain and interference management can be achieved simultaneously.

This book aims at antenna array enabled wireless communication technologies and deals with the most advanced theories and methods of array beamforming. The potentials and challenges of large-scale antenna array beamforming technologies in next-generation mobile communication and some important emerging application scenarios are discussed, closely combining with current development needs and trends of wireless communication technologies. A series of unique insights and innovative approaches are proposed and the possible further research directions are pointed out, which can provide important reference for researchers and communication engineers. The materials and results presented in this book are mainly from the authors' research groups. Since array beamforming technology is broad, there are still many topics that are not covered in this book. Also, since this book is about beamforming at transmitter (Tx), space-time coding is not specially considered. Moreover, this book focuses on emerging analog beamforming and hybrid beamforming technologies, while the conventional fully-digital beamforming strategies are briefly introduced in Chapter 1 as a preliminary. In addition, we only focus on narrow-band beamforming, while wide-band beamforming is not included in this book.

In this book, we start from the fundamentals of antenna array, introducing point-to-point beamforming and training technologies, and then present array beamforming enabled emerging technologies. The diagram of the relationship among all the chapters of the book is shown in Fig. 1.

Chapter 1 provides the fundamentals on wireless communications and antenna array technologies. Specifically, we start from the channel model and the signal model for single-antenna wireless communication systems, where the impacts of large-scale fading and small-scale fading on the channel gain are discussed. The narrowband fading models are introduced, including the Rayleigh fading in a non-line-of-sight environment, the Rician fading in a line-of-sight environment, and the Nakagami-distribution-based fading model. Then, the channel model and signal model are extended to multiple-input multiple-output (MIMO) systems, where the diversity gain and spatial multiplexing gain can be obtained. Whereafter, the capacity of the MIMO channel is discussed from the perspective of information theory.

In wireless communication systems, it is critical to capture or estimate channel state information (CSI) in real time. Accurate CSI estimation can effectively enhance the performance. Due to the large number of antennas equipped at the transceivers, the entry-wise channel estimation usually requires high pilot overhead and computational complexity. In this regard, Chapter 2 introduces the channel estimation methods based on codebook and the corresponding beam search methods. First, we propose four different codebooks, namely deactivation codebook, joint sub-array and deactivation codebook, enhanced codebook, and codebook for hybrid structures. Then, four different search methods to acquire CSI are introduced. Exhaustive search that searches the whole Tx/receiver (Rx) angle to find the strongest angle pairs may suffer high time cost when the number of antennas is large. Hierarchical search based on hierarchical codebook can effectively reduce time complexity. Especially, we discuss the design method of closed-form hierarchical codebook. Besides, compressed sensing based channel estimation can be used to increase the accuracy. Finally, a joint beam search and compressed sensing method is presented which combines the benefits of both hierarchical codebook and compressed sensing.

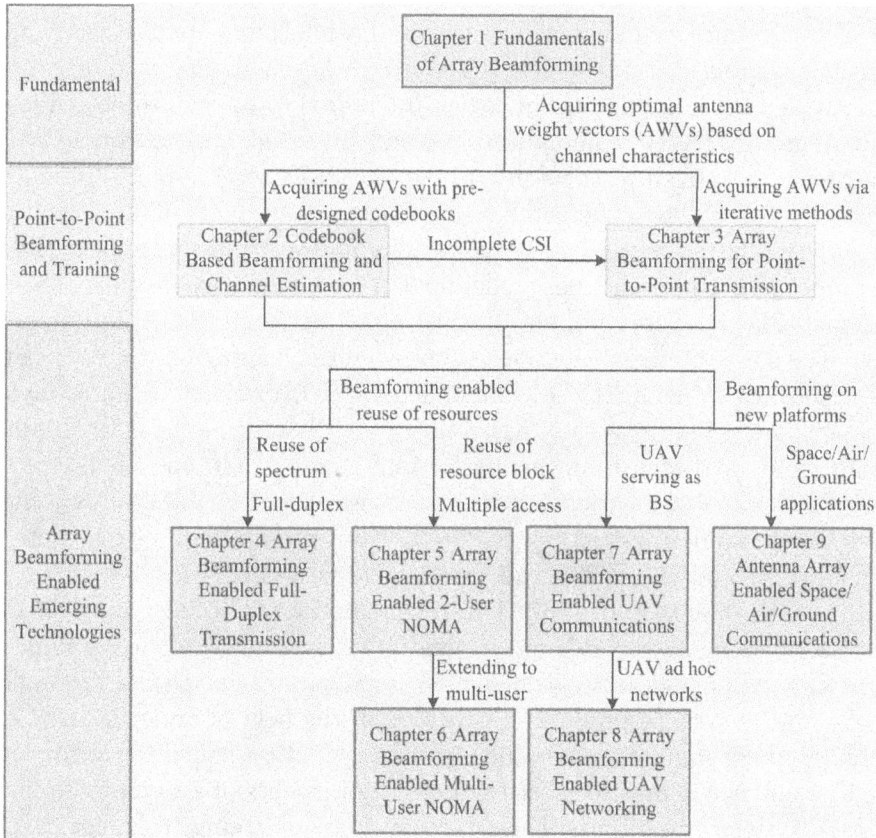

Figure 1 The diagram of this book.

With known CSI, antenna array beamforming can be applied in wireless communication scenarios to enhance the system performance, improve communication capacity, and increase connectivity. In Chapter 3, we introduce the methods of array beamforming for point-to-point transmission. A sub-optimal beamforming scheme is proposed to obtain the optimal antenna weight vectors via iterative eigenvalue decomposition (EVD) and a multipath grouping scheme is proposed to reduce overhead and increase system reliability. Then, a joint beamforming training scheme based on steering vectors is proposed and compared to a beamforming training scheme based on singular vectors. Chapter 4 introduces array-enabled full-duplex (FD) point-to-point transmission, where beamforming approaches for FD communication without constant-modulus (CM) constraint and considering CM constraint are discussed, respectively. Besides, beamforming for FD relay is proposed. On this basis, in Chapters 5 and 6, we further discuss non-orthogonal multiple access (NOMA) that allows multiple users to be served in the same time-frequency-code domain, and distinguish them in the power domain. Array beamforming enabled NOMA technology is emphasized, including the typical two-user system and the general multi-user system. The coupling mechanism of beamforming and power allocation is revealed.

We also introduce and analyze other array beamforming enabled emerging technologies. As unmanned aerial vehicle (UAV) technologies develop fast, large scale antenna array and beamforming technologies are desired to be applied on UAVs to help with air-to-ground (A2G) communications and flying ad hoc networks (FANETs). In Chapter 7, we introduce the scenarios where UAVs serve ground users as aerial base stations (BSs). The channel characteristics of A2G communications are quite different from that in ground communications and are more complex. Hence, we provide an overview of a channel model for UAV communications first. Then, three dimensional (3D) beamforming and flexible beam coverage of UAVs equipped with antenna arrays are discussed. Besides, we provide the solutions for UAV positioning and beamforming to enhance the communication performance for both single UAV systems and multi-UAV systems, respectively. In Chapter 8, the crucial issues and potentials for array beamforming enabled FANET are introduced. We discuss the potential technologies and solutions for the key issues arising for UAV ad-hoc networks, including the network architecture, link establishment and maintenance, integration of the sub-6 GHz and millimeter-wave bands, and network security.

With the rapid development of wireless communication technologies, the demand for all-time, all-domain, and full-space network services has exploded, and new communication requirements have been put forward on various space/air/ground platforms. Chapter 9 aims to provide an overview of the field of antenna array enabled space/air/ground communications and networking. Various emerging technologies facilitated by antenna arrays to meet the new requirements of space/air/ground communication systems are discussed. Enabled by these emerging technologies, the distinct characteristics, challenges, and solutions for space communications, airborne communications, and ground communications are reviewed.

In summary, this book studies in depth the application of array beamforming in channel estimation, transmission, multiple access, UAV communications, and space/air/ground communications. We hope that the book can provide some useful enlightenment and references for future wireless communication research.

This monograph was supported in part by the National Natural Science Foundation of China (NSFC) under grant numbers 61827901 and 62171010, and the Beijing Natural Science Foundation under grant number L212003. The content of this monograph is a compilation of the authors' research achievement over the years, which has been partially published in academic papers. The authors of this monograph would like to thank the other co-authors of these publications for their contributions. The authors would also like to thank Songqi Cao, Guangsheng Li, and Ke Liu for their editorial works on this monograph.

Zhenyu Xiao, Lipeng Zhu, Lin Bai, and Xiang-Gen Xia
Beijing, April, 2022

List of Symbols

General

a	scalar a
$\mathbf{a}$	vector $\mathbf{a}$
$\mathbf{A}$	matrix $\mathbf{A}$
$\mathcal{A}$	set $\mathcal{A}$
$|a|$	absolute value of scalar a
$|\mathcal{A}|$	cardinality of set $\mathcal{A}$
e	natural constant
π	circular constant
j	imaginary unit
Φ	empty set
$\mathbb{R}$	set of real value
$\mathbb{C}$	set of complex value
$\mathbb{C}^{M \times N}$	an $M \times N$-dimension linear space in complex domain
$\mathcal{CN}\left(\boldsymbol{\Gamma}, \boldsymbol{\Sigma}\right)$	Gaussian distribution with mean $\boldsymbol{\Gamma}$ and covariance matrix $\boldsymbol{\Sigma}$
$\mathbb{E}\left(\cdot\right)$	expected value of a random variable
$\Re\left(\cdot\right)$	real part of a complex number/vector
$\Im\left(\cdot\right)$	imaginary part of a complex number/vector
$\angle\left(\cdot\right)$	phase of a complex number/vector
$\measuredangle(\mathbf{x})$	angle vector of $\mathbf{x}$ in radian
$d\left(\cdot\right)$	differential of a function
$\partial\left(\cdot\right)$	partial differential of a function
$\circ$	Hadamard product
$\otimes$	Kronecker product
$\lceil\cdot\rceil$	ceiling integer operation
$\langle\mathbf{x}, \mathbf{y}\rangle$	inner product in Hilbert space
$\delta\left[k\right]$	discrete impulse response function

Vector/Matrix-related symbols

$\left(\cdot\right)^{\mathrm{T}}$	transpose
$\left(\cdot\right)^{*}$	conjugate
$\left(\cdot\right)^{\mathrm{H}}$	conjugate transpose
$\left(\cdot\right)^{\dagger}$	pseudo inverse
$\left(\cdot\right)_{\mathrm{R}}^{\dagger}$	right pseudo inverse

$\mathrm{vec}\,(\cdot)$	matrix vec operator
$\|\mathbf{a}\|_0$	0-norm of vector $\mathbf{a}$
$\|\mathbf{a}\|_2$	2-norm of vector $\mathbf{a}$
$\|\mathbf{a}\|_\infty$	infinite norm of vector $\mathbf{a}$
$\|\mathbf{A}\|_{\mathrm{F}}$	Frobenius of matrix $\mathbf{A}$
$\mathrm{LpSingVect}(\mathbf{A})$	left principal singular vectors of $\mathbf{A}$
$\mathrm{RpSingVect}(\mathbf{A})$	right principal singular vectors of $\mathbf{A}$
$\mathrm{pEigVect}(\mathbf{A})$	principal eigenvectors of $\mathbf{A}$
$\mathbf{I}_N$	identity matrix
$[\mathbf{a}]_i$	the i-th entry of vector $\mathbf{a}$
$[\mathbf{A}]_{i,:}$	the i-th row of matrix $\mathbf{A}$
$[\mathbf{A}]_{:,j}$	the j-th column of matrix $\mathbf{A}$
$[\mathbf{A}]_{i,j}$	the entry in the i-th row and j-th column of matrix $\mathbf{A}$

List of Abbreviations

A2A	air-to-air
A2G	air-to-ground
A2S	air-to-satellite
ABC	artificial bee colony
ACO	ant colony optimization
ACS	adaptive compressed sensing
ADC	analog-to-digital converter
AE	antenna element
AIS	alternating interference suppression
AO	alternating optimization
AoA	angle of arrival
AoD	angle of departure
AP	access point
ASLN	area secure link number
ASR	achievable sum rate
AWGN	additive white Gaussian noise
AWV	antenna weight vector
AZF	approximate zero-forcing
B5G	beyond 5G
BC-PSO	boundary-compresses particle swarm optimization
BER	bit error rate
BLER	block-error rate
BMW-MS	beam widening with multi-RF-chain subarray
BS	base station
CAZAC	constant-amplitude-zero-autocorrelation
CDF	cumulative distribution function
CDMA	code division multiple access
CGA	conjugate gradient algorithm
CM	constant-modulus
CMA	constant-modulus algorithm
CoMP	coordinated multipoint
CS	compressed sensing
CSI	channel state information
CSIR	channel state information at the receiver
CSIT	channel state information at the transmitter
CTS	clear-to-send

D2D	device-to-device
DAC	digital-to-analog converter
DEACT	beam pattern of the deactivation
DFS	Doppler frequency shift
DFT	discrete Fourier transform
DMI	directed matrix inversion
DN	destination node
DoF	degree of freedom
DPS	double phase shifter
EE	energy-efficiency
EVD	eigenvalue decomposition
EVM	error vector magnitude
FANET	flying ad-hoc networks
FD	full-duplex
FDD	frequency division duplex
FDMIMO	fully-digital MIMO
FM	frequency modulation
FRAB	finite resolution analog beamforming
FSK	frequency shift keying
GBCM	geometry-based stochastic channel model
GDP	generalized detection probability
GEO	Geostationary Earth Orbit
GNSS	global navigation satellite system
GPS	global positioning system
HAP	high-altitude platform
HD	half-duplex
IA	interference alignment
IEVD	iterative eigenvalue decomposition
IoT	Internet of Things
IQ	in phase-quadrature
IRS	intelligent reflecting surface
ISI	inter-symbol interference
ITU	International Telecommunication Union
JAR	joint achievable rate
JTR-BF	joint Tx/Rx beamforming
KD	Kronecker decomposition
KKT	Karush-Kuhn-Tucker
LAP	low-altitude platform
LB-MMSE	lower bound based MMSE
LCS	low-complexity search
LEO	Low Earth Orbit
LMMSE	linear minimum mean square error
LMS	least mean square
LNA	low noise amplifier
LoS	line of sight

LS	least square
LS-CMA	least square CMA
LS-MIMO	large-scale MIMO
MAB	multi-armed bandit
MAC	media access control
MBA	multiple beam array
MDR	multipath decomposition and recovery
MEO	Medium Earth Orbit
MF	matched filter
MIMO	multiple-input multiple-output
MISO	multiple-input single-output
ML	maximal likelihood
MMSE	minimum mean square error
MOSINR	maximum output signal-to-interference plus noise ratio
MPC	multipath component
MPDR	minimum power distortionless response
MPG	multipath-grouping
MRC	maximum-ratio combining
MRT	maximum-ratio transmission
MS	mobile station
MSE	mean square error
MTP	maximal transmit power
MUD	multi-user detection
MUI	multi-user interference
MVDR	minimum variance distortionless response
NLoS	non-line of sight
NOMA	non-orthogonal multiple access
NSEE	network-wide secrecy energy efficiency
NST	network-wide secrecy throughput
OFDMA	orthogonal frequency-division multiple access
OMA	orthogonal multiple access
OMP	orthogonal matching pursuit
OSINR	output signal-to-interference plus noise ratio
OSTBC	orthogonal spatial-time encoding method
PA	power amplifier
PAPC	per-antenna power constraint
PEP	pairwise-error probability
PIC	parallel interference cancellation
PM	phase modulation
PS	phase shifter
PSO	particle swarm optimization
QoS	quality of service
QPSK	quadrature phase shift keying
R2D	relay-to-destination node
RB	resource block

RF	radio frequency
RLS	recursive least squares
RSMA	rate-splitting multiple access
RTS	request-to-send
RWV	radio frequency weight vector
Rx	receiver
S2D	source node-to-destination node
S2R	source node-to-relay
SAGIN	space-air-ground integrated network
SDMA	spatial division multiple access
SDN	software defined networking
SGD	stochastic gradient descent
SGV	a singular vector based training scheme
SI	self-interference
SIC	successive interference cancellation
SIMO	single-input multiple-output
SINR	signal-to-interference-plus-noise ratio
SISO	single-input single-output
SLNR	signal-to-leakage-plus-noise ratio
SMI	sample matrix inversion
SN	source node
SNR	signal-to-noise ratio
SPARSE	sparse reconstruction approach
SPDT	single-pole double-throw
SPS	single phase shifter
STV	steering vector based joint beamforming training scheme
SVD	singular value decomposition
SWAP	size, weight, and power
TDD	time division duplex
TDL	tapped delay line
TDMA	time division multiple access
Tx	transmitter
UAV	unmanned aerial vehicle
UCA	uniform circular array
UE	user equipment
ULA	uniform linear array
URA	uniform rectangular array
URLLC	ultra-reliable and low-latency communication
V2I	vehicle-to-infrastructure
V2N	vehicle-to-network
V2P	vehicle-to-pedestrain
V2V	vehicle-to-vehicle
V2X	vehicle-to-everything
WLAN	Wireless Local Area Network
WPAN	Wireless Personal Area Networks

WTS	wait-to-send
ZF	zero-forcing
ZMCSCG	zero-mean circularly-symmetric complex Gaussian
ZMSW	zero-mean spatially white
ZP	zero-padded

CHAPTER 1

Fundamentals of Array Beamforming

1.1 INTRODUCTION

Nowadays, communication technology is rapidly evolving. Compared with the past, the current communication technology has significantly improved in terms of both reliability and effectiveness. Nevertheless, the cornerstone of information theory does not change. Actually, all the researchers and engineers in this field have always been devoting to one target, approaching the Shannon bound, i.e.,

$$C = B \log_2(1 + \frac{S}{N}), \tag{1.1}$$

where B is the channel bandwidth and $\frac{S}{N}$ is the signal-to-noise ratio (SNR). This equation tells us two possible directions for improving the capacity of a communication system, i.e., increasing the bandwidth or increasing the SNR. As we all know, spectrum has always been a scarce resource. Hence, increasing the SNR becomes one of the most important topics for modern communication systems. On the one hand, the transmit power is usually limited and cannot be discretionarily increased for most systems. On the other hand, the noise power depends on the environment and is difficult to be reduced in daily used systems. Both directions will face a bottleneck in increasing the capacity of real-world communication systems.

With the birth of multiple-input multiple-output (MIMO) technologies, a new dimension for increasing the channel capacity is found, i.e., the spatial domain. In MIMO systems, by employing diversity and multiplexing gains, multiple channels are created in the spatial domain, and different streams can be transmitted simultaneously. Hence, the data rate is able to be increased manifold. Moreover, except for diversity and multiplexing, beamforming can also be achieved by MIMO systems. Beamforming refers to the idea of treating the antenna array formed by multiple antennas as an antenna in a MIMO system. Hence, the radiation of multiple antenna elements (AEs) interferes, and finally the superimposed radiation field forms a directional beam. In this regard, beamforming is able to make the energy more concentrated to specific directions. Thus, the power of received signals can be increased by introducing array beamforming.

DOI: 10.1201/9781003366362-1

Large-scale antenna arrays, including phased arrays and digital arrays, are with great potential to improve communication performance[1]. An antenna array consists of a set of antennas which work together as a single antenna[1]. They can form a desired radiation pattern by designing the type, number, spacing, and geometries of the elements[2]. Benefiting from the small wavelength of high-frequency signals, a large number of antennas can be equipped in a small area to realize high array gains[3,4,5]. Besides, by flexible beamforming, an adaptive antenna array with a proper signal processing method is able to improve spectrum utilization and solve many problems, including multipath interference, channel fading, etc. In fact, the concepts of antenna array technologies and array signal processing have a long developing history.

Back in 1901, Marconi tried to obtain diversity gain by employing multiple antennas at the transmitting side. He believed that the use of multi-antenna technology can combat the fading of the channel. According to[6], two way communications were achieved then. The first antenna array used two poles, where the two poles were a yard apart at the top and converged at the bottom, forming a planar fan-shaped aerial.

Later, array signal processing has been applied in the military field for a long time, extracting the angles of radar targets or formulating narrow beams for jamming/anti-jamming. For instance, the concept of phased arrays has been around since the 1930s, and actual systems were in place in the 1950s. According to[7], in 1971, the first experimental X-band phased array antenna consisting of 80 linear AEs was implemented, in which 64 AEs were connected to 64 active X-band modules. In 1973, the experimental S-band phased array radar was implemented. The first successful results in continuous detection of aircraft in pulse compression were obtained thanks to the radar. In 1976, the X-band two dimensional phased array radar consisting of 208 AEs was manufactured. Since 1985, conformal array which is able to conform with an arbitrary curved surface has been studied by the technical research and development institute (TRDI). For instance, the X-band hemi-spherical conformal antenna was manufactured, and tests and evaluation of its antenna were conducted from 1989 to 1990. However, phased arrays were mainly used in radar systems.

At the same time, MIMO technology has also developed rapidly. From 1960s to 1970s, MIMO technology was mainly used for combating crosstalk in single-user scenarios. Inspired by the Nyquist's problem aiming to combat the inter-symbol interference (ISI), Shnidman made the earliest contribution to MIMO detection, considering the equalization problem of a bandwidth-limited pulse modulation problem[8]. From 1980s to 1990s, MIMO systems were used for multi-user detection (MUD) during the prevalence of code division multiple access (CDMA) systems.

Array signal processing was combined with wireless communications in the 1990s for symbol detection in small-scale multi-antenna systems, e.g., around the mid-to-late 1990's, angle-based spatial division multiple access (SDMA) was deployed in Japan and Australia[9]. However, there are too many scatters and reflectors, resulting in too many multipath components (MPCs) at Rx whose angles of arrival (AoA) cannot be distinguished. Besides, the number of antennas is small, which causes that the MIMO system cannot support accurate AoA estimation[10]. Meanwhile, AT&T Bell Labs researchers completed the groundbreaking of MIMO technology for wireless communication systems[11,12]. In 1994, Paulraj et al. proposed the concept of MIMO system, which simultaneously used multiple antennas at transmitter (Tx) and receiver (Rx) of wireless communications to increase the capacity of the wireless channel. In 1995, Teladar analyzed the channel capacity of the MIMO system in a fading channel, and it was also studied by Foschini and Gans[13]. In 1996, Roy et al. proposed air division multiple access systems that could use directional antennas at the same frequency to serve users in different directions. In 1998, Alamouti proposed a simple orthogonal spatial-time encoding method (OSTBC) for wireless communications[14], and Tarokh et al. and Guey et al. discussed the spatial-time coding for MIMO systems[15,16]. In 2005, Goldsmith studied the MIMO broadcast channel capacity problem and proposed the dirty-paper coding technique[17]. The above works have received great attention from both academia and industry all over the world, which leads to the rapid development of MIMO technology.

Nowadays, massive MIMO can be a good implementation of array signal processing technology. On one hand, with a large number of antennas, narrower beams can be formed, which are able to eliminate interference between users effectively. Moreover, narrow beams can also compensate path loss with high beam gains. On the other hand, in frequency domain, the tendency is to the high frequency band because of the spectrum shortage in conventional microwave communications, e.g., millimeter-wave spectrum. Meanwhile, massive MIMO has also led to more new directions for future research, e.g., channel estimation, hybrid beamforming, interference control, etc. As aforesaid, wireless communications have entered a period of vigorous development. Among them, MIMO technology and array antenna technology are the cornerstones of the rapid development of wireless communication technology. In this chapter, we give a brief introduction on the antenna array and the signal processing of antenna array.

This chapter provides some fundamentals for wireless communications and antenna array systems. In the second section, the channel model and MIMO system are introduced. Specifically, we introduce the fading characteristics, including large-scale fading and small-scale fading. To combat fading, we present diversity, single-input multiple-output (SIMO) and multiple-input single-output (MISO) systems. Moreover, the singular value decomposition (SVD) and channel capacity are derived for MIMO system. In the third section, we give an introduction on antenna array models. Based on the plane wavefront assumption, we present the model of two-element antenna arrays, including the uniform linear arrays (ULA), and planar array, for transmitting/receiving signals. The array channel model and the Saleh-Valenzuela model are derived. Then we introduce antenna array structures, i.e., phased arrays, digital arrays, hybrid antenna arrays, and irregular antenna arrays. Finally, in the fourth section, beamforming methods including conventional beamforming methods, statistic beamforming methods, and adaptive beamforming methods are provided to give a brief introduction of classical beamforming methods.

1.2 MIMO SYSTEMS

A MIMO system refers to using multiple receive and transmit antennas at Tx and Rx. When multiple antennas are used at both Tx and Rx, not only spectral utilization, diversity gain, and array gain, but also the throughput of the system can be improved benefiting from the increasing of the spatial multiplexing capability of the MIMO channel.

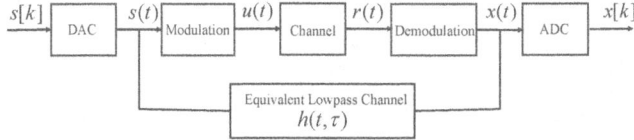

Figure 1.1 A wireless communication system.

When the spacing between the transceiver antennas is large, the channels between the different transceiver antennas can be regarded as independent. Consider the channel between the m-th transmit antenna and the n-th receive antenna, which form a single-input single-output (SISO) system. As shown in Fig. 1.1, the signal $s(t) = s_I(t) + js_Q(t)$ is the equivalent lowpass signal on the m-th transmit antenna, where $s_I(t)$ represents in-phase component and $s_Q(t)$ represents quadrature component. $u(t)$ represents a bandpass signal on the m-th transmit antenna at the carrier frequency f_c in the following form:

$$u(t) = \Re\left\{s(t)e^{j2\pi f_c t}\right\} = s_I(t)\cos(2\pi f_c t) - s_Q(t)\sin(2\pi f_c t). \qquad (1.2)$$

In order to analyze the wireless communication system with the equivalent lowpass response of the transmitted signal and the received signal, we define the lowpass channel impulse response between the m-th transmit antenna and the n-th receive antenna $h_{mn}(t, \tau)$ as follows, while making the analysis no longer depend on the carrier frequency f_c.

$$x(t) = h_{mn}(t, \tau) * s(t) + n(t), \qquad (1.3)$$

where $x(t) = \sum_{i=0}^{L(t)} \lambda_i(t)e^{j\phi_i}s(t - \tau_i(t))$, $L(t)$ represents the number of MPCs, $\lambda_i(t)$ represents the complex channel coefficient, ϕ_i depends on the delay and Doppler effect, and $\tau_i(t)$ represents the delay of the i-th path. Hence, the lowpass channel impulse response between the m-th transmit antenna and the n-th receive antenna can be expressed as

$$h_{mn}(t, \tau) = \sum_{i=0}^{L(t)} \lambda_i(t)\delta(\tau - \tau_i(t)). \qquad (1.4)$$

In the next subsection, we will take the above $h_{mn}(t, \tau)$ as an example and explain the fading of the channel. Note that in a MIMO system, the channel can be expressed as a matrix, and $h_{mn}(t, \tau)$ is the n-th row and m-th column element of the channel matrix. For ease of writing, the following $h(t, \tau)$ represents any element of the channel matrix $\mathbf{H}$.

1.2.1 Fading and Diversity

In wireless channels, the propagation of electromagnetic waves will gradually decay with the distance traveled, and it will also be affected by interference, obstacles, and even the atmosphere. At the same time, these factors can change over time because the transceivers may be mobile and the channel is dynamic. In the following, we will introduce the propagation characteristics of $h(t, \tau)$, including large-scale propagation effects and small-scale propagation effects. Diversity that can compensate fading is analyzed.

1.2.1.1 Large-Scale Propagation Effects

In this part, we will introduce the large-scale propagation effects which occur over relatively large distances. Large-scale propagation effects include path loss and shadowing. Path loss is caused by the diffusion and propagation characteristics of the channel during the propagation of electromagnetic waves. While shadowing is caused by the obstacles between transceivers, where the obstacles attenuate electromagnetic waves by reflection, scattering, diffraction, and diffraction.

1) Path Loss

As afore mentioned, path loss reflects the attenuation of signal power in relation to the propagation distance. Linear path loss can be defined as

$$P_L = \frac{P_t}{P_r}, \tag{1.5}$$

where P_t is the transmit power and P_r is the receive power. Then the dB value of the linear path loss can be defined as

$$P_L \mathrm{dB} = 10 \log \frac{P_t}{P_r} \mathrm{dB}. \tag{1.6}$$

Since the channel can only attenuate the signal, we define the dB path gain as the negative of the dB path loss, i.e., $P_G = -P_L$.

As for path loss, a simple example is free-space path loss. Consider a signal transmitted through free space to Rx and assume there are no obstacles between Tx and Rx. Then, the signal received at the receiving antenna can be written as

$$r(t) = \Re \left\{ \frac{\lambda \sqrt{G_l} e^{jkr}}{4\pi r} s(t) e^{j2\pi f_c t} \right\}, \tag{1.7}$$

where $r(t)$ is the signal received by the receive antenna, r is the distance between Tx and Rx, λ and $k = 2\pi/\lambda$ are the wavelength and wave number, respectively, and $\sqrt{G_l}$ is the product of the transmit and receive antenna field radiation patterns in the line of sight (LoS) direction representing the antenna gain. Then, the ratio of receive to transmit power is

$$\frac{P_r}{P_t} = \left[\frac{\sqrt{G_l}\lambda}{4\pi r} \right]^2 . \tag{1.8}$$

As mentioned in (1.8), the linear path loss falls off inversely proportional to the square of the distance r. The receive power can be expressed in dBm as

$$P_r\text{dBm} = P_t\text{dBm} + 10\log(G_l) + 20\log(\lambda) - 20\log(4\pi) - 20\log(r). \tag{1.9}$$

Then, we can define the free-space path loss as

$$P_L\text{dB} = 10\log\frac{P_t}{P_r} = -10\log\frac{G_l\lambda^2}{(4\pi r)^2}. \tag{1.10}$$

2) Shadowing

The signal will encounter obstructions during the propagation in a wireless channel, which will cause random attenuation of the signal. Meanwhile, changes in reflector surfaces and scatters can also cause signal attenuation. There are many factors involved, including location, size, and dielectric properties of obstacles, and these factors are usually unknown. Thus, a statistical model can be built to describe shadowing. The most common model is log-normal shadowing. Moreover, this model has been confirmed by measured data to accurately describe the change in receive power in indoor and outdoor environments[18, 19].

In the log-normal shadowing model, the ratio of transmit to receive power, i.e., $\psi = P_t/P_r$ is assumed random with following log-normal distribution

$$p(\psi) = \frac{\xi}{\sqrt{2\pi}\sigma_{\psi_{\text{dB}}}\psi} \exp\left[-\frac{(10\log\psi - \mu_{\psi_{\text{dB}}})^2}{2\sigma_{\psi_{\text{dB}}}^2} \right], \psi > 0, \tag{1.11}$$

where $\xi = 10/\ln 10$, $\mu_{\psi_{\text{dB}}}$ is the mean of $\psi_{\text{dB}} = 10\log\psi$ in dB, and $\sigma_{\psi_{\text{dB}}}$ is the standard deviation of ψ_{dB} in dB. As for the mean value $\mu_{\psi_{\text{dB}}}$, in empirical measurements, since the measurement of empirical path loss already includes the average for shadow fading, the mean value $\mu_{\psi_{\text{dB}}}$ is equal to the empirical path loss. While in the analytical model, the mean value should take into account path loss as well as shadowing caused by obstacles.

From (1.11), the mean value of linear path loss can be obtained as

$$\mu_\psi = \mathbb{E}[\psi] = \exp\left[\frac{\mu_{\psi_{\text{dB}}}}{\xi} + \frac{\sigma_{\psi_{\text{dB}}}^2}{2\xi^2} \right]. \tag{1.12}$$

Convert the linear path loss to log mean as

$$10\log\mu_\psi = \mu_{\psi_{\text{dB}}} + \frac{\sigma_{\psi_{\text{dB}}}^2}{2\xi^2}. \tag{1.13}$$

The performance in log-normal shadowing is typically parameterized by the log mean $\mu_{\psi_{\text{dB}}}$, defined as the average dB path loss in dB[20]. Hence, ψ in dB follows a normal distribution with mean value $\mu_{\psi_{\text{dB}}}$ and variance $\sigma_{\psi_{\text{dB}}}$.

We can overlay a path loss model with a shadow fading model to describe both path loss and shadow fading. In the combined model, average path loss in dB can be described by path loss model. The normal distribution is used to reflect the random shadowing caused by obstacles in the propagation path. According to this model, the ratio of receive to transmit power in dB can be represented as

$$\frac{P_r}{P_t}\text{dB} = 10\log K - 10\gamma\log\frac{d}{d_0} - \psi_{\text{dB}}, \qquad (1.14)$$

where ψ_{dB} is a Gaussian-distributed random variable with mean zero and variance $\sigma_{\psi_{\text{dB}}}^2$, γ denotes the path loss exponent, d denotes the distance between Tx and Rx, d_0 denotes the reference distance in the far field of the antenna and K is a constant path loss factor.

1.2.1.2 Small-Scale Propagation Effects

In the process of electromagnetic wave propagation, in addition to the LoS path, other paths will be generated because of reflection, scattering, and diffraction. These MPCs can interfere with each other, affecting the received signal. Usually this change occurs at wavelength order of magnitude, which is a small distance. Hence, it is called the small-scale propagation effect. In fact, there are many MPCs in real-world communication scenarios, and deterministic models cannot reflect the true characteristics of the channel. Therefore, a multipath channel must be described using statistical methods.

When Tx transmits a single pulse, Rx receives a pulse train due to the multiple resolvable MPCs contained in the channel. Define the time delay between the first arrived signal component and the last arrived signal component as the delay spread. The delay spread can lead to significant distortion of the received signal. Another feature of multipath channels that affects the received signal is time variability. Since Tx or Rx will move, the different MPCs in the channel are constantly changing, and the signal received by the receiver will also change.

Assume that the transmit signal is $u(t) = \Re\{s(t)\}\cos(2\pi f_c t) - \Im\{s(t)\}\sin(2\pi f_c t)$ whose bandwidth is B_u at Tx. Then, the signal received by the receive antenna can be represented as

$$r(t) = \Re\left\{\sum_{l=0}^{L(t)-1} \lambda_l(t)u\left(t - \tau_l(t)\right)e^{j\left(2\pi f_c(t-\tau_l(t))+\phi_{D_l}\right)}\right\}, \qquad (1.15)$$

where $l = 0$ corresponds to the LoS path, $L(t)$ is the number of resolvable MPCs, and $\lambda_l(t)$, ϕ_{D_l} and $\tau_l(t)$ represent amplitude, Doppler phase shift, and corresponding delay, respectively. The amplitude $\lambda_l(t)$ is related to the path loss and shadowing of the path. The phase shift caused by the delay $\tau_l(t)$ is $e^{-j2\pi f_c\tau_l(t)}$. Doppler frequency shift of the n-th path can be represented as $f_{D_l}(t) = v\cos\theta_l(t)/\lambda$, where $\theta_l(t)$ is the angle between the direction of arrival at Rx and the direction of Rx movement. $\phi_{D_l} = \int_t 2\pi f_{D_l}(t)\mathrm{d}t$ represents the Doppler phase shift caused by the Doppler frequency shift. The received signal is the convolution of the transmitted signal and the channel impulse response. Thus, the channel impulse response can be represented as

$$h(\tau, t) = \sum_{l=0}^{L(t)-1} \lambda_l(t) e^{-j\phi_l(t)} \delta\left(\tau - \tau_l(t)\right). \tag{1.16}$$

Note that there are two time parameters t, τ in (1.16), where t is the moment when Rx observes the impulse response, and $(t - \tau)$ is the moment when Tx sends impulse pulses[20]. Noted that when $f_c \tau_l \gg 1$, a small delay change on the l-th path will cause a large change in phase. Rapid changes in phase on each path will result in violent interference phenomena, resulting in drastic changes in the received signal. This phenomenon is called fading.

1) Narrowband Fading Models

In narrowband fading model, we suppose that the delay spread T_m is far less than the inverse signal bandwidth B, i.e., $T_m \ll B$, which implies that the delay associated with the l-th MPC satisfies $\tau_l \leq T_m, \forall l$. Thus, for all the paths, $s(t - \tau_l) \approx s(t)$. Hence, the channel impulse response (1.16) can be simplified as

$$h(t) = \sum_{l=0}^{L(t)-1} \lambda_l(t) e^{-j\phi_l(t)}, \tag{1.17}$$

which implies that the channel impulse response is only related to the complex coefficients, but not related to the transmit signal $s(t)$. Thus, (1.3) can be simplified as

$$x(t) = h(t)s(t) + n(t). \tag{1.18}$$

Note that in MIMO system, the narrowband assumption holds. In (1.2), the received in-phase and orthogonal components $r_I(t)$ and $r_Q(t)$ can be represented as

$$\begin{aligned} r_I(t) &= \sum_{l=0}^{L(t)-1} \alpha_l(t) \cos \varphi_l(t), \\ r_Q(t) &= \sum_{l=0}^{L(t)-1} \alpha_l(t) \sin \varphi_l(t). \end{aligned} \tag{1.19}$$

If $L(t)$ is large, $r_I(t)$ and $r_Q(t)$ approximate the joint Gaussian stochastic process, benefiting from the Central Limit Theorem and the static and periodic properties of $\alpha_l(t)$ and $\varphi_l(t)$. Specially, when $\varphi_n(t)$ is uniformly distributed, $r_I(t)$ and $r_Q(t)$ are independent homologous Gaussian random variables.

When there is no LoS component in the signal, assume that the variance of the in-phase component and the quadrature component is σ^2. The envelope of the received signal is

$$z(t) = |r(t)| = \sqrt{r_I^2(t) + r_Q^2(t)}, \tag{1.20}$$

which follows the Rayleigh distribution, and its probability density distribution is

$$p_Z(z) = \frac{2z}{P_r} \exp\left[-z^2/P_r\right] = \frac{z}{\sigma^2} \exp\left[-z^2/\left(2\sigma^2\right)\right], \quad x \geq 0, \qquad (1.21)$$

where $P_r = \sum_n \mathbb{E}[\alpha_n^2] = 2\sigma^2$ is the average received signal power only considering path loss and shadowing. Substitute $z^2(t) = |r(t)|^2$ into (1.21) and the probability density distribution of receive power can be transformed to

$$p_{Z^2}(x) = \frac{1}{P_r}e^{-x/P_r} = \frac{1}{2\sigma^2}e^{-x/\left(2\sigma^2\right)}, \quad x \geq 0. \qquad (1.22)$$

It can be seen that the power of the received signal follows an exponential distribution with the mean value of $2\sigma^2$.

When there is a LoS component in the signal, then $r_I(t)$ and $r_Q(t)$ are not zero mean. In such a case, the received signal is a superposition of the Gaussian component and LoS component. The envelope of the signal follows the following Rician distribution[21]

$$p_Z(z) = \frac{z}{\sigma^2} \exp\left[\frac{-\left(z^2 + s^2\right)}{2\sigma^2}\right] I_0\left(\frac{zs}{\sigma^2}\right), \quad z \geq 0, \qquad (1.23)$$

where $s^2 = \alpha_0^2$ is the power of the LoS component and $2\sigma^2 = \sum_{m,n \neq 0} \mathbb{E}[\alpha_n^2]$ is the average power of the non-LoS (NLoS) MPCs. The function I_0 is the modified Bessel function of 0-th order. The average received power in the Rician fading is given by

$$P_r = \int_0^\infty z^2 p_Z(z) dx = s^2 + 2\sigma^2. \qquad (1.24)$$

Define $K = s^2/2\sigma^2$ as the fading parameter, which is the ratio of the power in the LoS component to the power in the NLoS MPCs. Fading parameter K indicates the severity of the fading. The smaller K is, the more severe the fading is. For $K = 0$, we have Rayleigh fading, while for $K = \infty$, there is no fading which indicates no MPCs and only the LoS path is obtained.

In addition to the Rayleigh and Rician distributions, the Nakagami distribution can also describe the fading of the channel. The Nakagami distribution can be written as

$$p_Z(z) = \frac{2m^m z^{2m-1}}{\Gamma(m) P_r^m} \exp\left[\frac{-mz^2}{P_r}\right], \quad m \geq 0.5, \qquad (1.25)$$

where P_r is the average received power, $\Gamma(.)$ is the Gamma function, and m denotes the fading parameter. For $m = 1$, we have Rayleigh fading and for $m = \infty$, there is no fading, which is the same as $K = \infty$ in Rician fading. When $m = (K + 1)^2/(2K + 1)$, (1.25) approximates the Rician distribution with parameter K.

2) Wideband Fading Model

Different from the narrowband fading model, when the signal bandwidth gradually increases to $T_m = B^{-1}$, the approximation $s(t - \tau_n(t)) = s(t)$ no longer holds. Thus, after the transmit signal reaches Rx with different delays, the phase difference make the signals cancel each other. In the frequency domain, it manifests as a frequency-selective channel. In the time domain, it manifests as distortion of the signal.

For wideband signals, because the MPC delay extension will distort the received signal, it is necessary to show not only the amplitude and phase changes in the channel statistical model, but also the influence of delay spread and channel time variability. As mentioned above, channel impulse response $h(t, \tau)$ has two time parameters, i.e., t reflects the time-varying characteristic and τ represents the impulse response associated with a given multipath delay. Take the Fourier transform of the channel impulse response $h(t, \tau)$ with respect to t as

$$S_c(\tau, \rho) = \int_{-\infty}^{\infty} c(\tau, t) e^{-j2\pi\rho t} dt, \tag{1.26}$$

where S_c is defined as the deterministic scattering function of the lowpass equivalent channel impulse response $h(t, \tau)$ and the frequency parameter ρ reflects the Doppler characteristic of the channel.

1.2.1.3 Diversity

Channels in wireless communications have a very complex fading characteristic, and this fading characteristic will greatly affect the performance of communication, especially a SISO system. Fig. 1.2 shows the bit error rate (BER) curve using quadrature phase shift keying (QPSK) modulation in additive white Gaussian noise (AWGN) versus Rayleigh fading channel of a SISO system.

As can be seen from Fig. 1.2, the BER of Rayleigh fading channel is much larger than that of AWGN channel. Due to the fading characteristics of the channel, the

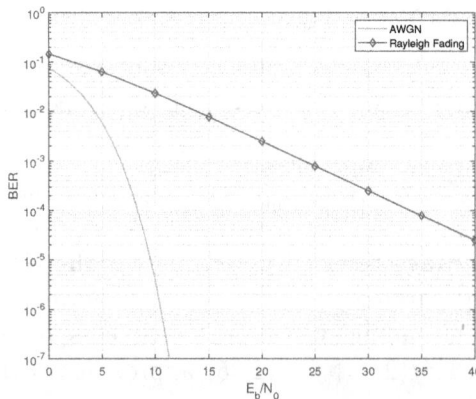

Figure 1.2 The comparison of BER between AWGN and Rayleigh fading channel with QPSK modulation.

Tx

h_0

h_1

Rx

#0

#1

h_{N_t-1} ⋮

#$N_t - 1$

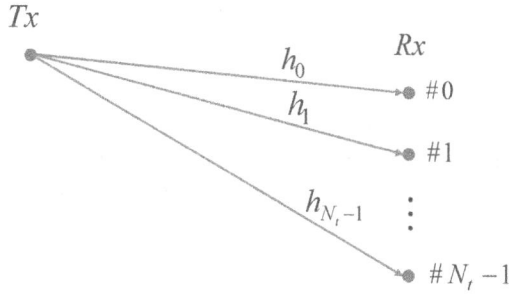

Figure 1.3 SIMO system.

instantaneous channel is in a state of deep fading which means a low SNR. The work of Tse and Viswanath[22] tell us that the channel in a state of deep fading is the main cause of the BER. Hence, we can employ diversity to overcome the impact of channel fading on the BER. Diversity refers to sending the same signal in a separate fading path. Since the probability of multiple channels being in deep fading at the same time is small, the degree of degradation of the received signal decreases after the receiver is merged, resulting a better BER performance of the system.

1.2.2 SIMO System

As shown in Fig. 1.3, the SIMO system is an example of utilizing diversity. SIMO systems combat fading by using multiple antennas at Rx to form multiple channels. If the spacing between adjacent receive antennas is large enough (a wavelength is enough), the channel fading from the transmit antenna to the receive antenna can be considered independent of each other.

The discrete channel model of the SIMO system can be represented as

$$\mathbf{x} = \mathbf{h}s + \mathbf{n}, \tag{1.27}$$

where $\mathbf{h} = [h_0, h_1, \cdots, h_{N_t-1}]^{\mathrm{T}}$ is the channel vector, $\mathbf{x} = [x_0, x_1, \cdots, x_{N_r-1},]$ is the lowpass received signal vector, and $\mathbf{n} = [n_0, n_1, \cdots, n_{N_r-1},]$ is the thermal noise vector, N_r is the number of receive antennas and x_i is the received signal on antenna i. In the SIMO system, the channel can be represented as a vector $\mathbf{h}$ instead of a matrix $\mathbf{H}$. Assume that Rx knows the channel information, i.e., the channel vector $\mathbf{h}$, which is known as channel state information at Rx (CSIR). In practice, by sending a pilot sequence for channel estimation, the channel gains can be easily obtained. Then, the maximal ratio combining (MRC) can be represented as

$$\mathbf{u} = \frac{\mathbf{h}}{\|\mathbf{h}\|_2}. \tag{1.28}$$

After combining, the output signal can be represented as

$$y = \mathbf{u}^{\mathrm{H}}\mathbf{x} = \frac{\mathbf{h}^{\mathrm{H}}\mathbf{h}}{\|\mathbf{h}\|_2}s + \frac{\mathbf{h}^{\mathrm{H}}\mathbf{n}}{\|\mathbf{h}\|_2}. \tag{1.29}$$

Then, the channel capacity of the SIMO system can be obtained

$$C = B \log_2 \left(1 + \frac{E_s}{N_0} \mathbf{h}^H \mathbf{h} \right) = B \log_2 \left(1 + \frac{E_s}{N_0} \sum_{i=0}^{N_r-1} |h_i|^2 \right), \tag{1.30}$$

where E_s represents average power required to send each symbol and $N_0 = \sigma_n^2 B$ is the noise power with B being the signal bandwidth.

1.2.3 MISO System

As shown in Fig. 1.4, a MISO system refers to a system which contains multiple transmit antennas at Tx and single receive antenna at Rx. The discrete model of a MISO system can be represented as

$$x = \mathbf{h}^T \mathbf{v} s + n, \tag{1.31}$$

where $\mathbf{h} = [h_0, h_1, \cdots, h_{N_t-1}]^T$ is the channel vector and $\mathbf{v}$ is the precoding vector. As the same with SIMO system, the channel can be represented as a vector $\mathbf{h}$ instead of a matrix $\mathbf{H}$. One of the problems that MISO systems need to consider is how to design precoding vectors to maximize the received SNR and make the system perform better. However, different from SIMO systems, in a MISO system, Tx may not know the channel state information (CSI). Under the condition where Tx does not know the CSI, an intuitive way is to divide the energy evenly among each antenna. In such a case, the precoding vector can be represented as

$$\mathbf{v} = \left[\frac{1}{\sqrt{N_t}}, \frac{1}{\sqrt{N_t}}, \cdots, \frac{1}{\sqrt{N_t}} \right]. \tag{1.32}$$

In this case, the channel capacity can be obtained as

$$C = B \log_2 \left(1 + \frac{E_s}{N_0} |\bar{h}|^2 \right), \tag{1.33}$$

where $\bar{h} = \frac{1}{N_t} \sum_{i=0}^{N_t-1} h_i$.

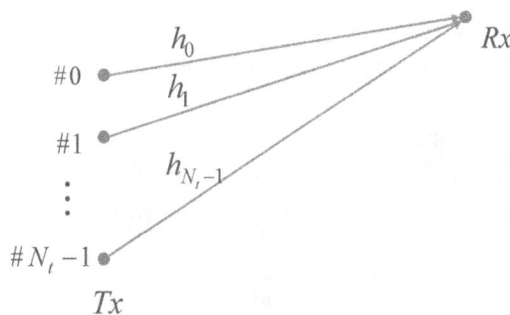

Figure 1.4 MISO system

An approach to improve the system performance is to employ space-time coding[15, 16, 23] at Tx, which does not require CSI at Tx. Let us see the simplest case of two Tx antennas and the Alamouti coding[23] is used. Then, in the first symbol period, s_0 and s_1 are transmitted to Rx from antenna 0 and antenna 1 simultaneously. In the second symbol period, $-s_1^*$ and s_0^* are transmitted to Rx from antenna 0 and antenna 1 simultaneously. Hence, the received signal can be represented as

$$
\begin{aligned}
x_0 &= h_0 s_0 + h_1 s_1 + n_0, \\
x_1 &= -h_0 s_1^* + h_1 s_0^* + n_1,
\end{aligned}
\tag{1.34}
$$

where x_0 and x_1 represent the received signals in the first symbol period and the second symbol period, respectively, n_0 and n_1 are the corresponding terms of additive noise in each symbol period. Write it in the form of matrix as

$$
\underbrace{\begin{pmatrix} x_0 \\ x_1^* \end{pmatrix}}_{\mathbf{x}} = \underbrace{\begin{pmatrix} h_0 & h_1 \\ h_1^* & -h_0^* \end{pmatrix}}_{\mathbf{H}} \underbrace{\begin{pmatrix} s_0 \\ s_1 \end{pmatrix}}_{\mathbf{s}} + \underbrace{\begin{pmatrix} n_0 \\ n_1^* \end{pmatrix}}_{\mathbf{n}},
\tag{1.35}
$$

where $\mathbf{H}$ is the equivalent channel matrix, which is a unitary matrix. Multiplying each side of the equation (1.35) from the left by $\mathbf{H}^{\mathrm{H}}$, we have

$$
\mathbf{x}' = \mathbf{H}^{\mathrm{H}} \mathbf{H} \mathbf{s} = \begin{pmatrix} |h_0|^2 + |h_1|^2 & 0 \\ 0 & |h_0|^2 + |h_1|^2 \end{pmatrix} \begin{pmatrix} s_0 \\ s_1 \end{pmatrix} + \mathbf{H}^{\mathrm{H}} \mathbf{n}.
\tag{1.36}
$$

Finally, the achievable data rate of Alamouti coded system can be obtained

$$
R = \log_2 \left(1 + \frac{E_s}{N_0} \frac{|h_0|^2 + |h_1|^2}{2} \right).
\tag{1.37}
$$

Thus, the Alamouti coding can achieve the channel capacity of the SIMO channel when the CSI is unknown. Since this book is about beamforming at Tx, space-time coding is not specially considered.

When the CSI is known at Tx, denoted as channel state information at the transmitter (CSIT), the precoding vector can be expressed as

$$
\mathbf{v} = \frac{\mathbf{h}^{\mathrm{H}}}{\|\mathbf{h}\|_2},
\tag{1.38}
$$

which is able to maximize the SNR at Rx. Then the channel capacity in this case can be obtained as

$$
C = B \log_2 \left(1 + \frac{E_s}{N_0} \mathbf{h}^{\mathrm{H}} \mathbf{h} \right) = B \log_2 \left(1 + \frac{E_s}{N_0} \sum_{i=0}^{N_t - 1} |h_i|^2 \right).
\tag{1.39}
$$

Note that this approach is also called the matched beamforming or conventional beamforming because the precoding vector makes the transmission in the direction of the matched channel. However, this approach requires knowledge of the CSI. Hence, in a time duplex system, a feedback from Rx is required. Nevertheless, in a frequency duplex system, the uplink and downlink channels may not be the same, which means the CSI at Tx is greatly reduced. Moreover, this approach will not be able to cancel the interference.

1.2.4 MIMO System

Systems with multiple antennas at both Tx and Rx are MIMO systems. In MIMO systems, both the receive antennas and the transmit antennas can be used for diversity gain. Moreover, by exploiting the structure of the channel gain matrix, independent paths can be obtained to transmit independent signals, which is called multiplexing.

1.2.4.1 Narrowband MIMO Model

In this subsection, we will extend the channel models (1.17) and (1.18) into a MIMO system. Assume that there are N_t transmit antennas and N_r receive antennas. Thus, the discrete signal channel can be expressed as

$$
\begin{bmatrix} x_1 \\ \vdots \\ x_{N_r} \end{bmatrix} = \begin{bmatrix} h_{11} & \cdots & h_{1N_t} \\ \vdots & \ddots & \vdots \\ h_{N_r1} & \cdots & h_{N_rN_t} \end{bmatrix} \begin{bmatrix} s_1 \\ \vdots \\ s_{N_t} \end{bmatrix} + \begin{bmatrix} n_1 \\ \vdots \\ n_{N_r} \end{bmatrix}, \tag{1.40}
$$

or simplified as $\mathbf{x} = \mathbf{H}\mathbf{s} + \mathbf{n}$, where $\mathbf{H}$ is the $N_r \times N_t$ channel response matrix and $\mathbf{n}$ is the $N_r \times 1$ noise vector. h_{mn} represents the channel gain from transmit antenna n to receive antenna m and its fading characteristic has been introduced in Section 1.2.1.

Similar to the MISO system. We can divide the MIMO systems into three types. The first one is to assume that Tx does not acquire CSI. The second one is to assume that Tx acquires CSI by implementing a feedback or exploiting reciprocal properties of propagation in time-division duplexing systems. The last one is that both Tx and Rx do not acquire the CSI. In such a case, some distributions on the channel matrix may be assumed. For instance, the zero-mean spatially white (ZMSW) model, which assumes the elements in matrix $\mathbf{H}$ to be independently identical distribution with zero mean, unit variance, and complex circularly symmetric Gaussian random variables, is commonly used.

1.2.4.2 Decomposition of the MIMO Channel

Consider a MIMO channel with N_t transmit antennas and N_r receive antennas. Assume that the channel matrix is acquired by both Tx and Rx. R_H represents the rank of matrix $\mathbf{H}$. Then, for any channel response matrix $\mathbf{H}$, we can obtain its SVD as

$$
\mathbf{H} = \mathbf{U}\mathbf{\Sigma}\mathbf{V}^\mathrm{H}, \tag{1.41}
$$

where the $N_r \times N_r$ matrix $\mathbf{U}$ and the $N_t \times N_t$ matrix $\mathbf{V}$ are unitary matrices and the $N_r \times N_t$ matrix $\mathbf{\Sigma}$ is a diagonal matrix of singular values $\{\lambda_i\}$ of $\mathbf{H}$, respectively. Specially, there are R_H nonzero singular values in $\mathbf{\Sigma}$.

According to the MIMO channel model $\mathbf{x} = \mathbf{H}\mathbf{s} + \mathbf{n}$, let $\tilde{\mathbf{x}} = \mathbf{U}^\mathrm{H}\mathbf{x}$ and $\mathbf{s} = \mathbf{V}\tilde{\mathbf{s}}$, we can obtain $\tilde{\mathbf{x}} = \mathbf{\Sigma}\tilde{\mathbf{s}} + \tilde{\mathbf{n}}$. Note that multiplying by a unitary matrix does not change the noise distribution, i.e., $\tilde{\mathbf{n}}$ is distributed identically with $\mathbf{n}$ as $\mathbf{n}$ is a multi-variant Gaussian noise. Since matrix $\mathbf{\Sigma}$ is a diagonal matrix, the channel is decomposed to several parallel channels, as shown in Fig. 1.5. Note that n_{min} is equal to R_H

representing the number of parallel channels. In a rich scattering environment, n_{min} is equal to min $\{N_t, N_r\}$. Other environments may lead to a low rank R_H, thus a low n_{min}. After decomposing the MIMO channel into parallel independent channels, independent data can be transmitted on these channels. The Gaussian noise $\tilde{n}$ can also be decomposed into n_{min} components, i.e., $\tilde{n} = [\tilde{n}_0, \tilde{n}_1, \cdots, \tilde{n}_{n_{min}-1}]^T$. Thus, the data rate of the MIMO system will be n_{min} times as much as that of the SISO system i.e., the multiplexing gain is n_{min}.

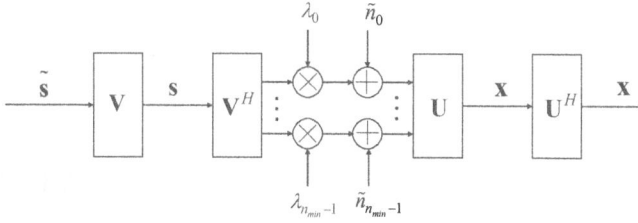

Figure 1.5 Decompose the MIMO channel into n_{min} parallel channels.

1.2.4.3   Channel Capacity of MIMO Channel

In this part, we assume that Rx acquires the knowledge of CSI. Under this assumption, the channel capacity can be obtained from maximizing the mutual information between the channel input vector **s** and output vector **x**

$$C = \max_{p(\mathbf{s})} I(\mathbf{S}, \mathbf{X}) = \max_{p(\mathbf{s})} [H(\mathbf{X}) - H(\mathbf{X}|\mathbf{S})], \tag{1.42}$$

where $I(\mathbf{S}, \mathbf{X})$ denotes the mutual information between **S** and **X**, $H(\mathbf{X})$ and $H(\mathbf{X}|\mathbf{S})$ denote the entropies of **X** and **X|S**, respectively. Note that $H(\mathbf{X}|\mathbf{S})$ represents the entropy in noise, which is independent of the signal. Thus, maximizing the entropy of **X** is to maximize the mutual information.

The entropy in **x** depends on the covariance matrix of **x**, and the covariance matrix can be represented as

$$\mathbf{R}_x = \mathbb{E}\left[\mathbf{x}\mathbf{x}^H\right] = \mathbf{H}\mathbf{R}_s\mathbf{H}^H + \mathbf{I}_{N_r}, \tag{1.43}$$

where $\mathbf{R}_s$ denotes the covariance matrix of **s**. It can be proved that when the input signal **s** is a zero-mean circularly-symmetric complex Gaussian (ZMCSCG) random vector, the output signal **x** is also a ZMCSCG random vector. And in such a case, the entropy in **x** is maximized with a power constraint $\text{Tr}(\mathbf{R}_x) = E_s/N_0$[11]. Hence, the mutual information can be obtained as

$$I(\mathbf{S}, \mathbf{X}) = \log_2 \det\left[\mathbf{I}_{N_r} + \mathbf{H}\mathbf{R}_s\mathbf{H}^H\right]. \tag{1.44}$$

Under the power constraint, the channel capacity of MIMO channel is

$$C = \max_{\mathbf{R}_s: \text{Tr}(\mathbf{R}_s)=(E_s/N_0)} \log_2 \det\left[\mathbf{I}_{N_r} + \mathbf{H}\mathbf{R}_s\mathbf{H}^H\right]. \tag{1.45}$$

Specifically, in a SISO system, the channel matrix degenerates into h, and the covariance matrix $\mathbf{R}_s$ of input signal becomes E_s/N_0 under the power constraint. Hence, the channel capacity of SISO system can be obtained as $C = B \log_2 \left(1 + (E_s/N_0) \cdot |h|^2\right)$. Extending the above equation to SIMO systems, the channel becomes a vector $\mathbf{h} = (h_1, h_2, \cdots, h_n)^T$. Hence, the channel capacity can be obtained as (1.30).

As for MISO system, the channel can be represented as $\mathbf{h} = (h_1, h_2, \cdots, h_n)$. If CSIT is obtained, the precoding vector can be represented as (1.38). Hence, the transmitted signal is

$$\mathbf{s} = \frac{\mathbf{h}^H}{\|\mathbf{h}\|_2} \cdot s. \tag{1.46}$$

Noted that under the power constraint, the power of the signal $\|\mathbf{s}\|_2^2$ equals to E_s. Substituting (1.46) into (1.45), the channel capacity of a precoded MISO system, i.e., (1.39) can be obtained.

Assuming the CSI is unknown at Tx and known at Rx, then it is impossible to optimize the power allocation at Tx to maximize channel capacity C for the precoded channel. If $\mathbf{H}$ conforms to the ZMSW model, its mean and variance are symmetrical for each antenna. Then, an intuitive way is to distribute Tx power evenly among the antennas at Tx. Under such an assumption, the covariance matrix of input signal $\mathbf{s}$ can be expressed as $\mathbf{R}_s = [E_s/(N_0 N_t)]\, \mathbf{I}_{N_t}$. In this case, channel mutual information can be maximized[11] and the maximized mutual information can be represented as

$$I(\mathbf{S}, \mathbf{X}) = \log_2 \det \left[\mathbf{I} + \frac{E_s}{N_t N_0} \mathbf{H} \mathbf{H}^H\right], \tag{1.47}$$

where $\mathbf{I}$ denotes the $N_r \times N_r$ identity matrix. Using SVD, we can obtain

$$\mathbf{I} = \sum_{i=0}^{n_{\min}-1} \log_2 \left(1 + \frac{E_i}{N_t N_0}\right), \tag{1.48}$$

where E_i/N_0 denotes the SNR of the i-th channel.

Noted that when the number of antennas at the transceiver is large, for fixed N_r, under the ZMSW model the law of large numbers implies that

$$\lim_{N_t \to \infty} \frac{1}{N_t} \mathbf{H} \mathbf{H}^H = \mathbf{I}. \tag{1.49}$$

Hence, the mutual information can be represented as

$$I(\mathbf{S}, \mathbf{X}) = N_r \log_2 \left(1 + \frac{E_s}{N_0}\right), \tag{1.50}$$

which is equal to the channel capacity of the SISO channel.

When the CSI is known at both Tx and Rx, the channel capacity can be reached by water-filling method. Assuming that the total transmit power is optimally allocated between the channels, the channel capacity of the MIMO system can be simplified to the sum of the channel capacities for each independent parallel. And the channel capacity can be obtained as

$$C = \sum_{i=0}^{n_{\min}-1} \log_2 \left(1 + \frac{E_i \lambda_i^2}{N_0}\right). \tag{1.51}$$

where λ_i is the i-th singular value of matrix $\mathbf{H}$. The optimal power allocation for each channel can be obtained by water-filling method, i.e., making the energy of each channel as[24]

$$E_i = \left(\mu - \frac{N_0}{\lambda_i^2}\right)^+, \quad \sum_i E_i = E_s, \tag{1.52}$$

where $(x)^+ = \max(x, 0)$.

The water-filling method compares the process of power allocation to the process of pouring water into a container. The bottom of the container is not flat, which has n_{min} different depths. The SNR of the channel corresponding to each depth is inverse. Power is poured into the container and the power on each channel is optimally allocated, as shown in Fig. 1.6.

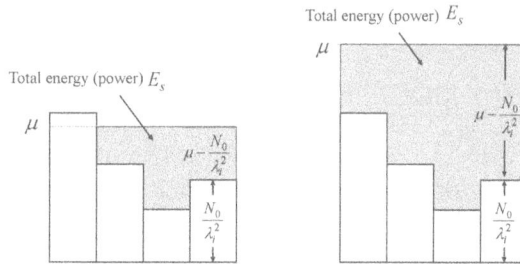

Figure 1.6 The power allocation in low SNR (left) and high SNR (right).

In practice, a common scenario is that the channel matrix undergoes a flat fading, i.e., the element h_{mn} changes overtime. In a time-varying channel, assume that the CSI is acquired at Rx but unknown at Tx. Hence, the transmit power is allocated evenly on each antenna at Tx. Define ergodic capacity as the maximum rate, averaged over all channel realizations that are only based on the distribution of $\mathbf{H}$[20]. Ergodic capacity can be represented as

$$C = \max_{\mathbf{R}_s : \mathrm{Tr}(\mathbf{R}_s)=(E_s/N_0)} \mathbb{E}_H \left[B \log_2 \det \left(\mathbf{I}_{N_r} + \mathbf{H}\mathbf{R}_x\mathbf{H}^{\mathbf{H}}\right)\right], \tag{1.53}$$

where the expectation is with respect to the distribution on the channel matrix $\mathbf{H}$. As mentioned above, the transmit power is allocated evenly on the antennas, i.e., $\mathbf{R}_x = \frac{E_s}{N_t N_0}\mathbf{I}$. Thus, the ergodic capacity turns into

$$C = \mathbb{E}_H \left[B \log_2 \det \left[\mathbf{I}_{N_r} + \frac{E_s}{N_t N_0}\mathbf{H}\mathbf{H}^{\mathbf{H}}\right]\right]. \tag{1.54}$$

When the CSI is known at both Tx and Rx, the transmit power is allocated by the water-filling method. The ergodic capacity is the average capacities with each channel realization. In a short-term power constraint, where in each channel realization the power is equal the average power constraint, the ergodic capacity can be obtained when the transmit power of each channel equals to average transmit power $\bar{E}$

$$
\begin{aligned}
C &= \mathbb{E}_H \left[\max_{\mathbf{R}_s : \mathrm{Tr}(\mathbf{R}_s) = (E_s/N_0)} B \log_2 \det \left(\mathbf{I}_{N_r} + \mathbf{H} \mathbf{R}_x \mathbf{H}^H \right) \right] \\
&= \mathbb{E}_H \left[\max_{E_i : \sum_i E_i \leq \bar{E}} B \sum_i \log_2 \left(1 + \frac{E_i \lambda_i^2}{N_0} \right) \right].
\end{aligned}
\tag{1.55}
$$

In a long-term power constraint, in each channel realization the power E_H is not necessarily the same. When the average power of each realization meets $\mathbb{E}_H[E_H] \leq \bar{E}$, the ergodic capacity can be represented as

$$
\begin{aligned}
C &= \max_{E_H : \mathbb{E}_H[E_H] = E_s} \mathbb{E} \left[\max_{\mathbf{R}_s : \mathrm{Tr}(\mathbf{R}_s) \leq (E_H/N_0)} B \log_2 \det \left(\mathbf{I}_{N_r} + \mathbf{H} \mathbf{R}_x \mathbf{H}^H \right) \right] \\
&= \max_{E_H : \mathbb{E}_H[E_H] \leq \bar{E}} \mathbb{E} \left[\max_{E_i : \sum_i E_i \leq E_H} B \sum_i \log_2 \left(1 + \frac{E_i \lambda_i^2}{N_0} \right) \right].
\end{aligned}
\tag{1.56}
$$

1.3 ANTENNA ARRAY MODELS

Aforementioned, in a MIMO system, when the transmit/receive antenna spacing is large, the channels between each transceiver antenna pair can be considered independent, so diversity technology can be used to combat channel fading. When antennas are very close to each other and are arranged in a certain regular manner, these antennas form an antenna array, where the radiation fields generated by all the AEs are coherently superimposed. Hence, the total radiation field has a high gain in a certain direction which is able to compensate the path loss. Meanwhile the interference in other directions can be eliminated. As shown in Fig. 1.7, the signal arrived at reference point P is enhanced while the interference at reference point Q is eliminated.

1.3.1 Plane Wavefront

Waves with equiphase plane are planar waves. In free space, the wave equiphase surface emitted by the wave source is usually a sphere, and when the distance from the wave source to the reference point is infinity, the equiphase surface of the received radiation wave is approximately a plane which is a plane wave. A plane wave can be written as

$$
\mathbf{x} = A \cos \left(\mathbf{k} \cdot \mathbf{r} - \omega t + \phi_0 \right),
\tag{1.57}
$$

where A is the amplitude of the signal, $\mathbf{k}$ is the wave vector that characterizes the direction of the wave propagation and $\mathbf{r}$ is the three-dimensional vector from source to reference point. The direction of the wave vector is defined as the direction of the wave velocity. Electromagnetic waves are transverse waves, with the amplitude direction perpendicular to the wave vector direction.

Figure 1.7 Beamforming

Note that the far-field can be expressed as

$$r > \frac{2L^2}{\lambda} \tag{1.58}$$

$$r \gg L \tag{1.59}$$

$$r \gg \lambda \tag{1.60}$$

where r is the distance from the the antenna array to the reference point and L is the length of the antenna array. In the communication system, this column of antennas is usually far from the signal source. And the radiation wave transmits in free space, whose propagation medium can be viewed as isotropic. The spatial signals will propagate in straight lines in the medium. Hence, when the spatial signal arrives at the antenna array, it can be seen as a plane wavefront. Then the different time delay of the array element can be determined by the geometry of the array and the direction of the spatial waves.

1.3.2 Two-Element Array

The simplest antenna array is the two-element array, as shown in Fig. 1.8. According to[25], the radiation field produced by a single dipole can be represented as

$$E_\theta \simeq j\eta \frac{kI_0 l e^{-jkr}}{4\pi r} \cos \Omega_E, \tag{1.61}$$

where Ω_E is the elevation angle of departure (AoD)/AoA, η is the intrinsic impedance of the medium, I_0 is the maximum current, and l is the length of the dipole. As shown in Fig. 1.8 (left), assume that the two dipoles do not interfere with each other. When they are placed side by side, the electric field at the reference point is a superposition of the two dipoles

$$\vec{E}_t = \vec{E}_0 + \vec{E}_1 = \hat{a}_{\Omega_E} j\eta \frac{I_0 l}{4\pi} \left\{ \frac{e^{-j[kr_0 - \beta/2]}}{r_0} \cos \Omega_{E0} + \frac{e^{-j[kr_1 + \beta/2]}}{r_1} \cos \Omega_{E1} \right\}, \tag{1.62}$$

where r_0 and r_1 denote the distance from dipole 0 and dipole 1 to the reference point, respectively, and β denotes is the difference in phase excitation between the elements. According to the far-field assumption (1.58), in phase variations:

$$\Omega_{E0} \simeq \Omega_{E1} \simeq \Omega_E,$$
$$r_0 \simeq r - \frac{d}{2}\cos\Omega_E, \tag{1.63}$$
$$r_1 \simeq r + \frac{d}{2}\cos\Omega_E,$$

where, for amplitude variations, we have

$$r_0 \simeq r_1 \simeq r, \tag{1.64}$$

and d denotes the space between dipole 0 and dipole 1. As shown in Fig. 1.8 (right), r_0 is parallel to r_1. Hence, under the assumption, equation (1.62) can be simplified as

$$E_t = \hat{\mathbf{a}}_{\Omega_E} j\eta \frac{kI_0 l e^{-jkr}}{4\pi r}\cos\Omega_E \left\{ 2\cos\left[\frac{1}{2}(kd\cos\Omega_E + \beta)\right]\right\}. \tag{1.65}$$

As can be seen from (1.65), the first part, $E_0 = E_{\Omega_E} \simeq j\eta(kI_0 l e^{-jkr}/4\pi r)\cos\Omega_E$, is the radiation field of a single dipole. The second part, $2\cos[(1/2)(kd\cos\Omega_E + \beta)]$, relates to the current ratio and the relative position between the two dipoles or the antennas, which is called array factor. It has been illustrated that the antenna array pattern is equal to the pattern of a single antenna multiplied by the array factor, which is called the pattern multiplication rule, expressed as

$$A(\Omega_E, \Omega_A) = A_{\text{single}} \cdot A_{\text{array}}, \tag{1.66}$$

where A_{single} denotes the field of a single antenna and A_{array} denotes the array factor. Ω_E and Ω_A, respectively, denote the elevation AoA/AoD and azimuth AoA/AoD. And the pattern multiplication rule can be extended to arrays with any number of antennas. Obviously, the array factor is independent of the axis direction of the AEs and is a function of the number of elements, their geometrical arrangement, their relative magnitudes, phases and spacings[25].

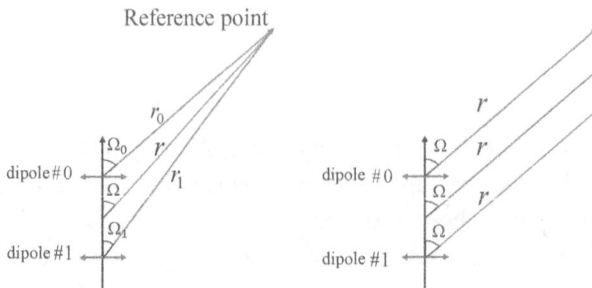

Figure 1.8 2-element array, **Left:** Near-field. **Right:** Far-field

Hence, as the point source is isotropic, when solving the array factor pattern, we can view each element in the antenna array as a point source.

Assume the currents in the two elements are in phase and the distance is $\lambda/2$. The beam pattern can be expressed as

$$A = 2\,|E_0|\cos\left(\frac{\pi}{2}\cos\Omega_E\right). \tag{1.67}$$

The antenna array pattern is shown in Fig. 1.9, the antenna array has no radiation along the axis, and the radiation along the vertical axis is the largest.

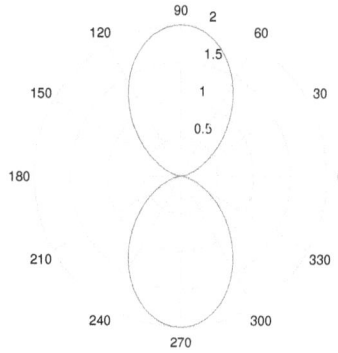

Figure 1.9 The 8-shaped pattern

When the phase difference of current in the elements is 90° and the distance is $\lambda/4$, we have

$$A = 2\,|E_0|\cos\left(\frac{\pi}{4}\cos\Omega_E \pm \frac{\pi}{4}\right). \tag{1.68}$$

The antenna array pattern is shown in Fig. 1.10, where the radiation field is unidirectional and the maximum radiation direction points to the element with current lag.

1.3.3 N-Element Array: Uniform Linear Array

ULA is a kind of antenna array where each AE has equal current amplitude, the phase increases or decreases in uniform proportion, and the antenna array is arranged on a straight line with equal spacing. An N element array is shown in Fig. 1.11. d denotes the distance between the adjacent AEs. We consider a scenario where the distance between the transceiver is much larger than the element spacing d, i.e., the far-field assumption holds. Then, electromagnetic waves arrive at Rx in the form of plane wave. The direction can be specified by azimuth angle Ω_A and the elevation angle Ω_E, and the beam direction vector is given by $\hat{\mathbf{r}} = [\sin\Omega_E\cos\Omega_A, \sin\Omega_E\sin\Omega_A, \cos\Omega_E]^{\mathrm{T}}$. The position vector of the n-th AE is given by $\mathbf{P}_n = [0, 0, (n-1)d]^{\mathrm{T}}$, $n \in \{1, 2, \cdots, N\}$. Thus, the difference of the wave path of the n-th AE and that of the first element

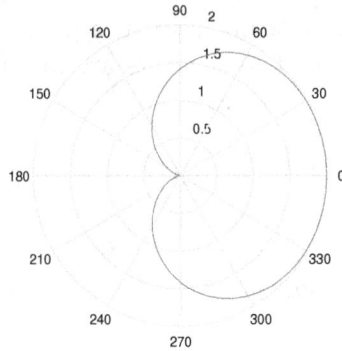

Figure 1.10 The cardioid pattern

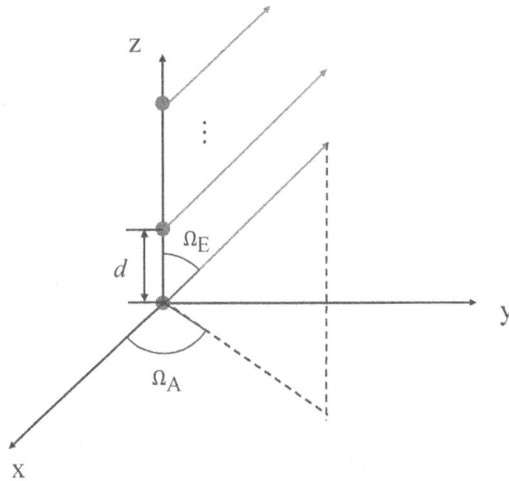

Figure 1.11 N-element array

is the inner product of the position vector and the beam direction vector can be expressed as

$$R_n - R_1 = -\hat{\mathbf{r}}^H \mathbf{P}_n = (n-1)d\cos\Omega_E. \tag{1.69}$$

Furthermore, if the bandwidth of the signal is much less than the carrier frequency, i.e., the signal is narrowband, the difference between the wave paths leads to a phase difference $k(n-1)d\cos\Omega_E$. Then we obtain the following steering vector for ULAs which will be introduced in Section 1.3.5.

$$\mathbf{a}(\Omega_E) = \frac{1}{\sqrt{N}}\left[1, e^{jkd\cos\Omega_E}, \cdots, e^{jk(N-1)d\cos\Omega_E}\right]^{\mathrm{T}}. \tag{1.70}$$

Assume that all the AEs have the same amplitude. There is a phase difference of β between two adjacent AEs. Considering all the AEs are point sources which are isotropic, the array factor can be represented as

$$A = \left(1 + e^{j\phi} + \cdots + e^{j(N-1)\phi}\right) = \sum_{n=0}^{N-1} e^{jn\phi}, \tag{1.71}$$

where $\phi = kd\cos\Omega_E + \beta$. It can also be written as

$$A = e^{j[(N-1)/2]\phi} \frac{\sin\left(\frac{N}{2}\phi\right)}{\sin\left(\frac{1}{2}\phi\right)}. \tag{1.72}$$

If the reference point is the physical center of the array, then the array factor is[25]

$$A = \frac{\sin\left(\frac{N}{2}\phi\right)}{\sin\left(\frac{1}{2}\phi\right)}. \tag{1.73}$$

And the normalized form of (1.73) can be represented as

$$A_n = \frac{1}{N} \frac{\sin\left(\frac{N}{2}\phi\right)}{\sin\left(\frac{1}{2}\phi\right)}. \tag{1.74}$$

When ϕ is small enough, in order to simplify the analysis, the following array factor can be obtained

$$A_n = \frac{1}{N} \frac{\sin\left(\frac{N}{2}\phi\right)}{\left(\frac{1}{2}\phi\right)}. \tag{1.75}$$

The maximum values of (1.75) can be obtained when (1.75) is in a $\sin(0)/0$ form, i.e.,

$$\frac{\phi}{2} = \frac{1}{2}(kd\cos\Omega_E + \beta)\Big|_{\Omega_E = \Omega_{Em}} = \pm m\pi \Rightarrow \Omega_{Em} = \arccos\left[\frac{\lambda}{2\pi d}(-\beta \pm 2m\pi)\right], \tag{1.76}$$

where $m = 1, 2, \cdots$. The array factor has only one maximum and occurs when $m = 0$. Thus, we can obtain the optimal angle Ω_{Em} as

$$\Omega_{Em} = \arccos\left(\frac{\lambda\beta}{2\pi d}\right). \tag{1.77}$$

The 3-dB point for the array factor of (1.75) occurs when Ω_E satisfies[25]

$$\frac{N}{2}\phi = \frac{N}{2}(kd\cos\Omega_E + \beta)\Big|_{\Omega_E = \Omega_{Eh}} = \pm 1.391 \Rightarrow \Omega_{Eh} = \arccos\left[\frac{\lambda}{2\pi d}\left(-\beta \pm \frac{2.782}{N}\right)\right]. \tag{1.78}$$

When the angles of the maximum value Ω_{Em} and the half-power value Ω_{Eh} are obtained, the half-power beamwidth can be represented as

$$\Theta_h = 2|\Omega_{Em} - \Omega_{Eh}|, \tag{1.79}$$

In addition to beamwidth, directivity can also be used to evaluate the concentration of the radiation. The directivity D can be defined as

$$D = \frac{U_{max}}{U_0}, \qquad (1.80)$$

where U_{max} is the maximum value of the radiation intensity and U_0 is the average value of the radiation intensity. For a normalized array factor (1.74) and (1.75), the maximum value U_{max} is equal to unity. And the average value U_0 can be obtained by performing surface integration in the radiative sphere and divided by 4π

$$U_0 = \frac{1}{4} \int_0^{2\pi} \int_0^{\pi} U \sin \Omega_E d\Omega_E d\Omega_A, \qquad (1.81)$$

where U denotes the radiation intensity.

1.3.3.1 Broadside Array

When the AEs of the ULA are in the same phase, as $\beta = 0$, the maximum radiation direction is $\cos \Omega_{Em} = 0$, $\Omega_{Em} = (2m+1)\pi$ (m is an integer), which means the maximum radiation direction is perpendicular to the antenna array axis, as shown in Fig. 1.12(left). In such a case, ULA is called broadside array. According to (1.75), the array factor can be represented as

$$A = \frac{\sin\left(\frac{Nkd}{2}\cos\Omega_E\right)}{N\left(\frac{kd}{2}\right)\cos\Omega_E}. \qquad (1.82)$$

When the distance between the antennas d increases, there may be some principal maxima in other directions, which is called the grating lobe. When the distance between the AEs is equal to multiples of a wavelength, the grating lobes appear, as shown in Fig. 1.12(right)

$$\phi = kd\cos\Omega_E + \beta|_{d=n\lambda, \beta=0} = 2\pi n \cos \Omega_E|_{\Omega_E=0, \Omega_E=\pi} = 2n\pi. \qquad (1.83)$$

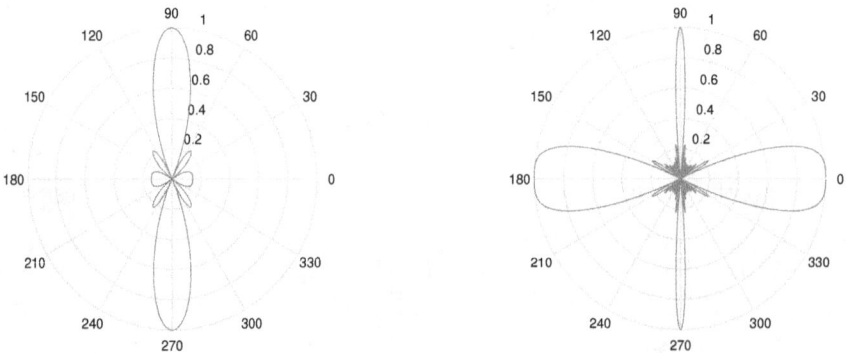

Figure 1.12 Broadside array pattern. **Left:**$d = \lambda/4$. **Right:**$d = \lambda$.

Thus, in addition to the maximum value in the desired direction $\Omega_E = \pi/2$, there are also additional maximum values along the axis $\Omega_E = 0$ and $\Omega_E = \pi$. Hence, in order to avoid the grating lobes, one way is to select the largest spacing between the AEs. The distance d is no larger than one wavelength $\lambda/2$.

For broadside array, according to the array factor in (1.82), the radiation intensity can be represented as

$$U(\Omega_E)^2 = [A_n]^2 = \left[\frac{\sin \left(\frac{Nkd}{2} \cos \Omega_E \right)}{N \left(\frac{kd}{2} \right) \cos \Omega_E} \right]^2. \tag{1.84}$$

Hence, the average radiation intensity is obtained by

$$U_0 = \frac{1}{2} \int_0^\pi \left[\frac{\sin \left(\frac{Nkd}{2} \cos \Omega_E \right)}{N \left(\frac{kd}{2} \right) \cos \Omega_E} \right]^2 \sin \Omega_E d\Omega_E \simeq \frac{\pi}{Nkd}. \tag{1.85}$$

Then, the directivity can be written as

$$D_0 \simeq \frac{Nkd}{\pi} = 2N \left(\frac{d}{\lambda} \right). \tag{1.86}$$

Let $L = (N-1)d$ denote the overall length of the array, then the directivity can be represented as

$$D_0 \simeq 2 \left(1 + \frac{L}{d} \right) \left(\frac{d}{\lambda} \right). \tag{1.87}$$

1.3.3.2 End-Fire Array

When the phase difference β of the adjacent antennas in the antenna array meets $\beta = kd$, the maximum radiation direction $\cos \Omega_{Em} = 1$, $\Omega_{Em} = 0$, which means the maximum radiation direction is parallel to the antenna array axis, as shown in Fig. 1.13(left). In such a case, the antenna array is called end-fire array. According to (1.75), the array factor can be represented as

$$A_n = \frac{\sin \left[\frac{Nkd}{2} (1 - \cos \Omega_E) \right]}{\frac{Nkd}{2} (1 - \cos \Omega_E)}. \tag{1.88}$$

It is the same as broadside array, when the distance d between the antennas increases, the grating lobe appears in addition to the main lobe, as shown in Fig. 1.13(right). The distance d is no larger than $\lambda/2$.

For an end-fire array, according to array factor (1.88), the radiation intensity can be represented as

$$U(\Omega_E)^2 = [A_n]^2 = \left[\frac{\sin \left[\frac{Nkd}{2} (1 - \cos \Omega_E) \right]}{\frac{Nkd}{2} (1 - \cos \Omega_E)} \right]^2. \tag{1.89}$$

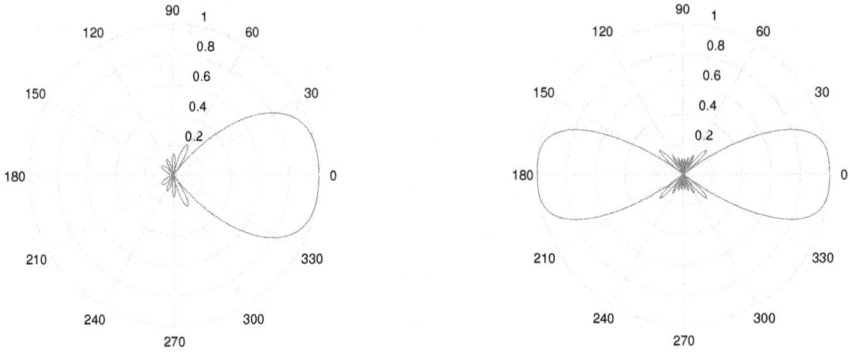

Figure 1.13 End-fire array pattern. **Left:**$d = \lambda/4$. **Right:**$d = \lambda/2$.

Hence, the average radiation intensity is obtained

$$U_0 = \frac{1}{2} \int_0^\pi \left[\frac{\sin\left[\frac{Nkd}{2}\left(1 - \cos\Omega_E\right)\right]}{\frac{Nkd}{2}\left(1 - \cos\Omega_E\right)} \right]^2 \sin\Omega_E d\Omega_E \simeq \frac{\pi}{2Nkd}. \tag{1.90}$$

Then the directivity can be written as

$$D_0 \simeq \frac{2Nkd}{\pi} = 4N\left(\frac{d}{\lambda}\right). \tag{1.91}$$

Let $L = (N-1)d$ denotes the overall length of the array, then the directivity can be represented as

$$D_0 \simeq 4\left(1 + \frac{L}{d}\right)\left(\frac{d}{\lambda}\right). \tag{1.92}$$

Noted that the directivity of the end-fire array (1.91) and (1.92) is twice that of the broadside array (1.86) and (1.87).

1.3.3.3 Phased (Scanning) Array

As mentioned above, the maximum radiation direction is $\Omega_{Em} = \beta/kd$, then we can change Ω_{Em} by changing the current phase difference β in adjacent AEs. And the phase difference β meets

$$\beta = kd\cos\Omega_{Em}. \tag{1.93}$$

Thus, by controlling the phase difference β in adjacent AEs, the maximum radiation direction can be changed in any direction. And this is the basic principle of a phased array. We take the number of AEs $n = 10$, the radiation direction $\Omega_{Em} = \pi/3$ and the distance between adjacent AEs, $d = \lambda/4$, as an example. Then the phase difference β equals to $\pi/4$. The array pattern is shown in Fig. 1.14.

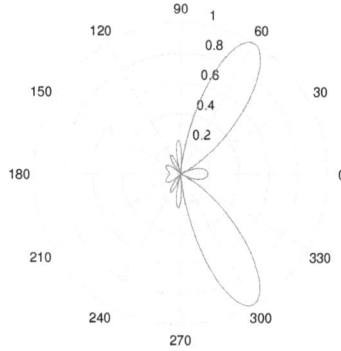

Figure 1.14 Phased array pattern, maximum radiation direction $\theta_m = \pi/3$

1.3.4 Planar Array

1.3.4.1 Uniform Rectangular Array

A uniform rectangular array (URA) refers to the equal current amplitude of each excitation AE, the phase changes in equal difference series, and the distance between the two adjacent elements is equal in a certain direction. As shown in Fig. 1.15, the AE spacing along the x, y direction is d_x, d_y, phase difference of β_x, β_y. Assume that there are M, N elements along the x, y directions, respectively. Hence, the position of the (m, n) element is $x_m = m d_x, y_n = n d_y$. Project the reference point to the xOy plane, and then connect to the origin. According to basic geometry, it is easy to know

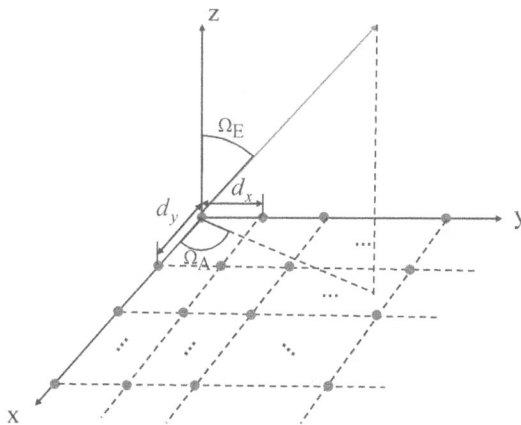

Figure 1.15 Uniform rectangular array

that the steering vectors of the URA is

$$\mathbf{a}_{\text{URA}} = \frac{1}{\sqrt{MN}} \left[1, \cdots, e^{jk[(M-1)d_x \sin \Omega_E \cos \Omega_A + (N-1)d_y \sin \Omega_E \sin \Omega_A]} \right], \qquad (1.94)$$

where Ω_E and Ω_A are the angle between the reference point and the z axis and the angle between the connection and x axis, respectively.

We suppose the AE current phase related to the element of $(0,0)$ position is $e^{[-j(m\beta_x + n\beta_y)]}$. As is the same with ULA, the electric wave radiated by the m-th antenna along x is $\phi_x = kd_x \sin\theta \cos\phi - \beta_x$, and the wave along y is $\phi_y = kd_y \sin \Omega_E \sin \Omega_A - \beta_y$. Array factor is the product of the two ULA array factors in (1.71), expressed as

$$\begin{aligned}
A &= A_x A_y \\
&= \left\{ \sum_{m=0}^{M-1} \left[e^{jmkd_x \sin \Omega_E \cos \Omega_A - \beta_x} \right] \right\} \cdot \left\{ \sum_{n=0}^{N-1} \left[e^{jnkd_y \sin \Omega_E \sin \Omega_A - \beta_y} \right] \right\} \\
&= \sum_{m=0}^{M-1} \sum_{n=0}^{N-1} e^{j(m\phi_x + n\phi_y)},
\end{aligned} \qquad (1.95)$$

According to (1.73), the normalized URA array factor can also be represented as

$$A_n = \frac{1}{M} \frac{\sin\left(\frac{M}{2}\phi_x\right)}{\sin\left(\frac{\phi_x}{2}\right)} \frac{1}{N} \frac{\sin\left(\frac{N}{2}\phi_y\right)}{\sin\left(\frac{\phi_y}{2}\right)}. \qquad (1.96)$$

As is the same with ULA, when the spacing between adjacent AEs is equal or larger than $\lambda/2$, the grating lobes will appear. To avoid the grating lobes, the same constraints need to be satisfied in URA, i.e., the spacings in the x- and y-directions, d_x and d_y must be less than $\lambda/2$.

The maximum radiation direction occurs when $\phi_x = 0, \phi_y = 0$, i.e.,

$$\begin{aligned}
kd_x \sin \Omega_E \cos \Omega_A - \beta_x &= 0, \\
kd_y \sin \Omega_E \sin \Omega_A - \beta_y &= 0.
\end{aligned} \qquad (1.97)$$

Note that the phases β_x and β_y are independent to each other. They can be adjusted and then the direction of the main lobe can be changed. In practice, the main lobes in the x- and y-directions are required to intersect and their maxima are directed toward the same direction. Thus, the 3D beamforming can be accomplished. By solving (1.97), we can obtain the maximum radiation direction $(\Omega_{Em}, \Omega_{Am})$ as

$$\tan \Omega_{Am} = \frac{\beta_y d_x}{\beta_x d_y}, \qquad (1.98)$$

$$\sin^2 \Omega_{Em} = \left(\frac{\beta_x}{kd_x}\right)^2 + \left(\frac{\beta_y}{kd_y}\right)^2. \qquad (1.99)$$

To produce only one mainlobe, $\sin^2 \Omega_E \leq 1$ is required. Then we can obtain

$$\left(\frac{\beta_x}{kd_x}\right)^2 + \left(\frac{\beta_y}{kd_y}\right)^2 \leq 1. \tag{1.100}$$

The constraints on $\beta_x, d_x, \beta_y, d_y$ can be obtained from (1.100).

Assume that the direction of the main beam is $(\Omega_{Em}, \Omega_{Am})$. And two planes are chosen to define the beamwidth. One is the elevation plane defined by $\Omega_A = \Omega_{A0}$ and the other is the plane which is perpendicular to the first plane. Θ and Φ denote the corresponding half-power beamwidths of the two planes, respectively. For a large array which is near broadside, the half-power beamwidth of the elevation plane is[26]

$$\Theta = \sqrt{\frac{1}{\cos^2 \Omega_{E0} \left[\Theta_{x0}^{-2} \cos^2 \Omega_{A0} + \Theta_{y0}^{-2} \sin^2 \Omega_{A0}\right]}}, \tag{1.101}$$

where Θ_{x0} and Θ_{y0} represent the beamwidths of a broadside linear array with M and N AEs, respectively. The half-power beamwidth Φ can be represented as[27]

$$\Psi = \sqrt{\frac{1}{\Theta_{x0}^{-2} \sin^2 \Omega_{A0} + \Theta_{y0}^{-2} \cos^2 \Omega_{A0}}}. \tag{1.102}$$

When $\beta_x = \beta_y = 0$, the current of each AE has the same phase excitation, the URA is broadside and beam points to the normal direction of the plane array. The directionality coefficient of URA array can be written as

$$D = 4\pi \frac{Md_x}{\lambda} \frac{Nd_y}{\lambda}. \tag{1.103}$$

When $d_x = d_y = \lambda/2$, we have

$$D = \pi MN \approx \pi D_x D_y, \tag{1.104}$$

where D_x and D_y are the directionality coefficients along x axis and y axis, respectively.

When the beam is scanned, the directionality coefficient is

$$D \approx \pi D_x D_y \cos \Omega_{Em}. \tag{1.105}$$

The directionality coefficient is related to Ω_{Em}, but not related to Ω_{Am}.

1.3.4.2 Uniform Circular Array

A uniform circular array (UCA) is an antenna array with angular symmetry properties, and the array is located on a circumference as shown in Fig. 1.16. Therefore, it is convenient to describe UCA using polar coordinates and Bessel functions. Suppose there is a UCA with radius a and N AEs. According to basic geometry, the steering vector of the UCA is

$$\mathbf{a}_{\text{UCA}} = \frac{1}{\sqrt{N}} \left[1, \cdots, e^{jkd\sin\Omega_E \cos\left[\Omega_A - \frac{2\pi}{N}(N-1)\right]}\right]. \tag{1.106}$$

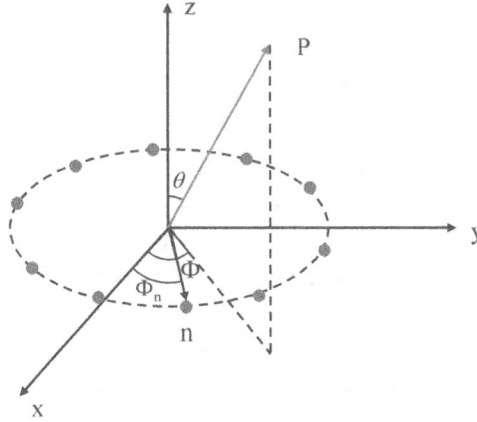

Figure 1.16 Uniform Circular Array

Hence, the array factor in the far-field is

$$A = \sum_{n=1}^{N} I_n e^{[jkd \sin \Omega_E \cos(\Omega_A - \Omega_{An}) + j\beta_n]},$$ (1.107)

where I_n is the current amplitude of the n-th element located in the angle $\Omega_{An} = \frac{2\pi}{N}n$, and β_n is the excitation phase of the n-th element. When the excitation currents are in phase and the main beam is in $(\Omega_{E0}, \Omega_{A0})$, β can be written as

$$\beta = -\frac{2\pi}{\lambda} d \sin \Omega_{E0} \cos(\Omega_{A0} - \Omega_{An}).$$ (1.108)

For convenience, we introduce some new parameters as follows

$$\rho = d \left[(\sin \Omega_E \cos \Omega_A - \sin \Omega_{E0} \cos \Omega_{A0})^2 + (\sin \Omega_E \sin \Omega_A - \sin \Omega_{E0} \sin \Omega_{A0})^2 \right]^{1/2},$$

$$\cos \xi = \frac{d}{\rho} (\sin \Omega_E \cos \Omega_A - \sin \Omega_{E0} \cos \Omega_{A0}).$$

(1.109)

Hence, the array pattern function is

$$A = \sum_{n=1}^{N} I_n e^{jk \cos(\xi - \Omega_{An})}.$$ (1.110)

For uniform excitation and equiangular distribution of circular arrays, i.e., $\Omega_{An} = 2\pi n/N$, the array pattern function can be expanded by the Fourier-Bessel function

$$A = NI \sum_{m=-\infty}^{\infty} J_{mN}(k\rho) e^{[jmN(\frac{\pi}{2} - \xi)]}.$$ (1.111)

Considering the horizontal array pattern function, the horizontal array pattern is the radiation field in the $\Omega_E = \pi/2$ plane. If the maximum direction of the beam is in the positive direction of x axis, let $\Omega_A = 0$. Then, we have

$$\beta = -kd\cos\left(\frac{2\pi n}{N}\right), \tag{1.112}$$

$$\rho = 2d\sin\frac{\Omega_A}{2} \quad (0 \le \Omega_A \le 2\pi), \tag{1.113}$$

$$\cos\xi = -\sin\frac{\Omega_A}{2} \quad \text{or} \quad \xi = \frac{\pi + \Omega_A}{2}. \tag{1.114}$$

Therefore, the horizontal array pattern function is

$$A_h = I_t \sum_{m=-\infty}^{\infty} e^{\left(-jmN\frac{\Omega_A}{2}\right)} J_{mN}\left(2\frac{2\pi}{\lambda}d\sin\frac{\Omega_A}{2}\right), \tag{1.115}$$

where, $I_t = NI$ is the total current amplitude of the circular array. The vertical array pattern is the radiation field in the $\Omega_E = 0$ plane. We have

$$\beta = 0, \tag{1.116}$$

$$\rho = d\sin\Omega_E, \tag{1.117}$$

$$\cos\xi = \cos\Omega_A \text{ or } \xi = \Omega_A. \tag{1.118}$$

Therefore, the vertical array pattern is

$$A_v = I_t \sum_{m=-\infty}^{\infty} e^{\left[jmN\left(\frac{\pi}{2}-\Omega_A\right)\right]} J_{mN}\left(kd\sin\Omega_E\right). \tag{1.119}$$

Then, we discuss the directionality coefficient of the UCA, which can be defined as

$$D = \frac{|A_{\max}|^2}{\frac{1}{4\pi}\int_0^{2\pi}\int_0^{\pi}|A|^2\sin\Omega_E d\Omega_E d\Omega_A}, \tag{1.120}$$

where $|A_{\max}|^2 = |A\left(\Omega_{E0}, \Omega_{A0}\right)|^2$ and $|A|^2$ can be expressed as

$$|A|^2 = AA^H = \sum_{m=1}^{N}\sum_{n=1}^{N} I_m I_n e^{j(\beta_m - \beta_n)} e^{jk\rho_{mn}\sin\Omega_E\cos(\Omega_A - \phi_{mn})}, \tag{1.121}$$

$$\rho_{mn} = \begin{cases} 2a\sin\left(\frac{\Omega_{Am}-\Omega_{An}}{2}\right) & m \ne n, \\ 0 & m = n. \end{cases} \tag{1.122}$$

$$\phi_{mn} = \arctan\left(\frac{\sin\Omega_{Am} - \sin\Omega_{An}}{\cos\Omega_{Am} - \cos\Omega_{An}}\right) \quad m \ne n. \tag{1.123}$$

Then we have

$$\int_0^{2\pi}\int_0^{\pi}|A|^2\sin\Omega_E d\Omega_E d\Omega_A$$
$$=4\pi\sum_{m=1}^{N}\sum_{n=1}^{N} I_m I_n e^{j(\beta_m - \beta_n)}\int_0^{\frac{\pi}{2}} J_0\left(\frac{2\pi}{\lambda}\rho_{mn}\sin\Omega_E\right)\sin\Omega_E d\Omega_E, \tag{1.124}$$

and

$$\int_0^{\frac{\pi}{2}} J_0\left(x \sin \Omega_E\right) \sin \Omega_E d\Omega_E = \left(\frac{\pi}{2}\right)^{\frac{1}{2}} \frac{J_{\frac{1}{2}}\left(x\right)}{x^{\frac{1}{2}}} = \frac{\sin x}{x}. \tag{1.125}$$

Hence, the directionality coefficient of UCA is

$$D = \frac{1}{W}\left|F\left(\Omega_{E0}, \Omega_{A0}\right)\right|^2, \tag{1.126}$$

where $W = \sum_{m=1}^N \sum_{n=1}^N I_m I_n \exp\left[j\left(\beta_m - \beta_n\right)\right] \frac{\sin k\rho_{mn}}{k\rho_{mn}}$. From (1.126), we can find that: 1) When $a \approx 7\lambda/8$, the directionality of UCA reaches maximum at $\Omega_{A0} = 0$; 2) When $a = \lambda/2, 7\lambda/4$, the directionality of UCA reaches maximum at $\Omega_{A0} = \pi/2$ and $\Omega_{A0} = 0$; 3) When $a = 3\lambda/4$, the directionality of UCA reaches maximum at $\Omega_{A0} = \pi/2$ and $\Omega_{A0} = 30°$.

1.3.5 Array Channel Model

Aforementioned, when using antenna array, beamforming can be performed, e.g., Fig. 1.13. When the ULA with N_r AEs is used only at Rx, then signal received by the n-th antenna at Rx can be expressed as (1.15). When the narrowband assumption in Section 1.2.1 and the far-field assumption in Section 1.3.1 hold, if only the LoS path is considered, the lowpass signal received by the n-th antenna can be expressed as

$$x(t) = \lambda(t)e^{j\phi_D}e^{-j2\pi f_c \tau_n(t)}s(t) + n(t), \tag{1.127}$$

where $\tau_n(t) = (r + d\cos\Omega_E)/c$ represents the time delay from the transmit antenna to the n-th receive antenna and r represents the distance from the transmit antenna to the 0-th receive antenna. Hence, the channel model of the n-th antenna can be expressed as

$$h_n(t) = \lambda(t)e^{j\phi_D}e^{-j2\pi f_c \tau_n(t)}. \tag{1.128}$$

Stacking the channel from all the N_r antennas, then the signal at Rx can be represented as $\mathbf{x} = \mathbf{h}s + \mathbf{n}$. The channel impulse response of the whole antenna array can be represented as

$$\mathbf{h}(t) = \lambda(t)e^{j\phi_D}e^{-jkr}\mathbf{a}_r(\Omega_E), \tag{1.129}$$

where the channel complex coefficient is assumed to be the same for all antenna pairs and $\mathbf{a}_r(\Omega_E)$ is the steering vector of the ULA and can be represented as

$$\mathbf{a}_r(\Omega_E) = \frac{1}{\sqrt{N_r}}\left[1, e^{jkd\cos\Omega_E}, \cdots, e^{jk(N-1)d\cos\Omega_E}\right]^{\mathrm{T}}, \tag{1.130}$$

which is the same as (1.70). Note that the steering vectors of URA and UCA can be represented as (1.94) and (1.106), respectively.

When the antennas are used only at Tx, then the signal can be represented as $x = \mathbf{h}^{\mathrm{H}}\mathbf{s} + n$. The channel impulse response can be represented as

$$\mathbf{h}(t) = \lambda(t)e^{j\phi_D}e^{-jkr}\mathbf{a}_t(\Omega_E). \tag{1.131}$$

When the ULAs are used both at Tx with N_t AEs and Rx with N_r AEs, the distance between the m-th transmit antenna to the n-th receive antenna is

$$r_{mn} = r + nd_r \cos \Omega_{Er} - md_t \cos \Omega_{Et}, \qquad (1.132)$$

where d_t, d_r denote the antenna spacing at Tx and Rx, respectively, and Ω_{Et}, Ω_{Er} denote the elevation AoD and elevation AoA, respectively. The channel impulse response between the m-th transmit antenna to the n-th receive antenna is

$$h_{mn} = \lambda(t)e^{j\phi_D}e^{-j2\pi\frac{r}{\lambda_c}}e^{jkd_t \cos \Omega_{Et}}e^{jkd_r \cos \Omega_{Er}}. \qquad (1.133)$$

Hence, the channel impulse matrix can be written as

$$\mathbf{H} = \sqrt{N_t N_r}\lambda(t)e^{j\phi_{D_l}-jkr}\mathbf{a}_r(\Omega_{Er})\mathbf{a}_t(\Omega_{Et})^{\mathrm{H}}, \qquad (1.134)$$

where $\mathbf{H}$ is a matrix with rank one and has a unique non-zero singular value $\lambda(t)\sqrt{N_r N_t}$. Thus, according to (1.51), the channel capacity can be obtained

$$C = \log\left(1 + \frac{E_s\lambda(t)^2 N_t N_r}{N_0}\right). \qquad (1.135)$$

When there are MPCs, the channel model can be written as

$$\mathbf{H} = \sqrt{N_t N_r} \sum_{l=0}^{L(t)-1} \lambda_l(t)e^{j\phi_{D_l}-jkr}\mathbf{a}_r(\Omega_{Er}, \Omega_{Ar})\mathbf{a}_t(\Omega_{Et}, \Omega_{At})^{\mathrm{H}}, \qquad (1.136)$$

where $L(t)$ is the number of MPCs.

1.3.6   Array Beamforming Structures

Because of the high path loss between space/air/ground, antenna arrays are usually exploited to provide high gain, narrow beam, low sidelobe level, etc. According to structures, antenna arrays can be classified as phased arrays, digital arrays, hybrid arrays, and irregular antenna arrays. Phased arrays exploit digital phase shifters and attenuators to change the beam quickly, which are flexible to track mobile users with high speed. Digital arrays implement amplitude and phase weighting in the digital domain, which can generate multiple adjustable beams simultaneously. To achieve the tradeoff between hardware complexity and beamforming performance, hybrid beamforming architectures are proposed and used, where both digital and analog beamformings are used simultaneously. Irregular antenna arrays are usually designed for wide coverage with well-designed sub-arrays.

1.3.6.1   Phased Arrays

Phased arrays are extensively exploited in radars, wireless communications, and electronic reconnaissance. The structure of phased array is shown in Fig. 1.17, where multiple antennas are connected to one radio frequency (RF) chain via phase shifters and attenuators.

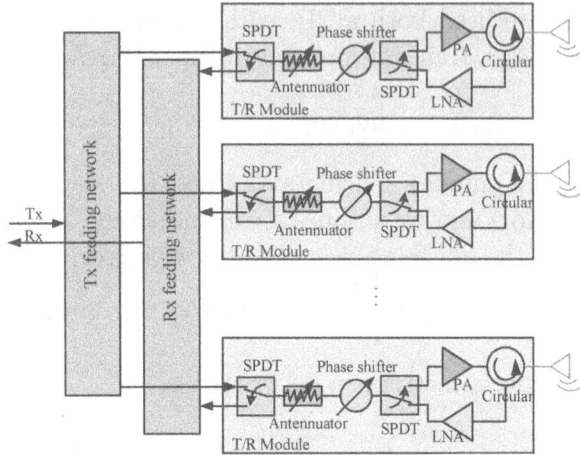

Figure 1.17 The structure of phased array.

Analog beamforming requires only one RF chain and is implemented by using phase shifters or switches in the analog domain. T/R modules are the core elements of the phased array, which are placed between antennas and the feeding network. T/R modules consist of circulars, power amplifiers (PAs), low noise amplifiers (LNAs), single-pole double-throw (SPDT) switches, phase shifters, and attenuators. The circulars and SPDT switches make it possible to reuse phase shifters and attenuators when transmitting and receiving are separate in time. The phase shifters and attenuators control and switch the beam pointing fast, and therefore the satellites with phased arrays can serve multiple wireless users simultaneously. However, only the phase of the signal can be adjusted at each AE, and thus less degree of freedoms (DoFs) are available.

1.3.6.2 *Digital Arrays*

Compared to phased arrays, digital arrays have a better dynamic range and are able to generate multiple beams simultaneously, which can serve users in different areas at the same time. The structure of a digital array is shown in Fig. 1.18, where each AE is connected to an independent RF chain. For each channel, analog to digital converters (ADCs) and digital to analog converters (DACs) are required to receive and transmit baseband signals, and digital beamforming is implemented in the baseband by digital weighting, which yields a high flexibility with sufficient DoFs to implement efficient precoding algorithms. However, a digital array is usually expensive and requires more power than a phased array. Thus, in theory, digital beamforming achieves higher performance compared to other architectures[28]. It can accommodate multi-stream transmission and distinguish signals simultaneously received from different directions. However, the digital beamforming architecture requires a dedicated RF chain for each AE. The corresponding hardware components, including ADCs, DACs, data converters, and mixers, entail a high hardware complexity and a large energy consumption. Currently, it is mainly exploited in shipborneintegrated electronic systems. Some digital arrays are also used in narrow band satellite communications now. However, with the development of technology, more and more digital phased arrays will be fabricated in space/air/ground communications.

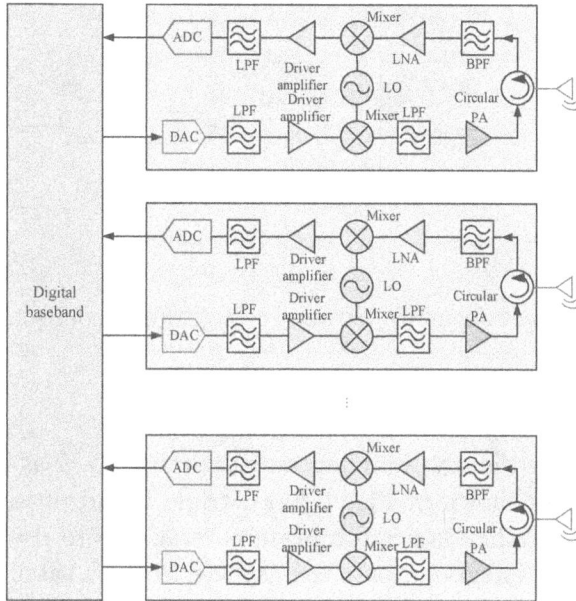

Figure 1.18 The structure of digital array.

1.3.6.3 Hybrid Antenna Arrays

To balance performance and cost, hybrid antenna arrays with analog-digital beamforming are proposed and fabricated for coverage in satellite communications[29]. A hybrid beamforming architecture uses a small number of RF chains and a large number of antennas to reduce cost and energy consumption while enabling multi-stream transmission to meet the overall performance requirements. There are two types of hybrid beamforming architectures, namely fully-connected and partially-connected. For the fully-connected hybrid beamforming architecture, as shown in Fig. 1.19(a), each RF chain connects with all AEs via phase shifters. And the fully-connected structure provides the full beamforming gain. As shown in Fig. 1.19(b), the partially-connected hybrid antenna array usually consists of $M \times N$ AEs, which contains M digital channels. Each digital channel is connected to an N-element analog phased sub-array. Thus, the hybrid antenna array can generate multiple digital beams with high gains and relatively low cost.

1.3.6.4 Irregular Antenna Arrays

For wide coverage, lots of irregular antenna arrays are designed and fabricated for spaceborne and airborne communications. Compared to conventional antenna arrays,

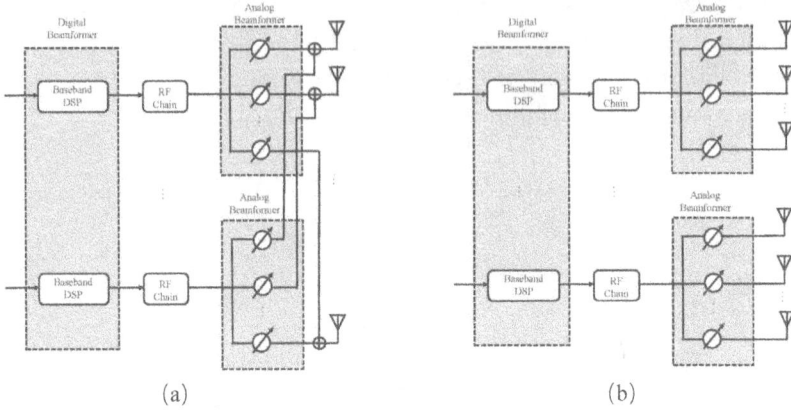

Figure 1.19 Hybrid beamforming Structure (a) Fully-connected. (b) Partially-connected.

they are usually divided in several independent parts to cover as much space as possible. For example, as shown in Fig. 1.20(a), triple round phased arrays in the X band are designed together to acquire beam coverage up to ±80°. In Fig. 1.20(b), multiple antennas are integrated on a hemisphere to track satellites in different direction by switching. This irregular antenna array provides a low-cost solution to the S-band satellite-ground digital datalink. Other irregular sparse antenna arrays, such as nested and co-prime arrays, are usually exploited in direction finding for wireless communications to provide high resolution.

Figure 1.20 Irregular antenna arrays for space/air/ground communications, (a) Triple round phased arrays in X-band. (b) Hemisphere antenna array in S-band.

1.4 ARRAY BEAMFORMING METHODS

In a MIMO system, by controlling the antenna weights, flexible beamforming can be used to overcome the high path loss caused by the long propagation distance. Moreover, beamforming technologies and array signal processing methods can improve the utilization rate of space, spectrum resources, and expand the system capacity while ensuring the communication quality. Although the directional pattern of antenna array is omnidirectional, the output beam gain can be focused in a particular direction by weighted summation, which is called "beam".

In order to suppress interference and noise, the beam pattern is supposed to have high main lobe and low side lobes and form a deep zero depression at the arrival direction of the interference. In other words, high signal-to-interference-plus-noise ratio (SINR) is desired. In this section, we discuss some classical beamforming methods, including conventional beamforming methods, statistic beamforming methods, and adaptive beamforming methods.

1.4.1 Conventional Beamforming Methods

In order to facilitate the understanding, we discuss a simple scenario first. We assume that the signal, interference and array are in the same plane. Thus, we can use a 1D angle to describe the direction of the signal, expressed as

$$\mathbf{a}_m = \mathbf{a}(\theta_m), \ m = 0, 1, \cdots, M-1, \tag{1.137}$$

where θ_m is the direction of the signal $s_m(t)$. If there is no interference signal, i.e., $M = 1$, the output signal of each array element is the desired signal with different phase shifts. In order to enhance the desired signal, we can perform phase shift operation on the output signals of each array element to make them have the same phase. Thus, the optimal weighted vector is

$$\mathbf{w}_{\mathrm{DAS}} = \mathbf{a}(\theta_0). \tag{1.138}$$

This is the conventional beamforming solution.

For more general scenarios, weighted vector design based on maximizing the output SNR can be utilized in conventional beamforming to make the main lobe steer to the direction of desired signal, i.e., θ_0. The optimal problem can be expressed as

$$\max_{\mathbf{w} \neq \mathbf{0}} \ \mathrm{OSNR}(\mathbf{w}) = \frac{\mathbf{w}^H (\mathbf{a}_0 \mathbf{a}_0^H) \mathbf{w}}{\mathbf{w}^H \mathbf{w}} \frac{\sigma_0^2}{\sigma^2}, \tag{1.139}$$

where $\mathbf{0}$ is the zero vector and $\frac{\sigma_0^2}{\sigma^2}$ is the input SNR, which will not affect the weighted vector.

According to (1.139), the norm of weighted vector $\mathbf{w}$ won't affect the output SNR. Therefore, the nonzero weighted vector that makes SNR($\mathbf{w}$) maximal is not unique. Note that

$$\mathrm{OSNR}_{\max} = \mathrm{OSNR}(\mathbf{w} \neq \mathbf{0}) = \mathrm{OSNR}(\frac{\mathbf{w}_{\max}}{\|\mathbf{w}_{\max}\|_2}), \tag{1.140}$$

where $\mathbf{w}_{\max}$ is one of the weighted vectors that makes SNR($\mathbf{w}$) maximal. (1.140) shows that there must be an optimal solution for the optimal problem (1.139) with unit norm. The unit-norm optimal solution can be acquired through solving the minimization problem with norm constraint as follow

$$\max_{\mathbf{w}} \ \mathbf{w}^H (\mathbf{a}_0 \mathbf{a}_0^H) \mathbf{w}, \\ \text{s.t. } \|\mathbf{w}\|_2^2 = 1. \tag{1.141}$$

Problem (1.141) can be solved using Lagrange Multiplier Algorithm. The optimal weighted vector is

$$\mathbf{w}_{\text{DAS}} = \frac{\mathbf{a}_0}{\mathbf{a}_0^H \mathbf{a}_0} = \frac{\mathbf{a}_0}{\|\mathbf{a}_0\|_2^2}. \tag{1.142}$$

1.4.2 Statistic Beamforming Methods

When a signal is received by an antenna array, the optimal output signals are measured in multiple ways according to the differences of the signals and environment. We assume that the received signal can be represented as $\mathbf{x} = [x_0, x_1, \cdots, x_{N-1}]^{\text{T}}$ and the signal $\mathbf{x}$ can be written as

$$\mathbf{x} = \mathbf{x}_d + \mathbf{x}_i + \mathbf{x}_n = \mathbf{x}_d + \mathbf{x}_{i,n}, \tag{1.143}$$

where $\mathbf{x}_d$ is the desired signal, $\mathbf{x}_i$ is the interference, $\mathbf{x}_n$ is the noise, and $\mathbf{x}_{i,n}$ is the combination of interference and noise, respectively. Then, $\mathbf{x}$ is shaped by the beamformer as

$$\begin{aligned} y &= w_1^* x_1 + w_2^* x_2 + \cdots + w_N^* x_N \\ &= w_1^* x_0 + w_2^* x_0 e^{jkd\cos\Omega_E} + \cdots + w_N^* x_0 e^{j(N-1)kd\cos\Omega_E} \\ &= x_0 \sum_{i=0}^{N-1} w_i^* e^{j(i-1)kd\cos\Omega_E} \\ &= \mathbf{w}^H \mathbf{x}, \end{aligned} \tag{1.144}$$

In this subsection, we discuss how to obtain the weight vector $\mathbf{w}$, so that the output signal y is as close as possible to the transmitted signal s according to the statistical optimal beamforming methods, including minimum variance distortionless response (MVDR), minimum power distortionless response (MPDR), minimum mean square error (MMSE), and maximum output signal-to-interference-plus-noise ratio (MOSINR).

1.4.2.1 MVDR and MPDR

In practice, it is sometimes not possible to accurately determine the form and the direction of the useful signal. Therefore, on the basis of assuming that the expected signal and the interference signal are both zero means; in order to better detect the useful signal and eliminate the background noise, the MVDR and MPDR criteria are proposed. As shown in (1.144), let $y = \mathbf{w}^H \mathbf{x}$ be the output of the beamformer, y_d is the output of the desired signal, y_i is the output of interference, and y_n is the output of noise, i.e., the output of the beamformer y can be written as

$$y = y_d + y_i + y_n = y_d + y_{i,n}. \tag{1.145}$$

For MVDR, we want to minimize the power of the output interference and noise, which can be expressed as

$$\mathbb{E}\left\{|y_{i,n}|^2\right\} = \mathbb{E}\left\{\left|\mathbf{w}^H\mathbf{x}_{i,n}\right|^2\right\} = \mathbf{w}^H\mathbb{E}\left\{\mathbf{x}_{i,n}\mathbf{x}_{i,n}^H\right\}\mathbf{w} = \mathbf{w}^H\mathbf{R}_{i,n}\mathbf{w}, \tag{1.146}$$

where $\mathbf{R}_{i,n}$ denotes the covariance matrix of noise and interference, expressed as

$$\mathbf{R}_{i,n} = \mathbf{R}_i + \mathbf{R}_n, \tag{1.147}$$

where $\mathbf{R}_i = \{\mathbf{x}_i\mathbf{x}_i^{\mathrm{H}}\}$ represents the covariance matrix of the interference signal and $\mathbf{R}_n = \{\mathbf{x}_n\mathbf{x}_n^{\mathrm{H}}\}$ represents the correlation matrix of noise.

For MPDR, we are supposed to minimize the power distortionless response of the signal, which can be expressed as

$$\mathbb{E}\left\{|y|^2\right\} = \mathbb{E}\left\{\left|\mathbf{w}^{\mathrm{H}}\mathbf{x}\right|^2\right\} = \mathbf{w}^{\mathrm{H}}\mathbb{E}\left\{\mathbf{x}\mathbf{x}^{\mathrm{H}}\right\}\mathbf{w} = \mathbf{w}^{\mathrm{H}}\mathbf{R}_x\mathbf{w}, \tag{1.148}$$

where $\mathbf{R}_x$ is the covariance matrix of received signal $\mathbf{x}$.

In order to obtain a response without distortion, the following constraint must be added

$$\mathbf{w}^{\mathrm{H}}\mathbf{a}_0 = 1, \tag{1.149}$$

where $\mathbf{a}_0$ is the steering vector with a unit norm value, expressed as

$$\mathbf{a}_0 = \frac{\mathbf{a}}{\|\mathbf{a}\|_2} = \frac{\mathbf{a}}{\sqrt{\mathbf{a}^{\mathrm{H}}\mathbf{a}}}, \tag{1.150}$$

where $\mathbf{a}$ is the steering vector. Then for MVDR, we can formulate the optimal problem as

$$\min_{\mathbf{w}_{\mathrm{MVDR}}} \mathbf{w}^{\mathrm{H}}\mathbf{R}_{i,n}\mathbf{w}, \\ s.t. \ \mathbf{w}^H\mathbf{a}_0 = 1. \tag{1.151}$$

For MPDR, we can formulate the optimal problem as

$$\min_{\mathbf{w}_{\mathrm{MPDR}}} \mathbf{w}^{\mathrm{H}}\mathbf{R}_x\mathbf{w}, \\ s.t. \ \mathbf{w}^H\mathbf{a}_0 = 1. \tag{1.152}$$

We can solve the above two problems using Lagrangian Multiplier Algorithm. First, we build an objective function as follows

$$J(\mathbf{w}, \lambda) = \mathbf{w}^{\mathrm{H}}\mathbf{R}_{i,n}\mathbf{w} + \lambda\left(\mathbf{w}^{\mathrm{H}}\mathbf{a}_0 - 1\right) + \lambda^{\mathrm{H}}\left(\mathbf{w}^{\mathrm{H}}\mathbf{a}_0 - 1\right), \tag{1.153}$$

where λ is the Lagrangian multiplier. Then, take the partial derivative of $J(\mathbf{w}, \lambda)$ respect to $\mathbf{w}^{\mathrm{H}}$ and λ, respectively, as follows

$$\frac{\partial J(\mathbf{w}, \lambda)}{\partial \mathbf{w}^{\mathrm{H}}} = \mathbf{R}_{i,n}\mathbf{w} + \lambda\mathbf{a}_0 = 0. \tag{1.154}$$

We can obtain that

$$\mathbf{w} = -\lambda\mathbf{R}_{i,n}^{-1}\mathbf{a}_0. \tag{1.155}$$

Substituting it into the constraint in (1.151), we can obtain that

$$\lambda = -\frac{1}{\mathbf{a}_0^{\mathrm{H}}\mathbf{R}_{i,n}^{-1}\mathbf{a}_0}. \tag{1.156}$$

Finally, we can obtain the weight vector as follows

$$\mathbf{w}_{\text{MVDR}} = \mathbf{R}_{i,n}^{-1} \left(\frac{\mathbf{a}_0}{\mathbf{a}_0^{\text{H}} \mathbf{R}_{i,n}^{-1} \mathbf{a}_0} \right). \tag{1.157}$$

In the same way, the weight vector of MPDR can be obtained as

$$\mathbf{w}_{\text{MPDR}} = \mathbf{R}_x^{-1} \left(\frac{\mathbf{a}_0}{\mathbf{a}_0^{\text{H}} \mathbf{R}_x^{-1} \mathbf{a}_0} \right). \tag{1.158}$$

It is shown in[30] that the MVDR, which utilizes $\mathbf{R}_{i,n}$ in its minimization criterion, outperforms the MPDR. However, the latter can exhibit an improved robustness to steering errors.

1.4.2.2 MMSE

The MMSE criterion requires that the mean square error of the output signal and the reference signal is minimized. Assume that the desired signal arrives from Ω_{E0} while the interference signals, $i_1, i_2, \cdots, i_M$, arrive from $\Omega_{E1}, \Omega_{E2}, \cdots, \Omega_{EM}$, respectively. All the signals, including the desired signal and the interference signals, are received by Rx. Each receive antenna has a weight w_n, as shown in Fig. 1.21, where x_r is the reference signal.

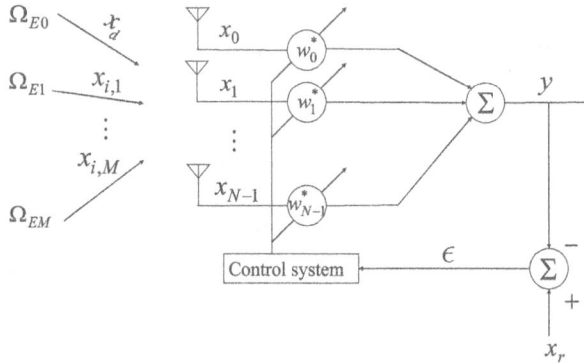

Figure 1.21 The mean square error adaptive system.

Note that the reference signal x_r needs to have a strong correlation with the desired signal x_d, but not with the interference signal x_i. If the reference signal has a high correlation with the interference signal, the MMSE criterion cannot be used. Let ϵ be the error between the output signal $\mathbf{w}^{\text{H}}\mathbf{x}$ and the reference signal x_r, expressed as

$$\epsilon = x_r - \mathbf{w}^{\text{H}}\mathbf{x}. \tag{1.159}$$

Then, the mean square error of the error ϵ can be expressed as

$$J(\mathbf{w}) = \mathbb{E}\left\{ |\epsilon|^2 \right\} = \mathbb{E}\left\{ |x_r|^2 \right\} - \mathbf{w}^{\text{H}}\mathbf{r}_{xr} - \mathbf{r}_{xr}^{\text{H}}\mathbf{w} + \mathbf{w}^{\text{H}}\mathbf{R}_x\mathbf{w}, \tag{1.160}$$

where $\mathbf{r}_{xd} = \mathbb{E}\{x_r^*\mathbf{x}\}$ is the covariance vector of the reference signal and the received signal and $\mathbf{R}_x$ is the same as that in (1.148). To obtain the optimal solution under the MMSE criterion, we need to take the derivative of $J(\mathbf{w})$ respect to $\mathbf{w}^H$ and let it equal to zero

$$\frac{\mathrm{d}J(\mathbf{w})}{\mathrm{d}\mathbf{w}^H} = -\mathbf{r}_{xr} + \mathbf{R}_x\mathbf{w} = 0. \tag{1.161}$$

Then, the optimal weight vector using MMSE method can be obtained as follows

$$\mathbf{w}_{\text{MMSE}} = \mathbf{R}_x^{-1}\mathbf{r}_{xr}. \tag{1.162}$$

1.4.2.3 MOSINR

Output signal-to-interference-plus-noise ratio (OSINR) is one of the main performance indices of the beamformer. MOSINR is to maximize OSINR, i.e., requiring in the output signal the desired signal to interference and noise signal power ratio be the maximum. Among them, the desired signal output power σ_d^2 and interference plus noise signal output power $\sigma_{i,n}^2$ is defined as

$$\sigma_d^2 = \mathbb{E}\{|\mathbf{w}^H\mathbf{x}_d|^2\} = \mathbf{w}^H\mathbf{R}_d\mathbf{w}, \tag{1.163}$$

$$\sigma_{i,n}^2 = \mathbb{E}\{|\mathbf{w}^H\mathbf{x}_{i,n}|^2\} = \mathbf{w}^H\mathbf{R}_{i,n}\mathbf{w}, \tag{1.164}$$

where $\mathbf{R}_d = \mathbb{E}\{|\mathbf{x}_d\mathbf{x}_d^H|^2\}$ represents the covariance matrix of the desired signal and $\mathbf{R}_{i,n}$ is defined in (1.147). Then, OSINR γ can be written as

$$\gamma = \frac{\sigma_d^2}{\sigma_{i,n}^2} = \frac{\mathbf{w}^H\mathbf{R}_d\mathbf{w}}{\mathbf{w}^H\mathbf{R}_{i,n}\mathbf{w}}, \tag{1.165}$$

By differentiating (1.165) with respect to $\mathbf{w}$ and equating it to zero, the maximum value of γ can be obtained. We can transform (1.165) to

$$\mathbf{R}_d\mathbf{w} = \gamma\mathbf{R}_{i,n}\mathbf{w}, \tag{1.166}$$

or

$$\mathbf{R}_{i,n}^{-1}\mathbf{R}_d\mathbf{w} = \gamma\mathbf{w}, \tag{1.167}$$

Equation (1.167) is an eigenvector equation and OSINR γ is an eigenvalue of this equation. Therefore, the maximum OSINR is equal to the maximum eigenvalue γ_{max} of the Hermitian matrix $\mathbf{R}_{i,n}^{-1}\mathbf{R}_d$. The eigenvector corresponding to the maximum eigenvalue is the optimal weight vector $\mathbf{w}_{\text{MOSINR}}$.

1.4.3 Adaptive Beamforming Methods

In practical applications, channel conditions may change over time. Hence, in adaptive beamforming, adaptive antenna array is able to adjust its weight factor in real time according to the signal environment based on the optimal filtering criteria. To obtain the expression of each optimal solution, the adaptive algorithm essentially estimates the statistic of the signal in real time, and then substitutes the estimated statistic into the corresponding optimal solution expression or iterative formula to obtain the current optimal weight vector. The adaptive antenna system architecture is shown in Fig. 1.22. In this subsection, we introduce several conventional adaptive beamforming methods, including least mean square (LMS), sample matrix inversion (SMI), recursive least squares (RLS), conjugate gradient algorithm (CGA), and constant modulus algorithm (CMA).

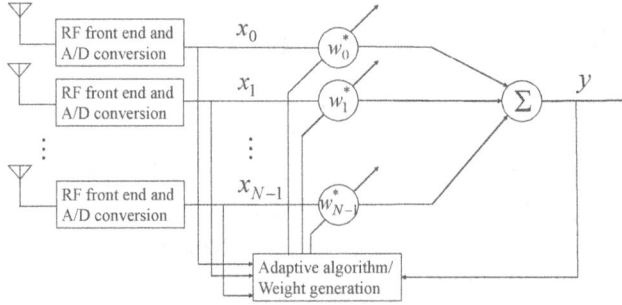

Figure 1.22   Adaptive antenna system architecture

1.4.3.1   LMS

The LMS algorithm is an iterative adaptive algorithm based on the steepest descent optimization method. The iterative formula can be expressed as

$$
\begin{aligned}
\mathbf{w}(k+1) &= \mathbf{w}(k) + \frac{1}{2}\mu \left[-\nabla_{\mathbf{w}} \left(\mathbb{E}\left\{ \epsilon^2(k) \right\} \right) \right]\Big|_{\mathbf{w}=\mathbf{w}(k)} \\
&= \mathbf{w}(k) + \mu \left(\mathbf{r}_{xr} - \mathbf{R}_x \mathbf{w}(k) \right),
\end{aligned}
\tag{1.168}
$$

where μ is a constant called a step size, $\nabla_{\mathbf{w}}$ is the gradient of the performance surface, $\mathbf{R}_x(k) = \mathbb{E}\{\mathbf{x}(k)\mathbf{x}^{\mathrm{H}}(k)\}$ is the $\mathbf{R}_x$ in the k-th snapshot, $\mathbf{r}_{xr}(k) = \mathbb{E}\{x_r^*(k)\mathbf{x}(k)\}$ is the $\mathbf{r}_{xr}$ in the k-th snapshot, $\mathbb{E}\{\epsilon(k)^2\}$ is the mean square error of the error ϵ in the k-th snapshot defined in (1.160).

The estimated values of $\mathbf{R}_x$ and $\mathbf{r}_{xr}$ are

$$
\begin{aligned}
\hat{\mathbf{R}}_x(k) &= \mathbf{x}(k)\mathbf{x}^{\mathrm{H}}(k), \\
\hat{\mathbf{r}}_{xr}(k) &= x_r^*(k)\mathbf{x}(k),
\end{aligned}
\tag{1.169}
$$

where $x_r(k)$ is a known training sequence. Thus, the iterative formula of the LMS algorithm can be obtained

$$
\begin{aligned}
\mathbf{w}(k+1) &= \mathbf{w}(k) + \mu \left[\mathbf{x}(k)x_r^*(k) - \mathbf{x}(k)\mathbf{x}^{\mathrm{H}}(k)\mathbf{w}(k) \right] \\
&= \mathbf{w}(k) + \mu \mathbf{x}(k) \left[x_r^*(k) - y^*(k) \right] \\
&= \mathbf{w}(k) + \mu \mathbf{x}(k)\epsilon^*(k).
\end{aligned}
\tag{1.170}
$$

In the LMS algorithm, the value of the step size μ has a very important influence on the performance of the algorithm, affecting the stability, convergence speed, and perturbation. According to[31], the well-known convergence condition for the LMS is

$$0 < \mu < \frac{2}{\lambda_{max}}, \tag{1.171}$$

where λ_{max} is the maximum eigenvalue of $\mathbf{R}_x$.

1.4.3.2 SMI

The SMI algorithm, also called directed matrix inversion (DMI), is mainly based on the MMSE criterion in (1.162). We can divide the received signal vector sequence into a continuous data block in units of several sample lengths. Then, $\mathbf{R}_x$ and $\mathbf{r}_{xr}$ are estimated in the units of data blocks. Assuming that the sample length of each data block is N, the received signal sequence in the first data block is $(\mathbf{x}(0), \mathbf{x}(1), \cdots, \mathbf{x}(N-1))$, and in the $(k+1)$-th data block, it is $(\mathbf{x}(kN+0), \mathbf{x}(kN+1), \cdots, \mathbf{x}(kN+N-1))$. Thus, the statistic is estimated for each data block in turn, and the corresponding weight vector can update in turn.

For the $(k+1)$-th data block, the estimations value of $\mathbf{R}_x$ and $\mathbf{r}_{xr}$ are

$$\hat{\mathbf{R}}_x(k) = \frac{1}{N} \sum_{i=0}^{N-1} \mathbf{x}(kN+i)\mathbf{x}^{\mathrm{H}}(kN+i),$$

$$\hat{\mathbf{r}}_{xr}(k) = \frac{1}{N} \sum_{i=0}^{N-1} \mathbf{x}(kN+i)x_r(kN+i). \tag{1.172}$$

Replacing the corresponding statistics in (1.162) by the estimates in (1.172), we can obtain the weight vector of the k-th data block as follows:

$$\mathbf{w}(k) = \hat{\mathbf{R}}_x^{-1}(k)\hat{\mathbf{r}}_{xr}(k). \tag{1.173}$$

Note that $\mathbf{w}(k)$ does not represent the weight of the k-th sampling moment, but the weight of the k-th data block.

1.4.3.3 RLS

The RLS algorithm is also mainly based on the MMSE criterion in (1.162). It does not divide the received signal vector sequence into data blocks like SMI but uses a data observation interval in the form of a sliding window. The estimations of $\mathbf{R}_x$ and $\mathbf{r}_{xd}$ are updated for each sampled signal vector that is newly received. For the k-th sample, we have

$$\hat{\mathbf{R}}_x(k) = \sum_{i=1}^{k} \lambda^{k-i}\mathbf{x}(i)\mathbf{x}^{\mathrm{H}}(i),$$

$$\hat{\mathbf{r}}_{xr}(k) = \sum_{i=1}^{k} \lambda^{k-i}\mathbf{x}(i)x_r(i), \tag{1.174}$$

where λ is called forgetting factor and needs to meet $0 \le \lambda \le 1$. There is a correlation between the statistics and weights before and after the update. Therefore, the operation of directly taking the inverse of the matrix can be replaced by an iterative operation, which can be expressed as

$$
\begin{aligned}
\hat{\mathbf{R}}_x(k) &= \lambda \hat{\mathbf{R}}_x(k-1) + \mathbf{x}(k)\mathbf{x}^{\mathrm{H}}(k), \\
\hat{\mathbf{r}}_{xr}(k) &= \lambda \hat{\mathbf{r}}_{xr}(k-1) + x_r^*(k)\mathbf{x}(k),
\end{aligned}
\tag{1.175}
$$

where (1.174) is divided into two terms, namely the sum of the first $i = k - 1$ terms and the last term $i = k$. According to the Woodbury identity, we have

$$
\hat{\mathbf{R}}_x^{-1}(k) = \lambda^{-1}\left[\hat{\mathbf{R}}_x^{-1}(k-1) - \mathbf{q}(k)\mathbf{x}^{\mathrm{H}}(k)\hat{\mathbf{R}}_x^{-1}(k-1)\right],
\tag{1.176}
$$

where

$$
\mathbf{q}(k) = \frac{\lambda^{-1}\hat{\mathbf{R}}_x^{-1}(k-1)\mathbf{x}(k)}{1 + \lambda^{-1}\mathbf{x}^{\mathrm{H}}(k)\hat{\mathbf{R}}_x^{-1}(k-1)\mathbf{x}(k)}.
\tag{1.177}
$$

Then, the weight vector update equation is

$$
\begin{aligned}
\mathbf{w}(k) &= \hat{\mathbf{R}}_x^{-1}\hat{\mathbf{r}}_{xr}(k) \\
&= \mathbf{w}(k-1) + \mathbf{q}(k)\epsilon(k).
\end{aligned}
\tag{1.178}
$$

Note that the introduction of a forgetting factor in the RLS algorithm is necessary. Because in the time-varying channel, the farther the data is from the current moment, the lower the current channel state and its correlation, and the less the sampled signal value contributes to the estimation of the weight vector. Moreover, the value of the forgetting factor can be changed depending on the speed of the time-varying channel change. According to[32], when the channel time-change rate is slow, a large forgetting factor can be selected. Otherwise, a small forgetting factor is selected.

1.4.3.4 CGA

The same as SMI algorithm, the CGA also needs to divide the received signals into data blocks and generates a weight vector in each data block. The difference is that CGA iteratively solves the least squares solution in the data block by the steepest descending gradient search iteration equation, which is similar to the LMS algorithm. The set of equations to be solved is as follows:

$$
\begin{cases}
\mathbf{x}^{\mathrm{H}}(kN)\mathbf{w}_k + r(kN) = x_r^*(kN), \\
\mathbf{x}^{\mathrm{H}}(kN+1)\mathbf{w}_k + r(kN+1) = x_r^*(kN+1), \\
\vdots \\
\mathbf{x}^{\mathrm{H}}(kN+N-1)\mathbf{w}_k + r(kN+N-1) = x_r^*(kN+N-1).
\end{cases}
\tag{1.179}
$$

The error vector can be expressed as

$$
\mathbf{r}_k = \mathbf{x}_{r,k}{}^* - \mathbf{X}_k^{\mathrm{H}}\mathbf{w}_k,
\tag{1.180}
$$

where $\mathbf{r}_k = [r(kN), r(kN+1), \cdots, r(kN+N-1)]^{\mathrm{T}}$, $\mathbf{X}_k = [\mathbf{x}(kN), \cdots, \mathbf{x}(kN+N-1)]$ and $\mathbf{x}_{r,k} = [x_r(kN), x_r(kN+1), \cdots, x_r(kN+N-1)]^{\mathrm{T}}$. The purpose of CGA is to minimize $\mathbf{r}_k^{\mathrm{H}}\mathbf{r}$. The propose of CGA is to minimize $\mathbf{r}_k^{\mathrm{H}}\mathbf{r}$. Let $\mathbf{w}_k(k)$ denote the result of the k-th iteration within the $(k+1)$-th data block and initialize it as $\mathbf{w}_k(0)$. Then, the initial error vector can be expressed as

$$\mathbf{r}_k(0) = \mathbf{x}_{r,k}^* - \mathbf{X}_k^{\mathrm{H}}\mathbf{w}_k(0). \tag{1.181}$$

The initial gradient vector respect to $\mathbf{w}_k$ on the error plane $\mathbf{r}_k^{\mathrm{H}}\mathbf{r}_k$ is

$$\mathbf{g}_k(0) = \mathbf{X}_k\mathbf{r}_k(0). \tag{1.182}$$

Then, we can implement iterations by the following equations

$$\begin{aligned}
\mathbf{w}_k(n+1) &= \mathbf{w}_k(n) - \mu(n)\mathbf{g}_k(n) \\
\mathbf{r}_k(n+1) &= \mathbf{r}_k(n) + \mu(n)\mathbf{X}_k^{\mathrm{H}}\mathbf{g}_k(n) \\
\mathbf{g}_k(n+1) &= \mathbf{X}_k\mathbf{r}_k(n+1) - \alpha(n)\mathbf{g}_k(n),
\end{aligned} \tag{1.183}$$

where the step sizes are

$$\begin{aligned}
\mu(n) &= \frac{\mathbf{r}_k^{\mathrm{H}}(n)\mathbf{X}_k\mathbf{X}_k^{\mathrm{H}}\mathbf{r}_k(n)}{\mathbf{g}_k^{\mathrm{H}}(n)\mathbf{X}_k\mathbf{X}_k^{\mathrm{H}}\mathbf{g}_k(n)} \\
\alpha(n) &= \frac{\mathbf{r}_k^{\mathrm{H}}(n+1)\mathbf{X}_k\mathbf{X}_k^{\mathrm{H}}\mathbf{r}_k(n+1)}{\mathbf{r}_k^{\mathrm{H}}(n)\mathbf{X}_k\mathbf{X}_k^{\mathrm{H}}\mathbf{r}_k(n)}.
\end{aligned} \tag{1.184}$$

1.4.3.5 CMA

The CMA is applied to situations where the desired signal has a constant envelope, e.g., frequency modulation (FM), phase modulation (PM), frequency shift keying (FSK) or quasi-constant modulation, etc. First, we define the cost function as follows:

$$J_{p,q}(k) = \mathbb{E}\left\{\left|\left[|y(k)|^p - \sigma^p\right]\right|^q\right\}, \tag{1.185}$$

where $y(k) = \mathbf{w}^{\mathrm{H}}(k)\mathbf{x}(k)$ is the output signal of the beamformer, $\sigma = 1$ is a constant called constant modules factor, p, q are positive real numbers typically set as 1 or 2. Usually a particular value of σ, i.e., 1 or not, does not affect the performance and the corresponding algorithm is called (p, q) constant modules algorithm, i.e., $\mathrm{CMA}_{p,q}$.

Since the cost function of CMA (1.185) is non-linear and cannot be solved directly, the optimal solution can be obtained by an iterative method. There are many types of CMAs. The first is the stochastic gradient descent (SGD) CMA. The proposed algorithm is based on the steepest descent gradient search. When p and q are determined, the iterative formula of SGD-CMA can be represented as

$$\mathbf{w}(k+1) = \mathbf{w}(k) + \mu\mathbf{x}(k)\epsilon^*(k), \tag{1.186}$$

where μ is the step size. When $p = 1, q = 2$, we can obtain

$$\epsilon(k) = \frac{y(k)}{|y(k)|} - y(k), \tag{1.187}$$

and when $p = 2, q = 2$, we can obtain

$$\epsilon(k) = 2y(k)\text{sgn}\left(1 - |y(k)|^2\right). \tag{1.188}$$

Note that (1.187) and (1.188), i.e., $p = 1, q = 2$ and $p = 2, q = 2$ are most commonly used. The convergence performance of SGD-CMA highly depends on the initial value and the step size. On the one hand, if the step size is too small, the convergence speed may be very slow. On the other hand, if the step size is too large, the performance will be easily maladjusted.

The second is the least square CMA (LS-CMA). It is similar to the SMI algorithm above that is done in data blocks. Let $\mathbf{X} = [x_k(0), x_k(1), \cdots, x_k(N-1)]$ be the data block at the k-th moment with N sample data. In the static LS-CMA, different data blocks are completely independent. Then the output signal can be represented as $\mathbf{y} = [y_k(0), y_k(1), \cdots, y_k(N-1)]$. Let e_n denote the error, i.e., $e_n(\mathbf{w}) = |y(n)| - 1 = |\mathbf{w}(n)x(n)| - 1$. Take the accumulated sum of squares of the error as the cost function

$$C(\mathbf{w}) = \sum_{n=0}^{N-1} |e_n(\mathbf{w})|^2 = \|e(\mathbf{w})\|_2^2, \tag{1.189}$$

where $C(\mathbf{w})$ denotes the cost function in order not to be confused with the Jacobian matrix $\mathbf{J}(\mathbf{w})$. Our purpose is to minimize the cost function. By using the Taylor series expansion, we can obtain

$$C(\mathbf{w} + \Delta) \approx \|e(\mathbf{w}) + \mathbf{J}^{\text{H}}(\mathbf{w})\Delta\|_2^2, \tag{1.190}$$

where Δ denotes the offset vector when the weight vector $\mathbf{w}$ is updated and $\mathbf{J}^{\text{H}}(\mathbf{w})$ is the complex Jacobian matrix of error $e(\mathbf{w})$ and can be represented as

$$\mathbf{J}^{\text{H}}(\mathbf{w}) = [\nabla e_0(\mathbf{w}), \nabla e_1(\mathbf{w}), \cdots, \nabla e_{N-1}(\mathbf{w})]. \tag{1.191}$$

Find the gradient of (1.191) and set it to 0. Then we can obtain the iterative formula of the weight vector as follows:

$$\mathbf{w}(k+1) = \mathbf{w}(k) - \left[\mathbf{J}(\mathbf{w}(k))\mathbf{J}^{\text{H}}(\mathbf{w}(k))\right]^{-1} \mathbf{J}(\mathbf{w}(k))e(\mathbf{w}(k)). \tag{1.192}$$

When $p = 1, q = 2$, the iterative formula of the weight vector can be represented as

$$\mathbf{w}(k+1) = [\mathbf{X}\mathbf{X}^{\text{H}}]^{-1}\mathbf{X}\mathbf{\Gamma}_k^*, \tag{1.193}$$

where $\mathbf{\Gamma}_k$ is the complex limiting output vector

$$\mathbf{\Gamma}_k = \left[\frac{y_k(0)}{|y_k(0)|}, \frac{y_k(1)}{|y_k(1)|}, \cdots, \frac{y_k(N-1)}{|y_k(N-1)|}\right]^{\text{H}}. \tag{1.194}$$

1.5 SUMMARY

This chapter mainly includes the fundamentals of array beamforming. In Section 1.2, we first introduced the propagating characteristics of the electromagnetic waves and diversity methods to overcome the impact of fading on BER. Then, conventional transmission structures that can improve the channel performance, i.e., SIMO, MISO and MIMO, were introduced. In Section 1.3, we presented some commonly used antenna array models. Before introducing the array models, we introduced the principle of plane wave at the beginning to introduce the common assumptions of the antenna array, i.e., the far-field assumption, and then extended the two-element array to the ULA and planar array (URA and UCA). At last, the array channel model and different array structures were introduced. In Section 1.4, some array beamforming methods were introduced, including conventional beamforming methods, statistical beamforming methods, and adaptive beamforming methods.

Bibliography

[1] Zhenyu Xiao, Zhu Han, Arumugam Nallanathan, Octavia A. Dobre, Bruno Clerckx, Jinho Choi, Chong He, and Wen Tong. Antenna array enabled space/air/ground communications and networking for 6G. *arXiv preprint arXiv:2110.12610*, 2021.

[2] Ming Xiao, Shahid Mumtaz, Yongming Huang, Linglong Dai, Yonghui Li, Michail Matthaiou, George K Karagiannidis, Emil Björnson, Kai Yang, I Chih-Lin, et al. Millimeter wave communications for future mobile networks. *IEEE J. Select. Areas Commun.*, 35(9):1909–1935, 2017.

[3] Xinyu Gao, Linglong Dai, Han Shuangfeng, I Chih-lin, and Robert W. Heath. Energy-efficient hybrid analog and digital precoding for mmWave MIMO systems with large antenna arrays. *IEEE J. Select. Areas Commun.*, 34(4):998–1009, Apr. 2016.

[4] Zhenyu Xiao, Lipeng Zhu, Jinho Choi, Pengfei Xia, and Xiang-Gen Xia. Joint power allocation and beamforming for non-orthogonal multiple access (NOMA) in 5G millimeter wave communications. *IEEE Trans. Wireless Commun.*, 17(5):2961–2974, May 2018.

[5] Lipeng Zhu, Jun Zhang, Zhenyu Xiao, Xianbin Cao, Dapeng Oliver Wu, and Xiang-Gen Xia. Millimeter-wave NOMA with user grouping, power allocation and hybrid beamforming. *IEEE Trans. Wireless Commun.*, 18(11):5065–5079, Nov. 2019.

[6] Pmbir K. Bondyopadhyay. The first application of array antenna. In *Proceedings 2000 IEEE International Conference on Phased Array Systems and Technology (Cat. No. 00TH8510)*, pages 29–32. IEEE, 2000.

[7] Enji Rai, Shinkichi Nishimoto, Takeshi Katada, and Hideaki Watanabe. Historical overview of phased array antennas for defense application in japan. In *Proceedings of International Symposium on Phased Array Systems and Technology*, pages 217–221. IEEE, 1996.

[8] David A. Shnidman. A generalized nyquist criterion and an optimum linear receiver for a pulse modulation system. *Bell Syst. Tech. J.*, 46(9):2163–277, 1967.

[9] Richard H Roy III and Bjorn Ottersten. Spatial division multiple access wireless communication systems, May 7 1996. US Patent 5,515,378.

[10] Mingjin Wang, Feifei Gao, Shi Jin, and Hai Lin. An overview of enhanced massive MIMO with array signal processing techniques. *IEEE J. Sel. Top. Sign. Proces.*, 13(5):886–901, 2019.

[11] Emre Telatar. Capacity of multi-antenna gaussian channels. *Eur. Trans. Telecommun.*, 10(6):585–595, 1999.

[12] Gerard J. Foschini. Layered space-time architecture for wireless communication in a fading environment when using multi-element antennas. *Bell Labs Tech. J.*, 1(2):41–59, 1996.

[13] Gerard J. Foschini and Michael J. Gans. On limits of wireless communications in a fading environment when using multiple antennas. *Wirel. Pers. Commun.*, 6(3):311–335, 1998.

[14] Siavash M. Alamouti. A simple transmit diversity technique for wireless communications. *IEEE J. Select. Areas Commun.*, 16(8):1451–1458, 1998.

[15] Vahid Tarokh, Nambi Seshadri, and A. Robert Calderbank. Space-time codes for high data rate wireless communication: Performance criterion and code construction. *IEEE Trans. Inform. Theory*, 44(2):744–765, 1998.

[16] Jiann-Ching Guey, Michael P. Fitz, Mark R. Bell, and Wen-Yi Kuo. Signal design for transmitter diversity wireless communication systems over rayleigh fading channels. *IEEE Trans. Commun.*, 47(4):527–537, 1999.

[17] Nihar Jindal and Andrea Goldsmith. Dirty-paper coding versus TDMA for MIMO broadcast channels. *IEEE Trans. Inform. Theory*, 51(5):1783–1794, 2005.

[18] Robert G. Kouyoumjian and Prabhakar H. Pathak. A uniform geometrical theory of diffraction for an edge in a perfectly conducting surface. *Proc. IEEE*, 62(11):1448–1461, 1974.

[19] Joram Walfisch and Henry L. Bertoni. A theoretical model of UHF propagation in urban environments. *IEEE Trans. Antennas Propagat.*, 36(12):1788–1796, 1988.

[20] Andrea Goldsmith. *Wireless communications*. Cambridge university press, 2005.

[21] Stephen O. Rice. Mathematical analysis of random noise. *Bell Syst. Tech. J.*, 23(3):282–332, 1944.

[22] David Tse and Pramod Viswanath. *Fundamentals of Wireless Communication*. Cambridge university press, 2005.

[23] Siavash M. Alamouti. A simple transmit diversity technique for wireless communications. *IEEE J. Select. Areas Commun.*, 16(8):1451–1458, 1998.

[24] Thomas M. Cover. *Elements of Information Theory*. John Wiley & Sons, 1999.

[25] Constantine A. Balanis. *Antenna Theory: Analysis and Design*. John wiley & sons, 2015.

[26] Robert S. Elliott. Beamwidth and directivity of large scanning arrays. *first of two parts, Microw. J.*, pages 53–60, 1963.

[27] Ryan T. Bates. Mode theory approach to arrays. Technical report, MITRE CORP BEDFORD MA, 1966.

[28] Wonil Roh, Ji-Yun Seol, Jeongho Park, Byunghwan Lee, Jaekon Lee, Yungsoo Kim, Jaeweon Cho, Kyungwhoon Cheun, and Farshid Aryanfar. Millimeter-wave beamforming as an enabling technology for 5G cellular communications: Theoretical feasibility and prototype results. *IEEE Commun. Mag.*, 52(2):106–113, 2014.

[29] Deyi Peng, Ashok Bandi, Yun Li, Symeon Chatzinotas, and Björn Ottersten. Hybrid beamforming, user scheduling, and resource allocation for integrated terrestrial-satellite communication. *IEEE Trans. Veh. Technol.*, 70(9):8868–8882, 2021.

[30] Livnat Ehrenberg, Sharon Gannot, Amir Leshem, and Ephraim Zehavi. Sensitivity analysis of MVDR and MPDR beamformers. In *2010 IEEE 26-th Convention of Electrical and Electronics Engineers in Israel*, pages 000416–000420. IEEE, 2010.

[31] Yilun Chen, Yuantao Gu, and Alfred O. Hero. Sparse LMS for system identification. In *2009 IEEE International Conference on Acoust., Speech, Signal Processing*, pages 3125–3128. IEEE, 2009.

[32] Ahmed El Zooghby. *Smart Antenna Engineering*. Artech House, 2005.

CHAPTER 2

Codebook-Based Beamforming and Channel Estimation

2.1 INTRODUCTION

In Chapter 1, we introduced the channel model of array beamforming, i.e., the Saleh-Valenzuela model. As can be seen from the channel model, beamforming technology uses the radiation fields of antenna elements (AEs) to form a directional beam, which is different from conventional omni-directional antennas. Hence, for array beamforming enabled communication systems, we need to provide a mechanism that automatically finds the best path between the information source and sink. Moreover, in practice, the channel is usually dynamically changing, which means the paths may appear/disappear with time. Thus, we need to be able to switch beam to the next best path in case the current best path is lost. This procedure is also called beam training.

To automatically find the best path in changing environments, normally two types of antenna arrays that can be used, where one is the fully adaptive antenna array, and the other is the switched antenna array. For the fully adaptive antenna array, i.e., the digital beamforming structure, introduced in Chapter 1, the iterative antenna training/tracking algorithm requires that both the modulus and phases of the array are controllable. A fully adaptive antenna array is the most versatile antenna array and, at least in theory, is able to form an unlimited number of beam patterns. This is achievable by configuring the phase shifters on all the antenna branches with appropriate beamforming weights. In case the channel has changed (line-of-sight (LOS) being blocked for example), a new set of beamforming weights may be configured to select the current best path. Instead, the switched array with only the phases controllable which results in the constant-modulus (CM) constraint on their weights[1,2,3,4] offers lower complexity and is more attractive. A switched antenna array, on the other hand, assumes a structured codebook, e.g., a finite number of fixed, pre-defined beam patterns or combining strategies. By configuring which beam is to be selected, the switched antenna array (equipped with a proper antenna training/tracking algorithm, of course) is able to automatically find the best path between the information source and sink. In case the channel has changed, a new beam pattern may be selected. Typically an antenna array with CM coefficients is favored due to low implementation cost[3, 4, 5, 6, 7].

Meanwhile, a hybrid analog/digital precoding structure was also proposed to realize multi-stream/multi-user transmission[8, 9, 10], where a small number of ratio frequency (RF) chains are tied to a large antenna array. No matter whether the analog beamforming or the hybrid precoding structure is exploited, entry-wise estimation of channel state information (CSI) is time costly due to large-size antenna arrays, and a more efficient beam training approach is needed.

For the switched beamforming, it requires to search, based on a pre-designed beamforming vector codebook, in the transmitter (Tx)/receiver (Rx) angle domain to find an angle of arrival (AoA)/angle of departure (AoD) pair corresponding to a single strong multipath component (MPC). An exhaustive search algorithm may be used, which sequentially tests all the beam directions in the angle domain and finds the best pair of Tx/Rx beamforming codewords. This is conceptually straightforward. However, the overall search time is prohibitive, because the number of candidate beam directions is usually large for massive multiple-input multiple-output (MIMO) systems.

To improve search efficiency, a beamforming approach called hierarchical search[3, 11, 12] is adopted, where the beam search space (at the transmitter and receiver sides, respectively) is represented by a codebook containing multiple codewords. For example, a coarse beamforming vector codebook may be defined with a small number of coarse/low-resolution beams (or sectors) covering the intended angle range, while a fine beamforming vector codebook may be defined with a large number of fine/high-resolution beams covering the same intended angle range, and that a coarse beam may have the same/similar coverage as that of multiple fine beams together. A divide-and-conquer search may then be carried out across the hierarchy of codebooks, by finding the best sector (or coarse beam) first on the low-resolution codebook level, and then the best fine beam on the high-resolution codebook level, with the best high-resolution beam contained by the best low-resolution beam. The codebooks can be designed in advance, and the best transmit/receive beamforming vectors are found by searching through their respective codebooks.

Although the hierarchical search method is time efficient and can achieve a high detection rate to acquire an MPC, it is usually limited to acquiring only one single MPC. Extending it to multiple MPCs is nontrivial due to the limited angle resolution of a normal codebook, since the angle estimation error would result in a significant effect on the search for the remaining MPCs. Increasing the angle resolution of the codebook may help to search multiple MPCs, though it not only increases the training overhead, but also requires high-resolution phase shifters.

Different from the hierarchical search method, the compressed sensing (CS)-based approaches are open-looped, which means that the pilot overhead does not increase in the multi-user case. However, the performance of the CS-based schemes is highly dependent on the number of measurements. To achieve a satisfactory estimation performance, the training overhead is in fact considerably high.

Hence, in this chapter, we first propose four different codebooks which will be used in the hierarchical beamforming approach. With the aid of a codebook, we develop several fast yet accurate channel estimation and enhanced beam search methods for communication systems with large antenna arrays by exploiting the spatial sparsity.

2.2 CODEBOOK DESIGN

In this section, we introduce four different codebooks and give a performance evaluation between them, namely Deactivation Codebook, Joint Sub-Array and Deactivation Codebook, Enhanced Codebook, and Codebook for Hybrid Structures. Although several hierarchial search schemes have been proposed for beam search in both literatures and some standards, like IEEE 802.15.3c and IEEE 802.11ad, a unified criterion is needed first to judge whether a codebook is suitable or not. Thus, in this section, we first propose two basic criteria for designing a hierarchical codebook.

2.2.1 Two Criteria

First, we define the steering vector function $\mathbf{a}(\cdot)$ as in (1.70)

$$\mathbf{a}(N,\Omega) = \frac{1}{\sqrt{N}}[e^{j\pi 0\Omega}, \ e^{j\pi 1\Omega}, ..., e^{j\pi(N-1)\Omega}]^{\mathrm{T}}. \tag{2.1}$$

where N is the number of antennas, Ω is AoD or AoA.

Let $A(\mathbf{w}, \Omega)$ denote the beam gain of $\mathbf{w}$ along angle Ω, which is defined as

$$A(\mathbf{w},\Omega) = \sqrt{N}\mathbf{a}(N,\Omega)^{\mathrm{H}}\mathbf{w} = \sum_{n=1}^{N}[\mathbf{w}]_n e^{-\mathrm{j}\pi(n-1)\Omega}, \tag{2.2}$$

Let $\mathcal{CV}(\mathbf{w})$ denote the beam coverage in the angle domain of antenna weight vector (AWV) $\mathbf{w}$, which can be mathematically expressed as

$$\mathcal{CV}(\mathbf{w}) = \left\{\Omega \middle| |A(\mathbf{w},\Omega)| > \rho \max_{\omega}|A(\mathbf{w},\omega)|\right\}, \tag{2.3}$$

where ρ is a factor within $(0,1)$ to determine the beam coverage of $\mathbf{w}$. It is easy to find that the coverage is smaller as ρ becomes greater. When $\rho = 1/\sqrt{2}$, the beam coverage is the 3 dB coverage, and the beam width is the well-known 3 dB beam width. Different codebook design methods may have different values of ρ, and codewords with different beam widths in the same codebook may also have different values of ρ.

Hierarchical search is simply *layered* search, i.e., the AWVs within the codebook are *layered* according to their beam width. As shown in Fig. 2.1, AWVs with a lower layer have larger beam width. Let $\mathbf{w}(k,n)$ denote the n-th codeword (or AWV) in the k-th layer, the two criteria are presented as follows.

Criterion 1: The union of the beam coverage of all the codewords within each layer should cover the whole angle domain, i.e.,

$$\bigcup_{n=1}^{N_k} \mathcal{CV}(\mathbf{w}(k,n)) = [-1,1], \ k = 0, 1, ..., K, \tag{2.4}$$

where N_k is the number of codewords in the k-th layer, K is the maximal index of the layer (there are $K + 1$ layers in total).

Criterion 2: The beam coverage of an arbitrary codeword within a layer should be covered by the union of those of several codewords in the next layer, i.e.,

$$CV(\mathbf{w}(k,n)) \subseteq \bigcup_{m \in \mathcal{I}_{k,n}} CV(\mathbf{w}(k+1,m)), \quad k = 0, 1, ..., K - 1, \tag{2.5}$$

where $\mathcal{I}_{k,n}$ is the index set with indices of the codewords in the $(k+1)$-th layer for the n-th codeword in the k-th layer. For convenience, we call $\mathbf{w}(k,n)$ a parent codeword, and $\{\mathbf{w}(k+1,m)|m \in \mathcal{I}_{k,n}\}$ the child codewords of $\mathbf{w}(k,n)$.

It is clear that Criterion 1 guarantees the full coverage, i.e., there is no miss of any angle during the beam search, while Criterion 2 establishes a tree-fashion relationship between the codewords, which enables hierarchical search. If each parent codeword has M child codewords, all the codewords in the codebook constitute an M-way tree with respect to their beam coverage in the angle domain. In such a case, hierarchical search can be easily realized by using the tree search algorithm in both receiver and transmitter as follows[12, 13].

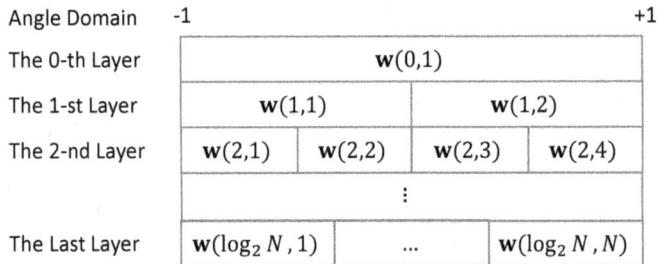

Angle Domain	-1			+1
The 0-th Layer	$\mathbf{w}(0,1)$			
The 1-st Layer	$\mathbf{w}(1,1)$		$\mathbf{w}(1,2)$	
The 2-nd Layer	$\mathbf{w}(2,1)$	$\mathbf{w}(2,2)$	$\mathbf{w}(2,3)$	$\mathbf{w}(2,4)$
		⋮		
The Last Layer	$\mathbf{w}(\log_2 N, 1)$	...	$\mathbf{w}(\log_2 N, N)$	

Figure 2.1 Beam coverage of a binary-tree structured codebook.

2.2.2 Antenna Deactivation Codebook

In this subsection, we design a codebook by using the deactivation method. This codebook contains two types of sub-codebooks, namely coarse and fine codebooks. It is noted that the coarse codebook and the coarse search are mandatory, while the fine codebook and the fine search are optional. In the case that only a rough beam direction is required, the fine codebook and fine search can be absent.

2.2.2.1 Coarse Codebook

In order to apply the binary search algorithm in coarse search, the coarse codebook should be hierarchical and organized into a binary tree. Based on the two criteria, let us set $M = 2$, which means each parent codeword has 2 child codewords. The structure of the coarse codebook is listed as Fig. 2.1. It is clear that there are $\log_2(N) + 1$ layers with indices from $k = 0$ to $k = \log_2(N)$, and the number of codewords in the

k-th layer $N_k = 2^k$. Here N denotes the number of antennas of an arbitrary array. Thus, $N = N_T$ at the transmitter and $N = N_R$ at the receiver. Besides, we have

$$CV(\mathbf{w}(k,n)) = CV(\mathbf{w}(k+1, 2n-1)) \cup CV(\mathbf{w}(k+1, 2n)),$$
$$k = 0, 1, ..., (\log_2(N) - 1), \quad n = 1, 2, 3, ..., 2^k. \tag{2.6}$$

In this book, we define

$$CV(\mathbf{a}(N, \Omega)) = \left[\Omega - \frac{1}{N}, \Omega + \frac{1}{N}\right], \tag{2.7}$$

which means that the steering vectors have a beam width $2/N$ centering at the steering angle[14]. In other words, within the beam coverage of $\mathbf{a}(N, \Omega)$, it has the maximal beam gain along the angle Ω, while the minimal beam gain along the angles $\Omega \pm 1/N$. Thus, we can compute the value of ρ for our codebook as

$$\rho = \left| \frac{\mathbf{a}(N, \Omega - 1/N)^H \mathbf{a}(N, \Omega)}{\mathbf{a}(N, \Omega)^H \mathbf{a}(N, \Omega)} \right| \text{ or } \left| \frac{\mathbf{a}(N, \Omega + 1/N)^H \mathbf{a}(N, \Omega)}{\mathbf{a}(N, \Omega)^H \mathbf{a}(N, \Omega)} \right|$$
$$= \frac{1}{N} \left| \sum_{n=1}^{N} e^{j\pi(n-1)/N} \right|. \tag{2.8}$$

Although the value of ρ depends on N, we have $\rho \approx 0.64$ given that N is large, e.g., $N \geq 8$. Even when N is small, ρ is still close to 0.64, e.g., $\rho = 0.65$ when $N = 4$.

Notice that the N codewords in the last layer cover an angle range $[-1, 1]$ in total, which means that all these codewords must have the narrowest beam width $2/N$ with different steering angles. In other words, the codewords in the last layer should be the steering vectors with angles evenly sampled within $[-1, 1]$. Consequently, we have $CV(\mathbf{w}(\log_2(N), n)) = [-1 + \frac{2n-2}{N}, -1 + \frac{2n}{N}]$, $n = 1, 2, ..., N$. With the beam coverage of the last-layer codewords, we can further obtain that of the codewords in the other layers in turn as an order of descending layer indices, i.e., obtain $CV(\mathbf{w}(\log_2(N) - 1, n))$, $CV(\mathbf{w}(\log_2(N) - 2, n))$, ..., $CV(\mathbf{w}(0, n))$ in turn. Finally, the beam coverage of all the codewords can be uniformly written as

$$CV(\mathbf{w}(k, n)) = [-1 + \frac{2n-2}{2^k}, -1 + \frac{2n}{2^k}], k = 0, 1, 2, ..., \log_2 N, \quad n = 1, 2, 3, ..., 2^k. \tag{2.9}$$

Comparing (2.9) with (2.7), it is clear that when

$$\mathbf{w}(k, n) = [\mathbf{a}(2^k, -1 + \frac{2n-1}{2^k})^T, \mathbf{0}_{(N-2^k) \times 1}^T]^T, \tag{2.10}$$

(2.9) is satisfied. From the formula above, we can know that there are only 2^k active antennas in the k-th layer ($k = 0, 2, ..., \log_2 N$) and other antennas are all truned off. This is why the name of this approach is deactivation. It is noted that the first layer ($k = 0$) only has one omnidirectional AWV $\mathbf{w}(0, 1)$. With the designed coarse codebook, the three principles are fulfilled. Fig. 2.2 shows an example of beam pattern of the deactivation (DEACT) approach for the case of $N = 128$. From this figure,

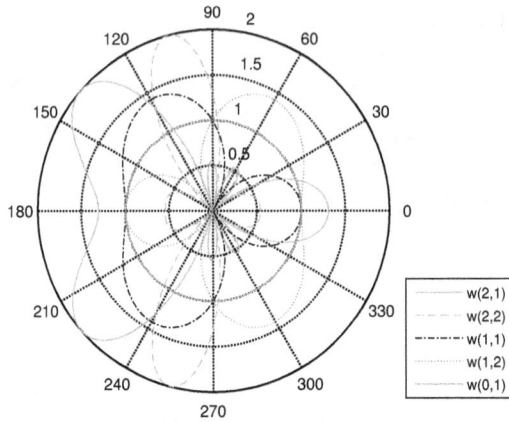

Figure 2.2 Beam patterns of $\mathbf{w}(2,1)$, $\mathbf{w}(2,2)$, $\mathbf{w}(1,1)$, $\mathbf{w}(1,2)$, and $\mathbf{w}(0,1)$ for the DEACT approach, where $N = 128$.

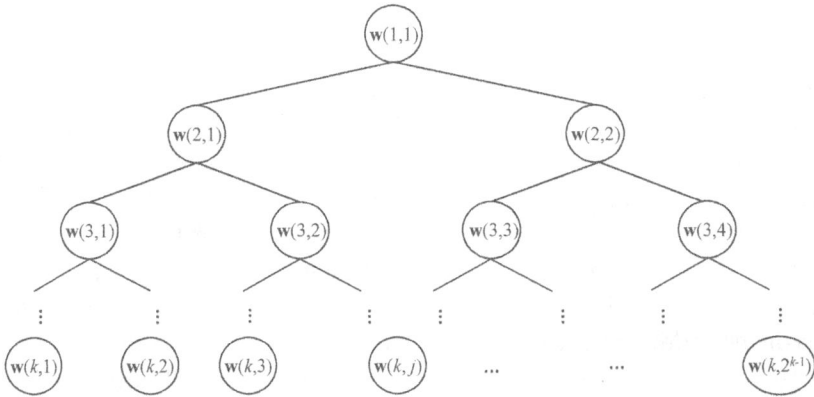

Figure 2.3 The binary-tree structure of the designed coarse codebook.

we find that the beam coverage of $\mathbf{w}(0,1)$ is just the union of those of $\mathbf{w}(1,1)$ and $\mathbf{w}(1,2)$, while the beam coverage of $\mathbf{w}(1,1)$ is just the union of those of $\mathbf{w}(2,1)$ and $\mathbf{w}(2,2)$.

Based on the coverage property, the coarse codebook can thus be organized into a binary tree structure shown in Fig. 2.3, which enables the binary search scheme.

2.2.2.2 Fine Codebook

We can observe that the AWVs of the last layer of the coarse codebook have the narrowest beam width. However, the angle resolution of the coarse codebook is $2/N$, which means that the maximal absolute angle error of the coarse codebook in the last layer is $1/N$. When a higher resolution is required, the coarse codebook can't meet the need. Without loss of generality, assume that we need to control the angle error to be lower than $1/(\alpha N)$, where $\alpha \geq 1$ represents the refinement factor. The resolution of the codebook should be no larger than $2/(\alpha N)$. Thus, the fine codebook should be

$$\mathbf{c}(i) = \mathbf{a}(N, -1 + \frac{2}{\alpha N}i), \; i = 1, 2, ..., \alpha N. \tag{2.11}$$

Fig. 2.4 shows the beams of the AWVs of the fine codebook in the case of $N = 4$ and $\alpha = 2$. It is observed that the fine codebook has αN AWVs with higher-resolution steering angles, and N AWVs of them are the same as those of the last layer of the coarse codebook, i.e.,

$$\mathbf{c}(\alpha i) = \mathbf{w}(N, i), \; i = 1, 2, ..., N. \tag{2.12}$$

Thus, in practice the size of a fine codebook is actually $(\alpha - 1)N$.

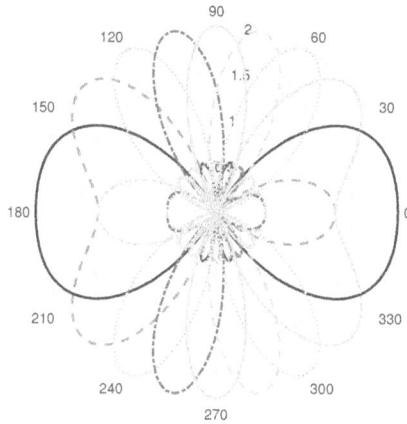

Figure 2.4 The beams of the AWVs of the fine codebook in the case of $N = 4$ and $\alpha = 2$.

2.2.3 Joint Sub-Array and Deactivation Codebook

From the previous subsection, it is noted that for the deactivation approach, when k is small, the number of active antennas is small, or even 1 when $k = 0$. This greatly limits the maximal total transmit power of a device. In general, we hope the number of active antennas is as large as possible, such that higher power can be transmitted, because per-antenna transmit power is usually limited. To achieve this target, we consider jointly using the sub-array and deactivation approach here. As the key of this approach is BeaM Widening via Single-RF Subarray, we term it BMW-SS.

The codebook with $M = 2$ has been designed in the previous codebook. We also want to design a codebook with the beam coverage shown in Fig. 2.1. According to (2.7), in the k-th layer, each codeword has a beam width of $2/2^k$. For the codewords of the last layer, we can also adopt the steering vectors according to (2.10). Compared with the codewords in the last layer, those in the lower layers have wider beams, and according to (2.9), codewords in the same layer have the same beam width but different steering angles. Thus, there are two basic tasks in the codebook design, namely to rotate the beam along required directions and to broaden the beam by required factors. We first introduce beam rotation.

2.2.3.1 Beam Rotation

Beam rotation can be realized according to the following theorem.

Theorem 2.2.1. $\mathcal{CV}(\mathbf{w} \circ \sqrt{N}\mathbf{a}(N, \psi)) = \mathcal{CV}(\mathbf{w}) + \psi$, where $\circ$ represents entry-wise product (a.k.a. Hadamard product), N is the number of elements of $\mathbf{w}$, ψ is an arbitrary angle. $\mathcal{A} + \psi$ is a new angle set with elements being those of the angle set $\mathcal{A}$ added by ψ.

Proof. Given an arbitrary N-element vector $\mathbf{w}$ and two arbitrary angles ψ and Ω, we want to prove that $\mathcal{CV}(\mathbf{w} \circ \sqrt{N}\mathbf{a}(N, \psi)) = \mathcal{CV}(\mathbf{w}) + \psi$, where $\mathcal{A} + \psi$ is a new angle set with elements being those of the angle set $\mathcal{A}$ added by ψ. Note that $\mathbf{w} \circ \sqrt{N}\mathbf{a}(N, \psi)$ is actually a new vector achieved based on $\mathbf{w}$ and the steering vector $\mathbf{a}(N, \psi)$. Let us first see the beam gain of this new vector.

$$
\begin{aligned}
& A(\mathbf{w} \circ \sqrt{N}\mathbf{a}(N, \psi), \Omega) \\
& \overset{(a)}{=} \sqrt{N}\mathbf{a}(N, \Omega)^{\mathrm{H}}(\mathbf{w} \circ \sqrt{N}\mathbf{a}(N, \psi)) \\
& \overset{(b)}{=} \sum_{n=1}^{N} [\mathbf{w}]_n e^{j\pi(n-1)\psi} e^{-j\pi(n-1)\Omega} \\
& = \sum_{n=1}^{N} [\mathbf{w}]_n e^{-j\pi(n-1)(\Omega - \psi)} \\
& \overset{(c)}{=} A(\mathbf{w}, \Omega - \psi),
\end{aligned}
\tag{2.13}
$$

where (a) and (c) are according to the definition of the beam gain in (2.2), while (b) is obtained according to definition of the entry-wise product.

Thus, we further have

$$
\begin{aligned}
& \mathcal{CV}(\mathbf{w} \circ \sqrt{N}\mathbf{a}(N, \psi)) \\
& \overset{(a)}{=} \{\Omega | \, |A(\mathbf{w} \circ \sqrt{N}\mathbf{a}(N, \psi), \Omega)| > \\
& \qquad\qquad \rho \max_{\omega} |A(\mathbf{w} \circ \sqrt{N}\mathbf{a}(N, \psi), \omega)|\} \\
& \overset{(b)}{=} \{\Omega | \, |A(\mathbf{w}, \Omega - \psi)| > \rho \max_{\omega} |A(\mathbf{w}, \omega - \psi)|\} \\
& \overset{(c)}{=} \{\Omega | \, |A(\mathbf{w}, \Omega - \psi)| > \rho \max_{\omega} |A(\mathbf{w}, \omega)|\} \\
& \overset{(d)}{=} \{\Omega_0 + \psi | \, |A(\mathbf{w}, \Omega_0)| > \rho \max_{\omega} |A(\mathbf{w}, \omega)|\} \\
& \overset{(e)}{=} \{\Omega_0 | \, |A(\mathbf{w}, \Omega_0)| > \rho \max_{\omega} |A(\mathbf{w}, \omega)|\} + \psi \\
& = \mathcal{CV}(\mathbf{w}) + \psi,
\end{aligned}
\tag{2.14}
$$

where (a) is according to the definition of beam coverage in (2.3), (b) is according to (2.13), (c) is based on the fact that the maxima of $|A(\mathbf{w}, \Omega - \psi)|$ does not depend on the angle offset ψ, (d) is obtained by letting $\Omega = \Omega_0 + \psi$, and (e) is obtained according to the definition of an angle set plus a single angle in Theorem 2.2.1. □

Theorem 2.2.1 implies that given an arbitrary codeword $\mathbf{w}$, we can rotate its beam coverage $\mathcal{CV}(\mathbf{w})$ by ψ with $\mathbf{w} \circ \sqrt{N}\mathbf{a}(N, \psi)$. This theorem helps to design all the other codewords in the same layer once one codeword in this layer is found. To explain this, we need to emphasize that all the codewords in the same layer have the same beam width but different steering angles according to (2.9), which means that the beam coverage of all the codewords can be assumed to have the same shape but different offsets in the angle domain. Thus, we can obtain one codeword based on another in the same layer as long as we know the angle gap between them according to Theorem 2.2.1. In particular, suppose we find the first codeword in the k-th layer $\mathbf{w}(k,1)$. According to (2.9), we do know that the angle gap between the n-th codeword in the k-th layer, i.e., $\mathbf{w}(k,n)$, and $\mathbf{w}(k,1)$ is $\frac{2n-2}{2^k}$, $n = 2, 3, ..., 2^k$. Then, we can obtain all the other codewords in this layer based on $\mathbf{w}(k,1)$ according to Theorem 2.2.1 (see Corollary 2.2.1 below).

Corollary 2.2.1. *Given the first codeword in the k-th layer $\mathbf{w}(k,1)$, all the other codewords in the k-th layer can be found through rotating $\mathbf{w}(k,1)$ by $\frac{2n-2}{2^k}$, $n = 2, 3, ..., 2^k$, respectively, i.e., $\mathbf{w}(k,n) = \mathbf{w}(k,1) \circ \sqrt{N}\mathbf{a}(N, \frac{2n-2}{2^k})$.*

Proof. To prove this corollary, we need to prove that, according to (2.9), when $\mathbf{w}(k,n) = \mathbf{w}(k,1) \circ \sqrt{N}\mathbf{a}(N, \frac{2n-2}{2^k})$, $\mathbf{w}(k,n) \in \mathcal{W}(N)$ and $\mathcal{CV}(\mathbf{w}(k,n)) = [-1 + \frac{2n-2}{2^k}, -1 + \frac{2n}{2^k}]$.
Since

$$[\mathbf{w}(k,n)]_i = [\mathbf{w}(k,1) \circ \sqrt{N}\mathbf{a}(N, \frac{2n-2}{2^k})]_i = [\mathbf{w}(k,1)]_i e^{j\pi(n-1)\frac{2n-2}{2^k}}, \qquad (2.15)$$

we have $|[\mathbf{w}(k,n)]_i| = |[\mathbf{w}(k,1)]_i|$. As $\mathbf{w}(k,1) \in \mathcal{W}(N)$, $\mathbf{w}(k,n) \in \mathcal{W}(N)$.
In addition, $\mathbf{w}(k,1)$ has a beam coverage $[-1, -1 + \frac{2}{2^k}]$. According to Theorem 2.2.1,

$$\mathcal{CV}(\mathbf{w}(k,n))$$
$$=\mathcal{CV}(\mathbf{w}(k,1) \circ \sqrt{N}\mathbf{a}(N, \frac{2n-2}{2^k}))$$
$$=\mathcal{CV}(\mathbf{w}(k,1)) + \frac{2n-2}{2^k} \qquad (2.16)$$
$$=[-1, -1 + \frac{2}{2^k}] + \frac{2n-2}{2^k}$$
$$=[-1 + \frac{2n-2}{2^k}, -1 + \frac{2n}{2^k}].$$

□

2.2.3.2 Beam Broadening

The remaining task is to broaden the beam for each layer. Given an N-element array, generally we would expect a beam width of $2/N$. Nevertheless, this beam width is in fact achieved by concentrating the transmit power at a specific angle Ω_0, i.e., by selecting AWV to maximize $|A(\mathbf{w}, \Omega_0)|$. Intuitively, if we design the AWV to disperse the transmit power along different widely-spaced angles, the beam width can be broadened. More specifically, if a large antenna array is divided into multiple sub-arrays, and these sub-arrays point at sufficiently-spaced directions, a wider beam can be shaped.

To illustrate this, let us separate the N-antenna array into S sub-arrays with N_S antennas in each sub-array, which means $N = SN_S$. In addition, letting $\mathbf{f}_m = [\mathbf{w}]_{(m-1)N_S+1:mN_S}$, we have $[\mathbf{f}_m]_n = [\mathbf{w}]_{(m-1)N_S+n}$, $m = 1, 2, ..., S$. $\mathbf{f}_m$ can be seen as the sub-AWV of the m-th sub-array. Therefore, the beam gain of $\mathbf{w}$ writes

$$
\begin{aligned}
A(\mathbf{w}, \omega) &= \sum_{n=1}^{N} [\mathbf{w}]_n e^{-\mathrm{j}\pi(n-1)\omega} \\
&= \sum_{m=1}^{S} \sum_{n=1}^{N_S} [\mathbf{w}]_{(m-1)N_S+n} e^{-\mathrm{j}\pi((m-1)N_S+n-1)\omega} \\
&= \sum_{m=1}^{S} \sum_{n=1}^{N_S} e^{-\mathrm{j}\pi(m-1)N_S\omega} [\mathbf{f}_m]_n e^{-\mathrm{j}\pi(n-1)\omega} \\
&= \sum_{m=1}^{S} e^{-\mathrm{j}\pi(m-1)N_S\omega} A(\mathbf{f}_m, \omega),
\end{aligned}
\tag{2.17}
$$

which means that the beam gain of $\mathbf{w}$ can be seen as the union of those of $\mathbf{f}_m$. According to (2.7), by assigning $\mathbf{f}_m = e^{\mathrm{j}\theta_m} \mathbf{a}(N_S, -1 + \frac{2m-1}{N_S})$, where $e^{\mathrm{j}\theta_m}$ can be seen as a scalar coefficient with unit norm for the m-th sub-array, the m-th sub-array has beam coverage $\mathcal{CV}(\mathbf{f}_m) = [-1 + \frac{2m-2}{N_S}, -1 + \frac{2m}{N_S}]$, $m = 1, 2, ..., S$. Hence, $\mathbf{w}$ has the beam coverage

$$
\mathcal{CV}(\mathbf{w}) = \bigcup_{m=1}^{S} \mathcal{CV}(\mathbf{f}_m) = [-1, -1 + \frac{2S}{N_S}] = [-1, -1 + \frac{2S^2}{N}],
\tag{2.18}
$$

i.e., the beam width has been broadened by S^2 by using the sub-array technique, where a broadening factor S comes from the number of sub-arrays, while another factor S results from the reduction factor of the sub-array size.

However, in the above process, the mutual effects between different sub-arrays are not taken into account. In the case of $\mathbf{f}_m = e^{\mathrm{j}\theta_m} \mathbf{a}(N_S, -1 + \frac{2m-1}{N_S})$, we have

$$
\begin{aligned}
A(\mathbf{w}, \omega) \mid \mathbf{f}_m &= e^{\mathrm{j}\theta_m} \mathbf{a}(N_S, -1 + \frac{2m-1}{N_S}) \\
&= \sqrt{N_S} \sum_{m=1}^{S} e^{-\mathrm{j}\pi(m-1)N_S\omega} e^{\mathrm{j}\theta_m} \times \mathbf{a}(N_S, \omega)^{\mathrm{H}} \mathbf{a}(N_S, -1 + \frac{2m-1}{N_S}).
\end{aligned}
\tag{2.19}
$$

As the steering vector has the properties that $\mathbf{a}(N_{\mathrm{S}}, -1 + \frac{2m-1}{N_{\mathrm{S}}})^{\mathrm{H}}\mathbf{a}(N_{\mathrm{S}}, -1 + \frac{2n-1}{N_{\mathrm{S}}}) = 0$ when $m \neq n$, the beam gain of $\mathbf{f}_m$ along the angle $-1 + \frac{2m-1}{N_{\mathrm{S}}}$ is affected little by $\mathbf{f}_n$. It is clear that $|A(\mathbf{w}, -1 + \frac{2m-1}{N_{\mathrm{S}}})| = \sqrt{N_{\mathrm{S}}}$ for $m = 1, 2, ..., S$, which means that the beam gains along angles $\omega = -1 + \frac{2m-1}{N_{\mathrm{S}}}$ are significant.

Additionally, to reduce the beam fluctuation, it is required that the intersection points in the angle domain of these coverage regions, i.e., $\omega = -1 + \frac{2n}{N_{\mathrm{S}}}$, $n = 1, 2, ..., S - 1$, also have high beam gain, and this can be realized by adjusting coefficients $e^{j\theta_m}$. Concretely, we have (2.21), where in (a) we have used the fact that $\mathbf{a}(N_{\mathrm{S}}, \omega_1)^{\mathrm{H}}\mathbf{a}(N_{\mathrm{S}}, \omega_2)$ is small and can be neglected when $|\omega_1 - \omega_2| > 2/N_{\mathrm{S}}$, in (b) we have exploited the condition that N_{S} is even in this paper. To maximize $|A(\mathbf{w}, \omega)|^2$, we face the problem

$$\underset{\Delta\theta}{\text{maximize}} \quad |f(N_{\mathrm{S}}, \Delta\theta)|^2, \tag{2.20}$$

which has a solution that $\Delta\theta = (2k - \frac{N_{\mathrm{S}}-1}{N_{\mathrm{S}}})\pi$, where $k \in \mathbb{Z}$. Thus, we may choose $\theta_m = -jm\frac{N_{\mathrm{S}}-1}{N_{\mathrm{S}}}\pi$, which satisfies $\Delta\theta = \pi$, to reduce the fluctuation of the beam.

$$
\begin{aligned}
A(\mathbf{w}, \omega) \mid &\{\mathbf{f}_m = e^{j\theta_m}\mathbf{a}(N_{\mathrm{S}}, -1 + \frac{2m-1}{N_{\mathrm{S}}}), \ \omega = -1 + \frac{2n}{N_{\mathrm{S}}}\} \\
=&\sqrt{N_{\mathrm{S}}}\sum_{m=1}^{S} e^{-j\pi(m-1)N_{\mathrm{S}}\omega}e^{j\theta_m}\mathbf{a}(N_{\mathrm{S}}, \omega)^{\mathrm{H}}\mathbf{a}(N_{\mathrm{S}}, -1 + \frac{2m-1}{N_{\mathrm{S}}}) \\
\overset{(a)}{\approx}&\sqrt{N_{\mathrm{S}}}e^{-j\pi(n-1)N_{\mathrm{S}}\omega}e^{j\theta_n}\mathbf{a}(N_{\mathrm{S}}, -1 + \frac{2n}{N_{\mathrm{S}}})^{\mathrm{H}}\mathbf{a}(N_{\mathrm{S}}, -1 + \frac{2n-1}{N_{\mathrm{S}}}) + \\
&\sqrt{N_{\mathrm{S}}}e^{-j\pi n N_{\mathrm{S}}\omega}e^{j\theta_{n+1}}\mathbf{a}(N_{\mathrm{S}}, -1 + \frac{2n}{N_{\mathrm{S}}})^{\mathrm{H}}\mathbf{a}(N_{\mathrm{S}}, -1 + \frac{2n+1}{N_{\mathrm{S}}}) \\
=&\frac{1}{\sqrt{N_{\mathrm{S}}}}e^{-j\pi(n-1)N_{\mathrm{S}}\omega}e^{j\theta_n} \times \left(\sum_{i=1}^{N_{\mathrm{S}}} e^{-j\pi(i-1)/N_{\mathrm{S}}} + e^{-j\pi N_{\mathrm{S}}\omega}e^{j(\theta_{n+1}-\theta_n)}\sum_{i=1}^{N_{\mathrm{S}}} e^{j\pi(i-1)/N_{\mathrm{S}}}\right) \\
\overset{(b)}{=}&\frac{1}{\sqrt{N_{\mathrm{S}}}}e^{-j\pi(n-1)N_{\mathrm{S}}\omega}e^{j\theta_n} \times \left(\sum_{i=1}^{N_{\mathrm{S}}} e^{-j\pi(i-1)/N_{\mathrm{S}}} + e^{j\Delta\theta}\sum_{i=1}^{N_{\mathrm{S}}} e^{j\pi(i-1)/N_{\mathrm{S}}}\right) \\
\overset{\Delta}{=}&\frac{1}{\sqrt{N_{\mathrm{S}}}}e^{-j\pi(n-1)N_{\mathrm{S}}\omega}e^{j\theta_n}f(N_{\mathrm{S}}, \Delta\theta),
\end{aligned}
$$

$$\tag{2.21}$$

In summary, by using the sub-array method and setting $\mathbf{f}_m = e^{-jm\frac{N_{\mathrm{S}}-1}{N_{\mathrm{S}}}\pi}\mathbf{a}(N_{\mathrm{S}}, -1 + \frac{2m-1}{N_{\mathrm{S}}})$, $m = 1, 2, ..., S$, we obtain a codeword $\mathbf{w}$ with a beam width $\frac{2S}{N_{\mathrm{S}}} = \frac{2S^2}{N}$. If we jointly using the sub-array and deactivation method, we may obtain codewords with beam width $\frac{2N_{\mathrm{A}}}{N_{\mathrm{S}}} = \frac{2SN_{\mathrm{A}}}{N}$ by setting as

$$
\mathbf{f}_m = \begin{cases} e^{-jm\frac{N_{\mathrm{S}}-1}{N_{\mathrm{S}}}\pi}\mathbf{a}(N_{\mathrm{S}}, -1 + \frac{2m-1}{N_{\mathrm{S}}}), & m = 1, 2, ..., N_{\mathrm{A}}, \\ \mathbf{0}_{N_{\mathrm{S}} \times 1}, & m = N_{\mathrm{A}} + 1, N_{\mathrm{A}} + 2, ..., S. \end{cases} \tag{2.22}
$$

where N_{A} is the number of active sub-arrays.

2.2.3.3 Codebook Generation

Recall that we need to design $\mathbf{w}(k, n)$ with beam width $2/2^k$ in the k-th layer.

When $k = \log_2(N)$, we have $\mathbf{w}(\log_2(N), n) = \mathbf{a}(N, -1 + \frac{2n-1}{N})$, $n = 1, 2, ..., N$.

When $k = \log_2(N) - \ell$, where $\ell = 1, 2, ..., \log_2(N)$, we follow the following procedures to compute $\mathbf{w}(k, n)$:

- Separate $\mathbf{w}(k, 1)$ into $S = 2^{\lfloor (\ell+1)/2 \rfloor}$ sub-arrays with $\mathbf{f}_m = [\mathbf{w}(k, 1)]_{(m-1)N_S+1:mN_S}$, where $\lfloor \cdot \rfloor$ is the flooring integer operation, $m = 1, 2, ..., S$;

- Set $\mathbf{f}_m$ as (2.22), where $N_A = S/2$ if ℓ is odd, and $N_A = S$ if ℓ is even;

- According to Corollary 2.2.1, we have $\mathbf{w}(k, n) = \mathbf{w}(k, 1) \circ \sqrt{N} \mathbf{a}(N, \frac{2(n-1)}{N})$, where $n = 2, 3, ..., 2^k$;

- Normalize $\mathbf{w}(k, n)$.

Fig. 2.5 shows an example of the beam pattern of the BMW-SS approach in the case of $N = 128$. From this figure, we find that the beam coverage of $\mathbf{w}(0, 1)$ is just the union of those of $\mathbf{w}(1, 1)$ and $\mathbf{w}(1, 2)$, while the beam coverage of $\mathbf{w}(1, 1)$ is just the union of those of $\mathbf{w}(2, 1)$ and $\mathbf{w}(2, 2)$. Comparing the beam pattern of DEACT shown in Fig. 2.2 with that in Fig. 2.5, it can be observed that although there are small-scale fluctuations for BMW-SS, the beams of BMW-SS appear flatter than those of DEACT within the covered angle.

On the other hand, for BMW-SS all the codewords either have all the antennas activated, or have half of them activated, which shows a significant advantage over DEACT in terms of the maximal total transmit power, especially for the low-layer codewords. Fig. 2.6 shows the comparison of beam patterns of BMW-SS, DEACT,

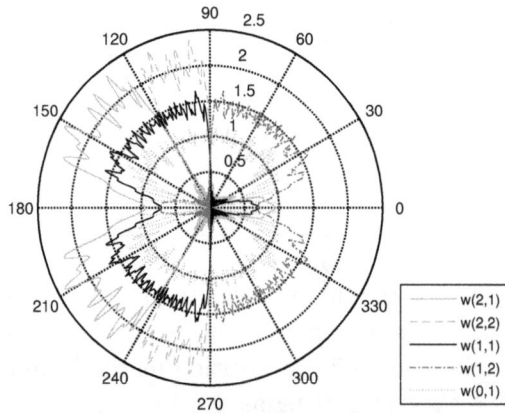

Figure 2.5 Beam patterns of $\mathbf{w}(2, 1)$, $\mathbf{w}(2, 2)$, $\mathbf{w}(1, 1)$, $\mathbf{w}(1, 2)$, and $\mathbf{w}(0, 1)$ for the BMW-SS approach, where $N = 128$.

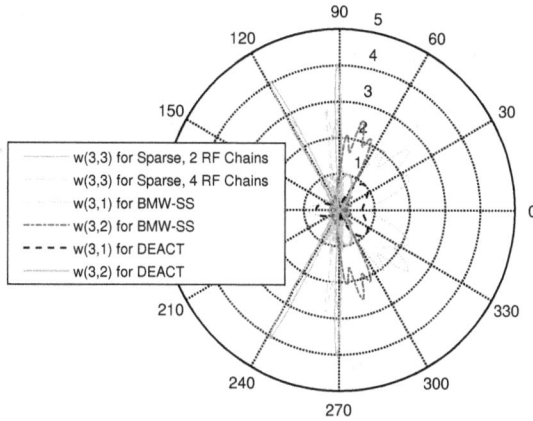

Figure 2.6 Comparison of the beam patterns of BMW-SS, DEACT, and the approach in[13] (termed as Sparse) with the per-antenna transmit power model, where $N = 32$. $L_d = 1$ for the Sparse approach.

and the approach in[13] (termed as Sparse) with the per-antenna transmit power model, where all the weights of active antennas have a unit amplitude. From this figure, we find that BMW-SS offers much higher beam gains than DEACT due to exploiting much greater number of active antennas. In addition, for the Sparse codebook, when the number of RF chains is small, there are deep sinks within the beam coverage of the wide-beam codewords, and the sink is more severe when the number of RF chains is smaller, which is in accordance with the results in[13]. Clearly, if the angle of departure (AoD) or angle of arrival (AoA) of an MPC is along the sink angle, it cannot be detected with the codeword, which results in miss detection of the MPC. By contrast, BMW-SS does not have such deep sinks.

It is noted that the corresponding hierarchical search of the designed codebook will eventually converge to a codeword of the last layer, i.e., a steering vector, at both ends. We can find that the angle resolution of the last layer is $2/N$. Thus, the designed codebook is just coarse codebook, while the corresponding search method is coarse search, like those in[12]. If a higher angle resolution is required, a fine codebook composed by steering vectors with a smaller sampling gap than $2/N$ is necessary. Details are referred to[12].

2.2.4 Enhanced Codebook

Although the two previous codebooks can be used for analog beamforming/combining devices[1], part of the antennas need to be switched off for some codewords, i.e., a strict

---

[1]Codebooks that can be used for analog beamforming/combining devices can be surely used for hybrid analog beamforming/combining devices.

CM constraint is not satisfied. This not only reduces the maximal transmit power but also requires an analog switch for each AE path, leading to additional cost and power consumption[15]. In this part, we propose an enhanced sub-array scheme to design a hierarchical codebook as shown in Fig. 2.1, with a single RF chain under a strict CM constraint. The designed codebook can be used for both analog and hybrid beamforming/combining devices.

2.2.4.1 Preliminaries of Codebook Design

According to Fig. 2.1, we have

$$\mathcal{CV}(\mathbf{w}(k,n)) = [-1 + \frac{2n-2}{2^k}, -1 + \frac{2n}{2^k}], k = 0, 1, 2, ..., \log_2 N, \ n = 1, 2, 3, ..., 2^k.$$

$$(2.23)$$

The task of codebook design is to design all the codewords $\mathbf{w}(k,n)$ satisfying the beam coverage in (2.23) under the CM constraint. The key is to design codewords with relatively wide beams. Suppose the desired beam coverage is $[\Omega_l, \ \Omega_u]$; then a good codeword should satisfy that the beam gain along the angles within the coverage is as high as possible, while that along the angles out of the coverage is as low as possible. Hence, an optimization problem can be formulated to design a codeword:

$$\underset{\mathbf{w}}{\text{minimize}} \quad \varepsilon \qquad (2.24a)$$

$$\text{subject to} \quad |A(\mathbf{w}, \Omega)| > 1, \ \Omega \in [\Omega_l, \ \Omega_u], \qquad (2.24b)$$

$$|A(\mathbf{w}, \Omega)| < \varepsilon, \ \Omega \notin [\Omega_l, \ \Omega_u], \qquad (2.24c)$$

$$|[\mathbf{w}]_1| = |[\mathbf{w}]_2| = ... = |[\mathbf{w}]_N|. \qquad (2.24d)$$

Since every $\Omega \in [\Omega_l, \ \Omega_u]$ should satisfy constraint (2.24b), there are infinite constraints for continuous Ω. We can sample the angle domain $[-1, \ 1]$ to obtain discrete Ω, such that an optimization problem with limited number of constraints is formulated. However, since most of the constraints are non-convex, the problem is difficult to solve. An optimal solution of this problem can be hardly found even with the exhaustive search method, because the size of $\mathbf{w}$, i.e., N, is too large in general. For instance, if we search over the possible phases of all the weight elements with a step of $\pi/18$, we need $(2\pi/(\pi/18))^N = 36^N$ tests, which is prohibitively high even for off-line computation. In such a case, sub-optimal heuristic methods are usually adopted to design appropriate codewords[11, 12, 13, 16]. In this book, we present an improved heuristic method for codebook design under strict CM constraint.

We first define the beam width of a steering vector with N antennas:

$$\mathcal{CV}(\mathbf{a}(N, \Omega)) = \left[\Omega - \frac{1}{N}, \Omega + \frac{1}{N}\right],$$

which means that the steering vectors have a beam width $2/N$ centering at the steering angle[14].

2.2.4.2 The Enhanced Sub-Array Scheme

Based on the preliminaries, we propose an enhanced sub-array scheme to design the codeword $\mathbf{w}(k,1)$ with beam coverage $[-1, -1 + \frac{2}{2^k}]$. The other codewords in the k-th layer, i.e., $\{\mathbf{w}(k,n) \mid n = 2, 3, ..., 2^k\}$ can be obtained by using Lemma 1 with $\mathbf{w}(k,1)$. The idea of the scheme is to divide the large array into several virtual sub-arrays, and let the sub-arrays steer to evenly-spaced angles within the beam coverage. A key difference of the enhanced sub-array scheme from the joint sub-array and deactivation method in[11] is that beam overlap is allowed in the enhanced scheme, while in[11] the steering angles must be sufficiently spaced. Beam overlap increases the mutual influence between adjacent sub-arrays, and thus calls for more delicate weight setting and optimization. In the following, we will introduce the enhanced sub-array scheme in detail.

1) The Number of Sub-Arrays

Firstly, we decide the number of sub-arrays that we need to use. Let an N-element antenna array be divided into S sub-arrays. Then each sub-array has N/S antennas, and the beam width of each sub-array is $2/(N/S) = 2S/N$. If the steering directions of the S sub-arrays are spaced by $2S/N$ in the cosine angle domain, the total beam width of the sub-arrays is $2S/N * S = 2S^2/N$[11], which is S^2 times of the beam width of a steering vector with N antennas. Hence, the broadening factor of the sub-array technique is S^2.

According to (2.23), the targeted beam width of the k-th layer codewords is $2/2^k = 2^{1-k}$. Hence, the number of sub-arrays for the k-th layer codewords satisfies $S_k = \sqrt{\frac{2^{1-k}}{2/N}} = \sqrt{2^{-k}N}$. A problem is that $\sqrt{2^{-k}N}$ is not necessarily an integer, and even $\sqrt{2^{-k}N}$ is an integer, it does not necessarily hold that N is an integer multiple of S. To address this issue, we can overlap the beam coverage of the sub-arrays, i.e., the angle space between adjacent sub-arrays can be less than $2S/N$. In such a case, the broadening factor may be less than S^2, and thus $S_k \geq \sqrt{2^{-k}N}$. Furthermore, we assume that both S_k and N are integer powers of 2. With this assumption, it can be obtained that the number of sub-arrays for the k-th layer codewords is

$$S_k = 2^{\lceil (\log_2 N - k)/2 \rceil}, \tag{2.25}$$

where $\lceil \cdot \rceil$ is the ceiling integer operation. It can be observed that the numbers of sub-arrays for the $\log_2 N$, $(\log_2 N - 1)$, $(\log_2 N - 2)$, $(\log_2 N - 3)$, $(\log_2 N - 4)$,...-th layer codewords are 1, 2, 2, 4, 4, ... With this setting, it is assured that $S_k \geq \sqrt{2^{-k}N}$ and N is an integer times of S_k.

2) The Weight Settings of the Sub-Arrays

Next, we need to set the AWVs of the S_k sub-arrays. For $\mathbf{w}(k,1)$, the steering angle space between adjacent sub-arrays is $\Delta = 2^{1-k}/S_k$, and the steering angles of the sub-arrays are

$$\omega_m = -1 + \frac{2m-1}{2}\Delta, \ m = 1, 2, ..., S_k. \tag{2.26}$$

Let $\mathbf{f}_m$ denote the AWV of the m-th sub-array. Considering the CM constraint, $\mathbf{f}_m$ can be expressed as

$$\mathbf{f}_m = \sqrt{\frac{N_S}{N}} e^{j\theta_m} \mathbf{a}(N_S, \omega_m), \tag{2.27}$$

where $N_S = N/S_k$ is the number of antennas of each sub-array, $e^{j\theta_m}$ is a co-phase factor between different sub-arrays. With these notations, $\mathbf{w}(k,1)$ is set as

$$[\mathbf{w}(k,1)]_{(m-1)N_S+1:mN_S} = \mathbf{f}_m, \ m = 1, 2, ..., S_k. \tag{2.28}$$

It is clear that with the setting in (2.28), $\mathbf{w}(k,1)$, it obeys the CM constraint and has a unit l_2-norm. The remaining issue is to determine the co-phases θ_m.

3) Co-Phase Optimization of the Sub-Arrays

To optimize the co-phases is challenging because there are two objectives for the codeword design. The first one is to maximize the beam gain along the main lobe direction, and the other is to minimize the gain fluctuation within the beam coverage. Even if we can formulate an optimization problem, the number of variables can be large when S_k is large, which means that the numerical search method may be of high computational complexity. In this subsection, we propose an intuitive approach to formulate an optimization problem, and we find a sub-optimal solution with closed form for the problem.

With the setting in (2.28) for $\mathbf{w}(k,1)$, it is guaranteed that the main power of the antenna array is within the beam coverage $[-1, -1 + \frac{2}{2^k}]$. Since the sub-arrays steer along ω_m, $m = 1, 2, ..., N_S$, the beam gains along ω_m would be large. To reduce the gain fluctuation, we hope that the intersection points in the angle domain of the beam regions of the sub-arrays, i.e., $\Omega_\ell = -1 + \ell\Delta$, $\ell = 1, 2, ..., N_S - 1$, also have high beam gains. As a result, we can formulate the following optimization problems

$$\underset{\theta_m}{\text{maximize}} \ |A(\mathbf{w}(k,1), \Omega_\ell)|, \ \ell = 1, 2, ..., N_S - 1, \tag{2.29}$$

where $\mathbf{w}(k,1)$ is shown in (2.28), and $\Omega_\ell = -1 + \ell\Delta$. Note that θ_m are involved in all the $(N_S - 1)$ optimization problems; thus it is almost impossible to find an optimal solution for all these problems. Fortunately, with some manipulations, we are able to find a sub-optimal solution with closed form. Here is the derivation process of the sub-optimal solution.

$$A(\mathbf{w}(k,1), \Omega_\ell) = \sum_{n=1}^{N} [\mathbf{w}(k,1)]_n e^{-j\pi(n-1)\Omega_\ell}$$

$$= \sum_{m=1}^{S_k} \sum_{n=1}^{N_S} [\mathbf{w}(k,1)]_{(m-1)N_S+n} e^{-j\pi((m-1)N_S+n-1)\Omega_\ell} \tag{2.30}$$

$$= \frac{N_S}{\sqrt{N}} \sum_{m=1}^{S_k} e^{-j\pi(m-1)N_S\Omega_\ell} e^{j\theta_m} \mathbf{a}(N_S, \Omega_\ell)^{\mathrm{H}} \mathbf{a}(N_S, \omega_m).$$

It is clear that it is still complicated to determine θ_m by optimizing the absolute beam gain in (2.30). Notice that $|\mathbf{a}(N_S, \omega_1)^H \mathbf{a}(N_S, \omega_2)|$ becomes smaller when $|\omega_1 - \omega_2|$ becomes greater from 0 to $2/N_S$, and can be neglected when $|\omega_1 - \omega_2| > 2/N_S$. This means that the two sub-arrays with steering angles closest to Ω_ℓ have the most significant effects on the beam gain along Ω_ℓ, while the sub-arrays with steering angles far from Ω_ℓ have much smaller effect on the beam gain along Ω_ℓ. This motivates us to consider only the two close sub-arrays when optimizing the beam gain for simplicity. Since $\Omega_\ell = -1 + \ell\Delta$, the two close steering angles are ω_ℓ and $\omega_{\ell+1}$. Consequently, we have

$$
\begin{aligned}
&A(\mathbf{w}(k,1), \Omega_\ell) \\
&\approx \frac{N_S}{\sqrt{N}} e^{-j\pi(\ell-1)N_S\Omega_\ell} e^{j\theta_\ell} \mathbf{a}(N_S, \Omega_\ell)^H \mathbf{a}(N_S, \omega_\ell) + \\
&\quad \frac{N_S}{\sqrt{N}} e^{-j\pi\ell N_S\Omega_{\ell+1}} e^{j\theta_{\ell+1}} \mathbf{a}(N_S, \Omega_\ell)^H \mathbf{a}(N_S, \omega_{\ell+1}) \\
&= \frac{1}{\sqrt{N}} e^{-j\pi(\ell-1)N_S\Omega_\ell} e^{j\theta_\ell} \times \\
&\quad \left(\sum_{i=1}^{N_S} e^{-j\pi(i-1)\Delta/2} + e^{j\pi N_S\Omega_\ell} e^{j(\theta_{\ell+1}-\theta_\ell)} \sum_{i=1}^{N_S} e^{j\pi(i-1)\Delta/2} \right).
\end{aligned}
\tag{2.31}
$$

Thus, we further obtain

$$
\begin{aligned}
&A(\mathbf{w}(k,1), \Omega_\ell) \\
&= \frac{1}{\sqrt{N}} e^{-j\pi(\ell-1)N_S\Omega_\ell} e^{j\theta_\ell} \times \left(e^{-j\pi(N_S-1)\Delta/4} \frac{\sin(-N_S\pi\Delta/4)}{\sin(-\pi\Delta/4)} \right. \\
&\quad \left. + e^{j\pi N_S\Omega_\ell} e^{j(\theta_{\ell+1}-\theta_\ell)} e^{j\pi(N_S-1)\Delta/4} \frac{\sin(N_S\pi\Delta/4)}{\sin(\pi\Delta/4)} \right) \\
&= \frac{1}{\sqrt{N}} e^{-j\pi(\ell-1)N_S\Omega_\ell} e^{j\theta_\ell} \times \frac{\sin(N_S\pi\Delta/4)}{\sin(\pi\Delta/4)} \\
&\quad \left(e^{-j\pi(N_S-1)\Delta/4} + e^{j\pi N_S\Omega_\ell} e^{j(\theta_{\ell+1}-\theta_\ell)} e^{j\pi(N_S-1)\Delta/4} \right).
\end{aligned}
\tag{2.32}
$$

It is clear that to optimize $|A(\mathbf{w}(k,1))|$, it should hold that

$$
2n\pi - \pi(N_S - 1)\Delta/4 = \pi N_S\Omega_\ell + (\theta_{\ell+1} - \theta_\ell) + \pi(N_S - 1)\Delta/4, \tag{2.33}
$$

where n is an arbitrary integer. Without loss of generality, we set $n = 0$. As N_S is also an integer power of 2, we have

$$
\theta_{\ell+1} - \theta_\ell = -\pi(N_S - 1)\Delta/2 - \pi N_S\ell\Delta. \tag{2.34}
$$

The final solution is

$$
\theta_m = -\pi m(N_S - 1)\Delta/2 - \pi N_S m(m-1)\Delta/2, \tag{2.35}
$$

where $\Delta = 2^{1-k}/S_k$.

4) Codebook Generation

In this part, we summary the codebook generation with the proposed enhanced sub-array technique.

Recall that we need to design $\mathbf{w}(k,n)$ with beam width $2/2^k$ in the k-th layer. For $k = 0, 1, 2, ..., \log_2 N$, we follow the following procedures to compute $\mathbf{w}(k,n)$:

- Separate $\mathbf{w}(k,1)$ into $S_k = 2^{\lceil (\log_2 N - k)/2 \rceil}$ sub-arrays; thus each sub-array has $N_S = N/S_k$ antennas;

- Set the AWVs of the S_k sub-arrays: for $m = 1, 2, ..., S_k$, set $[\mathbf{w}(k,1)]_{(m-1)N_S+1:mN_S} = \sqrt{\frac{N_S}{N}} e^{j\theta_m} \mathbf{a}(N_S, \omega_m)$, where θ_m is shown in (2.35), and $\mathbf{a}(N_S, \omega_m)$ is shown in (2.1);

- According to Corollary 2.2.1, we have $\mathbf{w}(k,n) = \mathbf{w}(k,1) \circ \sqrt{N} \mathbf{a}(N, \frac{2(n-1)}{N})$, $n = 2, 3, ..., 2^k$, where $\circ$ is the entry-wise product.

It is clear that there is no deactivation operation for all the codewords. Thus, unlike the deactivation method in[12] and the joint sub-array and deactivation method in[11], the above presented codebook does not require an on-off switch in each antenna branch and increases the maximal total transmit power. Fig. 2.7 shows the beam pattern comparison between the presented enhanced sub-array scheme (Presented) and the joint sub-array and deactivation method in[11] (JOINT), where we can find that for the 1st and the 3rd layer codewords, the enhanced scheme can achieve a significantly higher beam gain, due to no deactivation operation. Meanwhile, we can find that for the presented scheme, the beam width of $\mathbf{w}(1,1)$ is indeed roughly 2 times that of $\mathbf{w}(2,1)$, and 4 times that of $\mathbf{w}(3,1)$, which are in accordance with that in Fig. 2.1.

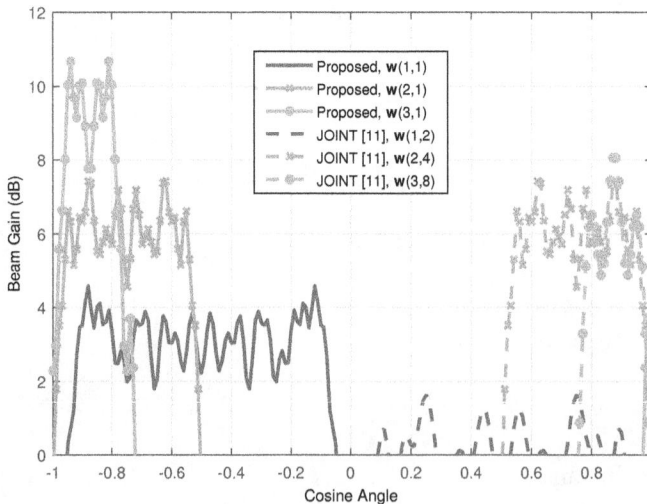

Figure 2.7 Beam Comparison between the presented enhanced sub-array scheme (Presented) and the joint sub-array and deactivation method in[11] (JOINT) with a uniform linear array (ULA), where $N = 64$.

2.2.5   Codebook for Hybrid Structures

Without loss of generality, we consider a point-to-point millimeter-wave system with a hybrid digital/analog precoding/combining structure[2], as shown in Fig. 2.8, where multiple RF chains[3] are tied to a half-wavelength spaced uniform linear array (ULA) at both Tx and Rx. Analog precoding/combining refers to signal operations in the analog RF (including only phase shift and signal summation) in comparison with those in the digital baseband. Due to the phase shifters, the analog precoding/combining matrices have CM constraints, which challenge the design of a proper beam pattern with desired beam width and steering angle. Relevant system parameters are listed below, where N_{ST} is the number of data streams.

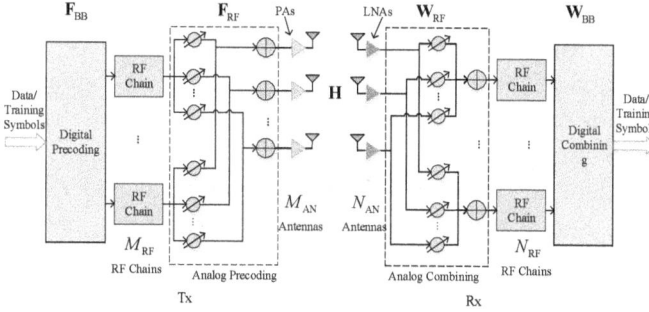

Figure 2.8   Illustration of a hybrid analog/digital precoding and combing structure with power amplifiers (PAs).

M_{RF}	The number of RF chains at the Tx.
M_{AN}	The number of antennas at the Tx.
N_{RF}	The number of RF chains at the Rx.
N_{AN}	The number of antennas at the Rx.
$\mathbf{F}_{BB}$	$M_{RF} \times N_{ST}$ digital precoding matrix at the Tx.
$\mathbf{F}_{RF}$	$M_{AN} \times M_{RF}$ analog precoding matrix at the Tx.
$\mathbf{W}_{BB}$	$N_{RF} \times N_{ST}$ digital combining matrix at the Rx.
$\mathbf{W}_{RF}$	$N_{AN} \times N_{RF}$ analog combining matrix at the Rx.
$\underline{\mathbf{F}}$	A Tx *composite codeword*, $\underline{\mathbf{F}} \triangleq (\mathbf{F}_{RF}, \mathbf{F}_{BB})$.
$\underline{\mathbf{f}}_i$	A Tx codeword, $\underline{\mathbf{f}}_i \triangleq (\mathbf{F}_{RF}, [\mathbf{F}_{BB}]_{:,i})$.
$\underline{\mathbf{W}}$	A Rx *composite codeword*, $\underline{\mathbf{W}} \triangleq (\mathbf{W}_{RF}, \mathbf{W}_{BB})$.
$\underline{\mathbf{w}}_i$	A Tx codeword, $\underline{\mathbf{w}}_i \triangleq (\mathbf{W}_{RF}, [\mathbf{W}_{BB}]_{:,i})$.

---

[2]The hybrid structure greatly reduces the hardware complexity in comparison with a fully digital precoding/combining structure, and the hybrid precoding/combining may need only partial instead of full CSI in millimeter-wave communication. However, the precoding performance of the hybrid structure is not as good as the fully digital structure.

[3]Note that a Tx RF chain consists of signal processing components after digital-to-analog converter (DAC) but before phase shifters, including filters, amplifiers, up-converter, etc. While a Rx RF chain consists of signal processing components after phase shifters but before analog-to-digital converter (ADC), including filters, amplifiers, down-converter, etc.

Tx codebook is a collection of composite codewords, and a Tx composite codeword is a precoding matrix pair $(\mathbf{F}_{\text{RF}}, \mathbf{F}_{\text{BB}})$, which can be seen as the composite of M_{RF} Tx codewords $(\mathbf{F}_{\text{RF}}, [\mathbf{F}_{\text{BB}}]_{:,i})_{i=1,2,\cdots,M_{\text{RF}}}$. The construction of the Rx codebook is similar to the Tx codebook. Basically, we have $M_{\text{RF}} \leq M_{\text{AN}}$ and $N_{\text{RF}} \leq N_{\text{AN}}$. In practical millimeter-wave systems, M_{RF} and N_{RF} are much smaller than M_{AN} and N_{AN}, respectively.

Without loss of generality, we adopt the same channel model as that in[11, 12, 13, 17, 18, 19], which is given by

$$\mathbf{H} = \sqrt{M_{\text{AN}} N_{\text{AN}}} \sum_{\ell=1}^{L} \lambda_\ell \mathbf{a}(N_{\text{AN}}, \Omega_\ell) \mathbf{a}(M_{\text{AN}}, \psi_\ell)^{\text{H}}, \qquad (2.36)$$

where λ_ℓ is the complex coefficient of the ℓ-th path, L is the number of MPCs, $\mathbf{a}(\cdot)$ is the steering vector function, Ω_ℓ and ψ_ℓ are the cosine AoD and cosine AoA of the ℓ-th path, respectively.

In this subsection, we want to design a hierarchical codebook with the beam coverage shown in Fig. 2.1 based on the hybrid structure shown in Fig. 2.8. We emphasize that we have both the per-antenna power constraint (PAPC) on PAs and the CM constraint on the analog precoding/combining matrices. Since a Rx codebook design can be the same as a Tx codebook design, we only proceed with Tx codebook design.

With the hybrid structure, an arbitrary codeword $\underline{\mathbf{w}} \triangleq (\mathbf{F}_{\text{RF}}, [\mathbf{F}_{\text{BB}}]_{:,i})$ shapes an AWV $\mathbf{w} = \mathbf{F}_{\text{RF}}[\mathbf{F}_{\text{BB}}]_{:,i}$, and the beam steering and coverage of $\underline{\mathbf{w}}$ are in fact reflected by $\mathbf{w}$. Hence, the codebook design here is to design $\underline{\mathbf{w}}(k,n)$ such that $\mathbf{w}(k,n)$ has the beam coverage $\mathcal{CV}(\mathbf{w}(k,n)) = \mathcal{CV}(\underline{\mathbf{w}}(k,n))$. *For convenience, we also call* $\mathbf{w}$ *a codeword,* but we emphasize that we want to design $\underline{\mathbf{w}} \triangleq (\mathbf{F}_{\text{RF}}, [\mathbf{F}_{\text{BB}}]_{:,i})$ rather than just $\mathbf{w}$ itself, because $\mathbf{w}$ is solely determined by $\underline{\mathbf{w}}$ but not vice versa.

Consequently, a codeword $\mathbf{w}$ has the following structure:

$$\mathbf{w} = \mathbf{F}_{\text{RF}} \mathbf{f}_{\text{BB}} = \sum_{j=1}^{M_{\text{RF}}} [\mathbf{f}_{\text{BB}}]_j [\mathbf{F}_{\text{RF}}]_{:,j}, \qquad (2.37)$$

where $\mathbf{f}_{\text{BB}} = [\mathbf{F}_{\text{BB}}]_{:,i}$, $|[\mathbf{F}_{\text{RF}}]_{:,j}| = \frac{1}{\sqrt{M_{\text{AN}}}} \mathbf{1}$ (the CM constraint). Note that the codewords belonging to the same composite codeword share the same $\mathbf{F}_{\text{RF}}$.

Given the target beam pattern of $\mathbf{w}(k,n)$ shown in Fig. 2.1, we need to design $(\mathbf{F}_{\text{RF}}, \mathbf{f}_{\text{BB}})$ for each $\mathbf{w}(k,n)$, which is challenging due to the CM constraint on $\mathbf{F}_{\text{RF}}$. In[13], this problem is solved by exploiting the sparse reconstruction approach (SPARSE). While in[16], the problem is further constrained by letting $|\mathbf{f}_{\text{BB}}| = \mathbf{1}$, i.e., the transmit power of each RF chain is the same. In such a case, a codeword is a combination of multiple RF vectors with equal power, and it is intuitive that by steering these RF vectors to equally spaced angles, a wide beam can be shaped. This is just the PS-DFT codebook proposed in[16].

In this subsection, we let $|\mathbf{f}_{\text{BB}}| = \mathbf{1}$ to simplify the problem, just the same as[16]. However, we propose different methods to design the codewords. In the following, we will first establish a generalized detection probability (GDP) metric to evaluate the quality of an arbitrary Tx codeword $\mathbf{w}$. Then we will design a hierarchical codebook with the target beam coverage shown in Fig. 2.1 with the codeword structure (2.37).

2.2.5.1 The GDP Metric

Given an arbitrary target codeword to cover an angle range $[\psi_0, \psi_0+B]$, it is clear that the best codeword should have constant absolute beam gain within the covered angle range (i.e., a totally flat beam pattern)[16]. However, due to the CM constraint on the analog precoding/combining matrices and the PAPC, an ideal codeword can be hardly designed. Hence, sub-optimal designs are of interest, and there have been different approaches[13, 16]. To the best of our knowledge, however, there is no particular metric to directly evaluate the quality of a codeword in the regime of millimeter-wave communications under PAPC.

Intuitively, good codewords should have flat beam patterns, and the mean square error (MSE) can be adopted to measure how flat a beam pattern is. Moreover, in millimeter-wave communications, the saturation power of a power amplifier (PA) is limited. We use maximal transmit power (MTP) to represent the total transmit power of all the AEs. In real-world communication systems, we would like to send signals with as high as possible MTP, which is, however, limited by the saturation power of the PA in each antenna branch branch. Thus, we need to carefully design the codewords such that a large MTP is allowed. For instance, DEACT is not of high quality, because a lot of AEs are turned off, which significantly lowers the MTP. In fact, under the PAPC, the MTP of an arbitrary codeword $\mathbf{w}$ is given by

$$\frac{P_{\text{MAX}}(\mathbf{w})}{\|\mathbf{w}\|_2^2} = \frac{P_{\text{PER}}}{\max\limits_{1\leq n\leq N}(|[\mathbf{w}]_n|^2)} \triangleq \frac{P_{\text{PER}}}{\|\mathbf{w}\|_\infty^2}, \qquad (2.38)$$

where P_{PER} is the saturation power for each antenna branch. It is clear that given fixed P_{PER}, $P_{\text{MAX}}(\mathbf{w})$ is maximized when $\mathbf{w}$ (with unit l_2-norm) has CM elements.

Both MSE and MTP may affect the quality of a codeword and their effects are different. In fact, these two metrics may be contradictory to each other, i.e., a codeword with small MSE tends to have a small MTP as well. We directly bridge the metric to the detection performance in beamforming training. During beamforming training, many Tx/Rx codeword pairs will be selected to detect the AoD/AoA of an MPC. When the AoD/AoA of the MPC locate within the coverage of the codewords, the detection probability is a direct and exact metric incorporating both MSE and MTP. Hence, we can derive the average detection probability, and generalize a metric based on the average detection probability for the Tx codewords.

Suppose that Tx transmits a training sequence with codeword $\mathbf{w}_T$, and Rx receives with codeword $\mathbf{w}_R$, i.e., $\mathbf{w}_T$ and $\mathbf{w}_R$ are fixed. The target beam coverage of $\mathbf{w}_T$ is $[\psi_0, \psi_0 + B]$. We want to develop a metric to evaluate the Tx codeword $\mathbf{w}_T$ based on the average detection probability of a single MPC.

Let $\mathbf{H}_0$ denote the channel response for the MPC to be detected, and it can be defined as

$$\mathbf{H}_0 = \sqrt{M_{\text{AN}}N_{\text{AN}}}\lambda\mathbf{a}(N_{\text{AN}}, \Omega)\mathbf{a}(M_{\text{AN}}, \psi)^{\text{H}}, \qquad (2.39)$$

where λ, Ω, and ψ denote the gain, AoA and AoD of the MPC, respectively. Following the channel model in (2.36), we assume $\lambda \sim \mathcal{CN}(0,1)$, Ω, and ψ are uniformly distributed within $[-1,1]$.

Given $\mathbf{H}_0$, the detection problem can be formulated as binary hypothesis testing given by[16]

$$y = \begin{cases} \mathbf{w}_\mathrm{R}^\mathrm{H}\mathbf{n} \sim \mathcal{CN}(0, N_0), & \mathcal{H}_0 \\ \sqrt{P}\mathbf{w}_\mathrm{R}^\mathrm{H}\mathbf{H}_0\mathbf{w}_\mathrm{T}s + \mathbf{w}_\mathrm{R}^\mathrm{H}\mathbf{n} \sim \mathcal{CN}(S, N_0), & \mathcal{H}_1 \end{cases} \tag{2.40}$$

where $\mathcal{H}_0$ and $\mathcal{H}_1$ represent the cases when the AoD locates outside and within of $[\psi_0, \psi_0 + B]$, respectively, $Y = \sqrt{P}\mathbf{w}_\mathrm{R}^\mathrm{H}\mathbf{H}_0\mathbf{w}_\mathrm{T}s$ denotes the received signal. Given a threshold ΓN_0, the instantaneous detection probability is given by

$$p_\mathrm{D}(\Gamma) = \Pr\{|(y|\mathcal{H}_1)|^2 > \Gamma N_0\} = \Pr\{|Y + n|^2 > \Gamma N_0\}, \tag{2.41}$$

where $n = \mathbf{w}_\mathrm{R}^\mathrm{H}\mathbf{n}$. To derive the average detection probability, we need to average $p_\mathrm{D}(\Gamma)$ on all the random variables. Note that Y depends on $\mathbf{H}_0$ and $\mathbf{H}_0$ depends on λ, Ω, and ψ. Hence, we need to average $p_\mathrm{D}(\Gamma)$ on n, λ, Ω, and ψ.

Let us first fix Ω and ψ and average $p_\mathrm{D}(\Gamma)$ on n and λ. Since $Y = \sqrt{P}\mathbf{w}_\mathrm{R}^\mathrm{H}\mathbf{H}_0\mathbf{w}_\mathrm{T}s$, when Ω and ψ are fixed, $\mathbf{H}_0$ has only one random parameter $\lambda \sim \mathcal{CN}(0,1)$. In such a case, Y can be seen as a zero-mean complex Gaussian variable, and $(Y + n) \sim \mathcal{CN}(0, (1 + \gamma)N_0)$, where γ denotes the average received signal-to-noise ratio (SNR) given by

$$\begin{aligned} \gamma &= \mathbb{E}_\lambda\left\{|\sqrt{P}\mathbf{w}_\mathrm{R}^\mathrm{H}\mathbf{H}_0\mathbf{w}_\mathrm{T}s|^2/N_0\right\} \\ &= \frac{PM_\mathrm{AN}N_\mathrm{AN}}{N_0}|\mathbf{w}_\mathrm{R}^\mathrm{H}\mathbf{a}(N_\mathrm{AN}, \Omega)\mathbf{a}(M_\mathrm{AN}, \psi)^\mathrm{H}\mathbf{w}_\mathrm{T}|^2 \\ &= \frac{P}{N_0}|A(\mathbf{w}_\mathrm{T}, \psi)|^2|A(\mathbf{w}_\mathrm{R}, \Omega)|^2, \end{aligned} \tag{2.42}$$

where $|A(\mathbf{w}_\mathrm{T}, \psi)|$ and $|A(\mathbf{w}_\mathrm{R}, \Omega)|$ are in fact Tx and Rx array gains depending on ψ and Ω, respectively. According to[20], $|Y + n|^2/N_0$ obeys Chi-square distribution, and its cumulative distribution function (CDF) is $F(y) = 1 - e^{-y/(1+\gamma)}$. Thus, we have

$$\bar{p}_{\mathrm{D}0}(\Gamma) = 1 - F(\Gamma) = e^{-\Gamma/(1+\gamma)}. \tag{2.43}$$

After averaging over λ and n, we need to further average $\bar{p}_{\mathrm{D}0}(\Gamma)$ in (2.43) on Ω and ψ to obtain the ultimate average detection probability. Note that as we only want to evaluate the quality of the Tx codeword $\mathbf{w}_\mathrm{T}$ with angle coverage $[\psi_0, \psi_0 + B]$, we can first get rid of the effects of the Rx codeword $\mathbf{w}_\mathrm{R}$ and AoA Ω. Consequently, we assume the RX gain is fixed for simplicity, and without loss of generality, let $|A(\mathbf{w}_\mathrm{R}, \Omega)|^2 = 1$. As a result, γ reduces to

$$\gamma = \frac{P}{N_0}|A(\mathbf{w}_\mathrm{T}, \psi)|^2. \tag{2.44}$$

And considering the MTP of $\mathbf{w}_{\mathrm{T}}$, the maximal received SNR is

$$
\begin{aligned}
\gamma_{\mathrm{MAX}} &= \frac{P_{\mathrm{MAX}}}{\epsilon^2 N_0} |A(\mathbf{w}_{\mathrm{T}}, \psi)|^2 \\
&= \frac{P_{\mathrm{PER}}}{\|\mathbf{w}_{\mathrm{T}}\|_\infty^2 \epsilon^2 N_0} |A(\mathbf{w}_{\mathrm{T}}, \psi)|^2 \\
&\triangleq \frac{\gamma_{\mathrm{PER}}}{\|\mathbf{w}_{\mathrm{T}}\|_\infty^2} |A(\mathbf{w}_{\mathrm{T}}, \psi)|^2,
\end{aligned}
\tag{2.45}
$$

where ϵ^2 is the pass loss, γ_{PER} denotes the per-antenna received SNR under the PAPC.

Consequently, the average detection probability is given by

$$
\begin{aligned}
\bar{p}_{\mathrm{D}}(\Gamma) &= \frac{1}{B} \int_{\psi_0}^{\psi_0+B} e^{-\Gamma/(1+\gamma_{\mathrm{MAX}})} d\psi \\
&= \frac{1}{B} \int_{\psi_0}^{\psi_0+B} \exp\left(-\frac{\Gamma}{1 + \frac{\gamma_{\mathrm{PER}}}{\|\mathbf{w}_{\mathrm{T}}\|_\infty^2} |A(\mathbf{w}, \psi)|^2}\right) d\psi.
\end{aligned}
\tag{2.46}
$$

We can see that $\bar{p}_{\mathrm{D}}(\Gamma)$ depends on both Γ and γ_{PER} in addition to the codeword $\mathbf{w}_{\mathrm{T}}$ itself. Hence, it cannot be directly used as a general metric. Note that the threshold Γ affects only the tradeoff between detection probability in hypothesis $\mathcal{H}_1$ and false-alarm probability in hypothesis $\mathcal{H}_0$[21]. A small Γ leads to high detection probability but higher false-alarm probability as well. In fact, the threshold itself does not affect the detection capability which involves both detection probability and false-alarm probability[21]. Based on this fact, we can just set $\Gamma = 1$ without loss of generality. When Γ is larger/smaller, detection probability will be lower/higher, but the comparison result of average detection probability between two different codewords basically maintains.

On the other hand, γ_{PER} may affect the comparison result of the average detection probability between two different codewords. Intuitively, when γ_{PER} is sufficiently high, the beam pattern (reflected by $|A(\mathbf{w}, \psi)|$ in (2.46)) is dominant, but when γ_{PER} is small, the maximal received SNR (reflected by $\|\mathbf{w}_{\mathrm{T}}\|_\infty^2$ in (2.46)) is dominant. Hence, different γ_{PER} should be set for different systems.

A possible way is to set a typical value of γ_{PER} based on the system settings. For instance, the saturation power of a PA (P_{PER}) can be set to 8 dBm[22,23]. According to the Friis formula, when the wavelength of the carrier frequency is 1 centimeter (30 GHz), and the Tx/Rx distance is 100 meters, the pass loss will be $20\log(4\pi \times 10000) = 102$ dB. Besides, when the bandwidth $B = 100$ MHz, the noise power can be computed as $N_0 = 10\log(\kappa T B) = 10\log(1.38 \times 10^{-23} \times 300 \times 10^8 \times 10^3) = -94$ dBm, where κ and T are the Boltzmann constant and ambient temperature, respectively. Hence, the per-antenna received SNR is $\gamma_{\mathrm{PER}} = 8 - 102 - (-94) = 0$ dB.

It is noteworthy that the above computation is rough. The spreading gain of the training sequence and the receive antenna gain are not included. On the other hand, millimeter-wave circuit losses, like insertion loss, noise figure, etc., are not considered. The overall per-antenna received SNR may have a dynamic range centered on 0 dB. In this paper, we prefer to set $\gamma_{PER} = 0$ dB for conciseness, but it should be clarified that other typical values close to 0 dB are also applicable. It will be shown in[11] that a small change of γ_{PER} does not affect the comparison result of two codewords.

Based on the above discussions, we propose the metric of GDP for an arbitrary N-entry Tx codeword $\mathbf{w}$ with unit l_2-norm and target coverage $[\psi_0, \psi_0 + B]$:

$$\xi(\mathbf{w}, \psi_0, B) = \frac{1}{B} \int_{\psi_0}^{\psi_0+B} \exp\left(-\frac{\|\mathbf{w}\|_\infty^2}{\|\mathbf{w}\|_\infty^2 + |A(\mathbf{w}, \psi)|^2}\right) d\psi, \qquad (2.47)$$

where $\|\mathbf{w}\|_\infty^2 = \max\limits_{1 \leq n \leq N}(|[\mathbf{w}]_n|^2)$.

Although (2.47) is defined for Tx codewords, it can also be used for Rx codewords, because small input fluctuation of low-noise amplifier (LNA) is also favored in millimeter-wave communications, where the linearity of LNA may not be perfect due to the high frequency and large signal bandwidth. In the case that the linearity of LNA is good enough, (2.47) can be modified by replacing $\|\mathbf{w}\|_\infty^2$ with constant 1 for Rx codewords. In this book, we use (2.47) for both Tx/Rx codeword designs. In addition, although in the derivation of (2.47) we have assumed Gaussian-distributed path gain, this metric can also be applied to cases when the path gain obeys other distributions or is even unknown. In such a case, the GDP metric in (2.47) is suboptimal.

It is clear that GDP depends on both $\|\mathbf{w}\|_\infty^2$ and $|A(\mathbf{w}, \psi)|^2$; thus in fact both MTP of $\mathbf{w}$ and MSE of the beam pattern ($|A(\mathbf{w}, \psi)|$) are incorporated in the GDP. According to (2.47), a good codeword should have elements with close amplitudes, such that the MTP will be higher. In addition, a good codeword should also have a flat beam pattern; thus deep sinks within the beam coverage should be avoided.

Moreover, one significance of GDP lies in that it enables a general optimization approach to design the codewords. In particular, if we want to design an arbitrary codeword $\mathbf{w}$ with target beam coverage $[\psi_0, \psi_0 + B]$, we can formulate the following problem

$$\underset{\mathbf{b}}{\text{maximize}} \quad \xi(\mathbf{w}(\mathbf{b}), \psi_0, B), \qquad (2.48a)$$

$$\text{subject to} \quad \text{Constraints on } \mathbf{b}, \qquad (2.48b)$$

where $\mathbf{b}$ is a parameter vector to be determined, and the constraints can include other desired structure constraints to simplify the search complexity in addition to the CM constraint. As it is difficult to further simplify the expression and obtain a closed-form GDP, it is thus difficult to obtain a solution of (2.48) through an analytical approach. Fortunately, by appropriately designing the structure constraints, the problem can be solved through numerical search methods. In fact, the proposed beam widening with multi-RF-chain subarray/low-complexity search (BMW-MS/LCS) codebook is just obtained with the optimization approach (cf. (2.53)) by adopting the sub-array structure.

Another significance of GDP lies in that it provides an additional way to compare two different codewords/codebooks besides simulation. With the same target beam coverage, a codeword with a higher GDP has better performance. For two different codebooks with the same coverage structure, its performance is basically determined by the codewords with the widest beams, because the widest beams have the lowest beam gain in general. Hence, by comparing the GDPs of the widest codewords of two different codebooks, we can evaluate which one is better.

2.2.5.2 Hierarchical Codebook Design

In this subsection, we present the BMW-MS approach to design a Tx hierarchical codebook based on multi-RF-chain sub-array technique[4]. In order to obtain the coefficients for each sub-array, we present two candidate solutions. The first one is a LCS solution to optimize the GDP metric, and the second one is a closed-form solution to pursue flat beam patterns. Hence, the BMW-MS approach with the two solutions are termed as BMW-MS/LCS and BMW-MS/closed-form, respectively.

It is noteworthy that when letting $|\mathbf{f}_{\mathrm{BB}}| = \mathbf{1}$ in (2.37) the structure of the Tx codeword can be further written as

$$\mathbf{w} = \sum_{i=1}^{M_{\mathrm{RF}}} \mathbf{v}_i, \qquad (2.49)$$

where $\mathbf{v}_i = [\mathbf{F}_{\mathrm{RF}}]_{:,i}$ is the radio frequency weight vector (RWV) of the i-th RF chain, and the phases of $\mathbf{f}_{\mathrm{BB}}$ have been absorbed into those of $\mathbf{v}_i$; thus we have in fact let $\mathbf{f}_{\mathrm{BB}} = \mathbf{1}$ here.

The BMW-MS Approach

A critical challenge to design the hierarchical codebook shown in Fig. 2.1 is beam widening, i.e., to design the low-layer codewords that have wide beam widths. Intuitively, if M_{RF} is sufficiently large, wide beams can be shaped by steering these RF RWVs toward equally spaced angles within the beam coverage. This is just the PS-DFT approach[16]. However, in practice M_{RF} may be rather small. In such a case, we consider to use the sub-array technique to shape a wide beam. In particular, a large RWV of each RF chain can be divided into multiple sub-vectors (called sub-arrays), and these sub-arrays can point at different directions, such that a wider beam can be shaped.

To illustrate this, let us separate the N-element RWV of each RF chain into S sub-arrays with N_{S} elements in each sub-array, which means $N = SN_{\mathrm{S}}$. In addition, letting $\mathbf{f}_{i,m} = [\mathbf{v}_i]_{(m-1)N_{\mathrm{S}}+1:mN_{\mathrm{S}}}$, we have $[\mathbf{f}_{i,m}]_n = [\mathbf{v}_i]_{(m-1)N_{\mathrm{S}}+n}$, $m = 1, 2, ..., S$, $n = 1, 2, ..., N_{\mathrm{S}}$, and $i = 1, 2, ..., M_{\mathrm{RF}}$. $\mathbf{f}_{i,m}$ can be seen as the sub-RWV of the m-th

---

[4]Rx codebook design is similar.

sub-array of the i-th RF chain. Therefore, the beam gain of $\mathbf{w}$ writes

$$
\begin{aligned}
A(\mathbf{w}, \omega) &= \sum_{n=1}^{N} \left[\sum_{i=1}^{M_{\mathrm{RF}}} \mathbf{v}_i \right]_n e^{-j\pi(n-1)\omega} \\
&= \sum_{n=1}^{N} \sum_{i=1}^{M_{\mathrm{RF}}} [\mathbf{v}_i]_n e^{-j\pi(n-1)\omega} \\
&= \sum_{m=1}^{S} \sum_{n=1}^{N_{\mathrm{S}}} \sum_{i=1}^{M_{\mathrm{RF}}} [\mathbf{v}_i]_{(m-1)N_{\mathrm{S}}+n} e^{-j\pi((m-1)N_{\mathrm{S}}+n-1)\omega} \quad (2.50) \\
&= \sum_{i=1}^{M_{\mathrm{RF}}} \sum_{m=1}^{S} \sum_{n=1}^{N_{\mathrm{S}}} e^{-j\pi(m-1)N_{\mathrm{S}}\omega} [\mathbf{f}_{i,m}]_n e^{-j\pi(n-1)\omega} \\
&= \sum_{i=1}^{M_{\mathrm{RF}}} \sum_{m=1}^{S} e^{-j\pi(m-1)N_{\mathrm{S}}\omega} A(\mathbf{f}_{i,m}, \omega),
\end{aligned}
$$

where we can find that the beam coverage of $\mathbf{w}$ can be controlled by controlling the $M_{\mathrm{RF}}S$ sub-arrays $\mathbf{f}_{i,m}$. It is noteworthy that the coefficient of the m-th sub-array is $e^{-j\pi(m-1)N_{\mathrm{S}}\omega}$. As the coefficient depends on m and ω, it induces coupling effect between different sub-arrays of the same RF chain. When the angle gap of two adjacent sub-arrays of the same RF chain is not wide enough, the coupling effect will be significant. In contrast, the coefficient does not depend on i. Hence, there is no coupling effect between different sub-arrays of different RF chains, which means that the steering angles of two sub-arrays of different RF chains can be close without affecting each other.

Based on the above observation, we present the BMW-MS approach for beam widening, i.e., to cover an arbitrary angle range $[\Omega_0, \Omega_0 + B]$ with M_{RF} RF chains, where each RF chain is decomposed into S sub-arrays, and the sub-RWVs $\mathbf{f}_{i,m}$ are set to steer along the angles

$$
\omega_{i,m} = \Omega_0 + (i - 1/2)\Delta\theta + (m - 1)M_{\mathrm{RF}}\Delta\theta, \quad (2.51)
$$

where $\Delta\theta = B/(M_{\mathrm{RF}}S)$, i.e., $\mathbf{f}_{i,m}$ satisfies

$$
\mathbf{f}_{i,m} = \sqrt{\frac{N_{\mathrm{S}}}{N}} e^{j\theta_{i,m}} \mathbf{a}(N_{\mathrm{S}}, \omega_{i,m}), \quad (2.52)
$$

where $\theta_{i,m}$ are phase parameters (in the angle domain instead of cosine angle domain) to be determined. Since the beam width of a sub-array is $2/N_{\mathrm{S}}$, $\Delta\theta$ should be no larger than $2/N_{\mathrm{S}}$; otherwise there will be sink between two adjacent sub-arrays.

An example of the beam patterns of the sub-arrays is shown in Fig. 2.9, where $N_{\mathrm{S}} = 8$, $M_{\mathrm{RF}} = S = 2$, $\Delta\theta = 2/N_{\mathrm{S}} = 0.25$, $B = 1$, and $\Omega_0 = -1$. The intuition of this approach is explained as follows. As we want to cover an angle interval of B, and there are $M_{\mathrm{RF}}S$ controllable sub-arrays in total, we can evenly steer these sub-arrays with an angle gap $\Delta\theta = B/(M_{\mathrm{RF}}S)$ over the desired angle range. Moreover, in order to reduce the coupling effect between adjacent sub-arrays of the same RF chain, we set their angle gaps as wide as possible; hence we try to set sub-arrays of different RF chains to steering along adjacent angles. For instance, the steering angles from small to large are $\omega_{1,1}$, $\omega_{2,1}$, $\omega_{1,2}$, $\omega_{2,2}$ in turn, such that $\omega_{1,1}$ and $\omega_{1,2}$, as well as $\omega_{2,1}$ and $\omega_{2,2}$, are largely spaced.

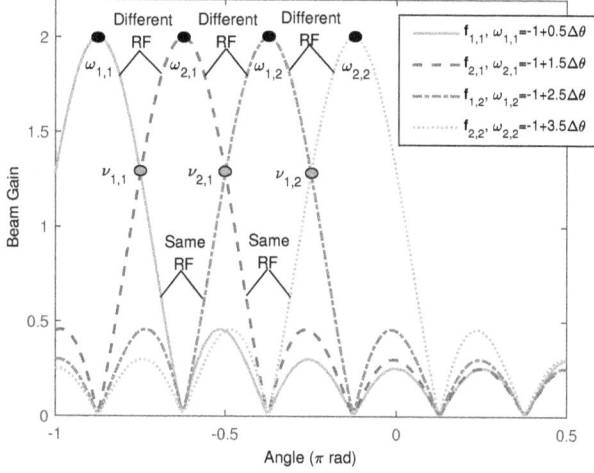

Figure 2.9 The beam patterns of the sub-arrays, where $N_S = 8$, $M_{RF} = S = 2$, $\Delta\theta = 2/N_S = 0.25$, $B = 1$, and $\Omega_0 = -1$.

2.2.5.3 Low-Complexity Search and Closed-Form Solutions

A remaining critical issue is to determine the coefficients $\theta_{i,m}$ for the BMW-MS approach in (2.52). We present two solutions as follows, i.e., an LCS and a closed form solutions.

1) A Low-Complexity Search Solution

According to (2.48), the following optimization problem can be formulated:

$$\underset{\theta_{i,m}}{\text{maximize}} \quad \xi(\mathbf{w}, \Omega_0, B), \tag{2.53a}$$

$$\text{subject to} \quad [\mathbf{v}_i]_{(m-1)N_S+1:mN_S} = \mathbf{f}_{i,m} = \sqrt{\frac{N_S}{N}} e^{j\theta_{i,m}} \mathbf{a}(N_S, \omega_{i,m}), \tag{2.53b}$$

which is a non-convex problem. Although the exhaustive grid search can be directly adopted to search over the feasible domains of $\theta_{i,m}$, it has a high computational complexity that grows exponentially with the total number of the sub-arrays. To lower the search complexity, we assume equal-difference phase sequences for the sub-arrays of the same RF chain and the sub-arrays of different RF chains, respectively, i.e., we let

$$\theta_{i,m} = m\phi_1 + i\phi_2, \tag{2.54}$$

where $\phi_1 \in [0, 2\pi]$ is the phase difference of the phase sequence for the sub-arrays of the same RF chain, and $\phi_2 \in [0, 2\pi]$ is the phase difference of the phase sequence for the sub-arrays of different RF chains. With this assumption, the problem (2.53) reduces to a 2-parameter search problem, which does not grow with the total number of sub-arrays. Thus, the computational complexity becomes affordable.

2) A Closed-Form Solution

According to (2.51), the sub-arrays are set to steer along $\omega_{i,m}$ with an angle gap $\Delta\theta$. This can only guarantee that the beam gains along these directions are high. To pursue a flat beam pattern, we also hope that the beam gains along the other angles between adjacent $\omega_{i,m}$ are high, such that the beam pattern is flatter. Thus, we can design $\theta_{i,m}$ to maximize the beam gains along the middle angles of adjacent $\omega_{i,m}$, i.e.,

$$\nu_{i,m} = \Omega_0 + i\Delta\theta + (m-1)M_{\mathrm{RF}}\Delta\theta, \tag{2.55}$$

where $im \neq M_{\mathrm{RF}}S$. Fig. 2.9 also shows the locations of $\nu_{i,m}$.

Since $\mathbf{f}_{i,m} = \sqrt{\frac{N_{\mathrm{S}}}{N}} e^{j\theta_{i,m}} \mathbf{a}(N_{\mathrm{S}}, \omega_{i,m})$, according to (2.2) the beam gain of $\mathbf{w}$ along angles $\nu_{i,m}$ can be derived as

$$A(\mathbf{w}, \nu_{k,n}) = \frac{N_{\mathrm{S}}}{\sqrt{N}} \sum_{i=1}^{M_{\mathrm{RF}}} \sum_{m=1}^{S} e^{-j\pi(m-1)N_{\mathrm{S}}\nu_{k,n}} e^{j\theta_{i,m}}$$
$$\mathbf{a}(N_{\mathrm{S}}, \nu_{k,n})^{\mathrm{H}} \mathbf{a}(N_{\mathrm{S}}, \omega_{i,m}). \tag{2.56}$$

It is clear that it is still complicated to determine $e^{\theta_{i,m}}$ by optimizing the absolute beam gain in (2.56). Notice that $|\mathbf{a}(N_{\mathrm{S}}, \omega_1)^{\mathrm{H}}\mathbf{a}(N_{\mathrm{S}}, \omega_2)|$ becomes smaller when $|\omega_1 - \omega_2|$ becomes greater from 0 to $2/N_{\mathrm{S}}$, and can be neglected when $|\omega_1 - \omega_2| > 2/N_{\mathrm{S}}$. This means that the two sub-arrays with steering angles closest to $\nu_{k,n}$ have the most significant effects on the beam gain along $\nu_{k,n}$, while the sub-arrays with steering angles far from $\nu_{k,n}$ have a little effect on the beam gain along $\nu_{k,n}$ (see Fig. 2.9). This motivates us to consider only the two close sub-arrays when optimizing the beam gain for simplicity. With this idea, we can finally obtain

$$\theta_{i,m} = \pi m(m-1)N_{\mathrm{S}}M_{\mathrm{RF}}\Delta\theta/2 - \pi(mM_{\mathrm{RF}} + i)(N_{\mathrm{S}} - 1)\Delta\theta/2, \tag{2.57}$$

where $\Delta\theta = B/(M_{\mathrm{RF}}S)$, $i = 1, 2, ..., M_{\mathrm{RF}}$, and $m = 1, 2, ..., S$.

Derivation: As shown in Fig. 2.9, there are two different types of positions of $\nu_{k,n}$. The first one is $\nu_{k,n}$ with $k = 1, 2, ..., M_{\mathrm{RF}} - 1$. The closest steering angles to it are $\omega_{k,n}$ and $\omega_{k+1,n}$, i.e., the two corresponding sub-arrays have adjacent RF indices and the same sub-array index. The other one is $\nu_{M_{\mathrm{RF}},n}$. The closest steering angles to it are $\omega_{M_{\mathrm{RF}},n}$ and $\omega_{1,n+1}$, i.e., the two corresponding sub-arrays have adjacent sub-array indices but the RF index switches from M_{RF} to

1. The beam gain of the first type of $\nu_{k,n}$ is derived as in (2.58)

$$
\begin{aligned}
&A(\mathbf{w}, \nu_{k,n})\big|_{k<M_{\mathrm{RF}},\ \mathbf{f}_{i,m}=e^{j\theta_{i,m}}\mathbf{a}(N_{\mathrm{S}},\omega_{i,m})} \\
&= \frac{N_{\mathrm{S}}}{\sqrt{N}} \sum_{i=1}^{M_{\mathrm{RF}}} \sum_{m=1}^{M_{\mathrm{S}}} e^{-j\pi(m-1)N_{\mathrm{S}}\nu_{k,n}} e^{j\theta_{i,m}} \mathbf{a}(N_{\mathrm{S}},\nu_{k,n})^{\mathrm{H}} \mathbf{a}(N_{\mathrm{S}},\omega_{i,m}) \\
&\approx \frac{N_{\mathrm{S}}}{\sqrt{N}} e^{-j\pi(n-1)N_{\mathrm{S}}\nu_{k,n}} e^{j\theta_{k,n}} \mathbf{a}(N_{\mathrm{S}},\nu_{k,n})^{\mathrm{H}} \mathbf{a}(N_{\mathrm{S}},\omega_{k,n}) + \\
&\quad \frac{N_{\mathrm{S}}}{\sqrt{N}} e^{-j\pi(n-1)N_{\mathrm{S}}\nu_{k,n}} e^{j\theta_{k+1,n}} \mathbf{a}(N_{\mathrm{S}},\nu_{k,n})^{\mathrm{H}} \mathbf{a}(N_{\mathrm{S}},\omega_{k+1,n}) \\
&= \frac{1}{\sqrt{N}} e^{-j\pi(n-1)N_{\mathrm{S}}\nu_{k,n}} e^{j\theta_{k,n}} e^{-j\pi(N_{\mathrm{S}}-1)\Delta\theta/4} \frac{\sin(-\pi\Delta\theta N_{\mathrm{S}}/4)}{\sin(-\pi\Delta\theta/4)} + \\
&\quad \frac{1}{\sqrt{N}} e^{-j\pi(n-1)N_{\mathrm{S}}\nu_{k,n}} e^{j\theta_{k+1,n}} e^{j\pi(N_{\mathrm{S}}-1)\Delta\theta/4} \frac{\sin(\pi\Delta\theta N_{\mathrm{S}}/4)}{\sin(\pi\Delta\theta/4)} \\
&= \frac{1}{\sqrt{N}} \frac{\sin(\pi\Delta\theta N_{\mathrm{S}}/4)}{\sin(\pi\Delta\theta/4)} e^{-j\pi(n-1)N_{\mathrm{S}}\nu_{k,n}} e^{j\theta_{k,n}} e^{-j\pi(N_{\mathrm{S}}-1)\Delta\theta/4} \\
&\quad \left(1 + e^{j(\theta_{k+1,n}-\theta_{k,n}+\pi(N_{\mathrm{S}}-1)\Delta\theta/2)}\right).
\end{aligned}
\tag{2.58}
$$

where we have used

$$
\begin{aligned}
\sum_{i=1}^{N} e^{j(i-1)\theta} &= \frac{1-e^{jN\theta}}{1-e^{j\theta}} = \frac{e^{jN\theta/2}(e^{-jN\theta/2}-e^{jN\theta/2})}{e^{j\theta/2}(e^{-j\theta/2}-e^{j\theta/2})} \\
&= e^{j(N-1)\theta/2} \frac{\sin(N\theta/2)}{\sin(\theta/2)}.
\end{aligned}
\tag{2.59}
$$

From (2.58) we can find that to optimize the absolute gain, we have

$$
\theta_{k+1,n} - \theta_{k,n} = -\pi(N_{\mathrm{S}}-1)\Delta\theta/2.
\tag{2.60}
$$

In addition, the beam gain of the other type of $\nu_{k,n}$ is derived as in (2.61),

$$
\begin{aligned}
&A(\mathbf{w}, \nu_{M_{\mathrm{RF}},n})\big|_{\mathbf{f}_{i,m}=e^{j\theta_{i,m}}\mathbf{a}(N_{\mathrm{S}},\omega_{i,m})} \\
&= \frac{N_{\mathrm{S}}}{\sqrt{N}} \sum_{i=1}^{M_{\mathrm{RF}}} \sum_{m=1}^{M_{\mathrm{S}}} e^{-j\pi(m-1)N_{\mathrm{S}}\nu_{k,n}} e^{j\theta_{i,m}} \mathbf{a}(N_{\mathrm{S}},\nu_{k,n})^{\mathrm{H}} \mathbf{a}(N_{\mathrm{S}},\omega_{i,m}) \\
&\approx \frac{N_{\mathrm{S}}}{\sqrt{N}} e^{-j\pi(n-1)N_{\mathrm{S}}\nu_{k,n}} e^{j\theta_{k,n}} \mathbf{a}(N_{\mathrm{S}},\nu_{k,n})^{\mathrm{H}} \mathbf{a}(N_{\mathrm{S}},\omega_{k,n}) + \\
&\quad \frac{N_{\mathrm{S}}}{\sqrt{N}} e^{-j\pi n N_{\mathrm{S}}\nu_{k,n}} e^{j\theta_{1,n+1}} \mathbf{a}(N_{\mathrm{S}},\nu_{k,n})^{\mathrm{H}} \mathbf{a}(N_{\mathrm{S}},\omega_{1,n+1}) \\
&= \frac{1}{\sqrt{N}} e^{-j\pi(n-1)N_{\mathrm{S}}\nu_{k,n}} e^{j\theta_{k,n}} e^{-j\pi(N_{\mathrm{S}}-1)\Delta\theta/4} \frac{\sin(-\pi\Delta\theta N_{\mathrm{S}}/4)}{\sin(-\pi\Delta\theta/4)} + \\
&\quad \frac{1}{\sqrt{N}} e^{-j\pi n N_{\mathrm{S}}\nu_{k,n}} e^{j\theta_{1,n+1}} e^{j\pi(N_{\mathrm{S}}-1)\Delta\theta/4} \frac{\sin(\pi\Delta\theta N_{\mathrm{S}}/4)}{\sin(\pi\Delta\theta/4)} \\
&= \frac{1}{\sqrt{N}} \frac{\sin(\pi\Delta\theta N_{\mathrm{S}}/4)}{\sin(\pi\Delta\theta/4)} e^{-j\pi(n-1)N_{\mathrm{S}}\nu_{k,n}} e^{j\theta_{k,n}} e^{-j\pi(N_{\mathrm{S}}-1)\Delta\theta/4} \\
&\quad \left(1 + e^{j[(\theta_{1,n+1}-\theta_{k,n}+\pi(N_{\mathrm{S}}-1)\Delta\theta/2)-\pi N_{\mathrm{S}}\nu_{k,n}]}\right).
\end{aligned}
\tag{2.61}
$$

where we can find that to optimize the absolute gain, we have

$$
\begin{aligned}
\theta_{1,n+1} - \theta_{M_{\mathrm{RF}},n} \\
= &- \pi(N_{\mathrm{S}} - 1)\Delta\theta/2 + \pi N_{\mathrm{S}}(M_{\mathrm{RF}}\Delta\theta + (n-1)M_{\mathrm{RF}}\Delta\theta) \qquad (2.62) \\
= &- \pi(N_{\mathrm{S}} - 1)\Delta\theta/2 + \pi N_{\mathrm{S}} n M_{\mathrm{RF}}\Delta\theta.
\end{aligned}
$$

Based on (2.60) and (2.62), we finally obtain (2.57).

2.2.5.4 Codebook Generation

Up to now we have assumed that M_{RF}, S, and N_{S} are known in advance. However, in practice M_{RF} is given by the system setting, while S and N_{S} are in fact determined by the beam width B of the codeword to be designed. In other words, S and N_{S} may be different for different codewords with different beam widths. Since when S is smaller, N_{S} will be bigger and a higher beam gain can be provided, S should be as small as possible. As $\Delta\theta = B/(M_{\mathrm{RF}}S) \le 2/N_{\mathrm{S}}$ and $N = SN_{\mathrm{S}}$, we can obtain

$$
S = \left\lceil \sqrt{BN/(2M_{\mathrm{RF}})} \right\rceil, \qquad (2.63)
$$

where $\lceil \cdot \rceil$ is the ceiling operation.

Recall that we need to design $\underline{\mathbf{w}}(k,n)$ instead of just $\mathbf{w}(k,n)$ itself. By exploiting the BMW-MS approach, we can design $\underline{\mathbf{w}}(k,n) \triangleq (\mathbf{F}_{\mathrm{RF}(k,n)}, \mathbf{f}_{\mathrm{BB}(k,n)} = \mathbf{1})$. Recall again that different codewords within the same composite codeword share the same $\mathbf{F}_{\mathrm{RF}}$, i.e., the same $\{\mathbf{v}_i\}_{i=1,2,...,M_{\mathrm{RF}}}$. This can be satisfied by using Corollary 2.2.1 in Section 2.2.3 for beam rotation.

According to Corollary 2.2.1, we have

$$
\begin{aligned}
\mathbf{w}(k,n) &= \mathbf{w}(k,1) \circ \sqrt{N}\mathbf{a}(N, \frac{2(n-1)}{M_{\mathrm{RF}}^k}) \\
&= \mathbf{F}_{\mathrm{RF}(k,1)}\mathbf{1} \circ \sqrt{N}\mathbf{a}(N, \frac{2(n-1)}{M_{\mathrm{RF}}^k}) \qquad (2.64) \\
&= \mathbf{F}_{\mathrm{RF}(k,1)} \sqrt{N}\mathbf{a}(N, \frac{2(n-1)}{M_{\mathrm{RF}}^k}),
\end{aligned}
$$

which means that all the codewords within the same layer can share the same $\mathbf{F}_{\mathrm{RF}}$.

In summary, the codebook is generated as follows, where $k = 1, ..., \log_{M_{\mathrm{RF}}}(M_{\mathrm{AN}})$, $N = M_{\mathrm{AN}}$.

- Split each RF chain into $S = \left\lceil \sqrt{B_k N/(2M_{\mathrm{RF}})} \right\rceil$ sub-arrays, where $B_k = 2/M_{\mathrm{RF}}^k$. Let the number of antennas of each sub-array be $N_{\mathrm{S}} = N/S$.

- Compute $\underline{\mathbf{w}}(k,1) = (\mathbf{F}_{\mathrm{RF}(k,1)} = \{\mathbf{v}_i\}_{i=1,2,...,M_{\mathrm{RF}}}, \mathbf{1})$, where $[\mathbf{v}_i]_{(m-1)N_{\mathrm{S}}+1:mN_{\mathrm{S}}} = \sqrt{\frac{N_{\mathrm{S}}}{N}}e^{j\theta_{i,m}}\mathbf{a}(N_{\mathrm{S}}, \omega_{i,m})$. $\omega_{i,m}$ is computed as (2.51) ($\Omega_0 = -1$, $B = 2/2^k$). $\theta_{i,m}$ can either be computed by solving (2.53) with the low-complexity search method (BMW-MS/LCS) or according to the closed-form expression (2.57) (BMW-MS/closed-form).

- Compute $\underline{\mathbf{w}}(k,n) = (\mathbf{F}_{\mathrm{RF}(k,1)}, \sqrt{N}\mathbf{a}(N, \frac{2(n-1)}{M_{\mathrm{RF}}^k}))$, where $n = 2, 3, ..., M_{\mathrm{RF}}^k$.

We emphasize that the main purpose of BMW-MS is to design a full hierarchical codebook shown in Fig. 2.1 with as less as possible RF chains, because in reality the number of available RF chains in a millimeter-wave device would be small, e.g., typically only 2, 4, or 8. In fact, we recommend to select $M_{\mathrm{RF}} = 2$ to realize BMW-MS, because fewer RF chains help to reduce the input/output fluctuation of the PAs, and with 2 RF chains BMW-MS can already achieve promising performance as we shall see from simulations later, but a larger number of RF chains can improve the efficiency of channel estimation.

In addition, we adopt a ULA model in this subsection. There are also other types of antenna arrays in practice, e.g., uniform plane array (UPA) and uniform circular array (UCA). Different from ULA, both UPA and UCA are 2-dimensional arrays with both azimuth and elevation steering angles. Although the idea, i.e., multi-RF-chain sub-array, behind the presented BMW-MS approach can also be used for the UPA and UCA models, the corresponding derivations are non-trivial and may be difficult. In particular, both the definitions of the steering vector in (2.1) and the beam gain in (2.2) must be redefined in a 2-dimensional angle domain if UPA or UCA is adopted, and all the derivations in Section 2.2.5 must be updated based on the new definitions of steering vector and beam gain. Subject to the structure of the 2-dimensional arrays, especially UCA, the derivations may be much more complicated than those for ULA in this book. For this reason, we think that the extension to UPA and ULA models would be an interesting research problem.

2.2.6 Performance Evaluation of Designed Hierarchical Codebooks

In this subsection, we evaluate the performance of the above designed hierarchical codebooks. For convenience, Antenna Deactivation codebook is termed as DEACT, Joint Sub-Array and Deactivation codebook is termed as JOINT, Enhanced codebook is termed as ENHANCE. We consider two different system models on the transmit power, namely total transmit power and per-antenna transmit power. The total transmit power signal model reflects the performance for the case when the transmit power on each antenna branch can be high enough, while the per-antenna transmit power signal model reflects the limit performance for the case when the transmit power on each antenna branch is limited[5]. The per-antenna transmit power model makes more sense in millimeter-wave communication, where the output power of a single PA is generally limited[1,24]. The activation/deactivation operations of a codebook are irrelevant to the power models. In particular, no matter which power model is adopted, the codewords of JOINT either have all or half of the antennas activated, those of DEACT have varying numbers of the antennas activated and the number may be quite small, while those of ENHANCE always have all the antennas activated.

Besides, in the simulations, both line of sight (LoS) and non-line of sight (NLoS) channel models are considered based on (2.36). For LoS channel, the first MPC has a constant coefficient and random AoD and AoA, while the other NLoS MPCs have complex Gaussian-distributed coefficients and random AoDs and AoAs[18,25]. The

---

[5]The limit performance here refers to the performance in the case when all the active antennas transmit with maximal power.

LoS MPC is generally much stronger than the NLoS MPCs. Both the LoS and NLoS channels are sparse in the angle domain, because the number of MPCs is much smaller than the numbers of the Tx/Rx antennas[13, 18, 25]. For all the codebooks, the hierarchical search method introduced in the next chapter is used. The performances of received SNR and success rate are all averaged on the instantaneous results of 10^4 random realizations of the LoS/NLoS channel.

We first discuss the two transmit power models that are used in performance evaluation. Letting s denote the transmitted symbol with unit power, the received signal is

$$y = \sqrt{P_{\text{tot}}} \mathbf{w}_R^H \mathbf{H} \mathbf{w}_T s + \mathbf{w}_R^H \mathbf{n}, \tag{2.65}$$

where P_{tot} is the *total transmit power* of all the active antennas, $\mathbf{w}_T$ and $\mathbf{w}_R$ are the transmit and receive AWVs, respectively, $\mathbf{H}$ is the channel matrix, $\mathbf{n}$ is the Gaussian noise vector with power N_0, i.e., $\mathbb{E}(\mathbf{n}\mathbf{n}^H) = N_0\mathbf{I}$. Let $\mathcal{W}(N)$ denote a set of vectors with N entries

$$\mathcal{W}(N) = \{\nu[\beta_1 e^{j\theta_1}, \beta_2 e^{j\theta_2}, ..., \beta_N e^{j\theta_N}]^T | \beta_i \in \{0,1\}, \theta_i \in [0, 2\pi), i = 1, 2, ..., N\}, \tag{2.66}$$

where ν is a normalization factor such that all the vectors have *unit power*. We can find that each entry of an arbitrary vector in $\mathcal{W}(N)$ has either an amplitude ν (activated) or is 0 (deactivated). Consequently, we have $\mathbf{w}_T \in \mathcal{W}(N_T)$, and $\mathbf{w}_R \in \mathcal{W}(N_R)$. It is noted that this signaling is based on the total transmit power, and we can further define the total transmission SNR as $\gamma_{\text{tot}} = P_{\text{tot}}/N_0$, and the received SNR with the total transmit power model as

$$\eta_{\text{tot}} = \gamma_{\text{tot}} |\mathbf{w}_R^H \mathbf{H} \mathbf{w}_T|^2. \tag{2.67}$$

The power gain under this model is

$$G_{\text{tot}} = \frac{\eta_{\text{tot}}}{\gamma_{\text{tot}}} = |\mathbf{w}_R^H \mathbf{H} \mathbf{w}_T|^2, \tag{2.68}$$

which is also the array gain.

On the other hand, in millimeter-wave communication the scaling abilities of PAs are generally limited. Thus, a per-antenna transmit power model is also with significance to characterize the best transmission ability of the transmitter, which is shown as

$$y = \sqrt{P_{\text{per}} N_{\text{Tact}}} \mathbf{w}_R^H \mathbf{H} \mathbf{w}_T s + \mathbf{w}_R^H \mathbf{n}, \tag{2.69}$$

where P_{per} is the per-antenna transmit power, N_{Tact} is the number of active antennas of $\mathbf{w}_T$, which varies with different $\mathbf{w}_T$. Also, we have $\mathbf{w}_T \in \mathcal{W}(N_T)$, and $\mathbf{w}_R \in \mathcal{W}(N_R)$. In addition, the per-antenna transmission SNR is defined as $\gamma_{\text{per}} = P_{\text{per}}/N_0$, and the received SNR with the per-antenna transmit power model is defined as

$$\eta_{\text{per}} = \gamma_{\text{per}} N_{\text{Tact}} |\mathbf{w}_R^H \mathbf{H} \mathbf{w}_T|^2. \tag{2.70}$$

The power gain under this model is

$$G_{\text{per}} = \frac{\eta_{\text{per}}}{\gamma_{\text{per}}} = N_{\text{Tact}} |\mathbf{w}_R^H \mathbf{H} \mathbf{w}_T|^2, \tag{2.71}$$

which includes both the transmit power gain equal to the number of active antennas N_{Tact} and the array gain $|\mathbf{w}_{\mathrm{R}}^{\mathrm{H}}\mathbf{H}\mathbf{w}_{\mathrm{T}}|^2$.

It is worth mentioning that the total and per-antenna transmit power models are suitable for cases when the scaling abilities of PA are high enough and limited, respectively. However, there is no difference for codebook design between these two models.

2.2.6.1 Total Transmit Power Model

In this part, the total transmit power signal model shown in (2.65) is used. With this model, the deactivation of antennas will not affect the total transmit power, i.e., the total transmit power is the same for the involved schemes.

Fig. 2.10 shows the received power during each search step with the JOINT, DEACT codebooks and ENHANCE under both LoS and NLoS channels, where $N_{\mathrm{T}} = N_{\mathrm{R}} = 64$, $L = 3$, $P_{\mathrm{tot}} = 1$ W, and $N_0 = 10^{-4}$ W, i.e., the SNR for beam training is sufficiently high, which means the length of the training sequence is sufficiently long. The upper bound is achieved by the exhaustive search method. For the LoS channel, the LoS component has 15 dB higher power than that of an NLoS MPC. From this figure, we can find that the received-power performance of these three codebooks is similar to each other. Under both channels, at the beginning, i.e., in the first two steps, DEACT behaves slightly better than JOINT, while in the following steps, JOINT and ENHANCE slightly outperform DEACT, until all methods achieve the same performance after the search process because they have the same last-layer codewords. Meanwhile, all approaches reach the upper bound under the LoS channel, while achieve a performance close to the upper bound under the NLoS channel. This is because under the LoS channel, the LoS component is the optimal MPC, and it is acquired by all JOINT, DEACT, ENHANCE and the exhaustive search. However, under the NLoS channel, JOINT, DEACT, and ENHANCE codebook acquire an

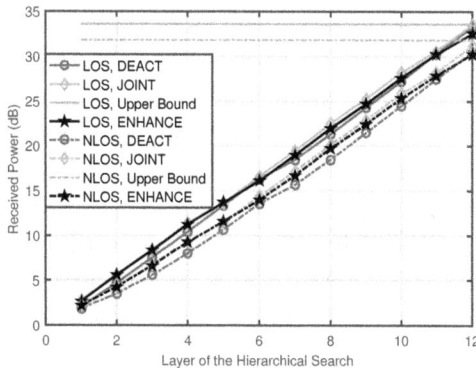

Figure 2.10 Received power during each search step with the JOINT and DEACT codebooks under both LoS and NLoS channels, where $N_{\mathrm{T}} = N_{\mathrm{R}} = 64$, $L = 3$, $P_{\mathrm{tot}} = 1$ W, and $N_0 = 10^{-4}$ W. Steps 1 to 6 are for Rx training, while Steps 7 to 12 are for Tx training.

arbitrary MPC of the L NLoS MPCs, which may not be the optimal one acquired by the exhaustive search.

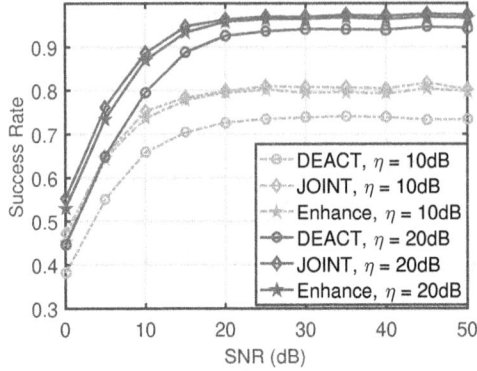

Figure 2.11 Success rate of hierarchical search with the JOINT, DEACT, and Sparse codebooks under LoS channel, where $N_T = N_R = 64$, $L = 3$. η is the power difference in dB between the LoS component and an NLoS MPC.

Fig. 2.11 shows the success rate of hierarchical search with JOINT, DEACT, and ENHANCE codebook under the LoS channel, where $N_T = N_R = 64$, $L = 3$. η is the power difference in dB between the LoS component and an NLoS MPC. From this figure, it is observed that both the transmission SNR γ_{tot} and η affect the success rate. For all the codebooks, the success rate improves as γ_{tot} increases. However, due to the mutual effect of MPCs (i.e., spatial fading), the success rate improves little when γ_{tot} is already high enough. Basically when η is bigger, the mutual effect is less, and the success rate is higher. Furthermore, we can find that the success rate with the JOINT codebook and the ENHANCE codebook are higher than that with the DEACT codebook. This is because the beams of the first two codebooks are flatter than those of the DEACT codebook; thus, they are more robust to the spatial fading. It is shown that the ENHANCE codebook can achieve a significant SNR gain at the mediate to low SNR range. This benefit over JOINT is due to that there is no deactivation operation on the antennas for all the codewords in ENHANCE. In contrast, some codewords of the JOINT codebook need to turn off half of the antennas, which results in a reduction of the maximal transmit power. However, at the high SNR regime, JOINT behaves slightly better than ENHANCE in this book, because the steering angle space between adjacent sub-arrays of ENHANCE is smaller than that of JOINT, which leads to severer fluctuation on the beam pattern.

Fig. 2.12 shows the success rate of hierarchical search with JOINT, DEACT, and ENHANCE codebooks under NLoS channel, where $N_T = N_R = 64$. From this figure, the same performance variation with respect to the transmission SNR γ_{tot} can be observed as that in Fig. 2.11. In addition, JOINT and ENHANCE codebook basically outperform DEACT, and the superiority depends on L. When $L = 1$, i.e., there is only one MPC, those codebooks all achieve a 100% success rate when γ_{tot} is high enough, because there is no spatial fading. In contrast, when $L > 1$, all these schemes can hardly achieve a 100% success rate, due to the mutual effect of multiple MPCs. But the success rate with the JOINT codebook and the ENHANCE codebook are higher than that with the DEACT codebook.

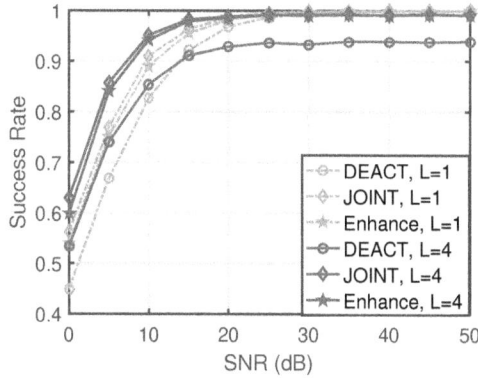

Figure 2.12 Success rate of hierarchical search with the JOINT, DEACT, and Sparse codebooks under the NLoS channel, where $N_T = N_R = 64$.

2.2.6.2 Per-Antenna Transmit Power Model

In this part, the per-antenna transmit power signal model is used to compare the limit performances of three codebooks with the same per-antenna transmit power. With this model, the deactivation of antennas will significantly affect the total transmit power. In particular, the total transmit power is lower if the number of active antennas is smaller.

Fig. 2.13 shows the received power during each search step with those three codebooks under both LoS and NLoS channels, where $N_T = N_R = 64$, $L = 3$, $P_{per} = 1$ W, and $N_0 = 10^{-4}$ W. The upper bound is achieved by the exhaustive search method. For the LoS channel, the LoS component has 15 dB higher power than that of an NLoS MPC. Comparing this figure with Fig. 2.10, we find a significant difference that with the per-antenna transmit power model JOINT and ENHANCE codebook have a distinct superiority over DEACT during the search process, especially at the beginning of the search process. The superiority is about 15 dB at the beginning, and it becomes less as the search goes on, until vanishes at the end of beam search, i.e., the three methods achieve the same received SNR after the search process. The superiority of the first two codebook result from the fact that the number of the active antennas for the codewords with wide beams is significantly greater than that for DEACT, and thus JOINT and ENHANCE codebook have a much higher total transmit power than DEACT when the per-antenna transmit power is the same.

Moreover, the increasing speed of received power is the same from Step 1 to Step 6 for the three schemes, but from Step 7 to Step 12, the increasing speed for JOINT varies, and that for DEACT becomes greater than that from Step 1 to Step 6. The increasing speed for the ENHANCE codebook is fixed. This is because with

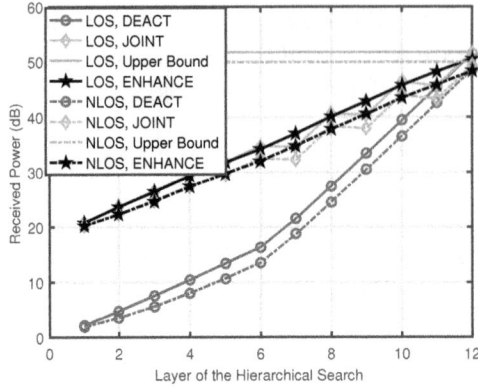

Figure 2.13 Received SNR during each search step with the JOINT and DEACT codebooks under both LoS and NLoS channels, where $N_T = N_R = 64$, $L = 3$, $P_{per} = 1$ W, and $N_0 = 10^{-4}$ W. Steps 1 to 6 are for Rx training, while Steps 7 to 12 are for Tx training.

per-antenna transmit power, there are two power gains during the search process according to (2.68), namely the array gain provided by narrowing the Tx/Rx beams and the total transmit power gain provided by increasing the number of active transmit antennas. For DEACT, there is only Rx array gain from Step 1 to Step 6, where Rx training is performed, while there are both Tx array gain and total transmit power gain from Step 7 to Step 12, where Tx training is performed; thus, the increasing speed of received power is greater from Step 7 to Step 12. For JOINT, there is also only Rx array gains from Step 1 to Step 6 for Rx training; thus the received power consistently increases with the same speed as DEACT. But from Step 7 to Step 12 for Tx training, although the Tx beam consistently becomes narrower, which means that Tx array gain is consistently improved, the number of active antennas alternatively changes between N_T and $N_T/2$, which means that the total transmit power may become larger or smaller. Hence, when both the Tx array gain and total transmit power increase, the received power improves with a speed the same as DEACT, while when the Tx array gain increases but the total transmit power decreases, the received SNR does not improve and may even decrease. For ENHANCE codebook, the number of active antennas is fixed, so there are only array gain from Step 1 to Step 12.

It is noted that the superiority of JOINT and ENHANCE codebook over DEACT at the beginning of the search process is with big significance for millimeter-wave communication, where per-antenna transmit power is generally limited. This superiority guarantees that with the first two codebooks, the success rate of beam search will be upgraded with the same transmission distance, or the transmission distance will be extended with the same success rate of beam search.

Figs. 2.14 and 2.15 show the success rates of hierarchical search with the three codebooks under LoS and NLoS channels, respectively. The same simulation conditions are adopted as those in Figs. 2.11 and 2.12, respectively, and the same results can be obtained from Figs. 2.14 and 2.15 as those from Figs. 2.11 and 2.12, respectively, except that the superiority of JOINT and ENHANCE codebook over DEACT becomes more significant in Figs. 2.14 and 2.15, which benefits from not only the fact that the beams of the two codebook are flatter than those of the DEACT codebook, but also that the number of the active antennas of the two codewords is basically much greater than that of DEACT, which offers much higher total transmit power. Also, Figs. 2.14 and 2.15 reveal that even with low per-antenna transmit power, the success rates of JOINT and ENHANCE codebook can be close to 100%, which is evidently better than those of DEACT and Sparse.

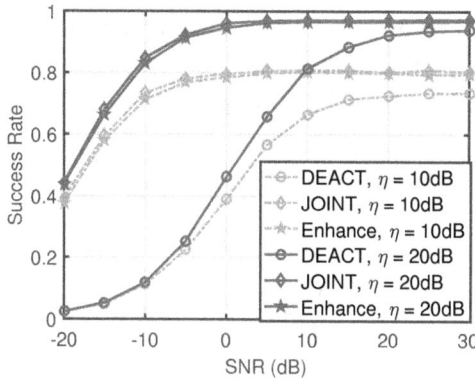

Figure 2.14 Success rate of hierarchical search with the JOINT and DEACT codebooks under the LoS channel, where $N_T = N_R = 64$, $L = 3$. η is the power difference in dB between the LoS component and an NLoS MPC.

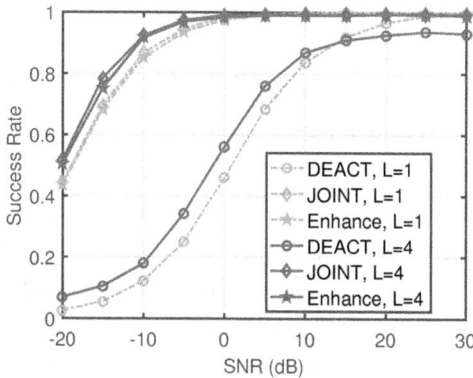

Figure 2.15 Success rate of hierarchical search with the JOINT and DEACT codebooks under NLoS channel, where $N_T = N_R = 64$.

2.3   BEAM SEARCH AND CHANNEL ESTIMATION

2.3.1   Exhaustive Beam Search

The exhaustive beam search scheme is straightforward, i.e., sequentially searching the whole Tx/Rx angle plane and finding the q_S (AoD AoA) pairs with the highest strengths. Therefore, the codebook for the exhaustive search consists of steering vectors with evenly sampled angles in the range of $[-1, 1]$, i.e., $\mathbf{a}(N, -1 + \frac{2i-1}{\alpha N})$, $i = 1, 2, ..., \alpha N$, where the sampling interval is $\frac{2}{\alpha N}$, and α is the *over-sampling factor*. The larger α is, the smaller the estimation errors of AoDs and AoAs are.

Regarding the considered system in Fig. 2.8, exhaustive search is realized by sequentially transmitting training sequences from the base station (BS) with codewords $\mathbf{a}(N_{\text{AN}}, -1 + \frac{2j-1}{\alpha N_{\text{AN}}})$ and receiving the training sequences at the mobile station (MS) with codewords $\mathbf{a}(M_{\text{AN}}, -1 + \frac{2i-1}{\alpha M_{\text{AN}}})$ for $i = 1, 2, ..., \alpha M_{\text{AN}}$ and $j = 1, 2, ..., \alpha N_{\text{AN}}$. Consequently, we can obtain the angle-domain matrix $\mathbf{G}$:

$$[\mathbf{G}]_{i,j} = \mathbf{a}(M_{\text{AN}}, -1 + \frac{2i-1}{\alpha M_{\text{AN}}})^{\text{H}} \mathbf{Ha}(N_{\text{AN}}, -1 + \frac{2j-1}{\alpha N_{\text{AN}}}),$$

$$i = 1, 2, ..., \alpha M_{\text{AN}}, \ j = 1, 2, ..., \alpha N_{\text{AN}}. \tag{2.72}$$

Afterwards, it is straightforward to find the q_S most significant peaks with $\mathbf{G}$ on the Tx/Rx angle plane. After finding the q_S most significant peaks, AoDs and AoAs of q_S most significant MPCs are determined. If we denote the time period of a training sequence as a time slot (i.e., a measurement), we need $\alpha^2 M_{\text{AN}} N_{\text{AN}}$ time slots to estimate $\mathbf{G}$ with the sequential search approach. Assuming that we have N_{RF} and M_{RF} RF chains at Tx and Rx, respectively, in each time slot we can estimate $N_{\text{RF}} M_{\text{RF}}$ elements of $\mathbf{G}$, by sending different orthogonal training sequences on the N_{RF} RF chains with different steering vectors at Tx, and receiving also with different steering vectors on the M_{RF} chains at Rx. Thus, the total time cost to estimate $\mathbf{G}$ is

$$T_{\text{SS}} = \frac{\alpha^2 M_{\text{AN}} N_{\text{AN}}}{N_{\text{RF}} M_{\text{RF}}}, \tag{2.73}$$

which is proportional to $M_{\text{AN}} N_{\text{AN}}$. While this method is feasible, the time cost would be significantly high for large-array devices.

2.3.2   Hierarchical Beam Search

To reduce the time cost of channel estimation, we present the hierarchical multi-beam search scheme, where a corresponding hierarchical codebook needs to be designed in advance. We have previously presented four codebooks. By using those codebooks, we can compute $\mathbf{w}(k, n)$. Based on the designed hierarchical codebook, we need to design a hierarchical multi-beam search scheme. The key of multi-beam search is how to cancel the effect of the already found beams in the on-going beam search. We present a method to shape an *already found channel response*, which can be subtracted when searching a new beam. The method has been reported in[26], and here we directly list the search algorithm in Algorithm 2.1. Then, we briefly illustrate the hierarchical beam search algorithm as follows. In this algorithm, we assume that there are M^k codewords with the same beam width but different steering angles in the k-th layer, where M is the *hierarchical factor* (cf.[26]) and for the over-sampling layer, the codewords $\mathbf{a}_i = \mathbf{a}(N, -1 + \frac{2i-1}{\alpha N})$, $i = 1, 2, ..., \alpha N$.

- *Search for the initial Tx/Rx codewords.* For every data stream, as the transmit power is generally limited, the beamforming training may not start from the 0-th layer, where the codeword is omni-directional and the gain is low. Instead, the beamforming training may need to start from a higher layer, e.g., the i_{LY}-th layer, to provide sufficient start-up beamforming gain. In this process, there are $M^{i_{\mathrm{LY}}}$ candidate codewords at both Tx and Rx. Thus, an exhaustive search over all Tx/Rx codeword pairs is adopted to search the best Tx/Rx codeword pair, which are treated as the parent codewords for the following search.

- *Hierarchical search.* In this process, a layered search is performed to refine the beam width step by step, until the most significant MPC is acquired at the K-th layer.

- *High-resolution search.* In this process, we uniformly sample the angle at the k-th layer into α pieces and acquire the optimal antenna weight vector by exhaustive search. In this process, we can also use a binary search to reduce the temporal complexity, which has been proposed in[12]. Then the channel response is updated.

2.3.3 Compressed Sensing for Channel Estimation

Both the LoS and NLoS channels are sparse in the angle domain, because the number of MPCs is much smaller than the numbers of the Tx/Rx antennas[11,13,18]. It is the reason why we can use Compressed Sensing method to acquire channel information.

To better illustrate the CS approach, we first adopt an analog beamforming/combining model, and the received symbol is expressed as

$$y = \sqrt{P}\mathbf{w}_{\mathrm{r}}^{\mathrm{H}}\mathbf{H}\mathbf{w}_{\mathrm{t}}s + \mathbf{n}, \tag{2.74}$$

where s is a transmitted symbol, P is the average transmit power, $\mathbf{w}_{\mathrm{r}}$ and $\mathbf{w}_{\mathrm{t}}$ are Rx/Tx AWVs, respectively, $\mathbf{H}$ is the channel matrix, and $\mathbf{n}$ is the Gaussian white noise. Let N_{r} and N_{t} denote the numbers of antennas at Rx and Tx, respectively. $\mathbf{w}_{\mathrm{r}}$ and $\mathbf{w}_{\mathrm{t}}$ are $N_{\mathrm{r}} \times 1$ and $N_{\mathrm{t}} \times 1$ vectors, respectively, with constant modulus and unit l_2-norm, i.e., $|\mathbf{w}_{\mathrm{r}}| = 1/\sqrt{N_{\mathrm{r}}}$ and $|\mathbf{w}_{\mathrm{t}}| = 1/\sqrt{N_{\mathrm{t}}}$. $|\cdot|$ denote the absolute value. In the case when a hybrid beamforming/combining structure is adopted at Rx and Tx, $\mathbf{w}_{\mathrm{r}}$, and $\mathbf{w}_{\mathrm{t}}$ will be the product of a digital beamforming/combining vector and an analog precoding/combining matrix with constant modulus[13]. Then a millimeter-wave channel can be expressed as[12,13,17,18,19,27]

$$\mathbf{H} = \sum_{\ell=1}^{L} \lambda_\ell \mathbf{a}_{\mathrm{r}}(\Omega_\ell)\mathbf{a}_{\mathrm{t}}^{\mathrm{H}}(\psi_\ell), \tag{2.75}$$

where λ_ℓ is the complex coefficient of the ℓ-th path and $\mathbb{E}\{\sum_{\ell=1}^{L}|\lambda_\ell|^2\} = N_{\mathrm{r}}N_{\mathrm{t}}$, L is the number of MPCs, $\mathbf{a}_{\mathrm{r}}(\cdot)$ and $\mathbf{a}_{\mathrm{t}}(\cdot)$ are Rx/Tx steering vector functions.

Algorithm 2.1: Hierarchical Multi-Beam Search Algorithm.

Input: $K = \max\{\log_M N_{AN}, \log_M M_{AN}\}$.
$i_{LY} = 2$. /*The initial layer index. It can be other values depending on practical requirements.*/
$\mathbf{H}_{fd} = 0$, /*The already found channel.*/

Output: The ℓth ($\ell = 1, 2, ..., q_S$) index pair is (I_ℓ, J_ℓ) within the over-sampling layer.

1: **for** $\ell = 1 : q_S$ **do**
2:    /*Search for the initial Tx/Rx codewords in the i_{LY}th layer.*/
3:    **for** $m = 1 : M^{i_{LY}}$ **do**
4:      **for** $n = 1 : M^{i_{LY}}$ **do**
5:       $y(m,n) = \mathbf{w}_{BS}(i_{LY}, n)^H[\sqrt{P}\mathbf{H}\mathbf{w}_{MS}(i_{LY}, m) + \mathbf{n}] - \mathbf{w}_{BS}(i_{LY}, n)^H\mathbf{H}_{fd}\mathbf{w}_{MS}(i_{LY}, m)$
6:      **end for**
7:    **end for**
8:    $(m_{MS} \ m_{BS}) = \underset{(m,n)}{\arg\max} |y(m,n)|$
9:    MS feeds back BS m_{BS}.

10:    /*Hierarchical refinement.*/
11:    **for** $k = i_{LY} + 1 : K$ **do**
12:      **for** $n = 1 : M$ **do**
13:       $y_{MS}(n) = \mathbf{w}_{BS}(k-1, m_{BS})^H[\sqrt{P}\mathbf{H}\mathbf{w}_{MS}(k, (m_{MS}-1)M + n) + \mathbf{n}] - \mathbf{w}_{BS}(k-1, m_{BS})^H\mathbf{H}_{fd}\mathbf{w}_{MS}(k, (m_{MS}-1)M + n)$
14:      **end for**
15:      $n_{MS} = \underset{n}{\arg\max} |y_{MS}(n)|$
16:      $m_{MS} = (m_{MS} - 1)M + n_{MS}$

17:      **for** $n = 1 : M$ **do**
18:       $y_{BS}(n) = \mathbf{w}_{BS}(k, (m_{BS}-1)M + n)^H[\sqrt{P}\mathbf{H}\mathbf{w}_{MS}(k, m_{MS}) + \mathbf{n}] - \mathbf{w}_{BS}(k, (m_{BS}-1)M + n)^H\mathbf{H}_{fd}\mathbf{w}_{MS}(k, m_{MS})$
19:      **end for**
20:      $n_{BS} = \underset{n}{\arg\max} |y_{BS}(n)|$
21:      $m_{BS} = (m_{BS} - 1)M + n_{BS}$
22:      MS feeds back BS m_{BS}.
23:    **end for**

24:    /*High-resolution refinement.*/
25:    **for** $m = (m_{MS} - 1)\alpha + 1 : m_{MS}\alpha$ **do**
26:      **for** $n = (m_{BS} - 1)\alpha + 1 : m_{BS}\alpha$ **do**
27:       $y(m,n) = \mathbf{w}_{BS}(i_{LY}, n)^H[\sqrt{P}\mathbf{H}\mathbf{w}_{MS}(i_{LY}, m) + \mathbf{n}] - \mathbf{w}_{BS}(i_{LY}, n)^H\mathbf{H}_{fd}\mathbf{w}_{MS}(i_{LY}, m)$
28:      **end for**
29:    **end for**
30:    $(I_\ell, J_\ell) = \underset{(m,n)}{\arg\max} |y(m,n)|$
31:    MS feeds back BS J_ℓ.

32:    /*Updating the already found channel response.*/
33:    $\mathbf{H}_{fd} = \mathbf{H}_{fd} + y(I_\ell, J_\ell)(\mathbf{g}_{I_\ell}^{MS})(\mathbf{g}_{J_\ell}^{BS})^H$
34: **end for**

35: **return** The ℓth ($\ell = 1, 2, ..., q_S$) index pair is (I_ℓ, J_ℓ) within the over-sampling layer.

Based on the signal model in (2.74), we may make multiple measurements with different Tx AWVs $[\mathbf{w}_{t1}, \mathbf{w}_{t2}, ..., \mathbf{w}_{tk_t}] \triangleq \mathbf{W}_t$ and Rx AWVs $[\mathbf{w}_{r1}, \mathbf{w}_{r2}, ..., \mathbf{w}_{rk_r}] \triangleq \mathbf{W}_r$, and we assume, without loss of generality, $s = 1$, then we observe the measurements

$$\mathbf{Y} = \sqrt{P}\mathbf{W}_r^H\mathbf{H}\mathbf{W}_t + \mathbf{N}, \tag{2.76}$$

where $\mathbf{N}$ is the noise matrix. By sampling the AoA/AoD domains with sufficiently high resolution δ, we can obtain $\mathbf{A}_r = [\mathbf{a}_r(-1+\frac{\delta}{2}), \mathbf{a}_r(-1+\frac{3\delta}{2}), ...]$ and $\mathbf{A}_t = [\mathbf{a}_t(-1+\frac{\delta}{2}), \mathbf{a}_t(-1+\frac{3\delta}{2}), ...]$. Then $\mathbf{H}$ can be approximately expressed as $\mathbf{H} = \mathbf{A}_r\mathbf{\Sigma}\mathbf{A}_t^H$, where $\mathbf{\Sigma}$ is a diagonal and sparse matrix with the diagonal entries corresponding to the channel coefficients λ_ℓ. Substituting $\mathbf{H}$ in (2.76) with this expression and vectorizing $\mathbf{Y}$, we have[28, 29]

$$\begin{aligned}
\text{vec}(\mathbf{Y}) &= \text{vec}(\sqrt{P}\mathbf{W}_r^H\mathbf{A}_r\mathbf{\Sigma}\mathbf{A}_t^H\mathbf{W}_t + \mathbf{N}) \\
&= \sqrt{P}\left((\mathbf{A}_t^H\mathbf{W}_t)^T \otimes (\mathbf{W}_r^H\mathbf{A}_r)\right)\text{vec}(\mathbf{\Sigma}) + \text{vec}(\mathbf{N}) \tag{2.77} \\
&\triangleq \sqrt{P}\mathbf{Q}\text{vec}(\mathbf{\Sigma}) + \text{vec}(\mathbf{N}),
\end{aligned}$$

where $\otimes$ is the Kronecker product. Since $\|\text{vec}(\mathbf{\Sigma})\|_0 = L \ll N_rN_t$, sparse recovery tools can be adopted to estimate $\mathbf{H}$, where the dictionary matrix $\mathbf{Q}$ can be obtained by randomly setting the Tx/Rx training AWVs in each measurement as $[\mathbf{w}_r]_k \in \{e^{j\theta}/\sqrt{N_r}\}$ and $[\mathbf{w}_t]_m \in \{e^{j\theta}/\sqrt{N_t}\}$ with uniformly distributed phase θ. Then the problem becomes

$$\text{vec}(\mathbf{\Sigma}^{CS}) = \underset{\mathbf{\Sigma}}{\arg\min} \quad \|\text{vec}(\mathbf{Y}) - \bar{\mathbf{Q}}\text{vec}(\mathbf{\Sigma})\|_2 \tag{2.78a}$$

$$\text{subject to} \quad \|\text{vec}(\mathbf{\Sigma})\|_0 = L, \tag{2.78b}$$

where $\text{vec}(\mathbf{\Sigma}^{CS})$ donates the estimated sparse matrix. (2.78) can be solved by exploiting the orthogonal matching pursuit (OMP) algorithm[28]. Then we can estimate the channel matrix $\mathbf{H}$ using $\mathbf{H} = \mathbf{A}_r\mathbf{\Sigma}\mathbf{A}_t^H$.

Note that as the number of candidate vectors, i.e., the number of columns of $\mathbf{Q}$, is large, the computational complexity of the CS approach is high. In addition, the total number of measurements is $T_{CS} = k_rk_t$. It was shown (and will also be shown later) that when T_{CS} is not large enough, the performance of the CS approach is not satisfactory[6][29].

2.3.4 Joint Beam Search and Compressed Sensing

First, we acquire the channel $\mathbf{H}_{fd}$ and the information about MPCs by using the hierarchical beam search algorithm that has been introduced in Section 2.3.2. The hierarchical codebook is shown in Fig. 2.16. Different from the hierarchical beam search, in this subsection, we only search the optimal MPCs at the front K layer. What we do is only a coarse search.

---

[6]The adaptive compressed sensing (ACS) scheme proposed in[13] can reduce the training overhead to some extent, but multiple RF chains are required to guarantee satisfactory performance.

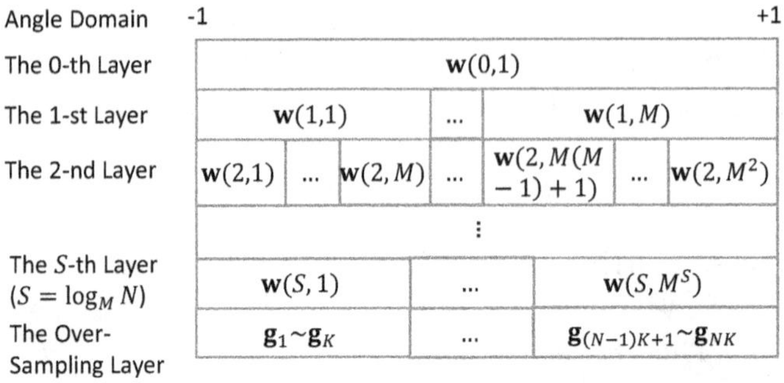

Angle Domain	-1							+1
The 0-th Layer	$\mathbf{w}(0,1)$							
The 1-st Layer	$\mathbf{w}(1,1)$		...		$\mathbf{w}(1,M)$			
The 2-nd Layer	$\mathbf{w}(2,1)$	...	$\mathbf{w}(2,M)$	...	$\mathbf{w}(2,M(M-1)+1)$	...	$\mathbf{w}(2,M^2)$	
				⋮				
The S-th Layer ($S = \log_M N$)	$\mathbf{w}(S,1)$			...		$\mathbf{w}(S,M^S)$		
The Over-Sampling Layer	$\mathbf{g}_1 \sim \mathbf{g}_K$			...		$\mathbf{g}_{(N-1)K+1} \sim \mathbf{g}_{NK}$		

Figure 2.16 Structure of a hierarchical codebook.

As we have estimated the channel $\mathbf{H}_{\mathrm{fd}}$, we have the following relation:

$$\mathbf{H} = \sum_{\ell=1}^{L} \lambda_\ell \mathbf{a}_{\mathrm{r}}(\Omega_\ell) \mathbf{a}_{\mathrm{t}}^{\mathrm{H}}(\psi_\ell) \approx \mathbf{H}_{\mathrm{fd}}. \tag{2.79}$$

To reconstruct the original channel $\mathbf{H}$, we need to estimate λ_ℓ, Ω_ℓ, and ψ_ℓ. Hence, we formulate the following problem

$$\underset{\lambda_\ell,\theta_\ell,\psi_\ell}{\text{minimize}} \quad \|\mathbf{H}_{\mathrm{fd}} - \sum_{\ell=1}^{L} \lambda_\ell \mathbf{a}_{\mathrm{r}}(\Omega_\ell) \mathbf{a}_{\mathrm{t}}^{\mathrm{H}}(\psi_\ell)\|_{\mathrm{F}}. \tag{2.80}$$

Then, analogous to the above CS approach, we could sample the AoA and AoD domains with a high resolution, i.e., an angle interval $2/(\alpha N_{\mathrm{r}})$ at Rx and $2/(\alpha N_{\mathrm{t}})$ at Tx, where α is the over-sampling factor, and we could obtain $\mathbf{A}_{\mathrm{r}} = [\mathbf{a}_{\mathrm{r}}(-1+\frac{1}{\alpha N_{\mathrm{r}}}), \mathbf{a}_{\mathrm{r}}(-1+\frac{3}{\alpha N_{\mathrm{r}}}), ..., \mathbf{a}_{\mathrm{r}}(-1+\frac{2\alpha N_{\mathrm{r}}-1}{\alpha N_{\mathrm{r}}})]$ and $\mathbf{A}_{\mathrm{t}} = [\mathbf{a}_{\mathrm{t}}(-1+\frac{1}{\alpha N_{\mathrm{t}}}), \mathbf{a}_{\mathrm{t}}(-1+\frac{3}{\alpha N_{\mathrm{t}}}), ..., \mathbf{a}_{\mathrm{t}}(-1+\frac{2\alpha N_{\mathrm{t}}-1}{\alpha N_{\mathrm{t}}})]$. This manipulation is applicable, but at the cost of a high computational complexity. In fact, by exploiting the search results in hierarchical beam search, we can significantly reduce the number of Rx and Tx candidate AWVs. Concretely, since the ℓ-th estimated AoA is $\hat{\Omega}_\ell = -1 + \frac{2n_{\mathrm{r}\ell}-1}{N_{\mathrm{r}}}$, the uncertainty range of the ℓ-th AoA should be $[\hat{\Omega}_\ell - \frac{2}{N_{\mathrm{r}}}, \hat{\Omega}_\ell + \frac{2}{N_{\mathrm{r}}}]$, which means that the candidate AoAs are the angle set obtained by sampling the angle range $[\hat{\Omega}_\ell - \frac{2}{N_{\mathrm{r}}}, \hat{\Omega}_\ell + \frac{2}{N_{\mathrm{r}}}]$ with an interval $2/(\alpha N_{\mathrm{r}})$. Consequently, the reduced Rx and Tx candidate AWVs are

$$\begin{aligned}
\bar{\mathbf{A}}_{\mathrm{r}} = \Big[& [\mathbf{a}_{\mathrm{r}}(\hat{\Omega}_1 - \frac{2}{N_{\mathrm{r}}} + \frac{2k}{\alpha N_{\mathrm{r}}})]_{k=0,1,...,2\alpha}, \\
& [\mathbf{a}_{\mathrm{r}}(\hat{\Omega}_2 - \frac{2}{N_{\mathrm{r}}} + \frac{2k}{\alpha N_{\mathrm{r}}})]_{k=0,1,...,2\alpha}, \cdots \\
& [\mathbf{a}_{\mathrm{r}}(\hat{\Omega}_L - \frac{2}{N_{\mathrm{r}}} + \frac{2k}{\alpha N_{\mathrm{r}}})]_{k=0,1,...,2\alpha} \Big],
\end{aligned} \tag{2.81}$$

and

$$\bar{\mathbf{A}}_\mathrm{t} = \Big[[\mathbf{a}_\mathrm{t}(\hat{\psi}_1 - \frac{2}{N_\mathrm{t}} + \frac{2k}{\alpha N_\mathrm{t}})]_{k=0,1,\dots,2\alpha},$$

$$[\mathbf{a}_\mathrm{t}(\hat{\psi}_2 - \frac{2}{N_\mathrm{t}} + \frac{2k}{\alpha N_\mathrm{t}})]_{k=0,1,\dots,2\alpha}, \dots \tag{2.82}$$

$$[\mathbf{a}_\mathrm{t}(\hat{\psi}_L - \frac{2}{N_\mathrm{t}} + \frac{2k}{\alpha N_\mathrm{t}})]_{k=0,1,\dots,2\alpha}\Big],$$

respectively, where $\hat{\psi}_\ell = -1 + \frac{2m_{\mathrm{t}\ell}-1}{N_\mathrm{t}}$. Then $\mathbf{H}$ can be approximately expressed as $\mathbf{H} = \bar{\mathbf{A}}_\mathrm{r}\boldsymbol{\Sigma}\bar{\mathbf{A}}_\mathrm{t}^\mathrm{H}$, where $\boldsymbol{\Sigma}$ is a diagonal and sparse matrix with the diagonal entries corresponding to the channel coefficients λ_ℓ, i.e., $\|\mathrm{vec}(\boldsymbol{\Sigma})\|_0 = L$. In a sequel,

$$\|\mathbf{H}_\mathrm{fd} - \sum_{\ell=1}^L \lambda_\ell \mathbf{a}_\mathrm{r}(\Omega_\ell)\mathbf{a}_\mathrm{t}(\psi_\ell)^\mathrm{H}\|_\mathrm{F} = \|\mathrm{vec}(\mathbf{H}_\mathrm{fd}) - \left(\bar{\mathbf{A}}_\mathrm{t}^* \otimes \bar{\mathbf{A}}_\mathrm{r}\right)\mathrm{vec}(\boldsymbol{\Sigma})\|_2$$
$$\stackrel{\Delta}{=} \|\mathrm{vec}(\mathbf{H}_\mathrm{fd}) - \bar{\mathbf{Q}}\mathrm{vec}(\boldsymbol{\Sigma})\|_2. \tag{2.83}$$

Hence, the problem (2.80) becomes

$$\underset{\lambda_\ell,\theta_\ell,\psi_\ell}{\mathrm{minimize}} \quad \|\mathrm{vec}(\mathbf{H}_\mathrm{fd}) - \bar{\mathbf{Q}}\mathrm{vec}(\boldsymbol{\Sigma})\|_2 \tag{2.84a}$$

$$\mathrm{subject\ to} \quad \|\mathrm{vec}(\boldsymbol{\Sigma})\|_0 = L, \tag{2.84b}$$

which is a standard sparse reconstruction problem and can be effectively solved by exploiting the OMP algorithm[28]. Note that an intrinsic difference between the problem shown in (2.84) and the one shown in (2.77) is that $\mathbf{Y}$ in (2.77) is measured by using random Tx/Rx AWVs, while $\mathbf{H}_\mathrm{fd}$ in (2.84) is obtained by using Algorithm 2.1 based on a hierarchical codebook.

In practice, the number of MPCs (i.e., L) is not known a priori. Besides, in some cases, it is not necessary to estimate all of the MPCs. In such cases, the number of MPCs in the presented multipath decomposition and recovery (MDR) approach, in both of the two stages, is set to $L = L_\mathrm{d}$, the desired number of MPCs. For instance, if we want to realize a 2-stream transmission, we only need to estimate $L_\mathrm{d} = 2$ MPCs, no matter how many MPCs the channel really has.

Since the sparse reconstruction stage does not need measurement, and the required feedback rate is small, the total number of measurements of the method is

$$T = L(4^{i_\mathrm{LY}} + 2(\log_2(N_\mathrm{r}) + \log_2(N_\mathrm{t}) - 2i_\mathrm{LY}) + 9). \tag{2.85}$$

i_LY is the initial layer in Algorithm 2.1. Note that this is the training overhead for an analog beamforming/combining structure. In the case of a hybrid structure, where parallel transmission of multiple-stream training sequences is available; the overhead will be further reduced.

2.3.5 Performance Evaluation

In this subsection, we evaluate the performance of the Joint Beam Search and Compressed Sensing approach. In the simulations, both LoS and NLoS channel models are considered based on (2.75). For the LoS channel, the first MPC has a constant coefficient and random AoD and AoA, while the other NLoS MPCs have complex Gaussian-distributed coefficients and random AoDs and AoAs[11, 18]. The LoS MPC is generally much stronger than the NLoS MPCs. For the NLoS channel, all the MPCs have complex Gaussian-distributed coefficients with the same variance and random AoDs and AoAs[11, 13, 18]. Both the LoS and NLoS channels are sparse in the angle domain because the number of MPCs is much smaller than the numbers of the Tx/Rx antennas[11, 13, 18]. Besides, $N_r = N_t = 32$ for all the simulations. The results are based on the average performance of 10^3 channel realizations.

First, we compare the performance of the Joint Beam Search and Compressed Sensing approach with the other alternatives in terms of success detection rate of MPCs and MSE of channel estimation, which is defined by MSE $= \mathbb{E}(\|\mathbf{H} - \hat{\mathbf{H}}\|_F^2)/(N_r N_t)$, where $\hat{\mathbf{H}}$ is the estimated channel matrix. The involved methods include the hierarchical search method and the CS method mentioned in this chapter, the ACS method in[13]. The hierarchical search method is to directly search multiple real MPCs using the same normal-resolution codebook designed in this chapter, and the ACS method exploits the codebook designed in[13].

Figs. 2.17 and 2.18 show the comparison results of success detection rate of MPCs and MSE of channel estimation, respectively, where $K = L = 2$. From these two figures we can find that the conventional hierarchical search method achieves poor performance. That is because the estimation error of AoAs and AoDs of the MPCs is significant due to the limited angle resolution of the codebook, which results in significant residual interference. Additionally, performances of the CS approach are highly dependent on the number of measurements, while those of the ACS approach are highly dependent on the number of RF chains. Only when the number of measurements and the number of RF chains are large enough, the CS and the ACS approaches, respectively, can achieve satisfactory performances; otherwise their performances will be not satisfactory. In comparison, the Joint Beam Search and Compressed Sensing

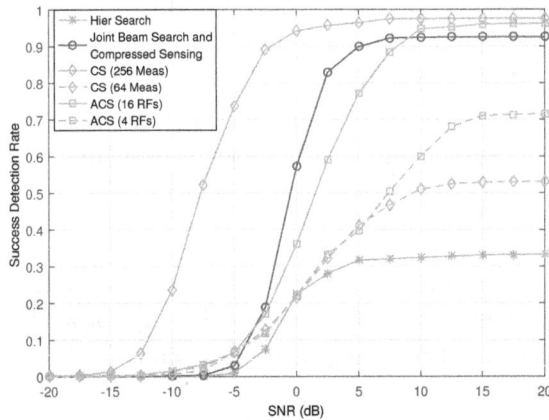

Figure 2.17 Comparison of the detection performance between different approaches, where $K = L = 2$. The success detection rate here means the rate to successfully detect all the MPCs. An NLoS channel model is adopted.

approach can achieve promising success detection rate and MSE performances, with only one single RF chain and a smaller number of measurements.

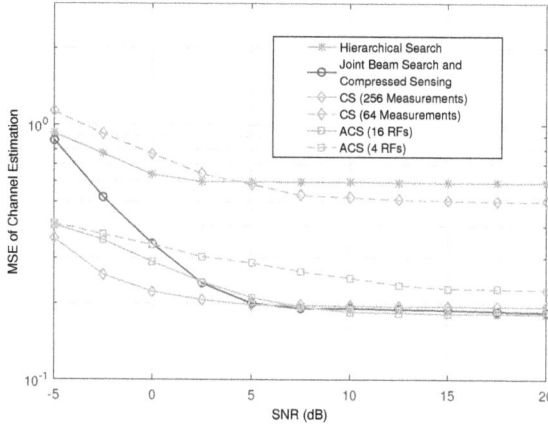

Figure 2.18 Comparison of the MSE performance between different approaches, where $K = L = 2$. An NLoS channel model is adopted.

In addition to the MSE performance, a more direct metric to evaluate the performance of millimeter-wave channel estimation is the relative gain loss of beamforming. For each method, the beamforming vectors at Tx and Rx are set to the right and left singular vectors of the estimated channel matrices, respectively, and then a practical beam gain can be obtained with the Tx/Rx beamforming vectors. The relative gain loss for a method is defined as the ideal beam gain subtracting the obtained practical beam gain in dB. Except the involved typical millimeter-wave channel estimation methods, there are also some baseline 1-stream beamforming methods, which do not need to estimate a full channel. For instance, the Exhaustive search method is to measure the channel on all AoA/AoD pairs ($K^2 N_r N_t$ pairs in total) and select the one with the maximum received energy. Another approach is that we make some measurements with random CM Tx/Rx AWVs, just the same as the CS method, and then we approximate the channel as a single path channel with a single AoA/AoD such that we can estimate them by using the maximal likelihood (ML) method.

Fig. 2.19 shows the comparison results of gain loss, where $K = L = 2$. Again we observe that the performance of the CS approach is highly dependent on the number of measurements, while those of the ACS approach are highly dependent on the number of RF chains. Only when the number of measurements and the number of RF chains are large enough, the CS and the ACS approaches can achieve satisfactory performances; otherwise their performances will not be satisfactory. In comparison, the Joint Beam Search and Compressed Sensing approach can achieve promising gain-loss performances, with only one single RF chain and a smaller number of measurements. These results show a well agreement with those from Fig. 2.18. Moreover, we can find that the two 1-stream methods, i.e., Exhaustive Search (1 Stream) and ML (1 Stream), behave poorer than MDR with high SNR while better with low SNR, this is because low SNR MDR induces additional noise when estimating more MPCs, while with high SNR MDR can make more accurate estimation of the other MPCs, thus, it obtains more channel energy by estimating more MPCs.

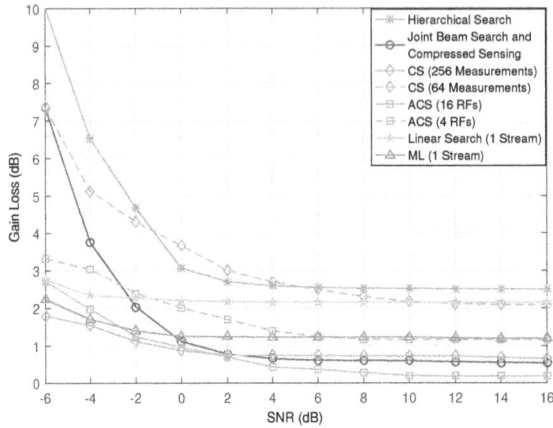

Figure 2.19 Comparison of the gain-loss performance between different approaches, where $K = L = 2$. The ML (1 Stream) method exploits 256 measurements. An NLoS channel model is adopted.

Lastly, we evaluate the training overhead that the proposed Joint Beam Search and Compressed Sensing approach requires and compare it with the other approaches. For a fair comparison, we consider only one-stream transmission, and do not count the overhead reduction that may be achieved via multi-stream transmission with multiple RF chains. Moreover, the training overhead is measured by the number of measurements for channel estimation, and for convenience, we assume $N_r = N_t = N$ and $L = 2$. As we know, the conventional least square (LS) requires at least $N_r N_t = N^2$ measurements. The proposed MDR approach requires T_{MDR} measurements as shown in (2.85), where S_0 can be typically set to 2. According to[13], the ACS approach requires $2^2 L^3 \log_2(N/L)$ measurements. For the pure CS approach, the required number of measurements is typically $\rho \log_2(N_{\mathrm{seq}})$[30], where N_{seq} is the length of the candidate row vectors in the dictionary matrix. In the context of this paper, $N_{\mathrm{seq}} = N_r N_t$. To guarantee a satisfactory performance, we set $\rho = 20$ here. Fig. 2.20 shows the comparison result, where we can find that the proposed MDR approach requires the least training overhead.

2.4 SUMMARY

In this chapter, we have introduced four different search methods and evaluated their performance. They are used to acquire channel information with minimum overhead. The exhaustive search method searches the whole Tx/Rx angle plane and finds the N_S (AoD AoA) pairs with the highest strengths. But when the number of antenna arrays is big, the time cost will be high. This method cost most time comparing with other three methods. Hierarchical beam search is based on the hierarchical codebook. This method searches only M AWV at every layer. At the last layer, named oversampling layer, we only search K elements. Comparing with the exhaustive search, the time

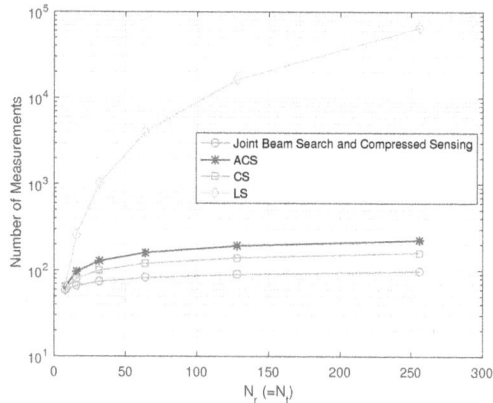

Figure 2.20 The comparison of required training overhead between different approaches.

complexity is effectively reduced. CS for channel estimation uses compressed sensing method to improve the estimation accuracy and reduce the training overload and exploits OMP algorithm to solve the optimization problem. Joint Beam Search and Compressed Sensing search has the advantages of both Hierarchical Beam Search and compressed sensing method. It uses hierarchical codebook to reduce the time cost of searching the channel information. And then, it uses compressed sensing method to optimize the channel information.

Bibliography

[1] Su-Khiong Yong, Pengfei Xia, and Alberto Valdes-Garcia. *60GHz Technology for Gbps WLAN and WPAN: from Theory to Practice*. Wiley, West Sussex, UK, 2011.

[2] Pengfei Xia, Huaning Niu, Jisung Oh, and Chiu Ngo. Practical antenna training for millimeter wave MIMO communication. In *IEEE Vehicular Technology Conference (VTC) 2008*, pages 1–5, Calgary, Canada, Oct. 2008. IEEE.

[3] Junyi Wang, Zhou Lan, Chang-woo Pyo, Tunçer Baykaş, Chin-Sum Sum, Azizur Rahman, Mohammad, Jing Gao, Ryuhei Funada, Fumihide Kojima, and Hiroshi Harada. Beam codebook based beamforming protocol for multi-Gbps millimeter-wave WPAN systems. *IEEE J. Select. Areas Commun.*, 27(8):1390–1399, Oct. 2009.

[4] Junyi Wang, Zhou Lan, Chin-Sean Sum, Chang-Woo Pyo, Jing Gao, Tunçer Baykaş, Azizur Rahman, Ryuhei Funada, Fumihide Kojima, and Ismail Lakkis. Beamforming codebook design and performance evaluation for 60GHz wideband WPANs. In *IEEE Vehicular Technology Conference Fall (VTC 2009-Fall)*, pages 1–6, Anchorage, AK, Sep. 2009. IEEE.

[5] Y Ming Tsang, Ada SY Poon, and Sateesh Addepalli. Coding the beams: Improving beamforming training in mmwave communication system. In *2011 IEEE Global Telecommunications Conference-GLOBECOM 2011*, pages 1–6. IEEE, 2011.

[6] Bin Li, Zheng Zhou, Weixia Zou, Xuebin Sun, and Guanglong Du. On the efficient beam-forming training for 60GHz wireless personal area networks. *IEEE Trans. Wireless Commun.*, 12(2):504–515, 2012.

[7] Zhenyu Xiao, Lin Bai, and Jinho Choi. Iterative joint beamforming training with constant-amplitude phased arrays in millimeter-wave communications. *IEEE Commun. Lett.*, 18(5):829–832, 2014.

[8] Ahmed Alkhateeb, Jianhua Mo, Nuria González-Prelcic, and W. Heath, Robert. MIMO precoding and combining solutions for millimeter-wave systems. *IEEE Commun. Mag.*, 52(12):122–131, Dec. 2014.

[9] Wonil Roh, Ji-Yun Seol, Jeongho Park, Byunghwan Lee, Jaekon Lee, Yungsoo Kim, Jaeweon Cho, Kyungwhoon Cheun, and Farshid Aryanfar. Millimeter-wave beamforming as an enabling technology for 5G cellular communications: theoretical feasibility and prototype results. *IEEE Commun. Mag.*, 52(2):106–113, Feb. 2014.

[10] Shu Sun, Theodore S Rappaport, W. Heath, Robert, Andrew Nix, and Sundeep Rangan. MIMO for millimeter-wave wireless communications: beamforming, spatial multiplexing, or both? *IEEE Commun. Mag.*, 52(12):110–121, Dec. 2014.

[11] Zhenyu Xiao, Tong He, Pengfei Xia, and Xiang-Gen Xia. Hierarchical codebook design for beamforming training in millimeter-wave communication. *IEEE Trans. Wireless Commun.*, 15(5):3380–3392, May 2016.

[12] Tong He and Zhenyu Xiao. Suboptimal beam search algorithm and codebook design for millimeter-wave communications. *Mobile Networks and Applications*, 20(1):86–97, Jan. 2015.

[13] Ahmed Alkhateeb, Omar El Ayach, Geert Leus, and W Heath, Robert. Channel estimation and hybrid precoding for millimeter wave cellular systems. *IEEE J. Sel. Top. Sign. Proces.*, 8(5):831–846, Oct. 2014.

[14] David Tse and Pramod Viswanath. *Fundamentals of Wireless Communication*. Cambridge University Press, New York, USA, 2005.

[15] Ehsan Adabi Firouzjaei. *Mm-Wave Phase Shifters and Switches*. PhD thesis, UC Berkeley, 2010.

[16] Song Noh, Michael D Zoltowski, and David J Love. Multi-resolution codebook based beamforming sequence design in millimeter-wave systems. In *IEEE Global Telecommunications Conference*, pages 1–6, San Diego, CA, USA, Dec. 2015.

[17] Omar El Ayach, Sridhar Rajagopal, Shadi Abu-Surra, Zhouyue Pi, and W. Heath, Robert. Spatially sparse precoding in millimeter wave MIMO systems. *IEEE Trans. Wireless Commun.*, 13(3):1499–1513, Mar. 2014.

[18] Sooyoung Hur, Taejoon Kim, David J Love, James V Krogmeier, Timothy A Thomas, and Amitava Ghosh. Millimeter wave beamforming for wireless backhaul and access in small cell networks. *IEEE Trans. Commun.*, 61(10):4391–4403, Oct. 2013.

[19] Jimmy Nsenga, Wim Van Thillo, François Horlin, Valery Ramon, André Bourdoux, and Rudy Lauwereins. Joint transmit and receive analog beamforming in 60 GHz MIMO multipath channels. In *IEEE International Conference on Communications (ICC)*, pages 1–5, Dresden, Germany, Jun. 2009. IEEE.

[20] John G Proakis. *Digital communications 5th Edition.* McGraw Hill Higher Education, New York, USA, 2007.

[21] Zhenyu Xiao, Changming Zhang, Depeng Jin, and Ning Ge. GLRT approach for robust burst packet acquisition in wireless communications. *IEEE Trans. Wireless Commun.*, 12(3):1127–1137, Mar. 2013.

[22] Yanyu Jin, Mihai AT Sanduleanu, Eduardo Alarcon Rivero, and John R Long. A millimeter-wave power amplifier with 25dB power gain and +8dBm saturated output power. In *European Solid State Circuits Conference (ESSCIRC)*, pages 276–279. IEEE, 2007.

[23] Yi Zhao and John R Long. A wideband, dual-path, millimeter-wave power amplifier with 20 dBm output power and pae above 15% in 130 nm SiGe-BiCMOS. *IEEE J. Solid-State Circuits*, 47(9):1981–1997, 2012.

[24] Kao-Cheng Huang and Zhaocheng Wang. *Millimeter Wave Communication Systems.* Wiley-IEEE Press, Hoboken, New Jersey, USA, 2011.

[25] Zhenyu Xiao, Xiang-Gen Xia, Depeng Jin, and Ning Ge. Iterative eigenvalue decomposition and multipath-grouping Tx/Rx joint beamformings for millimeter-wave communications. *IEEE Trans. Wireless Commun.*, 14(3):1595–1607, Mar. 2015.

[26] Zhenyu Xiao, Pengfei Xia, and Xiang-Gen Xia. Hierarchical multi-beam search for millimeter-wave MIMO systems. In *IEEE Vehicular Technology Conference (VTC Spring)*, pages 1–5. IEEE, 2016.

[27] Zhenyu Xiao, Lipeng Zhu, Jinho Choi, Pengfei Xia, and Xiang-Gen Xia. Joint power allocation and beamforming for non-orthogonal multiple access (NOMA) in 5G millimeter wave communications. *IEEE Trans. Wireless Commun.*, 17(5):2961–2974, May 2018.

[28] Junho Lee, Gye-Tae Gil, and Yong Hoon Lee. Exploiting spatial sparsity for estimating channels of hybrid MIMO systems in millimeter wave communications. In *IEEE Global Communications Conference*, pages 3326–3331. IEEE, 2014.

[29] Ahmed Alkhateeby, Geert Leusz, and W Heath, Robert. Compressed sensing based multi-user millimeter wave systems: How many measurements are needed? In *IEEE ICASSP*, pages 2909–2913, Apr. 2015.

[30] Waheed U Bajwa, Jarvis Haupt, Akbar M Sayeed, and Robert Nowak. Compressed channel sensing: A new approach to estimating sparse multipath channels. *Proceedings of the IEEE*, 98(6):1058–1076, Jun. 2010.

CHAPTER 3

Array Beamforming for Point-to-Point Transmission

3.1 INTRODUCTION

With accurate channel state information (CSI) estimation, the optimal antenna weight vectors (AWVs) at the transmitter (Tx)/receiver (Rx) can be found under well-known performance criteria, e.g., maximizing the receive signal-to-noise ratio (SNR)[1,2,3,4]. Then the AWVs at both ends are able to be appropriately set for beamforming to achieve array gain and compensate for high propagation attenuation, which is the *joint* Tx/Rx beamforming. However, when the number of antennas generally becomes large, the channel estimation becomes time-consuming in communication systems as more number of time slots are needed for transmitting pilots and performing feedback. In addition, the computational complexity is high, because the matrix decomposition, e.g., the singular value decomposition (SVD), is generally required.

In Chapter 2, we have presented some channel estimation methods based on codebook and the corresponding hierarchical codebook design solutions, i.e., exhaustive search, hierarchical beam search, Compressed Sensing for Channel Estimation and Joint Beam Search and Compressed Sensing search. Based on a pre-designed beamforming vector codebook, the training overload and computational complexity for channel estimation can be reduced.

Nevertheless, it is always required to consume the necessary time slots and computing resource to estimate the complete CSI at the Rx side and feed it back to the Tx side. In this regard, it is expected to exploring a more concise and effective solution, which can find reasonable AWVs to achieve high array gains but do not require full knowledge of CSI. This is so-called adaptive beamforming, where the desired AWVs at both ends can be found using iterative beamforming training between Tx and Rx. In this way, the knowledge of CSI implicated in the iterative training process can lead to finding a good beamforming vector, where we do not need to actually know the accurate CSI. This methodology is able to decrease the training overload and computational complexity compared to codebook-based channel estimation. Hence, in

DOI: 10.1201/9781003366362-3

this chapter, we present several solutions of adaptive beamforming to realize efficient point-to-point transmission.

It is worth noting that eigen decomposition based methods for adaptive beamforming have already been proposed[5, 6]. Different from the above strategies, in this chapter, a new sub-optimal adaptive beamforming scheme is first presented, which finds the AWVs via iterative eigenvalue decomposition (IEVD), provided that full CSI is available at both Tx and Rx. To make this sub-optimal scheme practically feasible in communications, a corresponding training approach is suggested to avoid the channel estimation and IEVD computation. The convergence analysis is also provided. Furthermore, in fast fading scenarios, a multipath-grouping (MPG) based beamforming scheme is presented to reduce overhead and increase system reliability. The scheme first groups the multipath components (MPCs) and then concurrently beamforms toward multiple steering angles of the grouped MPCs, so that both array gain and diversity gain are achieved. Owing to the MPG operation, the scheme guarantees a solution of AWV even when the number of MPCs is greater than that of the antennas at both ends. Pairwise-error probability (PEP) and diversity analyses are given. Then, in the third section, a joint beamforming training scheme (STV) based on steering vectors is presented, which uses the directional characteristics of the high-frequency band channel and compares its performance with the training scheme based on singular vectors (SGV).

3.2 SYSTEM AND CHANNEL MODELS

Without loss of generality, we consider a communication system with half-wavelength spaced uniform linear arrays (ULAs) of M and N elements at Tx and Rx, respectively[2], as shown in Fig. 3.1. A single radio frequency (RF) chain is tied to the ULA at Tx and Rx. At Tx, a single data stream is transmitted from multiple weighted antenna elements (AEs), and at Rx, signals from multiple AEs are weighted and combined to shape a single signal stream. It is noted that the system is half duplex or time-division duplex, i.e., a data stream can also be transmitted from Rx to Tx in the same frequency band but at a different time. According to the reported results of channel measurement for millimeter-wave communication[7, 8], only reflection contributes to generating MPCs besides the line of sight (LoS) component; scattering and diffraction effects are little due to the extremely small wave length of the communication. Thus, the MPCs in millimeter-wave communication have a directional feature, i.e., different MPCs have different physical transmit steering angles $\phi_{t\ell}$ and receive steering angles $\phi_{r\ell}$, as shown in Fig. 3.1. In fact,[9, 10, 11, 12] have reported such channel models. It is noted that although only reflection generates significant MPCs, the number of MPCs may be not always small, because there may be many good reflectors in communication. For instance, the walls, floor, ceiling, and metal objects are good reflectors for the indoor communication[8, 13, 14, 15], and the buildings are good reflectors for the outdoor communication[16, 17]. Moreover, the second-order reflection components may also have a significant strength[8, 13, 14, 15].

In order to achieve a high transmission speed, the bandwidth of communication is generally large. For instance, in both 60 GHz Wireless Local

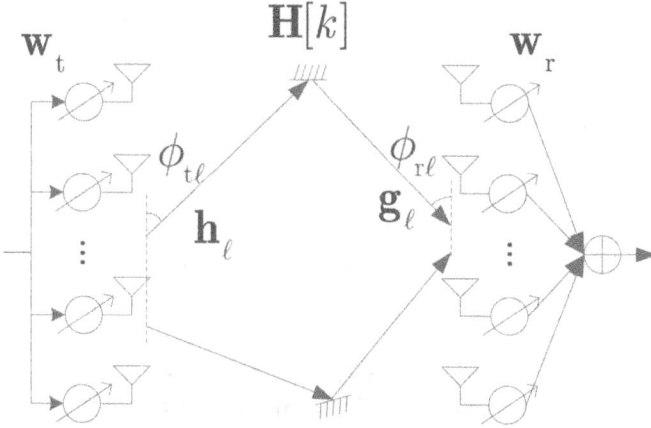

Figure 3.1 Illustration of the system.

Area Network (WLAN)[7,18,19,20,21,22] and millimeter-wave mobile broadband communications[23,24], the signal bandwidth is basically greater than 1 GHz, which means the equivalent symbol duration is less than 1 ns. That is to say, a difference of only 3 m in path distance would result in a path delay difference of about $3/(3 \times 10^8 \times 10^{-9}) = 10$ symbol durations. As in communication the propagation distance may be tens of meters for WLAN or hundreds of meters for mobile communications, the difference of path distances may be several or even tens of meters. Hence, a frequency-selective steering channel model is suitable for communication[1], which has been used in[4,22] and can be expressed as

$$\mathbf{H}[k] = \sum_{\ell=0}^{L-1} \hat{\mathbf{H}}_\ell \delta[k - \tau_\ell] = \sqrt{MN} \sum_{\ell=0}^{L-1} \mathbf{g}_\ell \lambda_\ell \mathbf{h}_\ell^{\mathrm{H}} \delta[k - \tau_\ell], \tag{3.1}$$

where $\hat{\mathbf{H}}_\ell = \sqrt{MN}\mathbf{g}_\ell\lambda_\ell\mathbf{h}_\ell^{\mathrm{H}}$, $(\cdot)^{\mathrm{H}}$ is the conjugate transpose operation, $\delta[k]$ is the discrete impulse response function, λ_ℓ and τ_ℓ are the channel coefficient, and delay of the ℓ-th MPC, respectively, and $\mathbf{g}_\ell$ and $\mathbf{h}_\ell$ are receive and transmit steering vectors of the ℓ-th MPC given by[9,10,11,12],

$$\mathbf{g}_\ell = \frac{1}{\sqrt{N}}[e^{j\pi 0\Omega_{r\ell}}, e^{j\pi 1\Omega_{r\ell}}, e^{j\pi 2\Omega_{r\ell}}, ..., e^{j\pi(N-1)\Omega_{r\ell}}]^{\mathrm{T}}, \tag{3.2}$$

and

$$\mathbf{h}_\ell = \frac{1}{\sqrt{M}}[e^{j\pi 0\Omega_{t\ell}}, e^{j\pi 1\Omega_{t\ell}}, e^{j\pi 2\Omega_{t\ell}}, ..., e^{j\pi(M-1)\Omega_{t\ell}}]^{\mathrm{T}}, \tag{3.3}$$

respectively, where $(\cdot)^{\mathrm{T}}$ is the transpose operator. Note that $\Omega_{t\ell}$ and $\Omega_{r\ell}$ represent the cosine transmit and receive angles of the ℓ-th MPC, and $\Omega_{t\ell} = \cos(\phi_{t\ell})$ and $\Omega_{r\ell} = \cos(\phi_{r\ell})$, respectively[25]. Therefore, $\Omega_{t\ell}$ and $\Omega_{r\ell}$ are within the range $[-1\ 1)$.

---

[1]All the beamforming methods presented in this book are only appropriate when the system bandwidth is sufficient to resolve all the MPCs with different steering angles.

For convenience, in the rest of this chapter, $\Omega_{t\ell}$ and $\Omega_{r\ell}$ are called transmit and receive angles, respectively.

It is assumed that $\mathbf{h}_\ell$ and $\mathbf{g}_\ell$ are quasi-static at Tx and Rx, which means that $\mathbf{h}_\ell$ and $\mathbf{g}_\ell$ vary slowly and can be well estimated at Tx and Rx, respectively. While λ_ℓ are independently and identically distributed complex Gaussian variables with zero mean and variance $1/L$. The basic evidence of this assumption is that the multipath directions, which determine $\mathbf{h}_\ell$ and $\mathbf{g}_\ell$, will change slowly with respect to node or scatter movement, but the phases of the multipath coefficients can change rapidly due to the short wave length of high-frequency band communication, leading to a fluctuation/fading in signal strength[13, 26, 27], i.e., $|\lambda_\ell|$. It is noted that in slow fading scenarios, λ_ℓ varies slowly compared with the transmission speed. Thus, $\mathbf{H}[k]$ varies slowly and can be well estimated, which means full CSI may be available at both Tx and Rx. However, in fast fading scenarios, λ_ℓ varies fast compared with the transmission speed, which means $\mathbf{H}[k]$ varies fast. Thus, the estimation of $\mathbf{H}[k]$ becomes frequent and time consuming, which may greatly degrade the system efficiency.

3.3 EIGENVALUE DECOMPOSITION BASED BEAMFORMING

For single-layer beamforming, it is known that, provided the CSI at both ends, the optimal AWVs at Tx/Rx can be found under well-known performance criteria, e.g., maximizing receive SNR[1, 2, 3, 4]. In narrow-band systems, it is well known that the optimal receive and transmit AWVs are left and right principal singular vectors of the channel matrix, respectively[1, 25]. However, in wideband communication which experiences frequency-selective channels, it is difficult to find a solution of the optimal transmit and receive AWVs. In[4], Nsenga et al. have proposed a sub-optimal scheme exploiting eigenvalue decomposition (EVD) and Schmidt decomposition within a high-dimensional space tensed by the transmit and receive AWVs. Although less CSI is required than full CSI, the channel estimation is still time-consuming, and the computations of EVD and Schmidt decomposition in the tensor space are also complicated due to the high dimension, which may limit the practical application of this scheme. On the other hand, in fast fading scenarios[2], these sub-optimal schemes that require CSI at both ends become infeasible due to frequent channel estimation or training, which are time-consuming. In order to reduce overhead and meanwhile increase system reliability, the scheme proposed by Park and Pan in[10] can be adopted, which utilizes the quasi-static steering angles of the MPCs in communications, and concurrently beamforms along multiple steering angles at both ends to achieve diversity gain in addition to array gain. This scheme is simple to implement and achieves full diversity, but it may be not efficient enough in array gain, and may be infeasible when the number of MPCs is larger than that of the antennas in either ends, because in such a case a solution of AWV may not exist.

---

[2]It is noted that a fast/slow fading channel means that the estimation of multi-antenna channel is required frequently/non-frequently in this book. For instance, in fast fading channel, the estimation may be required each several packets, but in slow fading channel, the estimation may be required each tens of packets.

In order to solve the above problems, in this section, we investigate Tx/Rx joint beamforming in communications. As the MPCs have different steering angles and independent fadings, beamforming aims at achieving array gain as well as diversity gain in this scenario. A sub-optimal beamforming scheme is presented to find the AWVs at Tx/Rx via iterative EVD, provided that full CSI is available at both Tx and Rx. To make this scheme practically feasible in communications, a corresponding training approach is suggested to avoid the channel estimation and iterative EVD computation. As in fast fading scenarios the training approach may be time-consuming due to frequent training, another beamforming scheme, which exploits the quasi-static steering angles in communications, is presented to reduce the overhead and increase the system reliability by MPG. The scheme first groups the MPCs and then concurrently beamforms toward multiple steering angles of the grouped MPCs, so that both array gain and diversity gain are achieved. Performance comparisons show that, compared with the corresponding state-of-the-art schemes, the iterative EVD scheme with the training approach achieves the same performance with a reduced overhead and complexity, while the MPG scheme achieves better performance with an approximately equivalent complexity.

The rest of this section is organized as follows. Firstly, we present the IEVD scheme and its training approach, and conduct the convergence analysis. Then, we give a brief description of the scheme proposed by Park and Pan, and then introduce the MPG scheme. Afterwards, we analyze the PEP and diversity performance. Finally, we present performance comparison.

By exploiting transmit and receive beamforming, the received signals $y[m]$ are expressed as

$$y[m] = \sqrt{\gamma} \sum_{\ell=0}^{L-1} \mathbf{w}_r^H \hat{\mathbf{H}}_\ell \mathbf{w}_t s[m - \tau_\ell] + \mathbf{w}_r^H \mathbf{n}, \tag{3.4}$$

where γ is the transmit SNR, which refers to the SNR without accounting the array gain achieved by beamforming in this book, $s[m]$ are the normalized transmitted information symbols, $\mathbf{w}_t$ and $\mathbf{w}_r$ are transmit and receive AWVs with unit l_2-norms, respectively, i.e., $\|\mathbf{w}_t\|_2 = \|\mathbf{w}_r\|_2 = 1$, L is the number of MPCs, $\mathbf{n}$ is the standard Gaussian complex noise vector.

3.3.1 Iterative EVD Scheme

In this section, we first introduce the IEVD scheme which requires full CSI. Afterwards, the training approach to obtain the sub-optimal AWVs in IEVD are described to avoid channel estimation and iterative EVD computation, which makes the presented IEVD scheme practically feasible. Finally, the analysis on the convergence of IEVD and the training approach is given.

3.3.1.1 Description of IEVD

For single-layer beamforming, provided full CSI, the optimal transmit and receive AWVs can be found under well-known performance criteria, e.g., maximizing receive SNR. We adopt the receive SNR as the criterion in this context, because we assume that the inter-symbol interference (ISI) caused by MPCs can be well addressed with equalization technology. In the case that there is no equalizer at Rx after beamforming, signal-to-interference-plus-noise ratio (SINR) may be a better criterion. With the criterion of receive SNR, under frequency-flat channels, the optimal AWVs are the principal singular vectors of the channel matrix, and they can be obtained by SVD on the channel matrix. However, under frequency-selective channels, the optimal AWVs aiming to maximizing receive SNR are difficult to find, and a feasible solution is still not available in the literature to the best of our knowledge. Although Nsenga's scheme[4], which exploits EVD and Schmidt decomposition within a high-dimensional space tensed by the transmit and receive AWVs, finds a sub-optimal solution, the channel estimation is still time-consuming, and the computations of EVD and Schmidt decomposition in the tensor space are also complicated due to the high dimension. Thus, in this subsection, we propose another sub-optimal solution.

The optimization problem is formulated as

$$\text{maximize} \quad \Gamma = \sum_{\ell=0}^{L-1} |\mathbf{w}_r^H \hat{\mathbf{H}}_\ell \mathbf{w}_t|^2,$$

$$\text{subject to} \quad \|\mathbf{w}_t\|_2 = \|\mathbf{w}_r\|_2 = 1. \tag{3.5}$$

where Γ is the power gain (or the array gain, and $\gamma\Gamma$ is the receive SNR). It is noted that in communications, there are two types of antenna arrays in practice that correspond to different constraints on AWV of one-layer beamforming. The first one is phased array with constant amplitude. In such a case all the elements of $\mathbf{w}_t$ or $\mathbf{w}_r$ should be the same constant, and only the phases can be adjusted. For instance,[28, 29] have adopted this model. While the other one is antenna array with both amplitude and phase adjustable, which has also be widely used in[4, 10, 22, 30], and there are also practical implementations of this model, e.g., in[31, 32]. In our model, both amplitude and phase are adjustable, i.e., the latter one is adopted.

In general, when we discuss the optimality of the solution to this problem, we implicitly assume that $\hat{\mathbf{H}}_\ell$ is known a priori, which corresponds to the slow fading case, where $\hat{\mathbf{H}}_\ell$ can be periodically estimated without too much overhead. However, in the fast fading case, the estimation of $\hat{\mathbf{H}}_\ell$ becomes frequent and the overhead used for the estimation becomes high, beamforming can only be conduct according to the quasi-static steering vectors, which will be discussed in the next section.

Given that $\hat{\mathbf{H}}_\ell$ is known a priori, when $L = 1$, the optimal AWVs are the principal singular vectors of $\hat{\mathbf{H}}_0$. When $L > 1$, the optimal AWVs are difficult to find. However, if given $\mathbf{w}_r$, we have

$$\Gamma = \sum_{\ell=0}^{L-1} \mathbf{w}_t^H \hat{\mathbf{H}}_\ell^H \mathbf{w}_r \mathbf{w}_r^H \hat{\mathbf{H}}_\ell \mathbf{w}_t = \mathbf{w}_t^H \left(\sum_{\ell=0}^{L-1} \hat{\mathbf{H}}_\ell^H \mathbf{w}_r \mathbf{w}_r^H \hat{\mathbf{H}}_\ell \right) \mathbf{w}_t. \tag{3.6}$$

The optimal $\mathbf{w}_t$ in such a case is the principal eigenvector of $\left(\sum_{\ell=0}^{L-1} \hat{\mathbf{H}}_\ell^H \mathbf{w}_r \mathbf{w}_r^H \hat{\mathbf{H}}_\ell \right)$. Similarly, given $\mathbf{w}_t$, we have

$$\Gamma = \sum_{\ell=0}^{L-1} \mathbf{w}_r^H \hat{\mathbf{H}}_\ell \mathbf{w}_t \mathbf{w}_t^H \hat{\mathbf{H}}_\ell^H \mathbf{w}_r = \mathbf{w}_r^H \left(\sum_{\ell=0}^{L-1} \hat{\mathbf{H}}_\ell \mathbf{w}_t \mathbf{w}_t^H \hat{\mathbf{H}}_\ell^H \right) \mathbf{w}_r. \tag{3.7}$$

The optimal $\mathbf{w_r}$ in such a case is the principal eigenvector of $\left(\sum_{\ell=0}^{L-1} \hat{\mathbf{H}}_\ell \mathbf{w_t} \mathbf{w_t}^H \hat{\mathbf{H}}_\ell^H\right)$. Based on this, we present the IEVD scheme in Algorithm 3.1. This scheme is a typical *alternating optimization* approach[33, 34], which does not guarantee the optimal solution, but is efficient to find a sub-optimal solution.

Algorithm 3.1: The IEVD Scheme.

---

1) Initialize:

  Randomly pick a normalized initial receive AWV $\mathbf{w_r}$.

2) Iteration:

  Iterate the following process ε times, then stop.

  Compute EVD on

$$\sum_{\ell=0}^{L-1} \hat{\mathbf{H}}_\ell^H \mathbf{w_r} \mathbf{w_r}^H \hat{\mathbf{H}}_\ell, \tag{3.8}$$

  and set the principal eigenvector to $\mathbf{w_t}$.

  Compute EVD on

$$\sum_{\ell=0}^{L-1} \hat{\mathbf{H}}_\ell \mathbf{w_t} \mathbf{w_t}^H \hat{\mathbf{H}}_\ell^H, \tag{3.9}$$

  and set the principal eigenvector to $\mathbf{w_r}$.

3) Result:

  $\mathbf{w_t}$ is the transmit AWV, and $\mathbf{w_r}$ is the receive AWV.

3.3.1.2 The Training Approach

It is clear that the IEVD scheme relies on a priori CSI. Since there are totally $N \times M \times L$ coefficients in the multipath channel matrices, the conventional channel estimation which estimates these coefficients one by one is rather time-consuming. Although when the path number is much smaller than the antenna number, the CSI can be fully characterized by the angle of departure (AoD), angle of arrival (AoA), and fading coefficient of each path, the angle estimation of AoD and AoA may also be complicated. Thus, only in slow fading scenarios, the IEVD scheme may be applicable. In fact, even in slow fading scenarios, the time-costly channel estimation may still significantly degrade the system efficiency due to the overhead used for channel estimation. In addition, the iterative EVD consumes much computation resource. Thus, the channel estimation and EVD computation make the IEVD scheme not so attractive even in slow fading scenarios. To address this problem, we present the training approach to obtain the AWVs in the IEVD scheme, which is based on the well-known "power method"[35], i.e., the principal eigenvector of an arbitrary $N \times N$

Algorithm 3.2: The Training Approach.

---

1) Initialize:

Randomly pick a normalized initial AWV $\mathbf{w}_r$ at the destination.

2) Iteration:

Iterate the following process ε times, then stop.

The destination transmits training sequences to the source: Keep transmitting training sequences with the same AWV $\mathbf{w}_r$ at the destination over M slots, one sequence in a slot. Meanwhile, use identity matrix $\mathbf{I}_M$ as the receive AWVs at the source, i.e., the i-th column of $\mathbf{I}_M$ as the receive AWV at the i-th slot. Ignoring the noise and decoding the training sequences, in the i-th slot we receive

$$r_i[\ell] = \mathbf{e}_i^{\mathrm{H}} \hat{\mathbf{H}}_\ell^{\mathrm{H}} \mathbf{w}_r, i = 1, 2, ..., M, \ell = 1, 2, ..., L, \qquad (3.10)$$

where $\mathbf{e}_i$ is the i-th column of $\mathbf{I}_M$. Then we arrive at

$$\mathbf{r}[\ell] = \{r_i[\ell]\}_{i=1,2,...,M} = \mathbf{I}_M^{\mathrm{H}} \hat{\mathbf{H}}_\ell^{\mathrm{H}} \mathbf{w}_r = \hat{\mathbf{H}}_\ell^{\mathrm{H}} \mathbf{w}_r, \qquad (3.11)$$

$$\mathbf{R}_{\mathrm{S}} = \sum_{\ell=0}^{L-1} \mathbf{r}[\ell](\mathbf{r}[\ell])^{\mathrm{H}}. \qquad (3.12)$$

Normalize $\mathbf{R}_{\mathrm{S}}^K \mathbf{e}_1$ and set the result to $\mathbf{w}_t$ as a new AWV in source.

The source transmits training sequences to the destination: Keep transmitting training sequences with the same AWV $\mathbf{w}_t$ at the source over N slots, one sequence in a slot. Meanwhile, use identity matrix $\mathbf{I}_N$ as the receive AWVs at the destination, i.e., the j-th column of $\mathbf{I}_N$ as the receive AWV at the j-th slot. Ignoring the noise and decoding the training sequences, in the j-th slot, we receive

$$\bar{r}_j[\ell] = \bar{\mathbf{e}}_j^{\mathrm{H}} \hat{\mathbf{H}}_\ell \mathbf{w}_t, j = 1, 2, ..., N, \ell = 1, 2, ..., L, \qquad (3.13)$$

where $\bar{\mathbf{e}}_j$ is the j-th row of $\mathbf{I}_N$. Then we arrive at

$$\bar{\mathbf{r}}[\ell] = \{\bar{r}_j[\ell]\}_{j=1,2,...,N} = \mathbf{I}_N^{\mathrm{H}} \hat{\mathbf{H}}_\ell \mathbf{w}_t = \hat{\mathbf{H}}_\ell \mathbf{w}_t, \qquad (3.14)$$

$$\mathbf{R}_{\mathrm{D}} = \sum_{\ell=0}^{L-1} \bar{\mathbf{r}}[\ell](\bar{\mathbf{r}}[\ell])^{\mathrm{H}}. \qquad (3.15)$$

Normalize $\mathbf{R}_{\mathrm{D}}^K \bar{\mathbf{e}}_1$ and set the result to $\mathbf{w}_r$ as a new AWV in destination.

3) Result:

$\mathbf{w}_t$ is the transmit AWV, and $\mathbf{w}_r$ is the receive AWV.

---

matrix $\mathbf{X}$ can be approximately computed by normalizing $\mathbf{X}^{m_{\text{Loop}}}\mathbf{w}$, where $\mathbf{w}$ is an arbitrary vector, given that m_{Loop} is sufficiently large.

The training approach utilizes the reciprocal feature of the channel, i.e., given $\mathbf{H}[k]$ as the forward channel response from source Tx to destination Rx [3], as shown in (3.1), the backward channel response from destination to source is $(\mathbf{H}[k])^{\text{H}}$ [4]. Based on the simplified approach to compute EVD as well as the reciprocal channel, the training approach is introduced in Algorithm 3.2. With this training approach, the channel estimation and iterative EVD computation are bypassed, which greatly increases the practical feasibility of the IEVD scheme.

There are various stopping rules for IEVD and its training approach. A simple one is to stop after a certain number of iterations, e.g., ε iterations used in Algorithms 3.1 and 3.2. In fact, these two schemes converge fast and basically only need 2 or 3 iterations, which will be shown later. Another one is to compute Γ after the n-th iteration and get $\Gamma^{(n)}$ according to (3.5). When $\Gamma^{(n)}/\Gamma^{(n-1)} < \mu$, stop the iteration, where μ is a predefined threshold slightly greater than 1, e.g., $\mu = 1.05$.

It is noted that the training approaches share the same iteration principle with IEVD. Thus, they have the same convergence properties, which will be analyzed in the next subsection.

3.3.1.3 Convergence Analysis

The convergence of IEVD, as well as its training approach, will be evident after proving the following theorem.

Theorem 3.3.1. *Let $\Gamma^{(n)}$ denote the value of Γ after the n-th iteration. Then $\{\Gamma^{(n)}|n = 1, 2, ...\}$ is a non-descending sequence, i.e., $\Gamma^{(n+1)} \geq \Gamma^{(n)}$.*

Proof. Firstly, we have

$$\Gamma^{(n)} = \sum_{\ell=0}^{L-1} |\left(\mathbf{w}_{\text{r}}^{(n)}\right)^{\text{H}}\hat{\mathbf{H}}_{\ell}\mathbf{w}_{\text{t}}^{(n)}|^2 = \left(\mathbf{w}_{\text{t}}^{(n)}\right)^{\text{H}}\left(\sum_{\ell=0}^{L-1}\hat{\mathbf{H}}_{\ell}^{\text{H}}\mathbf{w}_{\text{r}}^{(n)}\left(\mathbf{w}_{\text{r}}^{(n)}\right)^{\text{H}}\hat{\mathbf{H}}_{\ell}\right)\mathbf{w}_{\text{t}}^{(n)}, \quad (3.16)$$

where $\mathbf{w}_{\text{t}}^{(n)}$ and $\mathbf{w}_{\text{r}}^{(n)}$ are the transmit and receive AWVs after the n-th iteration. According to Algorithm 3.1, $\mathbf{w}_{\text{t}}^{(n+1)}$ is the principal eigenvector of $\left(\sum_{\ell=0}^{L-1}\hat{\mathbf{H}}_{\ell}^{\text{H}}\mathbf{w}_{\text{r}}^{(n)}\left(\mathbf{w}_{\text{r}}^{(n)}\right)^{\text{H}}\hat{\mathbf{H}}_{\ell}\right)$, i.e., the optimal solution to maximize $\Gamma^{(n)}$ given $\mathbf{w}_{\text{r}}^{(n)}$. Let

$$\Gamma_0^{(n+1)} = \left(\mathbf{w}_{\text{t}}^{(n+1)}\right)^{\text{H}}\left(\sum_{\ell=0}^{L-1}\hat{\mathbf{H}}_{\ell}^{\text{H}}\mathbf{w}_{\text{r}}^{(n)}\left(\mathbf{w}_{\text{r}}^{(n)}\right)^{\text{H}}\hat{\mathbf{H}}_{\ell}\right)\mathbf{w}_{\text{t}}^{(n+1)}, \quad (3.17)$$

then we have $\Gamma_0^{(n+1)} \geq \Gamma^{(n)}$.

---

[3]In the training process, Tx also needs to receive with the same antenna array, and Rx also needs to transmit with the same antenna array. Thus, we use the source and destination instead.

[4]Strictly speaking, the reciprocal channel of $(\mathbf{H}[k])$ should be $(\mathbf{H}[k])^{\text{T}[1]}$, but $(\mathbf{H}[k])^{\text{H}}$ is also usually used instead for convenience[36]. They are equivalent in beamforming design.

Let us write $\Gamma_0^{(n+1)}$ in another form

$$\Gamma_0^{(n+1)} = \left(\mathbf{w}_r^{(n)}\right)^H \left(\sum_{\ell=0}^{L-1} \hat{\mathbf{H}}_\ell \mathbf{w}_t^{(n+1)} \left(\mathbf{w}_t^{(n+1)}\right)^H \hat{\mathbf{H}}_\ell^H\right) \mathbf{w}_r^{(n)}. \tag{3.18}$$

Since $\left(\mathbf{w}_r^{(n+1)}\right)$ is the principal eigenvector of $\left(\sum_{\ell=0}^{L-1} \hat{\mathbf{H}}_\ell \mathbf{w}_t^{(n+1)} \left(\mathbf{w}_t^{(n+1)}\right)^H \hat{\mathbf{H}}_\ell^H\right)$, i.e., the optimal solution to maximize $\Gamma_0^{(n+1)}$ given $\left(\mathbf{w}_t^{(n+1)}\right)$, we have $\Gamma^{(n+1)} \geq \Gamma_0^{(n+1)}$, because

$$\begin{aligned}\Gamma^{(n+1)} &= \sum_{\ell=0}^{L-1} |\left(\mathbf{w}_r^{(n+1)}\right)^H \hat{\mathbf{H}}_\ell \mathbf{w}_t^{(n+1)}|^2 \\ &= \left(\mathbf{w}_r^{(n+1)}\right)^H \left(\sum_{\ell=0}^{L-1} \hat{\mathbf{H}}_\ell \mathbf{w}_t^{(n+1)} \left(\mathbf{w}_t^{(n+1)}\right)^H \hat{\mathbf{H}}_\ell^H\right) \mathbf{w}_r^{(n+1)}.\end{aligned} \tag{3.19}$$

Therefore, we have $\Gamma^{(n+1)} \geq \Gamma_0^{(n+1)} \geq \Gamma^{(n)}$. ☐

According to (3.5), $\Gamma < \infty$. Thus, $\{\Gamma^{(n)}|n = 1, 2, ...\}$ converges to a sub-optimal value, which guarantees the convergence of IEVD and its training approach.

The convergence rates of IEVD and its training approach are fast. From the iteration process, it can be observed that a temporary sub-optimal transmit and receive AWV can be directly found in each iteration, rather than progressing by only a small step toward the ultimate directions like Newton's method. Fig. 3.2 shows a simple example, which is a surface of $z = -(x - 1)^2 - (y - 1)^2$. With the *alternating optimization* method, like IEVD, the optimal solution can be quickly achieved. That is, given x, to optimize y, and given the optimized y, to optimize x. However, with

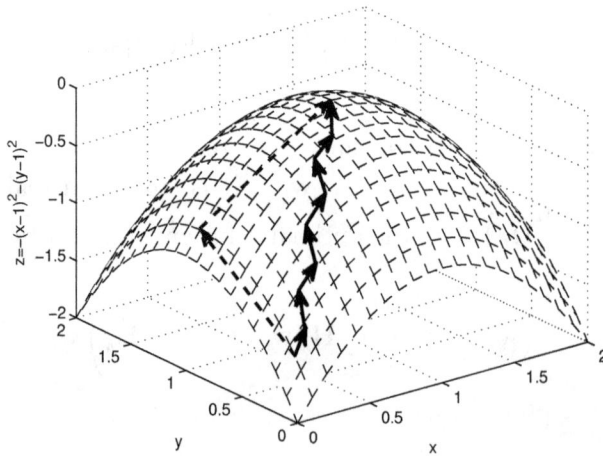

Figure 3.2 Illustration of the convergence process of IEVD. The arrows with solid and dash lines represent that of Newton's method and IEVD, respectively.

Newton's method, many iterations are required to reach the optimal solution, because each time only a small step can be made to approach the optimal solution.

3.3.2 Multipath Grouping Scheme

It is clear that the sub-optimal schemes, including both Nsenga's scheme and the presented IEVD scheme, require time-consuming estimation of CSI and complicated EVD computation. The training approach for IEVD reduces overhead and computational complexity, thus, it is more practical in slow fading scenarios where λ_ℓ varies slowly. However, in fast fading scenarios, even the training approach may not be practically applicable, not to mention Nsenga's and IEVD, which are CSI-required, because the training process and CSI estimation will need to be launched frequently due to fast variation of the channel, which will greatly decrease the system efficiency.

In fact, it is not possible to optimize the instantaneous SNR in fast fading scenarios due to the unknown instantaneous CSI. In such a case, average SNR may be a reasonable optimization objective, but it mingles with diversity gain, i.e., a higher average SNR may lead to a lower diversity gain. We hope to achieve both high array gain and diversity gain, so as to improve both average SNR and robustness of the system.

It is noted that in fast fading scenarios, λ_ℓ varies fast, but $\mathbf{g}_\ell$ and $\mathbf{h}_\ell$, the receive and transmit steering vectors, can still be well estimated and treated known a priori, because they vary slowly as mentioned in Section 3.2. By utilizing this feature, Park and Pan proposed a scheme to achieve diversity gain in addition to array gain, without a priori knowledge of λ_ℓ. Although their scheme is simple and achieves full diversity, it may be not efficient in array gain, and may be infeasible when the number of MPCs is larger than that of the antennas at either ends. Hence, we propose an improved diversity scheme for beamforming by exploiting MPG. In this section, we will first briefly introduce Park and Pan's scheme (Park-Pan), and then introduce the MPG scheme.

3.3.2.1 *The Diversity Scheme by Park and Pan*

The main beamforming approach for communications is to beamform toward only the direction of a single path to achieve array gain, but diversity gain cannot be achieved. In[10], multiple beams, which steer at different multipath directions, are concurrently shaped at both Tx and Rx to achieve beam diversity. It is noted that a beamforming scheme steering at different multipath directions means that the AWVs at both ends are set to have gains on the steering vectors of these MPCs. For instance, if Tx and Rx beamform toward only the direction of the k-th MPC, the AWVs are $\mathbf{w}_t = \mathbf{h}_k$ and $\mathbf{w}_r = \mathbf{g}_k$, respectively. While if Tx and Rx beamform toward multiple MPCs, the gains along the steering vectors of these MPCs need to be set first at Tx and Rx, respectively, and then the AWVs can be computed according to these gains.

Letting α_ℓ and β_ℓ denote the gains along the directions of the ℓ-th MPC at Rx and Tx, respectively, we have

$$y[m] = \sqrt{MN\gamma} \sum_{\ell=0}^{L-1} \alpha_\ell \beta_\ell \lambda_\ell s[m - \tau_\ell] + \mathbf{w}_r^H \mathbf{n}, \tag{3.20}$$

where

$$\alpha_\ell = \mathbf{w}_r^H \mathbf{g}_\ell, \quad \beta_\ell = \mathbf{h}_\ell^H \mathbf{w}_t. \tag{3.21}$$

As in the considered fast fading scenario λ_ℓ are assumed to be unavailable at both ends, but $\mathbf{g}_\ell$ and $\mathbf{h}_\ell$ are known a priori at Rx and Tx, respectively, the receive and transmit antenna gains along the directions of different MPCs can be simply and naturally set as $\alpha_\ell = \beta_\ell = 1$ for $\ell = 0, 1, ..., L - 1$.

Let $\mathbf{b}$ and $\mathbf{a}$ denote the transmit and receive gain vectors, respectively. We have

$$\mathbf{b} = [\beta_0, \beta_1, ..., \beta_{L-1}]^T \text{ and } \mathbf{a} = [\alpha_0, \alpha_1, ..., \alpha_{L-1}]^T. \tag{3.22}$$

The receive and transmit AWVs are consequently achieved as

$$\mathbf{w}_r = (\bar{\mathbf{G}}^H)_R^\dagger \mathbf{a}^* = \bar{\mathbf{G}}\left(\bar{\mathbf{G}}^H \bar{\mathbf{G}}\right)^{-1} \mathbf{a}^*, \tag{3.23}$$

and

$$\mathbf{w}_t = (\bar{\mathbf{H}}^H)_R^\dagger \mathbf{b} = \bar{\mathbf{H}}\left(\bar{\mathbf{H}}^H \bar{\mathbf{H}}\right)^{-1} \mathbf{b}, \tag{3.24}$$

respectively, where $(\cdot)^*$ is the conjugation operation, $(\cdot)_R^\dagger$ represents the right pseudo-inverse,

$$\bar{\mathbf{G}} = [\mathbf{g}_0, \mathbf{g}_1, ..., \mathbf{g}_{L-1}], \text{ and } \bar{\mathbf{H}} = [\mathbf{h}_0, \mathbf{h}_1, ..., \mathbf{h}_{L-1}]. \tag{3.25}$$

The final transmit and receive AWVs require a normalization on the obtained $\mathbf{w}_t$ and $\mathbf{w}_r$, respectively, for the unit l_2-norm constrain.

Park-Pan achieves full diversity and is simple to implement in the case of $L \leq \min(\{M, N\})$. In such a case, $\bar{\mathbf{G}}^H$ and $\bar{\mathbf{H}}^H$ are both row-rank matrices, and $(\bar{\mathbf{G}}^H)_R^\dagger$ and $(\bar{\mathbf{H}}^H)_R^\dagger$ exist, which guarantees a solution of AWV. However, when $L > \min(\{M, N\})$, $(\bar{\mathbf{G}}^H)_R^\dagger$ or $(\bar{\mathbf{H}}^H)_R^\dagger$ may not exist, because $\left(\bar{\mathbf{G}}^H \bar{\mathbf{G}}\right)^{-1}$ or $\left(\bar{\mathbf{H}}^H \bar{\mathbf{H}}\right)^{-1}$ in (3.23) may not exist, which means that the solution of AWV may not exist. Therefore, this scheme is not applicable in such a case.

Additionally, Park-Pan may be not effective when the MPCs have close steering angles. To illustrate this, let us look at the equation in the singular-vector space at Tx, i.e.,

$$\bar{\mathbf{H}}^H \mathbf{w}_t = \mathbf{b} \Rightarrow \bar{\mathbf{\Sigma}}^H \bar{\mathbf{U}}^H \mathbf{w}_t = \bar{\mathbf{V}}^H \mathbf{b}, \tag{3.26}$$

where $\bar{\mathbf{H}} = \bar{\mathbf{U}}\bar{\mathbf{\Sigma}}\bar{\mathbf{V}}^H$ is the SVD of $\bar{\mathbf{H}}$, $\bar{\mathbf{U}} = \{\bar{\mathbf{u}}_i\}_{i=1,2,...,M}$, $\bar{\mathbf{V}} = \{\bar{\mathbf{v}}_i\}_{i=1,2,...,L}$, $\bar{\mathbf{\Sigma}} = \text{diag}([\rho_1, \rho_2, ..., \rho_L])$. Letting $\tilde{\mathbf{w}}_t = \bar{\mathbf{U}}^H \mathbf{w}_t = \{\tilde{w}_{ti}\}_{i=1,2,...,M}$ and $\tilde{\mathbf{b}} = \bar{\mathbf{V}}^H \mathbf{b} = \{\tilde{b}_i\}_{i=1,2,...,L}$, we have $\tilde{w}_{ti} = \tilde{b}_i/\rho_i^*$, $i = 1, 2, ..., L$. Note that $|\tilde{w}_{ti}|^2$ and ρ_i^2 denote the transmit power allocated in the direction of $\bar{\mathbf{u}}_i$ and the power gain in this direction, respectively. Naturally, in order to achieve diversity gain and array gain, the transmit power should be evenly allocated in multiple directions with good gains. However, for

Park-Pan, the power allocation is ineffective, because less power is allocated in the direction with a greater gain ρ_i. When the steering angles of MPCs are largely spaced, the steering vectors are approximately orthogonal. Thus, the condition number of $\bar{\mathbf{H}}$, i.e., ρ_1/ρ_L, is small, and Park-Pan achieves good performance. However, when MPCs have close steering vectors, the condition number of $\bar{\mathbf{H}}$ will be very large, which means most of the power is allocated in the direction of $\bar{\mathbf{u}}_L$, which has the smallest gain. In such a case, the achieved array gain will be poor.

3.3.2.2 MPG Scheme

It is found that Park-Pan may be infeasible when there are too many MPCs, and may be not effective when there are MPCs with close steering angles, which makes the condition number of $\bar{\mathbf{H}}$ bad. Consequently, it probably benefits to group the MPCs with close steering angles to shape a single equivalent MPC, which will maintain a good condition number, because different groups of equivalent MPCs are basically widely spaced. This is the motivation of the MPG scheme. As this equivalent MPC represents all the MPCs within the corresponding group in AWV computations, the MPG operation can not only avoid the ineffectiveness caused by these MPCs with close angles, but also reduces the number of involved MPCs. In this subsection, we present the MPG beamforming scheme. As shown in Fig. 3.3, in our scheme the MPCs are grouped according to their transmit and receive steering angles at Tx and Rx, respectively. For each non-empty group, an equivalent steering vector is defined, and the corresponding antenna gain is set.

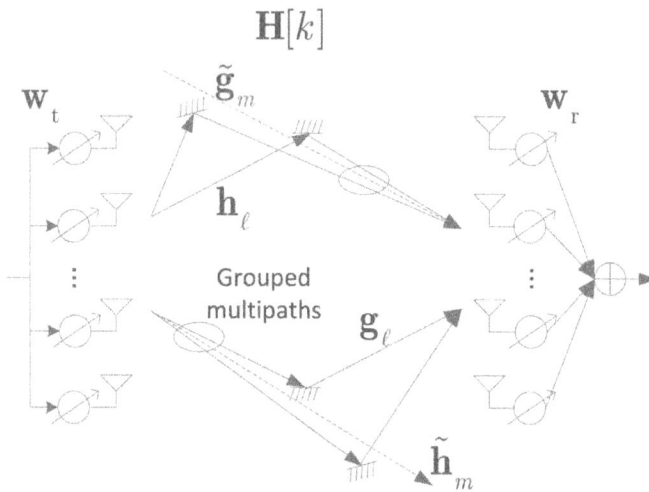

Figure 3.3 Illustration of the MPG schemes.

1) Transmit MPG:

An MPG scheme includes both transmit and receive MPCs. Let us look at Tx first. The angle range of the transmit steering vector is $[-1, 1)$. Uniformly divide the angle range into M segments with a spacing $2/M$, where the angle range of the i-th segment is $[-1 + 2(i-1)/M, \; -1 + 2i/M]$. As each MPC corresponds to a transmit steering vector, they are grouped into M groups according to their steering angles $\Omega_{t\ell}$. If $\Omega_{t\ell}$ fall within the i-th angle segment, the corresponding MPCs are grouped into the i-th group. For each group that has at least one MPC in it, i.e., non-empty, an equivalent steering vector is defined. Subsequently, when calculating the transmit AWV, each non-empty group is represented by its corresponding equivalent steering vector.

As an equivalent steering vector represents all the MPCs within the corresponding group in AWV computations, it must have a significant correlation with these MPCs, i.e., the inner product between the equivalent steering vector and the steering vector of any one MPC in this group is significant. There are many approaches to define the equivalent steering vectors for the non-empty groups. For example, a natural way is to define a new steering vector with a steering angle being the average angle of the steering angles of the MPCs within this group. However, this method requires estimation of all the steering angles of the MPCs, which is with high complexity and not practical. Therefore, we suggest an approach that is natural but simple to implement to define the equivalent steering vectors.

Let $\{N_j^{(i)}|j = 1, 2, ..., N_i\}$ denote the indices of the transmit steering vectors $\mathbf{h}_\ell$ falling within the i-th non-empty group, the equivalent steering vector for this group is defined as

$$\tilde{\mathbf{h}}_i = \left(\sum_{j=1}^{N_i} \mathbf{h}_{N_j^{(i)}} \right) \Big/ \left\| \sum_{j=1}^{N_i} \mathbf{h}_{N_j^{(i)}} \right\|_2. \tag{3.27}$$

The corresponding antenna gain in the direction of this equivalent steering vector is simply set as $\tilde{\beta}_i = 1$.

For the groups that have no multipath in them, i.e., empty, there are no equivalent steering vectors or antenna gains. Hence, the total number of equivalent transmit steering vectors, N_T, is no larger than M. Lastly, the transmit AWV is obtained as

$$\mathbf{w}_\mathrm{t} = \tilde{\mathbf{H}} \big(\tilde{\mathbf{H}}^\mathrm{H} \tilde{\mathbf{H}} \big)^{-1} \tilde{\mathbf{b}}, \tag{3.28}$$

where

$$\tilde{\mathbf{H}} = [\tilde{\mathbf{h}}_0, \tilde{\mathbf{h}}_1, ..., \tilde{\mathbf{h}}_{N_\mathrm{T}-1}], \tag{3.29}$$

and

$$\tilde{\mathbf{b}} = [\tilde{\beta}_0, \tilde{\beta}_1, ..., \tilde{\beta}_{N_\mathrm{T}-1}]^\mathrm{T} = [1, 1, ..., 1]^\mathrm{T}. \tag{3.30}$$

The final transmit AWV requires a normalization on the obtained $\mathbf{w}_\mathrm{t}$ for the unit l_2-norm constraint.

It is noted that as $N_\mathrm{T} \leq M$, $(\tilde{\mathbf{H}}^\mathrm{H}\tilde{\mathbf{H}})^{-1}$ exists almost surely. In other words, the transmit multipath grouping leads to a solution of transmit AWV almost surely.

2) Receive MPG:

Analogously, at Rx we also conduct multipath grouping. Uniformly divide the angle range into N segments with a spacing $2N$, where the angle range of the i-th segment is $[-1 + 2(i-1)/N \ -1 + 2i/N)$. As each MPC also corresponds to a receive steering vector, they are grouped into N groups according to their steering angles $\Omega_{r\ell}$. If $\Omega_{r\ell}$ fall within the i-th angle segment, the corresponding MPCs are grouped into the i-th group. For each group that has at least one MPC in it, i.e., non-empty, an equivalent steering vector is similarly defined. Let $\{M_j^{(i)} | j = 1, 2, ..., M_i\}$ denote the indices of the receive steering vectors $\mathbf{g}_\ell$ falling within the i-th non-empty group, the equivalent steering vector for this group is defined as

$$\tilde{\mathbf{g}}_i = \left(\sum_{j=1}^{M_i} \mathbf{g}_{M_j^{(i)}} \right) \Big/ \left\| \sum_{j=1}^{M_i} \mathbf{g}_{M_j^{(i)}} \right\|_2. \tag{3.31}$$

The corresponding antenna gain in the direction of this equivalent steering vector is simply set as $\tilde{\alpha}_i = 1$.

For the groups that have no multipath in them, there are no equivalent steering vectors or antenna gains. Hence, the total number of equivalent transmit steering vectors, M_R, is no larger than N. Lastly, the receive AWV is obtained as

$$\mathbf{w}_r = \tilde{\mathbf{G}} \left(\tilde{\mathbf{G}}^H \tilde{\mathbf{G}} \right)^{-1} \tilde{\mathbf{a}}^*, \tag{3.32}$$

where

$$\tilde{\mathbf{G}} = [\tilde{\mathbf{g}}_0, \tilde{\mathbf{g}}_1, ..., \tilde{\mathbf{g}}_{M_R-1}], \tag{3.33}$$

and

$$\tilde{\mathbf{a}} = [\tilde{\alpha}_0, \tilde{\alpha}_1, ..., \tilde{\alpha}_{M_R-1}]^T = [1, 1, ..., 1]^T. \tag{3.34}$$

The final receive AWV requires a normalization on the obtained $\mathbf{w}_r$ for the unit l_2-norm constraint.

Likewise, since $M_R \leq N$, $(\tilde{\mathbf{G}}^H \tilde{\mathbf{G}})^{-1}$ exists almost surely, which leads to a solution of receive AWV almost surely.

3) Realization:

The realization of the MPG scheme seems to require angle estimation at Tx and Rx, which is complicated. In fact, the peak of the spatial spectrum from a Bartlett beamformer can be used instead of more complicated AoD and AoA estimations. Take the transmit MPG for instance. Predefine M unit vectors $\mathbf{v}_i$, $i = 1, 2, ..., M$ as

$$\mathbf{v}_i = \frac{1}{\sqrt{M}} [e^{j\pi 0*(-1+(2i-1)/M)}, e^{j\pi 1*(-1+(2i-1)/M)},$$
$$e^{j\pi 2*(-1+(2i-1)/M)}, ..., e^{j\pi(M-1)*(-1+(2i-1)/M)}]^T. \tag{3.35}$$

For an arbitrary MPC with transmit angle $\Omega_{t\ell}$, its angle locates in the i-th segment, i.e., $[-1 + 2(i-1)/M \ -1 + 2i/M]$, is equivalent to that its steering vector $\mathbf{h}_\ell$ satisfies $|\mathbf{v}_i^H\mathbf{h}_\ell| > |\mathbf{v}_j^H\mathbf{h}_\ell|_{j\neq i}$. Hence, the alternative approach to realize MPG is to estimate the steering vectors $\mathbf{h}_\ell$, $\ell = 1, 2, ..., L$, at Tx, and for each steering vector $\mathbf{h}_\ell$, find the index of the unit vectors that maximizes $\arg_i |\mathbf{v}_i^H\mathbf{h}_\ell|$.

The index is the group number of $\mathbf{h}_\ell$. The receive MPC can be realized in the same way.

There are many ways to estimate the transmit and receive steering vectors, which are much simpler than to estimate full CSI. One way is to estimate them by the first iteration process of Algorithm 3.2. In the first iteration process, we can obtain $\mathbf{r}[\ell]$ and $\bar{\mathbf{r}}[\ell]$, respectively. In fact,

$$\mathbf{r}[\ell] = \hat{\mathbf{H}}_\ell^H\mathbf{w}_r = \left(\sqrt{NM}\lambda_\ell\mathbf{g}_\ell^H\mathbf{w}_r\right)\mathbf{h}_\ell, \tag{3.36}$$

and

$$\bar{\mathbf{r}}[\ell] = \hat{\mathbf{H}}_\ell\mathbf{w}_t = \left(\sqrt{NM}\lambda_\ell\mathbf{h}_\ell^H\mathbf{w}_t\right)\mathbf{g}_\ell. \tag{3.37}$$

Thus, by normalizing $\mathbf{r}[\ell]$ and $\bar{\mathbf{r}}[\ell]$, we get $\mathbf{h}_\ell$ and $\mathbf{g}_\ell$ at Tx (the source in Algorithm 3.2) and Rx (the destination in Algorithm 3.2), respectively.

It is noted that to use MPG in practice, there are many practical system issues to be considered, and a critical one is array calibration. To do accurate spectrum estimation, the phase and amplitude responses of each antenna need to be known. Although these could be measured ahead of time since there is only one active RF chain and the multiple channels are passive, the calibration of the responses of these channels can be a rather tricky issue because the gain is adjustable on each channel. As array calibration is not the focus of this book, relevant literatures, such as[37] and[38], are referred for further considerations.

3.3.2.3 Performance Analysis

Since PEP is an extensively used metric to reflect both array gain and diversity gain[25], we adopt it in this context. It is clear that no matter whether MPG or Park-Pan is adopted, an equivalent single-input single-output (SISO) multipath fading channel is observed after the single-layer beamforming, which yields

$$h[k] = \sum_{m=0}^v h_m\delta[k-m], \tag{3.38}$$

where $v \geq \max(\{\tau_\ell|\ell = 0, 1, ..., L-1\})$, and

$$h_m = \begin{cases} \mathbf{w}_r^H\hat{\mathbf{H}}_\ell\mathbf{w}_t, & m = \tau_\ell, \\ 0, & \text{otherwise.} \end{cases} \tag{3.39}$$

With the above SISO model (3.38), single-carrier block transmission, such as the zero-padded (ZP) block transmission[39], is adopted to evaluate the system performance. With the ZP, the minimum mean square error (MMSE) Rx can be used to collect the multipath diversity[39], and the PEP is upper bounded as

$$
\begin{aligned}
P\{\mathbf{s}_A \to \mathbf{s}_B\} &\lesssim \mathbb{E}_{\lambda_\ell}\left(Q\left(\frac{d\sqrt{\gamma \sum_{\ell=0}^{L-1}|h_{\tau_\ell}|^2}}{\sqrt{2N_0}}\right)\right) \\
&= \mathbb{E}_{\lambda_\ell}\left(Q\left(d\sqrt{\frac{\gamma}{2}\sum_{\ell=0}^{L-1}|\mathbf{w}_r^H\mathbf{g}_\ell\lambda_\ell\mathbf{h}_\ell^H\mathbf{w}_t|^2}\right)\right),
\end{aligned}
\tag{3.40}
$$

where $\mathbb{E}_{\lambda_\ell}$ is the expectation on the channel coefficients λ_ℓ which are fast fading, $Q(x)$ is the Q function, and d is the minimum distance of the signal constellation. Note that the ZP based block transmission achieves the best performance among all block based transmission systems for an SISO channel[39].

Since when $a > 1$, the Q function $Q(a)$ has the upper bound $e^{-a^2/2}$[25], when γ is large enough, we further have

$$
\begin{aligned}
P\{\mathbf{s}_A \to \mathbf{s}_B\} &\leq \mathbb{E}_{\lambda_\ell}\left(\exp\left(-d^2\frac{\gamma}{4}\sum_{\ell=0}^{L-1}|\mathbf{w}_r^H\mathbf{g}_\ell\lambda_\ell\mathbf{h}_\ell^H\mathbf{w}_t|^2\right)\right) \\
&= \mathbb{E}_{\lambda_\ell}\left(\prod_{\ell=0}^{L-1}\exp\left(-\frac{d^2\gamma}{4}|\mathbf{w}_r^H\mathbf{g}_\ell\lambda_\ell\mathbf{h}_\ell^H\mathbf{w}_t|^2\right)\right) \\
&= \prod_{\ell=0}^{L-1}\mathbb{E}_{\lambda_\ell}\left(\exp\left(-\frac{d^2\gamma}{4}|\mathbf{w}_r^H\mathbf{g}_\ell\mathbf{h}_\ell^H\mathbf{w}_t|^2|\lambda_\ell|^2\right)\right) \\
&= \prod_{\ell=0}^{L-1}\left(\frac{1}{1+d^2\gamma|\mathbf{w}_r^H\mathbf{g}_\ell\mathbf{h}_\ell^H\mathbf{w}_t|^2/4}\right) \triangleq P_{\mathrm{UB}},
\end{aligned}
\tag{3.41}
$$

which is suitable for both MPG and Park-Pan under the assumption that the AoD and AoA are uniformly distributed. Note that the difference in the AWV setting will lead to different PEP performance between these two schemes.

According to the definition in[25], the diversity gain D can thus be derived as

$$
D = -\lim_{\gamma\to\infty}\frac{\log P_{\mathrm{UB}}}{\log\gamma} = -\lim_{\gamma\to\infty}\frac{\log\prod_{\ell=0}^{L-1}\left(\frac{1}{1+d^2\gamma|\mathbf{w}_r^H\mathbf{g}_\ell\mathbf{h}_\ell^H\mathbf{w}_t|^2/2}\right)}{\log\gamma} = L,
\tag{3.42}
$$

which means that both MPG and Park-Pan achieve full diversity.

3.3.3 Performance Comparison

In this section, we conduct extensive comparisons between IEVD and Nsenga's scheme, both of which are sub-optimal schemes that require real-time full or partial CSI, as well as MPG and Pank-Pan, both of which are simpler schemes and feasible even in fast fading scenarios with only the steering vectors known a priori.

This section is organized as follows. First, we investigate the convergence rate of the presented IEVD scheme, as well as that of its training approach. Afterwards, we compare the performances of these schemes and discuss the complexity issue.

3.3.3.1 Convergence Rates of IEVD and Its Training Approach

The convergence rates of IEVD and its training approach are shown in Fig. 3.4 with $M = N = 8$. Recall that L is the number of MPCs, and K is the power of the square matrices to approximate the corresponding EVD. These curves are obtained via simulations over 10^5 randomly realized channels according to the model in (3.1). In these realizations, the steering angles $\Omega_{t\ell}$ and $\Omega_{r\ell}$, which determine the steering vectors, obey a uniform distribution within $[-1, 1)$; while the MPC coefficients $\{\lambda_\ell\}$ follow a complex Gaussian distribution with zero mean and variance $1/L$. In each channel realization, we obtain an instantaneous power gain Γ. The final average power gain is computed by averaging the 10^5 instantaneous power gains.

It is observed from Fig. 3.4 that both IEVD and its training approach converge fast. As the initial AWVs are randomly generated, there is basically no power gain when the number of iterations is 0. When $L = 1$, i.e., there is only one significant MPC, both IEVD and its training approach converge with only 1 iteration, even with $K = 1$. When $L > 1$, IEVD converges with only two iterations, while its training approach converges depending on K. It is clear that the training approach converges faster when K is greater, which means a higher computational complexity. Fortunately, when $K = 2$, the training approach can basically achieve convergence with 2 iterations. A small K and a fast convergence rate increase the practically applicability of the training approach.

On the other hand, it is found that as L increases, the obtained average gain becomes smaller, which means that the array gain becomes lower. However, as L

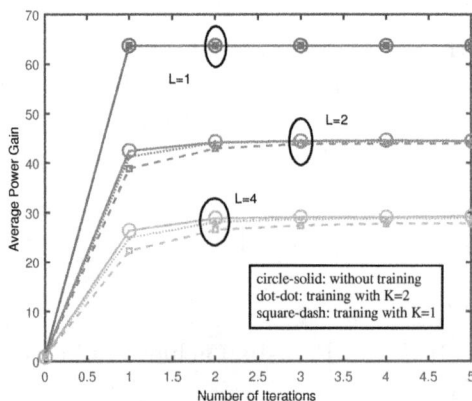

Figure 3.4 Convergence rates of IEVD and its training approach. $M = N = 8$. L is the number of MPC, and K is the power of the square matrices to approximate the corresponding EVD.

increases, more diversity gain is achieved. The metric of power gain cannot reflect diversity gain. Thus, we need to adopt alternative metrics to evaluate the performance.

3.3.3.2 Performance Comparisons

We next want to obtain and compare the numerical results of the upper bounded PEP for the four involved schemes, namely IEVD, Nsenga's, MPG, and Park-Pan. It is noted that the training approach of IEVD has the same performance with IEVD in the convergence state given that K is large enough, e.g., $K \geq 2$, and the iteration number is 3 in all the simulations in this subsection. The PEP upper bounds are achieved by Monte Carlo simulation under different scenarios according to (3.40), which is more accurate than (3.41). The steering angles $\Omega_{t\ell}$ and $\Omega_{r\ell}$ are either deterministic or random. For deterministic steering angles, only the channel coefficients λ_ℓ are random. Each realization of λ_ℓ determines an instantaneous bounded PEP according to (3.40), and an average bounded PEP is obtained by adopting their mean. For random steering angles, each realization of $\Omega_{t\ell}$, $\Omega_{r\ell}$, and λ_ℓ determines an instantaneous bounded PEP, and an average bounded PEP (termed PEP for short hereafter) is achieved by the same way. In the simulations, $M = N = 8$. Quadrature phase shift keying (QPSK) modulation is adopted; thus, $d = \sqrt{2}$.

We first investigate the case that the MPCs have the same transmit steering angle, but different receive steering angles. This case corresponds to the scenario that Tx is far away from Rx, and there are many reflectors around Rx. Afterwards, we investigate the case that the MPCs have different transmit and receive steering angles. This case corresponds to the scenario that Tx is not far away from Rx with many reflectors nearby.

Figs. 3.5 (left) and 3.5 (right) depict the PEP comparisons between the schemes with the same single steering angle at Tx and variable steering angles at Rx, with

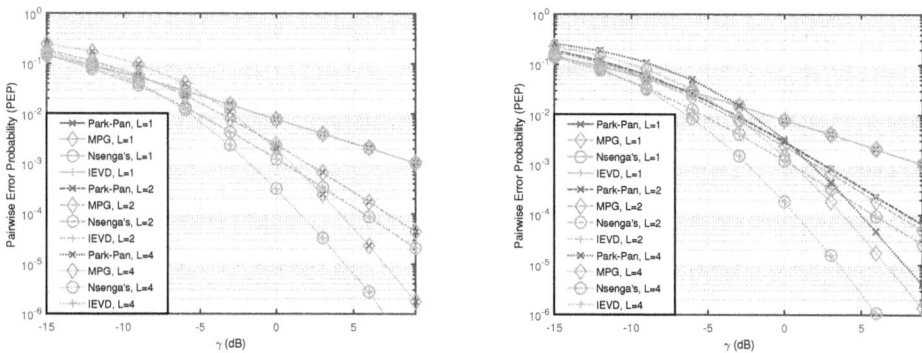

Figure 3.5 **Left**: PEP comparisons between the schemes with the same single steering angle at Tx and variable steering angles at Rx. The steering angles at Rx are deterministic and equally spaced. **Right**: PEP comparisons between the schemes with the same single steering angle at Tx and variable steering angles at Rx. The steering angles at Rx are random and uniformly distributed within $[-1\ 1)$.

deterministic and random steering angles, respectively. In the deterministic case, the steering angles are equally spaced, i.e., $\Omega_{t\ell} = \Omega_{r\ell} = -1 + 2(\ell - 1)/L$. While in the random case, the steering angles obey a uniform distribution within $[-1, 1)$.

Regarding the case of deterministic steering angles, all the four schemes achieve full diversity. The PEP performance of MPG is exactly the same as that of Park-Pan, as shown in Fig. 3.5 (left). This is because when $L \leq \min(\{M, N\})$, all the MPCs fall within different groups at Rx, and thus there is actually no such operations that multiple MPCs are grouped into a single one; consequently, the equivalent receive steering vectors of MPG are the same as those of Park-Pan. Also, there is, in fact, no MPG operation at Tx, because the MPCs have the same one transmit steering angle. On the other hand, the PEP performance of IEVD is exactly the same as that of Nsenga's, and achieves a better array gain than that of MPG and Park-Pan, which shows that both sub-optimal schemes are effective. When $L = 1$, the four schemes have the same performance, which is optimal, because there is only one steering angle at Tx and Rx.

Regarding the case of random steering angles, all four schemes also achieve full diversity. It is observed that, in such a case, the MPG scheme shows superiority over Park-Pan in array gain, as shown in Fig. 3.5 (right). As the steering angles are randomly generated at Rx, MPCs with close receive steering angles are grouped into a single equivalent MPC in the MPG scheme. Thus, the superiority of MPG versus Park-Pan indicates that the multipath grouping operation achieves an array gain, and as L increases, the superiority becomes more significant. On the other hand, IEVD also achieves the same performance as Nsenga's and has an increasingly better array gain than MPG and Park-Pan.

Figs. 3.6 (left) and 3.6 (right) depict the PEP comparisons between the schemes with variable steering angles at Tx and Rx, with deterministic and random steering angles, respectively.

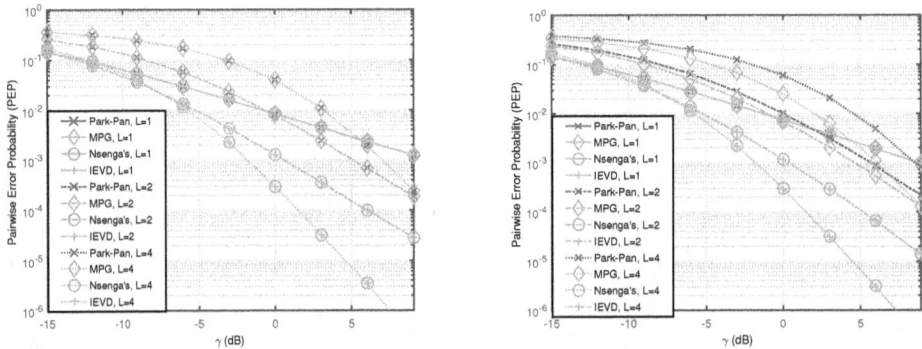

Figure 3.6 **Left**: PEP comparisons between the schemes with variable steering angles at Tx and Rx. The steering angles are deterministic and equally spaced at both ends. **Right**: PEP comparisons between the schemes with variable steering angles at Tx and Rx. The steering angles are random and uniformly distributed within $[-1, 1)$ at both ends.

It is found that in both deterministic and random cases, the same results can be observed as that from Figs. 3.5 (left) and 3.5 (right), respectively. That is, all the schemes achieve full diversity. Besides, MPG and Park-Pan achieve the same performance in the deterministic case, because there is actually no MPG operation in the MPG scheme. But in the random case, MPG achieves a better array gain than Park-Pan, which benefits from the MPG operation. IEVD and Nsenga's achieve the same performance in both cases, and have an increasingly superiority over MPG and Park-Pan as L increases. Comparing Fig. 3.5 (left) with Fig. 3.6 (left), and Fig. 3.5 (right) with Fig. 3.6 (right), it is found that the two optimal schemes almost achieve the same performance in the same one transmit steering angle case, and the variable transmit steering angles case, but MPG and Park-Pan have an increasing loss in array gain as L increases, which further shows the effectiveness of the sub-optimal schemes.

These results are all obtained through the bounded PEP curves. To demonstrate the rational of them, the block-error rate (BLER) curves, are obtained via simulation with deterministic and random steering angles as shown in Figs. 3.7 (left) and 3.7 (right), respectively. In the simulations, the block size is 32, and the ZP length is 8. A single BLER is obtained based on the simulation of 10^7 transmissions and receptions of a block with randomly realized channels, which guarantees that the BLER curves are accurate. From these two figures, the same results can be observed and concluded as that from Figs. 3.6 (left) and 3.6 (right).

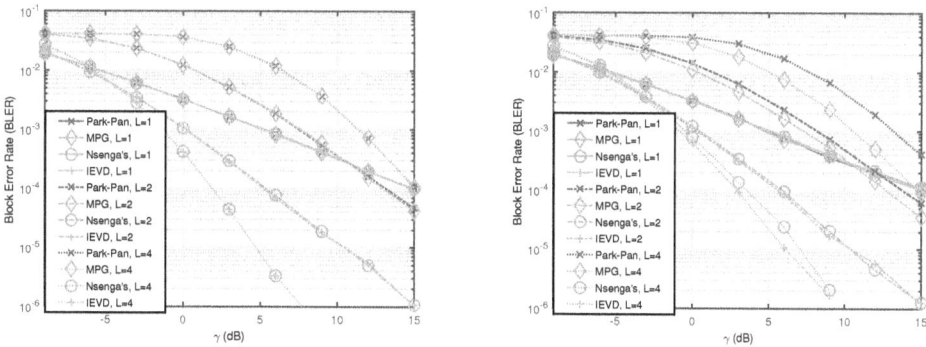

Figure 3.7 **Left**: BLER comparisons between the schemes with variable steering angles at Tx and Rx. The steering angles are deterministic and equally spaced at both ends. **Right**: BLER comparisons between the schemes with variable steering angles at Tx and Rx. The steering angles are random and uniformly distributed within $[-1, 1)$ at both ends.

As aforementioned, Park-Pan is infeasible when $L > \min(\{M, N\})$, but the two sub-optimal schemes and MPG are applicable. A natural way to make Park-Pan feasible even when $L > \min(\{M, N\})$ is to randomly select M and N MPCs at Tx and Rx to compute the AWVs. Park-Pan with such a process is termed as Park-Pan*. Figs. 3.8 (left) and 3.8 (right) show their PEP performances in the case of $L > \min(\{M, N\})$, with deterministic equally-spaced and uniformly-distributed random steering angles, respectively. It is observed again that in both cases the two sub-optimal schemes achieve the same performance, which is better than that of MPG and Park-Pan*. With deterministic angles, as shown in Fig. 3.8 (left), MPG almost achieves the same performance as Park-Pan*. While with random angles, as shown in Fig. 3.8 (right), MPG achieves a better array gain in the case of $L = 10$, and an approximately equivalent array gain to Park-Pan* in the case of $L = 20$. In addition, both MPG and Park-Pan* achieve full diversity. These results show that, in the presence of massive MPCs, MPG is still feasible and effective. In addition, Park-Pan can be made feasible by randomly selecting M and N MPCs, instead of L MPCs, at Tx and Rx to compute the AWVs.

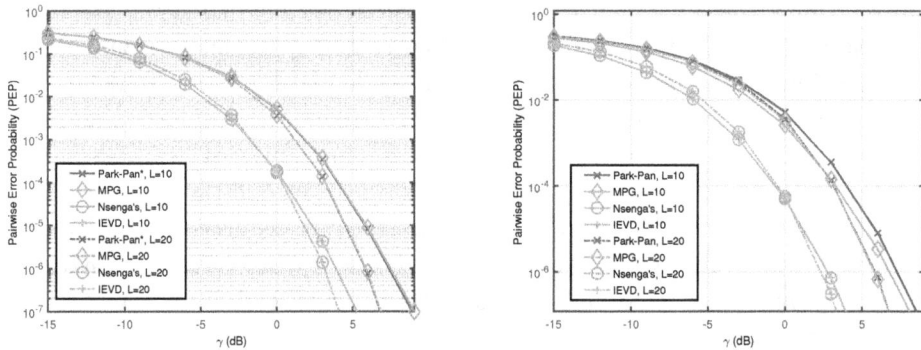

Figure 3.8 **Left**: PEP comparisons between the schemes with variable steering angles at Tx and Rx, where the number of MPCs is large. The steering angles are deterministic and equally spaced at both ends. Park-Pan* denotes Park-Pan with the process that randomly selecting M and N MPCs at Tx and Rx to compute the AWVs. **Right**: PEP comparisons between the schemes with variable steering angles at Tx and Rx, where the number of MPCs is large. The steering angles are random and uniformly distributed within $[-1, 1)$ at both ends. Park-Pan* denotes Park-Pan with the process that randomly selecting M and N MPCs at Tx and Rx to compute the AWVs.

3.3.3.3 Complexity Issue

As aforementioned, the two sub-optimal schemes, i.e., IEVD and Nsenga's, require real-time CSI. Thus, they are only applicable in slow fading scenarios. In fast fading scenarios, the MPG and Park-Pan schemes can be adopted, which both require only the quasi-static steering vectors $\mathbf{h}_\ell$ and $\mathbf{g}_\ell$, rather than the fast varying coefficients λ_ℓ, at Tx and Rx, respectively.

Compared with Nsenga's, the presented IEVD achieves exactly the same performance, but has a reduced overhead and implementation complexity. The overhead of channel estimation for Nsenga's scheme is $M \times N$ training sequences[4], while that for IEVD with the training approach is $2(M+N)$ training sequences, where the iteration number is set to 2. It is clear that the training overhead is greatly reduced by IEVD with the training approach, especially when the number of antennas is large at Tx and Rx. Moreover, IEVD with the training approach only needs to compute matrix multiplication, which has a much lower complexity than Nsenga's scheme, where the computations of EVD and Schmidt decomposition in the tensor space are required. In brief, IEVD with the training approach achieves the same performance with Nsenga's scheme, with a reduced overhead and complexity.

On the other hand, from the performance evaluations it can be found that, compared with Park-Pan, MPG achieves a better array gain. Besides, MPG does not increase overhead, because both MPG and Park-Pan only require the quasi-static steering vectors at Tx and Rx. In addition, although MPG requires grouping operation and computations of equivalent steering vectors, which are not required by Park-Pan, it needs a lower-dimensional matrix inverse than Park-Pan, i.e., $(\tilde{\mathbf{H}}^H\tilde{\mathbf{H}})^{-1}$ and $(\tilde{\mathbf{G}}^H\tilde{\mathbf{G}})^{-1}$ have lower dimensions than $\left(\bar{\mathbf{H}}^H\bar{\mathbf{H}}\right)^{-1}$ and $\left(\bar{\mathbf{G}}^H\bar{\mathbf{G}}\right)^{-1}$, respectively, due to the MPG operation. In summary, MPG achieves better performance than Park-Pan with an approximately equivalent complexity.

3.4   BEAM SPACE TRANSMISSION

Most adaptive beamforming training schemes adopt the same state-of-the-art approach, i.e., finding the best singular vector via iterative training without a priori CSI at both ends[1, 40]. This SGV requires that both the amplitudes and phases of the AWV are adjustable. On the other hand, in communications, phased arrays are usually implemented with the approach that only phases of the antenna branches are adjustable; the amplitudes are set constant to simplify the design and reduce the power consumption of phased arrays[2, 41, 42]. In fact, even in general multiple-input multiple-output (MIMO) systems[43], antenna branches with constant amplitude are also an optimization objective to reduce implementation complexity[44, 45]. In such a case, SGV becomes infeasible due to the constant modulus (CM) phased array. The schemes proposed in[44, 45] cannot be used here either, because these schemes are designed for transmitting beamforming with full or quantized a priori CSI at only the source device, but for communications with CM phased arrays, joint beamforming is required without a priori CSI at both the source and destination devices.

In this section, a steering vector based joint beamforming training scheme (STV), which exploits the directional feature of high-frequency band communication channels, is presented. Performance comparisons show that for LoS channels, both STV and SGV have fast convergence rates and achieve the optimal array gain. On the other hand, for non-LoS (NLoS) channels, STV achieves a faster convergence rate at the cost of a slightly lower array gain than SGV, which can also achieve the optimal array gain. The rest of this section is organized as follows. First introduce the system and channel model. Then introduce the SGV scheme and the STV scheme, and finally carry out performance evaluation and comparison.

Given the transmitting AWV $\mathbf{w}_t$ and the receiving AWV $\mathbf{w}_r$, where $\|\mathbf{w}_t\|_2 = \|\mathbf{w}_r\|_2 = 1$, the received signal y is given by

$$y = \mathbf{w}_r^H \mathbf{H} \mathbf{w}_t s + \mathbf{w}_r^H \mathbf{n}, \tag{3.43}$$

where s is the transmitted symbol, $\mathbf{n}$ is the noise vector. The target of beamforming training is to find appropriate transmitting and receiving AWVs to obtain a high receiving SNR, which is given by

$$\gamma\Gamma = |\mathbf{w}_\mathrm{r}^\mathrm{H}\mathbf{H}\mathbf{w}_\mathrm{t}|^2/\sigma^2, \tag{3.44}$$

where σ^2 is the noise power.

3.4.1 Singular-Vector Based Scheme

Let us introduce the SGV scheme first. It is known that the optimal AWVs to maximize $\gamma\Gamma$ is the principal singular vectors of the channel matrix $\mathbf{H}$[1,40][5]. Denote the SVD of $\mathbf{H}$ as

$$\mathbf{H} = \mathbf{U}\Sigma\mathbf{V}^\mathrm{H} = \sum\nolimits_{k=1}^{K} \rho_k\mathbf{u}_k\mathbf{v}_k^\mathrm{H}, \tag{3.45}$$

where $\mathbf{U}$ and $\mathbf{V}$ are unitary matrices with column vectors (singular vectors) $\mathbf{u}_k$ and $\mathbf{v}_k$, respectively, Σ is an $N \times M$ rectangular diagonal matrix with non-negative real values ρ_k on the diagonal, i.e., $\rho_1 \geq \rho_2 \geq ... \geq \rho_K \geq 0$ and $K = \min(M, N)$. The optimal AWVs are $\mathbf{w}_\mathrm{t} = \mathbf{v}_1$ and $\mathbf{w}_\mathrm{r} = \mathbf{u}_1$.

In the common case that $\mathbf{H}$ is unavailable, iterative beamforming training can be adopted to find the optimal AWVs. According to[1,40],

$$\mathbf{H}^{2m_\mathrm{Loop}} \triangleq (\mathbf{H}^\mathrm{H}\mathbf{H})^{m_\mathrm{Loop}} = \sum_{k=1}^{K} \rho_k^{2m_\mathrm{Loop}}\mathbf{v}_k\mathbf{v}_k^\mathrm{H}, \tag{3.46}$$

which can be obtained by m_Loop iterative trainings utilizing the reciprocal feature of the channel. When m_Loop is large,

$$\mathbf{H}^{2m_\mathrm{Loop}} \approx \rho_1^{2m_\mathrm{Loop}}\mathbf{v}_1\mathbf{v}_1^\mathrm{H}. \tag{3.47}$$

Thus, the optimal transmitting and receiving AWVs can be obtained by normalizing $\mathbf{H}^{2m_\mathrm{Loop}}\mathbf{w}_\mathrm{t}$ and $\mathbf{H} \times \mathbf{H}^{2m_\mathrm{Loop}}\mathbf{w}_\mathrm{r}$, respectively.

The SGV scheme is described in Algorithm 3.3. The iteration number ε depends on a practical channel response, which will be shown in Section 3.4.3. It is clear that although the SGV scheme is effective, it is required that both the amplitudes and phases of AWV are adjustable, which cannot be satisfied when CM phased arrays are used, where only phases are adjustable.

3.4.2 Steering-Vector Based Scheme

In fact, the SGV scheme is a general one suitable for an arbitrary channel $\mathbf{H}$. It does not use the specific feature of high-frequency band communications channels. In high-frequency band communications, the channel has a directional feature. The vectors $\{\mathbf{g}_\ell\}$ and $\{\mathbf{h}_\ell\}$ are CM steering vectors, not orthogonal bases, but in (3.45), $\{\mathbf{u}_k\}$ and $\{\mathbf{v}_k\}$ are strict non-CM orthogonal bases. Nevertheless, according to[25],

---

[5]The SVD on $\mathbf{H}$ gives a set of orthogonal transmitting and receiving AWV pairs, as well as the energies projected to these AWV pairs.

Algorithm 3.3: The SGV Scheme

---

1) Initialize:

Pick an initial transmitting AWV $\mathbf{w}_t$ at the source device. This AWV may be chosen either randomly or with the approach specified in Section 3.4.3.

2) Iteration: Iterate the following process ε times, and then stop.

Keep sending with the same transmitting AWV $\mathbf{w}_t$ at the source device over N slots. Meanwhile, use identity matrix $\mathbf{I}_N$ as the receiving AWVs at the destination device, i.e., the nth column of $\mathbf{I}_N$ as the receiving AWV at the nth slot. Consequently, we receive a vector

$$\mathbf{w}_r = \mathbf{I}_N^H \mathbf{H} \mathbf{w}_t + \mathbf{I}_N^H \mathbf{n}_r = \mathbf{H} \mathbf{w}_t + \mathbf{n}_r, \qquad (3.48)$$

where $\mathbf{n}_r$ is the noise vector. Normalize $\mathbf{w}_r$.

Keep sending with the same transmitting AWV $\mathbf{w}_r$ at the destination device over M slots. Meanwhile, use identity matrix $\mathbf{I}_M$ as the receiving AWVs at the source device. Consequently, we receive a new vector

$$\mathbf{w}_t = \mathbf{I}_M^H \mathbf{H}^H \mathbf{w}_r + \mathbf{I}_M^H \mathbf{n}_t = \mathbf{H}^H \mathbf{w}_r + \mathbf{n}_t, \qquad (3.49)$$

where $\mathbf{n}_t$ is the noise vector. Normalize $\mathbf{w}_t$.

3) Result:

$\mathbf{w}_t$ is the AWV at the source device, and $\mathbf{w}_r$ is the AWV at the destination device.

$|\mathbf{g}_m^H \mathbf{g}_n|$ and $|\mathbf{h}_m^H \mathbf{h}_n|$ are approximately equal to zero given that $|\Omega_{rm} - \Omega_{rn}| \geq 1/N$ and $|\Omega_{tm} - \Omega_{tn}| \geq 1/M$, respectively, i.e., the receiving and transmitting angles can be resolved by the arrays, which is the common case in high-frequency band communications. Consequently, as a sub-optimal approach, the steering vectors of the *strongest* MPC can be adopted as the transmitting and receiving AWVs at the source and destination devices, which leads to the presented STV scheme. The advantage of STV is that the elements of the steering vector have a constant envelope, which is suitable for the devices with CM phased arrays. Moreover, although the transmitting and receiving angles are required to be resolved by the arrays in the following analysis, the STV scheme can work even when there exists angles that cannot be resolved, because two or more MPCs associated with sufficiently close angles that cannot be resolved actually build a single equivalent MPC.

Assuming that $\mathbf{H}$ is available in advance, the background of STV is presented as follows. Using the directional feature of high-frequency band communications channels, we have

$$\mathbf{H}^{2m_{\mathrm{Loop}}} \approx \sum_{\ell=1}^{L} |\sqrt{MN}\lambda_\ell|^{2m_{\mathrm{Loop}}} \mathbf{h}_\ell \mathbf{h}_\ell^{\mathrm{H}}, \tag{3.50}$$

for a positive integer m. Suppose the k-th MPC is the strongest one. For $\ell \neq k$,

$$|\lambda_\ell|^{2m_{\mathrm{Loop}}} / |\lambda_k|^{2m_{\mathrm{Loop}}}, \tag{3.51}$$

exponentially decreases. This means that the contribution to the matrix product $\mathbf{H}^{2m_{\mathrm{Loop}}}$ from the other $L-1$ MPCs exponentially decreases, compared with the strongest one. Therefore, we have

$$\lim_{m_{\mathrm{Loop}} \to \infty} \mathbf{H}^{2m_{\mathrm{Loop}}} = |\sqrt{MN}\lambda_k|^{2m_{\mathrm{Loop}}} \mathbf{h}_k \mathbf{h}_k^{\mathrm{H}}. \tag{3.52}$$

Thus, for given a sufficiently large m and an arbitrary initial transmitting AWV $\mathbf{w}_{\mathrm{t}}$, we have

$$\mathbf{H}^{2m_{\mathrm{Loop}}} \mathbf{w}_{\mathrm{t}} = |\sqrt{MN}\lambda_k|^{2m_{\mathrm{Loop}}} \mathbf{h}_k \mathbf{h}_k^{\mathrm{H}} \mathbf{w}_{\mathrm{t}} = \left(|\sqrt{MN}\lambda_k|^{2m_{\mathrm{Loop}}} \mathbf{h}_k^{\mathrm{H}} \mathbf{w}_{\mathrm{t}} \right) \mathbf{h}_k,$$

which is $\mathbf{h}_k$ multiplied by a complex coefficient. It is noted that $\mathbf{h}_k$ is a constant-envelope steering vector. Hence, the desired transmitting AWV can be obtained by the signature estimation[6] where

$$\mathbf{e}_{\mathrm{t}} = \exp(\mathrm{j}\angle(\mathbf{H}^{2m_{\mathrm{Loop}}} \mathbf{w}_{\mathrm{t}}))/\sqrt{M}, \tag{3.53}$$

is to be estimated. Here, $\angle(\mathbf{x})$ represents the angle vector of $\mathbf{x}$ in radian. In fact, the signature estimation can be carried out by the entry-wise normalization on $\mathbf{H}^{2m_{\mathrm{Loop}}} \mathbf{w}_{\mathrm{t}}$.

In addition, we have

$$\mathbf{H} \times \mathbf{H}^{2m_{\mathrm{Loop}}} \mathbf{w}_{\mathrm{t}} = \left(\lambda_k \sqrt{MN} |\lambda_k \sqrt{MMN}|^{2m_{\mathrm{Loop}}} \mathbf{h}_k^{\mathrm{H}} \mathbf{w}_{\mathrm{t}} \right) \mathbf{g}_k. \tag{3.54}$$

Thus, the desired receiving AWV can be obtained by the signature estimation of

$$\mathbf{e}_{\mathrm{r}} = \exp(\mathrm{j}\angle(\mathbf{H} \times \mathbf{H}^{2m_{\mathrm{Loop}}} \mathbf{w}_{\mathrm{t}}))/\sqrt{N}. \tag{3.55}$$

It is clear that given full CSI, the AWVs steering along the *strongest* MPC in both ends can be obtained. In practical communications, however, $\mathbf{H}$ is basically unavailable at both ends; thus, we present the joint iterative beamforming training process of STV, which is shown in Fig. 3.9, and the corresponding algorithm is described in Algorithm 3.4. The iteration number ε depends on a practical channel response. According to the simulation results in Section 3.4.3, $\varepsilon = 2$ or 3 can basically guarantee convergence.

---

[6]There are other approaches to obtain the desired AWV here. The presented one is a simple one in implementation.

Algorithm 3.4: The STV Scheme

---

1) Initialize:

Pick an initial transmitting AWV $\mathbf{w}_t$ at the source device. This AWV may be chosen either randomly or with the approach specified in Section 3.4.3.

2) Iteration: Iterate the following process ε times, and then stop.

Keep sending with the same transmitting AWV $\mathbf{w}_t$ at the source device over N slots. Meanwhile, use discrete Fourier Transform (DFT) matrix $\mathbf{F}_N$ as the receiving AWVs at the destination device, i.e., the n-th column of $\mathbf{F}_N$ as the receiving AWV at the n-th slot. Note that $\mathbf{I}_N$ cannot be used for the receiving AWVs here, due to its non-constant-envelope entries, but other unitary matrices with constant-envelope entries are feasible. Consequently, we receiving a vector

$$\mathbf{w}_r = \mathbf{F}_N^H \mathbf{w}_{Ht} + \mathbf{F}_N^H \mathbf{n}_r, \qquad (3.56)$$

where $\mathbf{n}_r$ is the noise vector. Estimate the signature $\mathbf{e}_r$ as

$$\mathbf{e}_r = \exp(j\angle(\mathbf{F}_N \mathbf{w}_r))/\sqrt{N}, \qquad (3.57)$$

and assign $\mathbf{e}_r$ to $\mathbf{w}_r$.

Keep sending with the same transmitting AWV $\mathbf{w}_r$ at the destination device over M slots. Meanwhile, use DFT matrix $\mathbf{F}_M$ as the receiving AWVs at the source device. Consequently, we receiving a new vector

$$\mathbf{w}_t = \mathbf{F}_M^H \mathbf{H}^H \mathbf{w}_r + \mathbf{F}_M^H \mathbf{n}_t, \qquad (3.58)$$

where $\mathbf{n}_t$ is the noise vector. Estimate the signature $\mathbf{e}_t$ as

$$\mathbf{e}_t = \exp(j\angle(\mathbf{F}_M \mathbf{w}_t))/\sqrt{M}, \qquad (3.59)$$

and assign $\mathbf{e}_t$ to $\mathbf{w}_t$.

3) Result:

$\mathbf{w}_t$ is the AWV at the source device, and $\mathbf{w}_r$ is the AWV at the destination device.

It is noted that STV is tailored for communications devices with CM phased arrays based on SGV. Thus, STV and SGV have common features, e.g., both schemes need iteration. However, their mathematical fundamentals are different. SGV is to find the principal singular vectors of the channel matrix $\mathbf{H}$, which is optimal and applicable for arbitrary channels, while STV is to find the CM steering vectors of the *strongest* MPC by exploiting the directional feature, which is sub-optimal and only

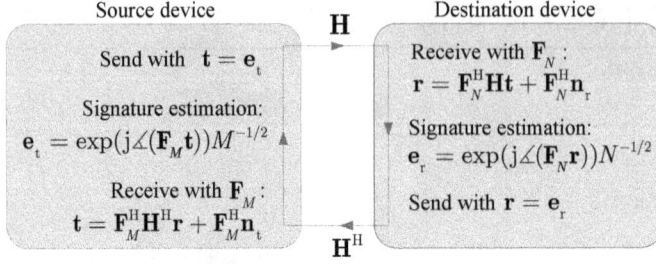

Figure 3.9 Process of the presented STV scheme.

feasible under high-frequency band communications channels. Thus, in each iteration, STV requires signature estimation, which is to estimate the CM steering vector of the *strongest* MPC. Meanwhile, in order to make STV feasible for CM phased arrays, it adopts the DFT matrices in transmitting and receiving training sequences, because the entries of them have a constant envelope.

3.4.3 Performance Evaluation

In this section, we evaluate the performances of STV, including array gain and convergence rate, and compare them with those of SGV via simulations. In all the simulations, the channel is normalized as

$$\mathbb{E}(\sum_{\ell=1}^{L} |\lambda_\ell|^2) = 1. \tag{3.60}$$

The transmitting SNR is thus $\gamma_t = 1/\sigma^2$, and the array gain becomes the ratio of receiving SNR to transmitting SNR, i.e.,

$$\eta = \gamma\Gamma/\gamma_t = |\mathbf{w}_r^H \mathbf{H} \mathbf{w}_t|^2. \tag{3.61}$$

The initial transmitting AWVs in the two schemes are selected under the principle that its power is evenly projected on the M basis vectors of the receiving matrices at the source, i.e., $\mathbf{I}_M$ and $\mathbf{F}_M$, respectively. Thus, the initial transmitting AWV for SGV is $\mathbf{1}_M/\sqrt{M}$, while that for STV is a normalized constant-amplitude-zero-autocorrelation (CAZAC) sequence with length M.

The array gain is empirically found using the ratio of the average receiving SNR to the average transmitting SNR over 10^3 realizations of channels. Furthermore, the SVD upper bound is obtained by averaging the squares of the principal singular values of these channel realizations. Channel realizations are generated under the Rician and Rayleigh fading models for LoS and NLoS channels, respectively. For LoS channels, the power of the LoS MPC is $|\lambda_1|^2 = 0.7692$, and the average powers of the NLoS MPCs are

$$\mathbb{E}(\{|\lambda_\ell|^2\}_{\ell=2,3,4}) = [0.0769 \ 0.0769 \ 0.0769]. \tag{3.62}$$

For NLoS channels,

$$\mathbb{E}(\{|\lambda_\ell|^2\}_{\ell=1,2,3,4}) = [0.25 \ 0.25 \ 0.25 \ 0.25]. \tag{3.63}$$

The transmitting and receiving steering angles are randomly generated within $[0\ 2\pi)$ in each realization.

Fig. 3.10 shows the achieved array gains of SGV and STV under LoS and NLoS channels, respectively, with different numbers of iterations, where $M = N = 16$. Fig. 3.11 shows the comparison of convergence rates between SGV and STV under LoS and NLoS channels with a high transmitting SNR, i.e., 25 dB, in the cases of $M = N = 16$ and $M = N = 32$, respectively. From Fig. 3.10 (left) and Fig. 3.11, it is found that under the LoS channel, both schemes achieve fast convergence rates and approach the optimal array gain, i.e., the SVD upper bound. From Fig. 3.10 (right) and Fig. 3.11, it is observed that under the NLoS channel, both schemes have slower convergence rates, and STV achieves a faster convergence rate at the cost of a slightly lower array gain than SGV that also approaches the SVD upper bound. It is noted that, although not shown in these figures, similar results are observed with a smaller or larger number of antennas.

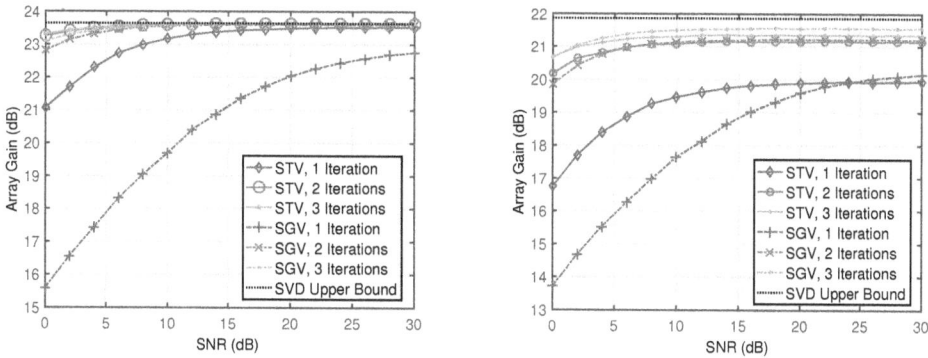

Figure 3.10 Comparison of array gain between SGV and STV under the LoS (left) and NLoS (right) channels with different numbers of iterations, where $M = N = 16$.

Explanations for these observations are as follows. Under the LoS channel, there is one and only one strong MPC, and the steering vectors of this MPC are almost the optimal AWVs. Thus, STV can achieve the optimal array gain. But under the NLoS channel, there are several MPCs with different steering angles (or steering vectors), and the STV scheme obtains one of them as an AWV, which is not optimal. Hence, STV cannot achieve the optimal array gain in such a case. On the other hand, since the SGV is based on the principal singular vector, it can surely achieve the SVD upper bound once convergence has been achieved. Besides, the fact that STV achieves a faster convergence rate in NLoS channels indicates that the signature estimation in each iteration of STV is more robust against noise, while the AWV estimation of SGV is more sensitive to noise.

In brief, although STV is tailored for high-frequency band communications devices with CM phased arrays, where SGV is infeasible, it has comparable performances to SGV in terms of the convergence rate and array gain under both LoS and NLoS channels. On the other hand, it is noted that a single iteration consumes $M + N$ training slots, which may significantly degrade the system efficiency, especially when the number of antennas is large. Hence, even if there is no CM constraint, i.e., both phase and amplitude are adjustable, and thus SGV is feasible, STV may still be favored in the case that the iteration number is constrained to be 1 or 2 to save training time, because it can achieve a higher array gain according to Fig. 3.11.

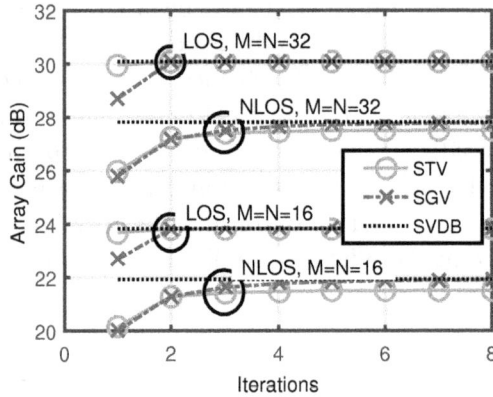

Figure 3.11 Comparison of convergence rates of array gains between SGV and STV under LoS and NLoS channels, where SVDB represents SVD bound.

3.5 SUMMARY

In this chapter, we introduced the system and channel models of the array beam-forming enabled point-to-point transmission, where two Tx/Rx joint beamforming schemes were presented. The first one is IEVD, which is sub-optimal and suitable for slow fading channel. To make this scheme practically feasible, a training approach has been suggested. It is demonstrated that IEVD and its training approach basically converge with only two iterations. Compared with the existing sub-optimal scheme, i.e., Nsenga's, IEVD with its training approach achieves the same performance, but reduced overhead and computational complexity. The other one is MPG suitable for fast fading channel, which groups the MPCs and concurrently beamforms toward multiple steering angles of the grouped MPCs. Compared with the existing alternative scheme, i.e., Park-Pan, MPG can work in the case that the number of MPCs is greater than that of antennas at Tx or Rx. Additionally, MPG achieves full diversity, the same as Park-Pan, but a better array gain than Park-Pan, when the number of MPCs is smaller than or around that of antennas at Tx and Rx, with an approximately equivalent complexity.

Since the existing SGV scheme cannot be used in communications with CM phased arrays, the STV scheme has been presented, which effectively exploits the directional feature of high-frequency band communications channels. Performance comparisons showed that under LoS channels, both the schemes achieve fast convergence rates and achieve the near-to-optimal array gain; under NLoS channels, STV achieves a faster convergence rate at the cost of a slightly lower array gain than SGV that can still approach the optimal array gain. In summary, the presented STV scheme is well-suited to high-frequency band communications with CM phased arrays. It has comparable performance to SGV in terms of convergence rate and array gain under both LoS and NLoS channels.

Bibliography

[1] Yang Tang, Branka Vucetic, and Yonghui Li. An iterative singular vectors estimation scheme for beamforming transmission and detection in MIMO systems. *IEEE Commun. Lett.*, 9(6):505–507, 2005.

[2] Junyi Wang, Zhou Lan, Chang-woo Pyo, Tuncer Baykas, Chin-sean Sum, Mohammad Azizur Rahman, Jing Gao, Ryuhei Funada, Fumihide Kojima, Hiroshi Harada, et al. Beam codebook based beamforming protocol for multi-gbps millimeter-wave WPAN systems. *IEEE J. Select. Areas Commun.*, 27(8):1390–1399, 2009.

[3] Bin Li, Zheng Zhou, Weixia Zou, Xuebin Sun, and Guanglong Du. On the efficient beam-forming training for 60GHz wireless personal area networks. *IEEE Trans. Wireless Commun.*, 12(2):504–515, 2012.

[4] Jimmy Nsenga, Wim Van Thillo, François Horlin, Valéry Ramon, André Bourdoux, and Rudy Lauwereins. Joint transmit and receive analog beamforming in 60 GHz MIMO multipath channels. In *2009 IEEE International Conference on Communications*, pages 1–5. IEEE, 2009.

[5] Xiqi Gao, Bin Jiang, Xiao Li, Alex B. Gershman, and Matthew R. McKay. Statistical eigenmode transmission over jointly correlated MIMO channels. *IEEE Trans. Inform. Theory*, 55(8):3735–3750, 2009.

[6] Shi Jin, Matthew R. McKay, Xiqi Gao, and Iain B. Collings. MIMO multichannel beamforming: SER and outage using new eigenvalue distributions of complex noncentral Wishart matrices. *IEEE Trans. Commun.*, 56(3):424–434, 2008.

[7] Su-Khiong Yong, Pengfei Xia, and Alberto Valdes-Garcia. *60GHz Technology for Gbps WLAN and WPAN: from Theory to Practice*. John Wiley & Sons, 2011.

[8] Alexander Maltsev, Roman Maslennikov, Alexey Sevastyanov, Artyom Lomayev, Alexey Khoryaev, Alexei Davydov, and Vladimir Ssorin. Characteristics of indoor millimeter-wave channel at 60 GHz in application to perspective WLAN system. In *Proceedings of the Fourth European Conference on Antennas and Propagation*, pages 1–5. IEEE, 2010.

[9] Nektarios Moraitis and Philip Constantinou. Indoor channel capacity evaluation utilizing ULA and URA antennas in the millimeter wave band. In *2007 IEEE 18th International Symposium on Personal, Indoor and Mobile Radio Communications*, pages 1–5. IEEE, 2007.

[10] Minyoung Park and Helen K. Pan. A spatial diversity technique for IEEE 802.11 ad WLAN in 60 GHz band. *IEEE Commun. Lett.*, 16(8):1260–1262, 2012.

[11] Vasanthan Raghavan and Akbar M. Sayeed. Sublinear capacity scaling laws for sparse MIMO channels. *IEEE Trans. Inform. Theory*, 57(1):345–364, 2010.

[12] Akbar M. Sayeed and Vasanthan Raghavan. Maximizing MIMO capacity in sparse multipath with reconfigurable antenna arrays. *IEEE J. Sel. Top. Sign. Proces.*, 1(1):156–166, 2007.

[13] Suiyan Geng, Jarmo Kivinen, Xiongwen Zhao, and Pertti Vainikainen. Millimeter-wave propagation channel characterization for short-range wireless communications. *IEEE Trans. Veh. Technol.*, 58(1):3–13, 2008.

[14] Hao Xu, Vikas Kukshya, and Theodore S. Rappaport. Spatial and temporal characteristics of 60-GHz indoor channels. *IEEE J. Select. Areas Commun.*, 20(3):620–630, 2002.

[15] Martin Jacob, Anton de Graauw, Maristella Spella, Pablo Herrero, Sebastian Priebe, Joerg Schoebel, and Thomas Kürner. Performance evaluation of 60 GHz WLAN antennas under realistic propagation conditions with human shadowing. In *2011 XXXth URSI General Assembly and Scientific Symposium*, pages 1–4. IEEE, 2011.

[16] Theodore S. Rappaport, Felix Gutierrez, Eshar Ben-Dor, James N. Murdock, Yijun Qiao, and Jonathan I. Tamir. Broadband millimeter-wave propagation measurements and models using adaptive-beam antennas for outdoor urban cellular communications. *IEEE Trans. Antennas Propagat.*, 61(4):1850–1859, 2012.

[17] Theodore S. Rappaport, Yijun Qiao, Jonathan I. Tamir, James N. Murdock, and Eshar Ben-Dor. Cellular broadband millimeter wave propagation and angle of arrival for adaptive beam steering systems. In *2012 IEEE Radio and Wireless Symposium*, pages 151–154. IEEE, 2012.

[18] Robert C. Daniels, James N. Murdock, Theodore S. Rappaport, and Robert W. Heath. 60 GHz wireless: Up close and personal. *IEEE Microwave*, 11(7):44–50, 2010.

[19] Kao-Cheng Huang and Zhaocheng Wang. *Millimeter Wave Communication Systems*, volume 29. John Wiley & Sons, 2011.

[20] Minyoung Park, Carlos Cordeiro, Eldad Perahia, and L. Lily Yang. Millimeter-wave multi-gigabit WLAN: Challenges and feasibility. In *2008 IEEE 19th International Symposium on Personal, Indoor and Mobile Radio Communications*, pages 1–5. IEEE, 2008.

[21] Eldad Perahia, Carlos Cordeiro, Minyoung Park, and L. Lily Yang. IEEE 802.11 ad: Defining the next generation multi-gbps wi-fi. In *2010 7th IEEE Consumer Communications and Networking Conference*, pages 1–5. IEEE, 2010.

[22] Zhenyu Xiao. Suboptimal spatial diversity scheme for 60 GHz millimeter-wave WLAN. *IEEE Commun. Lett.*, 17(9):1790–1793, 2013.

[23] Farooq Khan and Zhouyue Pi. MmWave mobile broadband (mmb): Unleashing the 3–300GHz spectrum. In *34th IEEE Sarnoff Symposium*, pages 1–6. IEEE, 2011.

[24] Zhouyue Pi and Farooq Khan. An introduction to millimeter-wave mobile broadband systems. *IEEE Commun. Mag.*, 49(6):101–107, 2011.

[25] David Tse and Pramod Viswanath. *Fundamentals of Wireless Communication*. Cambridge University Press, 2005.

[26] Theodore S. Rappaport et al. *Wireless Communications: Principles and Practice*, volume 2. Prentice Hall PTR New Jersey, 1996.

[27] Paul Marinier, Gilles Y. Delisle, and Charles L. Despins. Temporal variations of the indoor wireless millimeter-wave channel. *IEEE Trans. Antennas Propagat.*, 46(6):928–934, 1998.

[28] Omar El Ayach, Sridhar Rajagopal, Shadi Abu-Surra, Zhouyue Pi, and Robert W. Heath. Spatially sparse precoding in millimeter wave MIMO systems. *IEEE Trans. Wireless Commun.*, 13(3):1499–1513, 2014.

[29] Zhenyu Xiao, Lin Bai, and Jinho Choi. Iterative joint beamforming training with constant-amplitude phased arrays in millimeter-wave communications. *IEEE Commun. Lett.*, 18(5):829–832, 2014.

[30] Pengfei Xia, Huaning Niu, Jisung Oh, and Chiu Ngo. Practical antenna training for millimeter wave MIMO communication. In *2008 IEEE 68th Vehicular Technology Conference*, pages 1–5. IEEE, 2008.

[31] Sohrab Emami, Robert F. Wiser, Ershad Ali, Mark G. Forbes, Michael Q. Gordon, Xiang Guan, Steve Lo, Patrick T. McElwee, James Parker, Jon R. Tani, et al. A 60GHz CMOS phased-array transceiver pair for multi-gb/s wireless communications. In *2011 IEEE International Solid-State Circuits Conference*, pages 164–166. IEEE, 2011.

[32] Stephane Pinel, Padmanava Sen, Saikat Sarkar, Bevin G. Perumana, Debasis Dawn, David Yeh, Francesco Barale, Matthew Leung, Eric Juntunen, Praveen Babu Vadivelu, et al. 60ghz single-chip CMOS digital radios and phased array solutions for gaming and connectivity. *IEEE J. Select. Areas Commun.*, 27(8):1347–1357, 2009.

[33] Qiang Li, Mingyi Hong, Hoi-To Wai, Ya-Feng Liu, Wing-Kin Ma, and Zhi-Quan Luo. Transmit solutions for MIMO wiretap channels using alternating optimization. *IEEE J. Select. Areas Commun.*, 31(9):1714–1727, 2013.

[34] Rodrigo Caiado De Lamare and Raimundo Sampaio-Neto. Adaptive reduced-rank equalization algorithms based on alternating optimization design techniques for MIMO systems. *IEEE Trans. Veh. Technol.*, 60(6):2482–2494, 2011.

[35] Roger A. Horn and Charles Richard Johnson. *Matrix Analysis*. Cambridge University Press, 2012.

[36] Bettagere Nagaraja Bharath and Chandra R. Murthy. Channel training signal design for reciprocal multiple antenna systems with beamforming. *IEEE Trans. Veh. Technol.*, 62(1):140–151, 2012.

[37] Will P.M.N. Keizer. Fast and accurate array calibration using a synthetic array approach. *IEEE Trans. Antennas Propagat.*, 59(11):4115–4122, 2011.

[38] Boon Poh Ng, Joni Polili Lie, Meng Hwa Er, and Aigang Feng. A practical simple geometry and gain/phase calibration technique for antenna array processing. *IEEE Trans. Antennas Propagat.*, 57(7):1963–1972, 2009.

[39] Shuichi Ohno. Performance of single-carrier block transmissions over multipath fading channels with linear equalization. *IEEE Trans. Signal Processing*, 54(10):3678–3687, 2006.

[40] Pengfei Xia, Su-Khiong Yong, Jisung Oh, and Chiu Ngo. A practical SDMA protocol for 60 GHz millimeter wave communications. In *2008 42nd Asilomar Conference on Signals, Systems and Computers*, pages 2019–2023. IEEE, 2008.

[41] Alberto Valdes-Garcia, Sean T. Nicolson, Jie-Wei Lai, Arun Natarajan, Ping-Yu Chen, Scott K. Reynolds, Jing-Hong Conan Zhan, Dong G. Kam, Duixian Liu, and Brian Floyd. A fully integrated 16-element phased-array transmitter in sige bicmos for 60-GHz communications. *IEEE J. Solid-State Circuits*, 45(12):2757–2773, 2010.

[42] Emanuel Cohen, Claudio Jakobson, Shmuel Ravid, and Dan Ritter. A thirty two element phased-array transceiver at 60GHz with rf-if conversion block in 90nm flip chip CMOS process. In *2010 IEEE Radio Frequency Integrated Circuits Symposium*, pages 457–460. IEEE, 2010.

[43] Lin Bai and Jinho Choi. Lattice reduction-based MIMO iterative receiver using randomized sampling. *IEEE Trans. Wireless Commun.*, 12(5):2160–2170, 2013.

[44] Xiayu Zheng, Yao Xie, Jian Li, and Petre Stoica. MIMO transmit beamforming under uniform elemental power constraint. *IEEE Trans. Signal Processing*, 55(11):5395–5406, 2007.

[45] Jungwon Lee, Rohit U. Nabar, Jihwan P. Choi, and Hui-Ling Lou. Generalized co-phasing for multiple transmit and receive antennas. *IEEE Trans. Wireless Commun.*, 8(4):1649–1654, 2008.

CHAPTER 4

Array Beamforming Enabled Full-Duplex Transmission

4.1 INTRODUCTION

In Chapters 3 and 4, we have introduced methods of point-to-point beamforming, i.e., switched beamforming based on codebooks and adaptive beamforming via iterations, which are able to find the optimal antenna weight vectors (AWV) and obtain considerable beam gains. With considerable beam gains, the strength of signals and signal-to-noise ratio (SNR) at receive (Rx) can be improved. As a result, the data rate can be further improved.

However, this benefit may be offset to full-duplex (FD) transmissions by applying conventional half-duplex (HD) strategies at the transmitter (Tx), i.e., time division duplex (TDD) or frequency division duplex (FDD), which are not able to fully exploit the time-frequency domain and thus limit the spectrum efficiency. In particular, with TDD sufficient Tx/Rx switching guard time should be reserved to make sure the Tx/Rx circuits work normally. Moreover, when considering multiple access, there may be significant protocol overhead[1]. Even for bi-directional point-to-point transmission, there is still a lot of handshaking overhead[1]. While with FDD a large guard band should be arranged to make sure the interference leakage is small enough.

To this end, FD communications, where transmission and reception occur in the same time-frequency resource block is explored to achieve the reuse of carrier frequencies and may double the spectrum efficiency[2,3]. Moreover, FD provides more efficient and flexible access strategies for multiple access[3], and may not need handshaking for bi-directional point-to-point transmission, which can further increase the practical efficiency of communication.

Let us compare the complexity of an FD node and that of a regular node with FDD. There are a Tx radio frequency (RF) chain and a Tx antenna array, as well as a Rx RF chain and a Rx antenna array, at both the FD node and the regular FDD node. The only difference is that at the FD node the Rx needs to mitigate self-interference (SI), while at the regular FDD node, the Rx does not need to mitigate SI. For FD communication, SI cancellation can be done by using the beamforming technology presented in this chapter, which only needs to control the AWVs and almost does

DOI: 10.1201/9781003366362-4

not increase the system complexity. In brief, if beamforming technology is adopted to mitigate the SI, the complexity of an FD node is similar to that of a regular FDD node.

However, array beamforming enabled FD communication faces particular challenges. Naturally, the most critical issue in FD is also SI, which is the transmitted signal received by the local Rx at the same node and needs to be canceled. Especially for systems with large antenna arrays, specific antenna settings are required to enable FD transmission, which may increase the hardware complexity. We introduce two possible antenna configurations as shown in Fig. 4.1. There are one Tx RF chain and one Rx RF chain in both configurations. In Fig. 4.1(a) Tx/Rx share the same antenna array by using circulators. The function of the circulator is to separate the Tx signal path from the Rx signal path, and meanwhile, let the Tx/Rx signals be transmitted/received through the antenna connecting to it. In Fig. 4.1(b) Tx/Rx are equipped with separate antenna arrays. We compare these two configurations as follows.

Firstly, we compare the capability of SI suppression between the two configurations. For the configuration of sharing the same antenna array in Fig. 4.1(a), SI is suppressed by a circulator. The isolation of a circulator is dependent on the quality of the circulator. In the high-frequency band, the isolation is still limited in general and needs further improvement. For instance, the isolation of the circulator designed for millimeter-wave systems in[4] is only 18 dB, less than that in the micro-wave band in general. On the other hand, even if the circulator is good enough, the signal transmitted from an arbitrary antenna element (AE) will also result in significant SI to its neighbor AEs, because the distances between these AEs are quite small, basically at a level of the wavelength of the carrier. In contrast, with separate antennas as that in Fig. 4.1(b), SI can be suppressed by blocking the signals between Tx/Rx antenna arrays and/or locating the Tx/Rx antennas far away from each other. Moreover, with separate arrays, the parameters of Tx/Rx array positioning may also be exploited as additional degrees of freedom (DoFs) to mitigate SI and optimize the system performance, as will be discussed in detail later.

Then, we compare the cost and area of the two configurations. It is clear that the configuration with separate Tx/Rx arrays requires an additional array, while it does

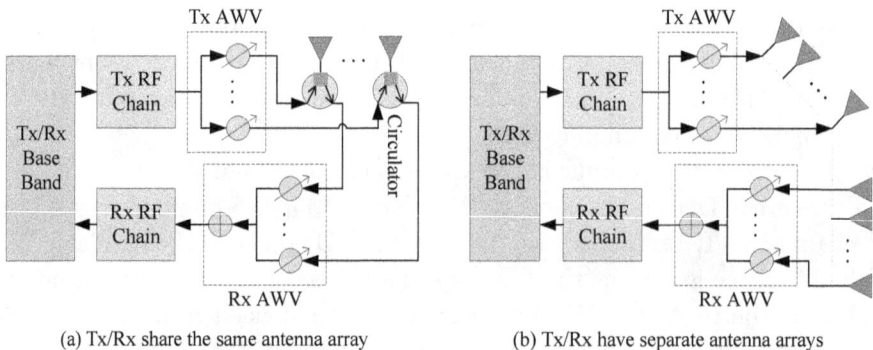

(a) Tx/Rx share the same antenna array   (b) Tx/Rx have separate antenna arrays

Figure 4.1 Possible antenna configurations of an FD node.

not need circulators. It is hard to say whether an antenna array or the circulators have a higher cost. Even if an antenna array has a higher cost, it is affordable in general because it is only a small part of the total cost of the whole transceiver. In addition to the cost, area is also important for an FD device, especially for portable devices. Note that in the micro-wave band the size of a micro-wave antenna is relatively large[2]. Hence, using seperate Tx/Rx antennas will double the antenna area, which may be unfavored for small devices, while sharing the same antennas can save a large area. In contrast, in the high-frequency band the size of an antenna is small; thus an additional antenna array may occupy only a small area. For instance, for a 4×4 uniform planar array (UPA) with half wavelength spacing at 30 GHz, the required area is only about 2×2 cm^2. However, when the number of the AEs is too large, the required area may be also large.

Although FD transmission requires additional hardware cost for multiple antenna systems, array beamforming technologies provide new DoFs for SI mitigation in FD communication systems. Thanks to the directional feature of the beams, the SI can be suppressed in the space domain via Tx/Rx beamforming, where the AWVs can be customized to minimize the effective channel gain between Tx and Rx antennas at the FD nodes. In fact, without considering the constant modulus (CM) constraint of communication, the beamforming technology can even completely cancel SI, including Tx RF impairments, by using the zero-forcing (ZF) filtering. However, under the CM constraint, the SI may not be completely mitigated, and it becomes a challenge for FD communication to mitigate SI under the CM constraint. We will focus on these problems and solutions in the following sections.

The content of this chapter is arranged as follows. In the second section, the channel modeling of FD communication is introduced, and the SI cancellation in FD communication is discussed. Then we propose several sub-optimal solutions to the joint Tx/Rx beamforming (JTR-BF) problem for FD communications. First, an iterative algorithm, which iteratively maximizes the signal power with ZF SI (ZF-Max-Power), is presented, and its convergence is proven. Next, two closed-form solutions are derived under minimum mean square error (MMSE), ZF, and maximum-ratio transmission (MRT) criteria, namely a lower-bound-based MMSE solution (LB-MMSE) and ZF SI with MRT (SI-ZF-MRT), where iterations are not required. In the third section, we propose to deploy an FD relay to improve the end-to-end performance of a communication system. The corresponding optimization problems are established to maximize the achievable rate between two nodes. Simulation results show that the presented optimization method can approach the performance upper bound of FD relay system.

4.2 BEAMFORMING FOR FULL-DUPLEX POINT-TO-POINT TRANSMISSION

As SI is still significant in FD communication even with separate Tx/Rx antenna arrays, we need to mitigate it as much as possible for transceivers. In the existing micro-wave band FD systems, antenna cancellation, which suppresses signals between Tx/Rx antennas, RF cancellation, which subtracts SI in the RF domain, and baseband cancellation, which mitigates residual SI in the baseband, are three typical methods to cancel SI, and a combination of them usually achieves better performance[2,3]. In this section, we focus on SI mitigation in the spatial domain, where the SI can be suppressed via appropriate Tx/Rx beamforming.

4.2.1 System Model

An FD communication system used in this chapter consisting of two nodes, namely Node #1 and Node #2, is illustrated in Fig. 4.2. Each node is equipped with a transmit antenna array and a receive antenna array, and supports only one data stream[5,6]. We denote by n_{t1} and n_{t2} the numbers of AEs of the transmit arrays at Node #1 and Node #2, respectively, while by n_{r1} and n_{r2} those of the receive arrays at Node #1 and Node #2, respectively. In our model, Node #1 transmits signals to Node #2 and receives signals from Node #2 simultaneously; thus both nodes suffer from SI transmitted by the local Txs.

It is noteworthy that although we depict separate antenna arrays for the Tx/Rx chains in Fig. 4.2 (this structure is indeed common in FD wireless communications[3,7,8,9,10]), the Tx/Rx chains may also share the same antenna array[2,3]. Fortunately, our signal model is suitable for both cases, while the one with the same array can be seen as a particular case of the separate arrays from the signaling viewpoint, i.e., the two arrays completely coincide with each other. Note that the SI with a shared array may be even higher than that with separate arrays.

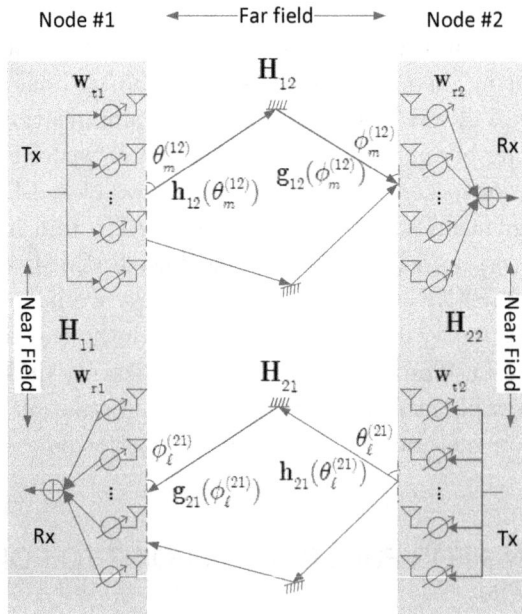

Figure 4.2 Illustration of the FD communication system.

4.2.2 Channel Model

As we can see from Fig. 4.2, there are two types of channels. The first one is the communication channel, which represents the channel for information-bearing signals exchanged between Node #1 and Node #2, i.e., $\mathbf{H}_{12}$ and $\mathbf{H}_{21}$, where $\mathbf{H}_{ij}$ represents the channel from Node #i to Node #j. The other one is the SI channel. Clearly, $\mathbf{H}_{11}$ and $\mathbf{H}_{22}$ are the SI channels.

4.2.2.1 Communication Channel

We first consider the model of communication channel. Since the distance between these two nodes is generally much greater than the wavelength of signal, the commonly used far-field channel model, which has a plane wavefront, is suitable for $\mathbf{H}_{12}$ and $\mathbf{H}_{21}$. According to channel measurement results for millimeter-wave communication[1, 11], mostly reflection contributes to generating multiple components (MPCs) besides the line of sight (LoS) component; scattering and diffraction effects are little due to the extremely short wavelength of millimeter-wave communication. Thus, the MPCs in millimeter-wave communication have a feature of directivity[6, 12, 13, 14, 15], i.e., different MPCs have different physical angles of departure (AoDs), i.e., $\theta_m^{(12)}$ and $\theta_\ell^{(21)}$, as well as angles of arrival (AoAs), e.g., $\phi_m^{(12)}$ and $\phi_\ell^{(21)}$, as shown in Fig. 4.2. The (narrow-band) communication channels can be expressed as[1] [6, 12, 13, 14, 15]

$$\mathbf{H}_{12} = \sqrt{n_{t1}n_{r2}} \sum_{m=1}^{M} \alpha_m \mathbf{g}_{12}(\phi_m^{(12)}) \mathbf{h}_{12}^{H}(\theta_m^{(12)}), \tag{4.1}$$

and

$$\mathbf{H}_{21} = \sqrt{n_{t2}n_{r1}} \sum_{\ell=1}^{L} \beta_\ell \mathbf{g}_{21}(\phi_\ell^{(21)}) \mathbf{h}_{21}^{H}(\theta_\ell^{(21)}), \tag{4.2}$$

where M and L are the numbers of MPCs, α_m and β_ℓ are the coefficients of MPCs, $\mathbf{g}_{12}(\phi_m^{(12)})$ and $\mathbf{g}_{21}(\phi_\ell^{(21)})$ are receive steering vectors, $\mathbf{h}_{12}(\theta_m^{(12)})$ and $\mathbf{h}_{21}(\theta_\ell^{(21)})$ are transmit steering vectors of $\mathbf{H}_{12}$ and $\mathbf{H}_{21}$, respectively. For uniform linear arrays (ULAs)[2] with half-wavelength spacing, these steering vectors are defined as (4.5), and they are all functions of the corresponding steering angles. For convenience, we have the following normalization:

$$\sum_{m=1}^{M} \mathbb{E}\{|\alpha_m|^2\} = \sum_{\ell=1}^{L} \mathbb{E}\{|\beta_\ell|^2\} = 1. \tag{4.3}$$

---

[1]In general, high-frequency band signals have a wide band, and thus a frequency selective channel may be suitable[16]. However, with beamforming only a very small number (or even only one) of strong MPCs may be searched out to form beams between Tx and Rx. As a result, the effect of delay spread may be substantially mitigated[17]. Due to this reason, a frequency flat channel model is extensively used in communication[6, 12, 13, 14, 15].

[2]Although ULA is adopted in this book, the developed schemes are also feasible for other types of arrays, like UPA or circular array, because different types of arrays affect only the channel matrices.

In the case of Tx/Rx sharing the same antenna array at a node, we have

$$\begin{cases} n_{r1} = n_{t1}, & n_{r2} = n_{t2}, & L = M, \\ \theta_m^{12} = \phi_m^{21}, & \theta_m^{21} = \phi_m^{12}, & \alpha_m = \beta_m, & m = 1, 2, ..., M, \end{cases} \tag{4.4}$$

i.e., parameters of $\mathbf{H}_{12}$ are the same as those of $\mathbf{H}_{21}$. However, since Tx/Rx have different RF chains[2,3], the beamforming and combining vectors are basically different.

Note that for communications there may be other models. For instance, in[18] a clustered channel model is adopted, where the channel includes several clusters, and a cluster consists of many MPCs with small angle differences. Different models are suitable for different communication circumstances. As our schemes do not exploit the specific feature of the communication channel, they can be used for different models.

$$\begin{aligned}
\mathbf{g}_{12}(\phi_m^{(12)}) &= \frac{\left[\exp\left(j\pi 0 \cos(\phi_m^{(12)})\right), ..., \exp\left(j\pi(n_{r2}-1)\cos(\phi_m^{(12)})\right)\right]^{\mathrm{T}}}{\sqrt{n_{r2}}}, \\
\mathbf{g}_{21}(\phi_\ell^{(21)}) &= \frac{\left[\exp\left(j\pi 0 \cos(\phi_\ell^{(21)})\right), ..., \exp\left(j\pi(n_{r1}-1)\cos(\phi_\ell^{(21)})\right)\right]^{\mathrm{T}}}{\sqrt{n_{r1}}}, \\
\mathbf{h}_{12}(\theta_m^{(12)}) &= \frac{\left[\exp\left(j\pi 0 \cos(\theta_m^{(12)})\right), ..., \exp\left(j\pi(n_{t1}-1)\cos(\theta_m^{(12)})\right)\right]^{\mathrm{T}}}{\sqrt{n_{t1}}}, \\
\mathbf{h}_{21}(\theta_\ell^{(21)}) &= \frac{\left[\exp\left(j\pi 0 \cos(\theta_\ell^{(21)})\right), ..., \exp\left(j\pi(n_{t2}-1)\cos(\theta_\ell^{(21)})\right)\right]^{\mathrm{T}}}{\sqrt{n_{t2}}}.
\end{aligned} \tag{4.5}$$

4.2.2.2 SI Channel

Next, we consider the strength of SI and the SI channel. An SI channel refers to the channel from the local Tx array to the local Rx array at the same node. It is different from the communication channel between two different nodes. In addition, the far-field range condition, i.e.,[19]

$$R \geq 2D^2/\lambda \tag{4.6}$$

does not hold in general for the LoS component of SI, where R is the distance between Tx and Rx arrays, D is the diameter of the antenna aperture, λ is the wavelength of the carrier. For instance, considering a half-wavelength spaced ULA with 32 elements, the far-field range should satisfy

$$R \geq 2(16\lambda)^2/\lambda = 1024\lambda, \tag{4.7}$$

which is basically too large for small-size devices like mobile phones or laptops even at the high-frequency band. Thus, the SI channel may have to use the near-field model, which has a spherical wavefront[19]. In such a case, the LoS path of the SI channel is highly dependent on the structures and relative positions of the Tx/Rx antenna arrays.

In addition to the LoS component, there are also NLoS components for the SI channel, due to the possible reflectors near the device (Fig. 4.3), i.e.,

$$\mathbf{H}_{\mathrm{SI}} = \mathbf{H}_{\mathrm{SI,L}} + \mathbf{H}_{\mathrm{SI,N}}. \tag{4.8}$$

Different from the LoS component, where the direct path between Tx/Rx antennas is short, the propagation distances via the NLoS paths are much longer, and can satisfy the far-field range condition in general[20]. Hence, the NLoS components may adopt the same model as the communication channel as above. Moreover, compared with the LoS components, the NLoS components experience much higher propagation loss as well as additional reflection loss. Hence, the strength of the NLoS components is much weaker than that of the LoS component. For these reasons, we put more attention to the LoS component of the SI channel.

Without loss of generality, we consider an antenna placement as shown in Fig. 4.3. The distance between the first elements of the two arrays is d, and the angle between these two ULAs is ω. With this antenna placement, the channel gain between the n-th receive antenna and the m-th Tx antenna, i.e., the coefficient corresponding to the n-th row and the m-th column of $\mathbf{H}_{\mathrm{SI,L}}$ (the SI channel matrix with a near-field model) is[21, 22]

$$[\mathbf{H}_{\mathrm{SI,L}}]_{nm} = h_{nm} = \frac{\rho}{r_{nm}} \exp\left(-j2\pi \frac{r_{nm}}{\lambda}\right), \tag{4.9}$$

where r_{nm} is the distance between the m-th element of the transmit array and the n-th element of the receive array, and ρ is a constant for power normalization.

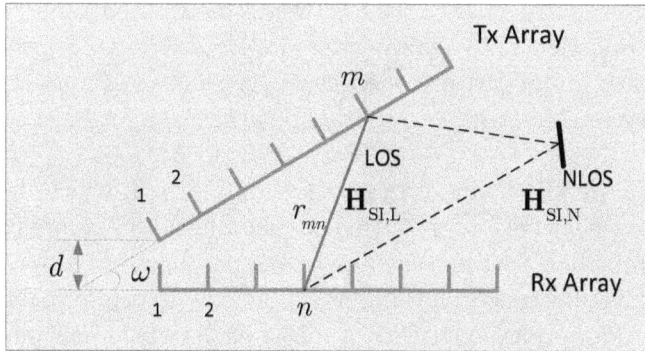

Figure 4.3 An illustration of Tx/Rx array positioning.

Interestingly, although the LoS-SI channel exploits the near-field model, from our model we find that when the number of AEs is large, the LoS-SI channel also shows spatial sparsity, which is similar to the millimeter-wave communication channel.

To show this feature, we set the Tx/Rx AWVs as steering vectors with (cosine) steering angles from -1 to 1, and obtain the beamforming gains

$$|\mathbf{a}(N, \alpha)^{\mathrm{H}} \mathbf{H}_{\mathrm{SI,L}} \mathbf{a}(N, \beta)|, \qquad (4.10)$$

with these angle combinations. Fig. 4.4 shows the results, where we can find that when the number of AEs is larger, the angle area, in which the beamforming gain is significant w.r.t. the largest beamforming gain, is smaller. When the number of elements is 32, as shown in the right-hand side figure, and the LoS-SI channel shows distinct spatial sparsity. In particular, the beamforming gain is significant only when $\beta - \alpha \approx \omega$, the angle of the Tx/Rx arrays; otherwise it is small.

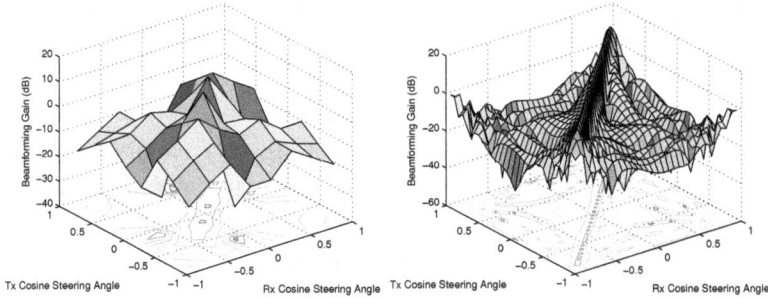

Figure 4.4 Spatial feature of the LoS-SI channel, where $\omega = 0$. The number of AEs is 8 for both the Tx/Rx arrays in the left-hand side figure, while 32 for the right-hand side figure.

This result implies that most of the SI power is distributed along only a few Tx/Rx steering angle pairs. Hence, by steering the Tx/Rx AWVs to other angle pairs, the SI can be efficiently mitigated provided that the number of antennas is sufficiently large, and the parameters, e.g., ω, can also be optimized for SI cancellation, as will be shown later.

As we can see, the SI channel can be very complicated, which includes both LoS component with a near-field propagation model and NLoS components with a far-field propagation model. Fortunately, due to the packaging of a portable FD device, the propagation circumstances within the device are basically stable. For instance, a mobile phone can be packaged with a metal shell, which can block the reflected signals from the back of the mobile phone and thus prevents the outside objects, e.g., hands, from affecting the SI channel. Hence, in general, an SI channel can be seen as slow-varying. As a consequence, we usually have enough time to make an accurate SI channel estimation.

4.2.3 FD Communication without CM Constraint

In order to realize FD in communications, we face a JTR-BF problem to maximize the Tx/Rx achievable rate. As this problem is non-convex, sub-optimal solutions are expected. The existing JTR-BF schemes proposed in communications are basically infeasible, because SI is not considered in these schemes [1,5,6,12,15,23,24,25]. Although the mitigation of loopback SI[26,27] as well as the utilization of loopback SI[28,29] are considered in multiple-input multiple-output (MIMO) relays, these methods cannot be employed in our setup, as their signal models are built for relay systems, but in this section a bi-directional FD transmission system is considered. In[30], digital beamforming to cancel SI in FD wireless communications was studied from an experimental perspective, where the JTR-BF with SI was not analytically investigated. Although some sub-optimal distributed solutions proposed for the K-user interference alignment (IA) problem[31,32,33,34], e.g., the maximum signal-to-interference-plus-noise ratio (max-SINR), maximum power (Max-Power), minimum leakage (Min-Leakage), and MMSE schemes[34], may be applicable, they are all iterative solutions designed for general K-user IA problems. Applying them to FD communications would result in high computational complexity for large antenna array. Moreover, the convergence of some of them is not yet proven in the literature[34] to the best of our knowledge.

In this section, we present several sub-optimal solutions to the JTR-BF problem for FD communications. Firstly, an iterative algorithm, which iteratively maximizes the signal power with ZF SI (ZF-Max-Power), is presented, and its convergence is proven. Next, two closed-form solutions are derived under MMSE, ZF, and MRT criteria, namely LB-MMSE and ZF SI with MRT (SI-ZF-MRT), where iterations are not required. Performance evaluations show that ZF-Max-Power approaches an upper bound on the joint achievable rate (JAR), and it needs only 2 iterations on average to achieve the convergence with random initial points. These performances of ZF-MAx-Power are almost the same as those of the best baseline for the IA problem[31,32,33,34], namely Max-SINR. However, the convergence of Max-SINR is unproven yet[34] to the best of our knowledge, and the computational complexity of ZF-Max-Power is significantly lower than that of Max-SINR since matrix inversion is not needed. The two closed-form solutions achieve sub-optimal performances to the upper bound depending on channel conditions. In addition, ZF-Max-Power and SI-ZF-MRT are robust against the geometry of Tx/Rx antenna arrays due to the operation of ZF SI, while LB-MMSE is not. ZF-Max-Power and LB-MMSE are robust against channel estimation errors, while SI-ZF-MRT is not. These results verify the feasibility of FD communication and the effectiveness of beamforming cancellation.

With the above system and channel models, the received signals at Node #1 and Node #2 are written as

$$y_1 = \mathbf{w}_{r1}^{\mathrm{H}} \left(\sqrt{P_{21}} \mathbf{H}_{21} \mathbf{w}_{t2} s_2 + \sqrt{P_{11}} \mathbf{H}_{11} \mathbf{w}_{t1} s_1 + \mathbf{n}_1 \right), \quad (4.11)$$

and

$$y_2 = \mathbf{w}_{r2}^{\mathrm{H}} \left(\sqrt{P_{12}} \mathbf{H}_{12} \mathbf{w}_{t1} s_1 + \sqrt{P_{22}} \mathbf{H}_{22} \mathbf{w}_{t2} s_2 + \mathbf{n}_2 \right), \quad (4.12)$$

respectively, where s_1 and s_2 are the transmitted symbols with unit power at Node #1 and Node #2, respectively, P_{21} and P_{12} are the average powers of the desired received signals, P_{11} and P_{22} are the average powers of the SI, $\mathbf{n}_1$, and $\mathbf{n}_2$ are the Gaussian white noise vectors with $\mathbb{E}\{\mathbf{n}_1\mathbf{n}_1^{\mathrm{H}}\} = \mathbf{I}_{n_{r1}}$ and $\mathbb{E}\{\mathbf{n}_2\mathbf{n}_2^{\mathrm{H}}\} = \mathbf{I}_{n_{r2}}$, respectively,

$\mathbf{w}_{t1}$ and $\mathbf{w}_{t2}$ are the transmit AWVs, and $\mathbf{w}_{r1}$ and $\mathbf{w}_{r2}$ are the receive AWVs. The two-norms of all these AWVs are normalized to 1.

With the transmit and receive AWVs, the JAR can be expressed as

$$R = \log_2 \left(1 + \frac{P_{21} |\mathbf{w}_{r1}^H \mathbf{H}_{21} \mathbf{w}_{t2}|^2}{1 + P_{11} |\mathbf{w}_{r1}^H \mathbf{H}_{11} \mathbf{w}_{t1}|^2} \right) + \log_2 \left(1 + \frac{P_{12} |\mathbf{w}_{r2}^H \mathbf{H}_{12} \mathbf{w}_{t1}|^2}{1 + P_{22} |\mathbf{w}_{r2}^H \mathbf{H}_{22} \mathbf{w}_{t2}|^2} \right),$$

(4.13)

where

$$P_{11} |\mathbf{w}_{r1}^H \mathbf{H}_{11} \mathbf{w}_{t1}|^2,$$

(4.14)

and

$$P_{22} |\mathbf{w}_{r2}^H \mathbf{H}_{22} \mathbf{w}_{t2}|^2,$$

(4.15)

are the average powers of SI at Nodes #1 and #2, respectively. The JTR-BF problem is formulated as

$$\underset{\mathbf{w}_{t1}, \mathbf{w}_{r1}, \mathbf{w}_{t2}, \mathbf{w}_{r2}}{\text{maximize}} \quad R \tag{4.16a}$$

$$\text{subject to} \quad \|\mathbf{w}_{t1}\|_2^2 = \|\mathbf{w}_{r1}\|_2^2 = 1, \tag{4.16b}$$

$$\|\mathbf{w}_{t2}\|_2^2 = \|\mathbf{w}_{r2}\|_2^2 = 1. \tag{4.16c}$$

As the main purpose of this section is to investigate the feasibility of FD communications and evaluate the performance of beamforming cancellation, the channel matrices and powers in (4.16) are assumed known a priori. In practice, these parameters can be estimated provided that the channel does not change too fast. For instance, as a communication channel has the feature of directivity and is sparse in the angle domain, schemes like AoD/AoA estimation[35], beam searching[14,25,36], iterative training[6,37], and even compressed sensing[38,39] can be adopted for the communication channel estimation. The estimation of the SI channel can be more straightforward, e.g., one can estimate a scalar channel coefficient between a single Tx/Rx antenna pair (i.e., the i-th transmit antenna and the j-th receive antenna) once at a time. After $n_{ri}n_{ti}$, $i = 1, 2$, measurements, the SI channel matrix can be estimated. Although $n_{ri}n_{ti}$ is large in communication systems, each measurement may require only one symbol thanks to the high strength of SI, rather than a long training sequence like those in[6,14,25,36,37,38]. Thus, the time cost of the SI channel estimation is yet affordable.

In addition, in our model both amplitudes and phases of the AWVs are controllable. Although CM array may be more common for the beamforming in communication due to a lower complexity[3] [6,14,25,36,38], the array structure without the CM constraint cannot be definitely ruled out, because it also arouses particular attention in both algorithm design [12,15,40,41] and implementation[42]. Moreover, the performance achieved with non-CM-constraint arrays can be seen as a bound achieved with the CM arrays, if the same scheme is adopted. Therefore, in this subsection we adopt the arrays without the CM constraint to investigate the feasibility of FD communication and evaluate the beamforming performance, letting the problem with CM constraint be further work.

---

[3] A CM array is, in fact, a phased array, where only the phases of the antenna weights are controllable as have been introduced in Chapter II.

4.2.3.1 The ZF-Max-Power Approach

It is clear that the JAR in (4.16) is not a concave function, and the equality constraints are not affine. Thus, (4.16) is not a convex/concave problem, and its globally optimal solution is hard to find. Consequently, we first give an upper bound JAR, and then propose the ZF-Max-Power approach in this section.

1) Upper Bound of the JAR

According to (4.13), since

$$P_{11}|\mathbf{w}_{r1}^{H}\mathbf{H}_{11}\mathbf{w}_{t1}|^{2} \geq 0, \tag{4.17}$$

and

$$P_{22}|\mathbf{w}_{r2}^{H}\mathbf{H}_{22}\mathbf{w}_{t2}|^{2} \geq 0, \tag{4.18}$$

an *upper bound function* on the JAR, denoted by R_{ub}, in the presence of SI can be easily obtained as

$$R \leq \log_2\left(1 + P_{21}|\mathbf{w}_{r1}^{H}\mathbf{H}_{21}\mathbf{w}_{t2}|^{2}\right) + \log_2\left(1 + P_{12}|\mathbf{w}_{r2}^{H}\mathbf{H}_{12}\mathbf{w}_{t1}|^{2}\right) \triangleq R_{\mathrm{ub}}. \tag{4.19}$$

The corresponding optimal AWVs for R_{ub} are

$$\begin{aligned} \mathbf{w}_{t1} &= \mathrm{RpSingVect}(\mathbf{H}_{12}), \ \mathbf{w}_{t2} = \mathrm{RpSingVect}(\mathbf{H}_{21}), \\ \mathbf{w}_{r1} &= \mathrm{LpSingVect}(\mathbf{H}_{21}), \ \mathbf{w}_{r2} = \mathrm{LpSingVect}(\mathbf{H}_{11}), \end{aligned} \tag{4.20}$$

where $\mathrm{LpSingVect}(\mathbf{X})$ and $\mathrm{RpSingVect}(\mathbf{X})$ represent the left and right principal singular vectors of $\mathbf{X}$, respectively. Thus, we have

$$R \leq R_{\mathrm{ub}} \leq R_{\mathrm{ub}}^{\star} < \infty, \tag{4.21}$$

where $R_{\mathrm{ub}}^{\star}$ is the maximum of the upper bound R_{ub}, and thus it is an upper bound on R.

2) The ZF-Max-Power Approach

As there are multiple coupled variables in (4.16), we consider using the alternating optimization approach[43] to obtain a sub-optimal solution of (4.16). The basic idea of alternating optimization (AO) is to alternately optimize a few parameters, and assume the other parameters being fixed and known[43]. In each round, a sub-problem with a few parameters is formulated and solved. Generally, this approach requires that the optimal solution to each sub-problem can be found. However, for the problem in (4.16), the AO approach cannot be directly used, because, as we can see, even given $\mathbf{w}_{r1}$ and $\mathbf{w}_{r2}$, the optimal $\mathbf{w}_{t1}$ and $\mathbf{w}_{t2}$ still cannot be easily found.

To make the AO approach feasible, we present the ZF-Max-Power scheme in this subsection. The motivation of this scheme is as follows. Since the SI is usually significant in FD communication, we can force the SI to zero and maximize the signal power. In particular, we add a constraint that the SI is completely mitigated. In such a case, we have $R = R_{ub}$, and the problem becomes

$$\underset{\mathbf{w}_{t1}, \mathbf{w}_{r1}, \mathbf{w}_{t2}, \mathbf{w}_{r2}}{\text{maximize}} \quad R_{ub} \tag{4.22a}$$

$$\text{subject to} \quad \|\mathbf{w}_{t1}\|_2^2 = \|\mathbf{w}_{r1}\|_2^2 = 1, \tag{4.22b}$$

$$\|\mathbf{w}_{t2}\|_2^2 = \|\mathbf{w}_{r2}\|_2^2 = 1, \tag{4.22c}$$

$$\mathbf{w}_{r1}^H \mathbf{H}_{11} \mathbf{w}_{t1} = \mathbf{w}_{r2}^H \mathbf{H}_{22} \mathbf{w}_{t2} = 0. \tag{4.22d}$$

Since we have added a new constraint to the original Problem (4.16), the solution of the new Problem (4.22) is sub-optimal to the original problem. To solve the new Problem (4.22), we need the following result.

Lemma 4.2.1. *Given an arbitrary set of linearly independent vectors $\{\mathbf{a}_i \in \mathbb{C}^{L \times 1}\}_{i=1,2,\ldots,N; \ N<L}$ and an arbitrary vector $\mathbf{a} \in \mathbb{C}^{L \times 1}$ that does not belong to the subspace spanned by $\{\mathbf{a}_i\}_{i=1,2,\cdots,N}$, the vector $\mathbf{b} \in \mathbb{C}^{L \times 1}$ that maximizes $|\langle \mathbf{b}, \mathbf{a} \rangle|^2$ with the constraints $\langle \mathbf{b}, \mathbf{a}_i \rangle = 0|_{i=1,2,\ldots,N}$ and $\|\mathbf{b}\|_2 = 1$ is[4]*

$$\mathbf{b} = \frac{\mathbf{a} - \sum_{i=1}^{N} \langle \mathbf{a}, \mathbf{v}_i \rangle \mathbf{v}_i}{\|\mathbf{a} - \sum_{i=1}^{N} \langle \mathbf{a}, \mathbf{v}_i \rangle \mathbf{v}_i\|_2}, \tag{4.23}$$

with $\mathbf{v}_1 = \mathbf{a}_1 / \|\mathbf{a}_1\|_2$ and

$$\mathbf{v}_i = \frac{\mathbf{a}_i - \sum_{j=1}^{i-1} \langle \mathbf{a}_i, \mathbf{v}_j \rangle \mathbf{v}_j}{\|\mathbf{a}_i - \sum_{j=1}^{i-1} \langle \mathbf{a}_i, \mathbf{v}_j \rangle \mathbf{v}_j\|_2}, \quad i > 1. \tag{4.24}$$

Proof. Since $\{\mathbf{a}_i \in \mathbb{C}^{L \times 1}\}_{i=1,2,\ldots,N; \ N<L}$ is a group of linear independent vectors, they can span an N-dimensional linear subspace denoted by $\mathbb{S} \triangleq \mathcal{S}\{\mathbf{a}_i|_{i=1,2,\ldots,N}\}$. Suppose $\{\mathbf{v}_i\}_{i=1,2,\ldots,N}$ is an orthogonal basis of the linear subspace $\mathbb{S}$, which can be easily obtained by using Gram-Schmidt orthogonalization shown in (4.24). Then constraint $\langle \mathbf{b}, \mathbf{a}_i \rangle = 0|_{i=1,2,\ldots,N}$ is equivalent to $\langle \mathbf{b}, \mathbf{v}_i \rangle = 0|_{i=1,2,\ldots,N}$.

Then, we can extend the set of vectors $\{\mathbf{v}_i\}_{i=1,2,\ldots,N}$ to an orthogonal basis of the L-dimensional linear space spanned by $\{\mathbf{v}_i\}_{i=1,2,\ldots,L}$. Vectors $\mathbf{a}, \mathbf{b} \in \mathbb{C}^{L \times 1}$ can be expressed as linear combinations of basis $\{\mathbf{v}_i\}_{i=1,2,\ldots,L}$ as

$$\begin{aligned} \mathbf{a} &= \alpha_1 \mathbf{v}_1 + \alpha_2 \mathbf{v}_2 + \ldots + \alpha_L \mathbf{v}_L, \\ \mathbf{b} &= \beta_1 \mathbf{v}_1 + \beta_2 \mathbf{v}_2 + \ldots + \beta_L \mathbf{v}_L, \end{aligned} \tag{4.25}$$

with

$$\begin{cases} \alpha_i = \langle \mathbf{a}, \mathbf{v}_i \rangle, \ i = 1, 2, \ldots, L, \\ \beta_i = \langle \mathbf{b}, \mathbf{v}_i \rangle, \ i = 1, 2, \ldots, L. \end{cases} \tag{4.26}$$

---

[4]An implicit assumption is that $\|\mathbf{a} - \sum_{i=1}^{N} \langle \mathbf{a}, \mathbf{v}_i \rangle \mathbf{v}_i\|_2$ is always not equal to zero. Otherwise, only the vector with all the elements being zero can satisfy constraints $\langle \mathbf{b}, \mathbf{a}_i \rangle = 0|_{i=1,2,\ldots,N}$, but $\|\mathbf{b}\|_2 = 1$ cannot be satisfied.

With the constraint $\langle \mathbf{b}, \mathbf{v}_i \rangle = 0|_{i=1,2,\ldots,N}$, we can obtain

$$\beta_1 = \beta_2 = \cdots = \beta_N = 0. \tag{4.27}$$

Therefore, vector $\mathbf{b}$ can be simplified as

$$\mathbf{b} = \beta_{N+1}\mathbf{v}_{N+1} + \beta_{N+2}\mathbf{v}_{N+2} + \ldots + \beta_L\mathbf{v}_L. \tag{4.28}$$

Then, $|\langle \mathbf{b}, \mathbf{a} \rangle|^2$ can be expressed as

$$
\begin{aligned}
&|\langle \mathbf{b}, \mathbf{a} \rangle|^2 \\
&= |(\beta_{N+1}\mathbf{v}_{N+1} + \beta_{N+2}\mathbf{v}_{N+2} + \ldots + \beta_L\mathbf{v}_L)^{\mathrm{H}}(\alpha_1\mathbf{v}_1 + \alpha_2\mathbf{v}_2 + \ldots + \alpha_L\mathbf{v}_L)|^2 \\
&= |\beta^*_{N+1}\alpha_{N+1} + \beta^*_{N+2}\alpha_{N+2} + \cdots + \beta^*_L\alpha_L|^2,
\end{aligned}
$$
$$\tag{4.29}$$

which can be maximized by using the well-known maximal ratio combining, i.e.,

$$
\begin{cases}
\beta_{N+1} = k\alpha_{N+1} = k\langle \mathbf{a}, \mathbf{v}_{N+1} \rangle, \\
\beta_{N+2} = k\alpha_{N+2} = k\langle \mathbf{a}, \mathbf{v}_{N+2} \rangle, \\
\ldots \\
\beta_L = k\alpha_L = k\langle \mathbf{a}, \mathbf{v}_L \rangle,
\end{cases}
\tag{4.30}
$$

where k is a constant which ensures a unit norm for $\mathbf{b}$, and thus, vector $\mathbf{b}$ can be written as

$$
\begin{aligned}
\mathbf{b} &= k\langle \mathbf{a}, \mathbf{v}_{N+1} \rangle \mathbf{v}_{N+1} + k\langle \mathbf{a}, \mathbf{v}_{N+2} \rangle \mathbf{v}_{N+2} + \ldots + k\langle \mathbf{a}, \mathbf{v}_L \rangle \mathbf{v}_L \\
&= k\mathbf{a} - \sum_{i=1}^{N} k\langle \mathbf{a}, \mathbf{v}_i \rangle \mathbf{v}_i.
\end{aligned}
\tag{4.31}
$$

According to constraint $\|\mathbf{b}\|_2 = 1$, we can obtain

$$\mathbf{b} = \frac{\mathbf{a} - \sum_{i=1}^{N} \langle \mathbf{a}, \mathbf{v}_i \rangle \mathbf{v}_i}{\|\mathbf{a} - \sum_{i=1}^{N} \langle \mathbf{a}, \mathbf{v}_i \rangle \mathbf{v}_i\|_2}. \tag{4.32}$$

This completes the proof.

$\square$

Specifically, when $N = 1$ the optimal $\mathbf{b}$ in Lemma 4.2.1 before normalization is

$$\mathbf{b}^\star = \mathbf{a} - \langle \mathbf{a}, \frac{\mathbf{a}_1}{\|\mathbf{a}_1\|_2} \rangle \frac{\mathbf{a}_1}{\|\mathbf{a}_1\|_2}, \tag{4.33}$$

and when $N = 2$ the optimal $\mathbf{b}$ in Lemma 4.2.1 before normalization is

$$\mathbf{b}^\star = \mathbf{a} - \langle \mathbf{a}, \frac{\mathbf{a}_1}{\|\mathbf{a}_1\|_2} \rangle \frac{\mathbf{a}_1}{\|\mathbf{a}_1\|_2} - \langle \mathbf{a}, \mathbf{b}_2 \rangle \mathbf{b}_2, \tag{4.34}$$

where

$$\mathbf{b}_2 = \frac{\mathbf{a}_2 - \langle \mathbf{a}_2, \frac{\mathbf{a}_1}{\|\mathbf{a}_1\|_2} \rangle \frac{\mathbf{a}_1}{\|\mathbf{a}_1\|_2}}{\|\mathbf{a}_2 - \langle \mathbf{a}_2, \frac{\mathbf{a}_1}{\|\mathbf{a}_1\|_2} \rangle \frac{\mathbf{a}_1}{\|\mathbf{a}_1\|_2}\|_2}. \tag{4.35}$$

Algorithm 4.1: The ZF-Max-Power Scheme.
---

1) Initialize:

Initialize the transmit AWVs as
$\mathbf{w}_{t1} = \text{RpSingVect}(\mathbf{H}_{12})$, $\mathbf{w}_{t2} = \text{RpSingVect}(\mathbf{H}_{21})$.

2) Iteration:

Iterate the following process N_S times, then stop.

Compute $\mathbf{w}_{r1}$ and $\mathbf{w}_{r2}$ according to (4.36) and (4.37), respectively; then normalize them.

Compute $\mathbf{w}_{t1}$ and $\mathbf{w}_{t2}$ according to (4.38) and (4.39), respectively; then normalize them.

3) Result:

$\mathbf{w}_{t1}$ and $\mathbf{w}_{t2}$ are the transmit AWVs, and $\mathbf{w}_{r1}$ and $\mathbf{w}_{r2}$ are the receive AWVs.

Let us go back to the ZF-Max-Power scheme. According to (4.33), given $\mathbf{w}_{t1}$ and $\mathbf{w}_{t2}$, the optimal $\mathbf{w}_{r1}$ and $\mathbf{w}_{r2}$ for (4.22) (before normalization) are

$$\mathbf{w}_{r1} = \mathbf{H}_{21}\mathbf{w}_{t2} - \langle \mathbf{H}_{21}\mathbf{w}_{t2}, \frac{\mathbf{H}_{11}\mathbf{w}_{t1}}{\|\mathbf{H}_{11}\mathbf{w}_{t1}\|_2} \rangle \frac{\mathbf{H}_{11}\mathbf{w}_{t1}}{\|\mathbf{H}_{11}\mathbf{w}_{t1}\|_2}, \tag{4.36}$$

and

$$\mathbf{w}_{r2} = \mathbf{H}_{12}\mathbf{w}_{t1} - \langle \mathbf{H}_{12}\mathbf{w}_{t1}, \frac{\mathbf{H}_{22}\mathbf{w}_{t2}}{\|\mathbf{H}_{22}\mathbf{w}_{t2}\|_2} \rangle \frac{\mathbf{H}_{22}\mathbf{w}_{t2}}{\|\mathbf{H}_{22}\mathbf{w}_{t2}\|_2}, \tag{4.37}$$

respectively.

Similarly, given $\mathbf{w}_{r1}$ and $\mathbf{w}_{r2}$, the optimal $\mathbf{w}_{t1}$ and $\mathbf{w}_{t2}$ for (4.22) (before normalization) are

$$\mathbf{w}_{t1} = \mathbf{H}_{12}^{\mathrm{H}}\mathbf{w}_{r2} - \langle \mathbf{H}_{12}^{\mathrm{H}}\mathbf{w}_{r2}, \frac{\mathbf{H}_{11}^{\mathrm{H}}\mathbf{w}_{r1}}{\|\mathbf{H}_{11}^{\mathrm{H}}\mathbf{w}_{r1}\|_2} \rangle \frac{\mathbf{H}_{11}^{\mathrm{H}}\mathbf{w}_{r1}}{\|\mathbf{H}_{11}^{\mathrm{H}}\mathbf{w}_{r1}\|_2}, \tag{4.38}$$

and

$$\mathbf{w}_{t2} = \mathbf{H}_{21}^{\mathrm{H}}\mathbf{w}_{r1} - \langle \mathbf{H}_{21}^{\mathrm{H}}\mathbf{w}_{r1}, \frac{\mathbf{H}_{22}^{\mathrm{H}}\mathbf{w}_{r2}}{\|\mathbf{H}_{22}^{\mathrm{H}}\mathbf{w}_{r2}\|_2} \rangle \frac{\mathbf{H}_{22}^{\mathrm{H}}\mathbf{w}_{r2}}{\|\mathbf{H}_{22}^{\mathrm{H}}\mathbf{w}_{r2}\|_2}, \tag{4.39}$$

respectively. Finally, the ZF-Max-Power scheme can be summarized as in Algorithm 4.1.

It is noted that there are various stopping rules for the ZF-Max-Power scheme. A simple one is to stop after a certain number of iterations, e.g., N_S iterations used in Algorithms 4.1. Another one is to compute R_{ub} after the n-th iteration and get $R_{\text{ub}}^{(n)}$ according to (4.13). When

$$R_{\text{ub}}^{(n)} / R_{\text{ub}}^{(n-1)} < \mu, \tag{4.40}$$

stop the iteration, where μ is a predefined threshold slightly greater than 1, e.g., $\mu = 1.05$.

3) Convergence Analysis and Complexity Comparison

To prove the convergence of the ZF-Max-Power scheme, we need the following theorem.

Theorem 4.2.1. *Let $R^{(n)}$ denote the value of R after the n-th iteration. Then $\{R^{(n)}|n = 1, 2, ...\}$ is a non-descending sequence, i.e., $R^{(n+1)} \geq R^{(n)}$.*

Proof. Under the constraints in (4.22), the JAR becomes

$$R(\mathbf{w}_{t1}, \mathbf{w}_{t2}, \mathbf{w}_{r1}, \mathbf{w}_{r2})$$
$$= \log_2\left(1 + \varepsilon_{21}|\mathbf{w}_{r1}^H \mathbf{H}_{21} \mathbf{w}_{t2}|^2\right) + \log_2\left(1 + \varepsilon_{12}|\mathbf{w}_{r2}^H \mathbf{H}_{12} \mathbf{w}_{t1}|^2\right) \qquad (4.41)$$
$$\triangleq R_{ub},$$

which shows that R is a function of $\mathbf{w}_{t1}$, $\mathbf{w}_{t2}$, $\mathbf{w}_{r1}$, and $\mathbf{w}_{r2}$.

Let $\mathbf{w}_{t1}^{(n)}$ and $\mathbf{w}_{t2}^{(n)}$ denote the transmit AWVs after the n-th iteration, and $\mathbf{w}_{r1}^{(n)}$ and $\mathbf{w}_{r2}^{(n)}$ the receive AWVs after the n-th iteration. Then we have

$$\begin{cases} R^{(n)} = R(\mathbf{w}_{t1}^{(n)}, \mathbf{w}_{t2}^{(n)}, \mathbf{w}_{r1}^{(n)}, \mathbf{w}_{r2}^{(n)}), \\ R^{(n+1)} = R(\mathbf{w}_{t1}^{(n+1)}, \mathbf{w}_{t2}^{(n+1)}, \mathbf{w}_{r1}^{(n+1)}, \mathbf{w}_{r2}^{(n+1)}). \end{cases} \qquad (4.42)$$

According to Algorithms 4.1, as well as (4.36) and (4.37), we have

$$R(\mathbf{w}_{t1}^{(n)}, \mathbf{w}_{t2}^{(n)}, \mathbf{w}_{r1}^{(n+1)}, \mathbf{w}_{r2}^{(n+1)}) \geq R(\mathbf{w}_{t1}^{(n)}, \mathbf{w}_{t2}^{(n)}, \mathbf{w}_{r1}^{(n)}, \mathbf{w}_{r2}^{(n)}) = R^{(n)}. \qquad (4.43)$$

Also, according to Algorithms 4.1, as well as (4.38) and (4.39), we have

$$R(\mathbf{w}_{t1}^{(n+1)}, \mathbf{w}_{t2}^{(n+1)}, \mathbf{w}_{r1}^{(n+1)}, \mathbf{w}_{r2}^{(n+1)}) = R^{(n+1)}$$
$$\geq R(\mathbf{w}_{t1}^{(n)}, \mathbf{w}_{t2}^{(n)}, \mathbf{w}_{r1}^{(n+1)}, \mathbf{w}_{r2}^{(n+1)}). \qquad (4.44)$$

Therefore, we have $R^{(n+1)} \geq R^{(n)}$. □

According to (4.21), $R \leq R_{ub}^\star < \infty$. Thus, $\{R^{(n)}|n = 1, 2, ...\}$ converges to a sub-optimal value, which guarantees the convergence of ZF-Max-Power.

On the other hand, the computational complexity of ZF-Max-Power is much lower than that of Max-SINR[34]. For simplicity, suppose $n_{t1} = n_{t2} = n_{r1} = n_{r2} = N$. According to Algorithm 4.1, in each iteration the main complexity of ZF-Max-Power lies in multiplications between a channel matrix and an AWV, as shown in (4.36), (4.37), (4.38), and (4.39). Hence, the computational complexity of ZF-Max-Power is roughly $\mathcal{O}(N^2)$, because there are about N^2 scalar multiplications for a multiplication of a channel matrix and an AWV. In contrast, according to[34] matrix inversion is required in each iteration of Max-SINR. Hence, Max-SINR has a computational complexity of $\mathcal{O}(N^3)$, which is significantly higher than that of ZF-Max-Power, especially in FD communication where N is large.

4.2.3.2 Closed-Form Solutions

In this section, we consider different criteria for the JTR-BF problem that provide us with closed-form solutions.

1) LB-MMSE

It is natural to perform receive and transmit beamforming separately when considering closed-form solutions. According to (4.13), an optimal solution can be achieved for receive beamforming by using the MMSE approach, which is equivalent to maximizing the SINR. However, for transmit beamforming, the optimal solution is hard to find. Thus, we show the way to optimize the lower bound on the JAR with MMSE, and the method is referred to as the LB-MMSE approach.

According to (4.13), we have

$$
\begin{aligned}
R &\geq \log_2 \left(\frac{P_{21}|\mathbf{w}_{r1}^{H}\mathbf{H}_{21}\mathbf{w}_{t2}|^2}{\mathbf{w}_{r1}^{H}\mathbf{w}_{r1} + P_{11}|\mathbf{w}_{r1}^{H}\mathbf{H}_{11}\mathbf{w}_{t1}|^2} \frac{P_{12}|\mathbf{w}_{r2}^{H}\mathbf{H}_{12}\mathbf{w}_{t1}|^2}{\mathbf{w}_{r2}^{H}\mathbf{w}_{r2} + P_{22}|\mathbf{w}_{r2}^{H}\mathbf{H}_{22}\mathbf{w}_{t2}|^2} \right) \\
&= \log_2 \left(\frac{P_{21}\mathbf{w}_{r1}^{H}\mathbf{H}_{21}\mathbf{w}_{t2}\mathbf{w}_{t2}^{H}\mathbf{H}_{21}^{H}\mathbf{w}_{r1}}{\mathbf{w}_{r1}^{H}\left(\mathbf{I} + P_{11}\mathbf{H}_{11}\mathbf{w}_{t1}\mathbf{w}_{t1}^{H}\mathbf{H}_{11}^{H}\right)\mathbf{w}_{r1}} \frac{P_{12}\mathbf{w}_{r2}^{H}\mathbf{H}_{12}\mathbf{w}_{t1}\mathbf{w}_{t1}^{H}\mathbf{H}_{12}^{H}\mathbf{w}_{r2}}{\mathbf{w}_{r2}^{H}\left(\mathbf{I} + P_{22}\mathbf{H}_{22}\mathbf{w}_{t2}\mathbf{w}_{t2}^{H}\mathbf{H}_{22}^{H}\right)\mathbf{w}_{r2}} \right) \\
&\triangleq R_1.
\end{aligned}
$$

(4.45)

First, we find $\mathbf{w}_{r1}$ and $\mathbf{w}_{r2}$ to maximize the lower bound R_1 by temporally treating $\mathbf{w}_{t1}$ and $\mathbf{w}_{t2}$ as fixed parameters. This sub-problem is actually to maximize SINR (or minimize MSE) at the two nodes.

Lemma 4.2.2. *Given a positive define Hermitian matrix* $\mathbf{R} \in \mathbb{C}^{M \times M}$ *and a vector* $\mathbf{a}$, *the solution to maximize*

$$
\frac{\mathbf{x}^{H}\mathbf{a}\mathbf{a}^{H}\mathbf{x}}{\mathbf{x}^{H}\mathbf{R}\mathbf{x}},
$$

(4.46)

over $\mathbf{x}$ *is* $\mathbf{x} = \mathbf{R}^{-1}\mathbf{a}$.

Proof. Since $\mathbf{R}$ is a positive define Hermitian matrix, $\mathbf{R}^{1/2}$ is also an invertible Hermitian matrix. Thus, we have

$$
\frac{\mathbf{x}^{H}\mathbf{a}\mathbf{a}^{H}\mathbf{x}}{\mathbf{x}^{H}\mathbf{R}\mathbf{x}} = \frac{\mathbf{x}^{H}\mathbf{a}\mathbf{a}^{H}\mathbf{x}}{\left(\mathbf{R}^{1/2}\mathbf{x}\right)^{H}\left(\mathbf{R}^{1/2}\mathbf{x}\right)}.
$$

(4.47)

Let $\tilde{\mathbf{x}} = \mathbf{R}^{1/2}\mathbf{x}$, the problem becomes to maximize

$$
\frac{\tilde{\mathbf{x}}^{H}\left(\mathbf{R}^{-H/2}\mathbf{a}\right)\left(\mathbf{R}^{-H/2}\mathbf{a}\right)^{H}\tilde{\mathbf{x}}}{\tilde{\mathbf{x}}^{H}\tilde{\mathbf{x}}}.
$$

(4.48)

By using Cauchy-Schwarz inequality, we can get

$$
\frac{\tilde{\mathbf{x}}^{H}\left(\mathbf{R}^{-H/2}\mathbf{a}\right)\left(\mathbf{R}^{-H/2}\mathbf{a}\right)^{H}\tilde{\mathbf{x}}}{\tilde{\mathbf{x}}^{H}\tilde{\mathbf{x}}} \leq \frac{\|\tilde{\mathbf{x}}\|_2^2 \|\mathbf{R}^{-H/2}\mathbf{a}\|_2^2}{\tilde{\mathbf{x}}^{H}\tilde{\mathbf{x}}},
$$

(4.49)

and the equality holds if and only if

$$\tilde{\mathbf{x}} = \mathbf{R}^{-1/2}\mathbf{a}, \tag{4.50}$$

i.e., the ratio is maximized for $\tilde{\mathbf{x}} = \mathbf{R}^{-1/2}\mathbf{a}$.

Therefore, the solution is $\tilde{\mathbf{x}} = \mathbf{R}^{-H/2}\mathbf{a} = \mathbf{R}^{-1/2}\mathbf{a}$, i.e., $\mathbf{x} = \mathbf{R}^{-1}\mathbf{a}$. $\qquad\square$

According to Lemma 4.2.2, the optimal $\mathbf{w}_{r1}$ and $\mathbf{w}_{r2}$ (before normalization) can be, respectively, found as

$$
\begin{aligned}
\mathbf{w}_{r1} &= \left(\mathbf{I} + P_{11}\mathbf{H}_{11}\mathbf{w}_{t1}\mathbf{w}_{t1}^{H}\mathbf{H}_{11}^{H}\right)^{-1}\mathbf{H}_{21}\mathbf{w}_{t2}, \\
\mathbf{w}_{r2} &= \left(\mathbf{I} + P_{22}\mathbf{H}_{22}\mathbf{w}_{t2}\mathbf{w}_{t2}^{H}\mathbf{H}_{22}^{H}\right)^{-1}\mathbf{H}_{12}\mathbf{w}_{t1}.
\end{aligned}
\tag{4.51}
$$

With these two receive AWVs, we further have (4.58), where inequalities (a) and (b) are based on Lemmas 4.2.3 and 4.2.4, respectively.

Lemma 4.2.3. *Given a positive define Hermitian matrix $\mathbf{R} \in \mathbb{C}^{M \times M}$ and a vector $\mathbf{a}$,*

$$\mathbf{a}^{H}\mathbf{R}^{-1}\mathbf{a} \geq \frac{(\mathbf{a}^{H}\mathbf{a})^2}{\mathbf{a}^{H}\mathbf{R}\mathbf{a}}. \tag{4.52}$$

Proof. Since $\mathbf{R} \in \mathbb{C}^{M \times M}$ is a positive define Hermitian matrix, it can be expressed as

$$\mathbf{R} = \sum_{i=1}^{M} \lambda_i \mathbf{v}_i \mathbf{v}_i^{H}, \tag{4.53}$$

where $\lambda_i > 0$ and $\{\mathbf{v}_i\}_{i=1,2,\dots,M}$ constitutes an orthogonal base in $\mathbb{C}^{M \times M}$. Thus, $\mathbf{a}$ can be expressed as

$$\mathbf{a} = \sum_{i=1}^{M} \alpha_i \mathbf{v}_i. \tag{4.54}$$

Besides, we have

$$\mathbf{R}^{-1} = \sum_{i=1}^{M} \frac{1}{\lambda_i} \mathbf{v}_i \mathbf{v}_i^{H}. \tag{4.55}$$

Thus,

$$
\begin{aligned}
\left(\mathbf{a}^{H}\mathbf{R}^{-1}\mathbf{a}\right)\left(\mathbf{a}^{H}\mathbf{R}\mathbf{a}\right) &= \left(\sum_{i=1}^{M} \frac{|\alpha_i|^2}{\lambda_i}\right)\left(\sum_{i=1}^{M} |\alpha_i|^2 \lambda_i\right) \\
&= \sum_{i=1}^{M} |\alpha_i|^4 + \sum_{i=1}^{M}\sum_{j=1}^{i-1} |\alpha_i|^2 |\alpha_j|^2 \left(\frac{\lambda_j}{\lambda_i} + \frac{\lambda_i}{\lambda_j}\right) \\
&\geq \sum_{i=1}^{M} |\alpha_i|^4 + 2\sum_{i=1}^{M}\sum_{j=1}^{i-1} |\alpha_i|^2 |\alpha_j|^2 \\
&= (\mathbf{a}^{H}\mathbf{a})^2.
\end{aligned}
\tag{4.56}
$$

$\qquad\square$

Lemma 4.2.4. *Given a positive define Hermitian matrix* $\mathbf{R} \in \mathbb{C}^{M \times M}$ *and a vector* $\mathbf{a}$ *with the formulation* $\mathbf{R} = \mathbf{I} + \varepsilon \mathbf{b} \mathbf{b}^{\mathrm{H}}$,

$$\frac{\mathbf{a}^{\mathrm{H}} \mathbf{a}}{\mathbf{a}^{\mathrm{H}} \mathbf{R} \mathbf{a}} \geq \frac{1}{1 + \varepsilon \mathbf{b}^{\mathrm{H}} \mathbf{b}}. \tag{4.57}$$

Proof. Let $\mathbf{v} = \mathbf{a}/\|\mathbf{a}\|_2$, and we have

$$
\begin{aligned}
R_1 =& \log_2 \left(P_{12} P_{21} \left(\mathbf{w}_{t2}^{\mathrm{H}} \mathbf{H}_{21}^{\mathrm{H}} \left(\mathbf{I} + P_{11} \mathbf{H}_{11} \mathbf{w}_{t1} \mathbf{w}_{t1}^{\mathrm{H}} \mathbf{H}_{11}^{\mathrm{H}} \right)^{-1} \mathbf{H}_{21} \mathbf{w}_{t2} \right) \right) + \\
& \log_2 \left(\mathbf{w}_{t1}^{\mathrm{H}} \mathbf{H}_{12}^{\mathrm{H}} \left(\mathbf{I} + P_{22} \mathbf{H}_{22} \mathbf{w}_{t2} \mathbf{w}_{t2}^{\mathrm{H}} \mathbf{H}_{22}^{\mathrm{H}} \right)^{-1} \mathbf{H}_{12} \mathbf{w}_{t1} \right) \\
\overset{(a)}{\geq}& \log_2 \left(P_{12} P_{21} \frac{\left(\mathbf{w}_{t2}^{\mathrm{H}} \mathbf{H}_{21}^{\mathrm{H}} \mathbf{H}_{21} \mathbf{w}_{t2} \right)^2}{\mathbf{w}_{t2}^{\mathrm{H}} \mathbf{H}_{21}^{\mathrm{H}} \left(\mathbf{I} + P_{11} \mathbf{H}_{11} \mathbf{w}_{t1} \mathbf{w}_{t1}^{\mathrm{H}} \mathbf{H}_{11}^{\mathrm{H}} \right) \mathbf{H}_{21} \mathbf{w}_{t2}} \right) + \\
& \log_2 \left(\frac{\left(\mathbf{w}_{t1}^{\mathrm{H}} \mathbf{H}_{12}^{\mathrm{H}} \mathbf{H}_{12} \mathbf{w}_{t1} \right)^2}{\mathbf{w}_{t1}^{\mathrm{H}} \mathbf{H}_{12}^{\mathrm{H}} \left(\mathbf{I} + P_{22} \mathbf{H}_{22} \mathbf{w}_{t2} \mathbf{w}_{t2}^{\mathrm{H}} \mathbf{H}_{22}^{\mathrm{H}} \right) \mathbf{H}_{12} \mathbf{w}_{t1}} \right) \\
\overset{(b)}{\geq}& \log_2 \left(P_{12} P_{21} \frac{\mathbf{w}_{t2}^{\mathrm{H}} \mathbf{H}_{21}^{\mathrm{H}} \mathbf{H}_{21} \mathbf{w}_{t2}}{\mathbf{w}_{t1}^{\mathrm{H}} \left(\mathbf{I} + P_{11} \mathbf{H}_{11}^{\mathrm{H}} \mathbf{H}_{11} \right) \mathbf{w}_{t1}} \frac{\mathbf{w}_{t1}^{\mathrm{H}} \mathbf{H}_{12}^{\mathrm{H}} \mathbf{H}_{12} \mathbf{w}_{t1}}{\mathbf{w}_{t2}^{\mathrm{H}} \left(\mathbf{I} + P_{22} \mathbf{H}_{22}^{\mathrm{H}} \mathbf{H}_{22} \right) \mathbf{w}_{t2}} \right) \\
=& \log_2 \left(P_{12} P_{21} \frac{\mathbf{w}_{t2}^{\mathrm{H}} \mathbf{H}_{21}^{\mathrm{H}} \mathbf{H}_{21} \mathbf{w}_{t2}}{\mathbf{w}_{t2}^{\mathrm{H}} \left(\mathbf{I} + P_{22} \mathbf{H}_{22}^{\mathrm{H}} \mathbf{H}_{22} \right) \mathbf{w}_{t2}} \frac{\mathbf{w}_{t1}^{\mathrm{H}} \mathbf{H}_{12}^{\mathrm{H}} \mathbf{H}_{12} \mathbf{w}_{t1}}{\mathbf{w}_{t1}^{\mathrm{H}} \left(\mathbf{I} + P_{11} \mathbf{H}_{11}^{\mathrm{H}} \mathbf{H}_{11} \right) \mathbf{w}_{t1}} \right) \triangleq R_2.
\end{aligned}
\tag{4.58}
$$

$$
\begin{aligned}
\frac{\mathbf{a}^{\mathrm{H}} \mathbf{a}}{\mathbf{a}^{\mathrm{H}} \mathbf{R} \mathbf{a}} &= \frac{\mathbf{a}^{\mathrm{H}} \mathbf{a}}{\mathbf{a}^{\mathrm{H}} \left(\mathbf{I} + \varepsilon \mathbf{b} \mathbf{b}^{\mathrm{H}} \right) \mathbf{a}} = \frac{1}{\mathbf{v}^{\mathrm{H}} \left(\mathbf{I} + \varepsilon \mathbf{b} \mathbf{b}^{\mathrm{H}} \right) \mathbf{v}} \\
&= \frac{1}{1 + \varepsilon |\mathbf{b}^{\mathrm{H}} \mathbf{v}|^2} \geq \frac{1}{1 + \varepsilon |\frac{\mathbf{b}^{\mathrm{H}} \mathbf{b}}{\|\mathbf{b}\|_2}|^2} = \frac{1}{1 + \varepsilon \mathbf{b}^{\mathrm{H}} \mathbf{b}}.
\end{aligned}
\tag{4.59}
$$

□

Next, let us find $\mathbf{w}_{t1}$ and $\mathbf{w}_{t2}$ to maximize the lower bound R_2. This sub-problem is equivalent to the following two optimization problems:

$$\max_{\mathbf{w}_{t1}} \frac{\mathbf{w}_{t1}^{\mathrm{H}} \mathbf{H}_{12}^{\mathrm{H}} \mathbf{H}_{12} \mathbf{w}_{t1}}{\mathbf{w}_{t1}^{\mathrm{H}} \left(\mathbf{I} + P_{11} \mathbf{H}_{11}^{\mathrm{H}} \mathbf{H}_{11} \right) \mathbf{w}_{t1}} \tag{4.60}$$

and

$$\max_{\mathbf{w}_{t2}} \frac{\mathbf{w}_{t2}^{\mathrm{H}} \mathbf{H}_{21}^{\mathrm{H}} \mathbf{H}_{21} \mathbf{w}_{t2}}{\mathbf{w}_{t2}^{\mathrm{H}} \left(\mathbf{I} + P_{22} \mathbf{H}_{22}^{\mathrm{H}} \mathbf{H}_{22} \right) \mathbf{w}_{t2}}. \tag{4.61}$$

These are generalized Rayleigh quotient problems, and the optimal transmit AWVs (before normalization) are

$$\mathbf{w}_{t1} = \text{pEigVect} \left(\left(\mathbf{I} + P_{11} \mathbf{H}_{11}^{\mathrm{H}} \mathbf{H}_{11} \right)^{-1} \mathbf{H}_{12}^{\mathrm{H}} \mathbf{H}_{12} \right) \tag{4.62}$$

and

$$\mathbf{w}_{t2} = \text{pEigVect} \left(\left(\mathbf{I} + P_{22} \mathbf{H}_{22}^H \mathbf{H}_{22} \right)^{-1} \mathbf{H}_{21}^H \mathbf{H}_{21} \right), \qquad (4.63)$$

respectively, where pEigVect($\mathbf{X}$) denotes the principal eigenvector of $\mathbf{X}$.

In brief, by exploiting the proposed LB-MMSE, the transmit AWVs are computed as (4.62) and (4.63), respectively. Based on the transmit AWVs, the receive AWVs (before normalization) are found as (4.51).

2) SI-ZF-MRT

In this scheme, we first consider transmit beamforming by adopting MRT, and then use ZF to suppress the SI for receive beamforming. This approach is referred to as SI-ZF-MRT.

By exploiting MRT for transmit beamforming, we have

$$\mathbf{w}_{t1} = \mathbf{H}_{12}^H \mathbf{w}_{r2}; \ \mathbf{w}_{t2} = \mathbf{H}_{21}^H \mathbf{w}_{r1}. \qquad (4.64)$$

To suppress the SI, we have

$$\mathbf{w}_{r1}^H \mathbf{H}_{11} \mathbf{H}_{12}^H \mathbf{w}_{r2} = 0 = \mathbf{w}_{r2}^H \mathbf{H}_{22} \mathbf{H}_{21}^H \mathbf{w}_{r1}. \qquad (4.65)$$

There are many solutions of $\mathbf{w}_{r1}$ and $\mathbf{w}_{r2}$ for (4.65). Among those, we want to find the receive AWVs to maximize JAR while satisfying (4.65). With (4.64) and (4.65), the JAR becomes

$$R = \log_2 \left(1 + P_{21} |\mathbf{w}_{r1}^H \mathbf{H}_{21} \mathbf{H}_{21}^H \mathbf{w}_{r1}|^2 \right) + \log_2 \left(1 + P_{12} |\mathbf{w}_{r2}^H \mathbf{H}_{12} \mathbf{H}_{12}^H \mathbf{w}_{r2}|^2 \right). \qquad (4.66)$$

There are two options to design $\mathbf{w}_{r1}$ and $\mathbf{w}_{r2}$. One is to first derive $\mathbf{w}_{r1}$ that maximizes (4.66) without considering the constraint (4.65); then derive $\mathbf{w}_{r2}$ to optimize (4.66) with (4.65) satisfied. With this option, we have

$$\mathbf{w}_{r1} = \text{pEigVect} \left(\mathbf{H}_{21} \mathbf{H}_{21}^H \right), \qquad (4.67)$$

and

$$\mathbf{w}_{r2} = \mathbf{a} - \langle \mathbf{a}, \frac{\mathbf{a}_1}{\|\mathbf{a}_1\|_2} \rangle \frac{\mathbf{a}_1}{\|\mathbf{a}_1\|_2} - \langle \mathbf{a}, \mathbf{b}_2 \rangle \mathbf{b}_2, \qquad (4.68)$$

where

$$\mathbf{a} = \text{pEigVect} \left(\mathbf{H}_{12} \mathbf{H}_{12}^H \right),$$
$$\mathbf{a}_1 = \mathbf{H}_{12} \mathbf{H}_{11}^H \mathbf{w}_{r1}, \qquad (4.69)$$

and

$$\mathbf{b}_2 = \frac{\mathbf{H}_{22} \mathbf{H}_{21}^H \mathbf{w}_{r1} - \langle \mathbf{H}_{22} \mathbf{H}_{21}^H \mathbf{w}_{r1}, \frac{\mathbf{a}_1}{\|\mathbf{a}_1\|_2} \rangle \frac{\mathbf{a}_1}{\|\mathbf{a}_1\|_2}}{\|\mathbf{H}_{22} \mathbf{H}_{21}^H \mathbf{w}_{r1} - \langle \mathbf{H}_{22} \mathbf{H}_{21}^H \mathbf{w}_{r1}, \frac{\mathbf{a}_1}{\|\mathbf{a}_1\|_2} \rangle \frac{\mathbf{a}_1}{\|\mathbf{a}_1\|_2} \|_2}. \qquad (4.70)$$

The other one is similar to the first one but with the positions of $\mathbf{w}_{r1}$ and $\mathbf{w}_{r2}$ exchanged. With this option, we have

$$\mathbf{w}_{r2} = \text{pEigVect} \left(\mathbf{H}_{12} \mathbf{H}_{12}^H \right), \qquad (4.71)$$

and

$$\mathbf{w}_{r1} = \mathbf{a} - \langle \mathbf{a}, \frac{\mathbf{a}_1}{\|\mathbf{a}_1\|_2} \rangle \frac{\mathbf{a}_1}{\|\mathbf{a}_1\|_2} - \langle \mathbf{a}, \mathbf{b}_2 \rangle \mathbf{b}_2, \tag{4.72}$$

where

$$\begin{aligned}
\mathbf{a} &= \text{pEigVect}\left(\mathbf{H}_{21}\mathbf{H}_{21}^{\text{H}}\right), \\
\mathbf{a}_1 &= \mathbf{H}_{11}\mathbf{H}_{12}^{\text{H}}\mathbf{w}_{r2},
\end{aligned} \tag{4.73}$$

and

$$\mathbf{b}_2 = \frac{\mathbf{H}_{21}\mathbf{H}_{22}^{\text{H}}\mathbf{w}_{r2} - \langle \mathbf{H}_{21}\mathbf{H}_{22}^{\text{H}}\mathbf{w}_{r2}, \frac{\mathbf{a}_1}{\|\mathbf{a}_1\|_2} \rangle \frac{\mathbf{a}_1}{\|\mathbf{a}_1\|_2}}{\|\mathbf{H}_{21}\mathbf{H}_{22}^{\text{H}}\mathbf{w}_{r2} - \langle \mathbf{H}_{21}\mathbf{H}_{22}^{\text{H}}\mathbf{w}_{r2}, \frac{\mathbf{a}_1}{\|\mathbf{a}_1\|_2} \rangle \frac{\mathbf{a}_1}{\|\mathbf{a}_1\|_2}\|_2}. \tag{4.74}$$

In brief, for SI-ZF-MRT, the two options are *i)* to find the receive AWVs according to (4.67) and (4.68), and then obtain the transmit AWVs according to (4.64); *ii)* to find the receive AWVs according to (4.71) and (4.72), and then obtain the transmit AWVs according to (4.64). Therefore, the one which has a higher JAR can be selected as the solution for SI-ZF-MRT.

Similar to SI-ZF-MRT, SI-ZF-maximum-ratio combining (MRC) is also applicable to the joint beamforming problem. With SI-ZF-MRC, receive beamforming is firstly performed by using MRC. Afterwards, transmit beamforming is carried out to maximize the JAR with the SI forced to zero. Since SI-ZF-MRC is similar to SI-ZF-MRT in formulation and performance, we do not present the details here.

3) Steering Beamforming

The conventional steering beamforming in communication can also be introduced here to compare with the alternatives. Steering beamforming does not require full channel information. Instead, it only requires the knowledge of the transmit and steering vectors for the most significant MPC of the communication channel, and does not consider the SI. Suppose the m-th and ℓ-th MPCs are the most significant ones from Node #1 to Node #2 and from Node #2 to Node #1, respectively. By using steering beamforming, the AWVs become

$$\begin{aligned}
\mathbf{w}_{t2} &= \mathbf{h}_{21}(\theta_\ell^{(21)}), \quad \mathbf{w}_{t1} = \mathbf{h}_{12}(\theta_m^{(12)}), \\
\mathbf{w}_{r1} &= \mathbf{g}_{21}(\phi_\ell^{(21)}), \quad \mathbf{w}_{r2} = \mathbf{g}_{12}(\phi_m^{(12)}).
\end{aligned} \tag{4.75}$$

By comparing the performance of steering beamforming with those of the presented schemes, we can see whether or not it is infeasible not to consider the SI in FD communication, and how much performance degradation it causes if we do not consider the SI in beamforming.

4.2.3.3 Simulation Results

In this part, we present the performances of all the involved schemes through numerical simulations, where the communication channel model and the SI channel model introduced in Section 4.2.1 are adopted. In all the evaluations, the near-field

SI channels are deterministic; while the far-field signal channels are random. Both LoS and non-LoS (NLoS) channels are considered for the communication channels. For NLoS channels, the transmit and receive steering angles are randomly generated within $[0, 2\pi)$, and the coefficients obey the circularly symmetric complex Gaussian distribution with the same average power. For LoS channels, the LoS component has a fixed coefficient and fixed transmit and receive steering angles, while the other NLoS components have random steering angles and coefficients with average power 15 dB lower than the LoS component. The total power of a generated channel obeys (4.3), and the total number of MPCs is 3 (We've also simulated with other numbers of MPCs, and similar results were obtained). For each curve in all the figures in this section, we have generated 1000 realizations with the LoS or NLoS channel models, and computed the average JAR based on these realizations. Moreover, we have considered both types of array settings in the evaluations, namely Tx/Rx have separate antenna arrays and the same antenna array at a node. In all the simulations, $n_{t1} = n_{r1} = n_{t2} = n_{r2} = 32$.

Firstly, we consider the JAR and convergence performances of the ZF-Max-Power scheme with random initial transmit AWVs, which are shown in Fig. 4.5 with relevant parameters listed in the caption. Both cases of separate arrays and the same array are included under LoS and NLoS channels. As ZF-Max-Power is an iterative method, we compare it with Max-SINR, which achieves the best performance within the typical solutions for the IA problem[34]. It is observed that ZF-Max-Power achieves a sub-optimal performance close to the upper bound after convergence, under both LoS and NLoS channels, especially with separate arrays. The slight superiority of the case with separate arrays is due to that the different channel parameters of $\mathbf{H}_{12}$ and $\mathbf{H}_{21}$ provide more degrees of spatial freedoms for beamforming than the case of the

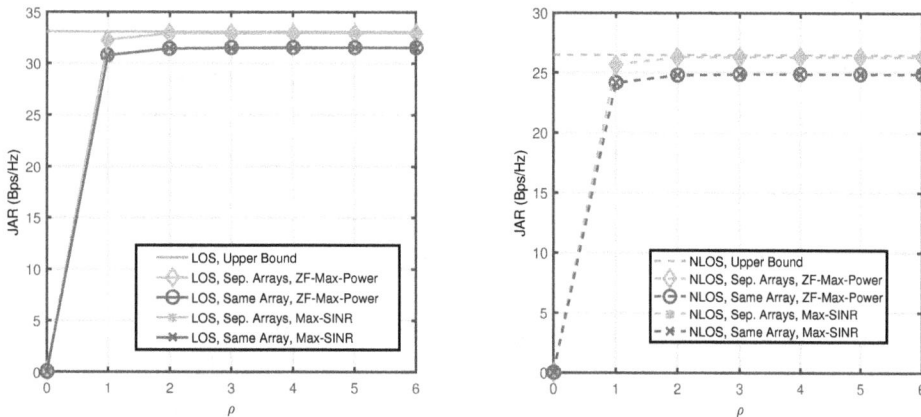

Figure 4.5 JAR and convergence performances of ZF-Max-Power with random initial transmit AWVs (Left: LoS channel, Right: NLoS channel). $P_{11} = P_{22} = 40$ dB. For LoS channels, $P_{12} = P_{21} = 20$ dB, while for NLoS channels, $P_{12} = P_{21} = 10$ dB. For the case of separate arrays, $d/\lambda = 1$ and $\omega = \pi/6$ rad, while for the case of sharing the same array, $d/\lambda = 0$ and $\omega = 0$ rad.

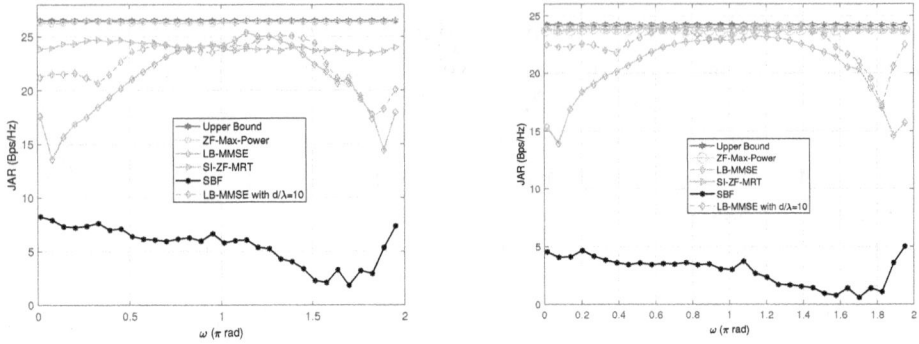

Figure 4.6 JAR performance of the involved schemes with respect to varying ω under LoS (the left-hand figure) and NLoS (the right-hand figure) channels in the case of separate arrays. $P_{11} = P_{22} = 40$ dB, $P_{12} = P_{21} = 10$ dB, $d/\lambda = 2$. For LB-MMSE, the JAR with $d/\lambda = 10$ is also plotted.

same array, where the channel parameters are the same. Moreover, the convergence speed of ZF-Max-Power is fast in all cases. Basically, only 2 iterations are required to achieve convergence. Interestingly, under the considered scenario, ZF-Max-Power achieves almost the same JAR and convergence performances as Max-SINR. However, it is noteworthy that ZF-Max-Power is a centralized iterative approach, where the iteration is performed at a certain node, and does not exploit channel reciprocity. By contrast, Max-SINR is a distributed iterative approach, where the iteration is performed at all distributed nodes by exploiting channel reciprocity. As a consequence, the convergence of ZF-Max-Power can be proven, while that of Max-SINR is still unproven in the literature[34] to the best of our knowledge. Moreover, ZF-Max-Power has a significantly lower computational complexity than Max-SINR.

Next, let us see the JAR performances of the presented schemes with respect to varying ω, d, i.e., the geometry of the arrays, when separate arrays are exploited at a node. Fig. 4.6 shows the JAR performances with respect to varying ω under LoS and NLoS channels, respectively, where SI is fixed. Fig. 4.7 (left) shows the JAR performances with respect to d under LoS channels. Similar results can be observed under NLoS channels. As we can see, SI is assumed fixed in Fig. 4.7 (left), which may be not practically reasonable, because in practice d significantly affects SI. However, in order to adequately evaluate the effects of d on the JAR performance, we assume a fixed SI in Fig. 4.7 (left). For rigorousness, we also assume varying SI in Fig. 4.7 (right), which is in accordance with the practice, where SI deteriorates with d^2. It is noteworthy that the strength of SI is typically much higher than that of the desired signal, because SI comes from the local Tx at the same node, while the desired signal comes from the remote Tx at the other node. Relevant parameters for these figures are listed in the corresponding captions.

By comparing these figures with each other, it can be observed that:

i) ZF-Max-Power is robust against ω, d, and SI, and approaches the upper bound in all these cases. This is because ZF-Max-Power not only forces SI to zero,

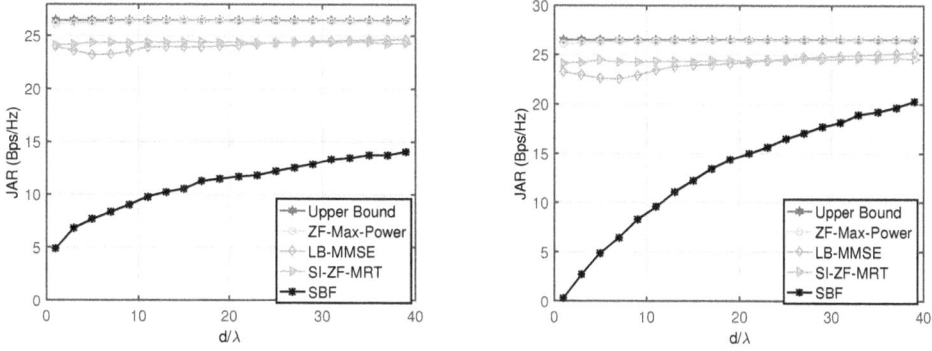

Figure 4.7 JAR performance of the involved schemes with respect to varying d under LoS channels in the case of separate arrays, $\omega = \pi$ rad. In the left-hand figure SI is assumed fixed, i.e., $P_{11} = P_{22} = 40$ dB, $P_{12} = P_{21} = 10$ dB; while in the right-hand figure SI varies with d/λ, i.e., $P_{11} = P_{22} = 60 - 20\log(d/\lambda)$ dB, $P_{12} = P_{21} = 10$ dB.

but also iteratively maximizes signal power. Thus, it achieves compelling performance that is insensitive to the geometry of the Tx/Rx arrays and SI.

ii) SI-ZF-MRT is also robust against ω, d, and SI, thanks to its ZF filtering to SI. In addition, it also achieves an acceptable performance, which is close to the upper bound. It is noted that the JAR gap between SI-ZF-MRT and the upper bound is greater under LoS channels than that under NLoS channels. This phenomenon can be explained by referring to (4.68), where $\mathbf{w}_{r2}$ is, in fact, set within an $(n_{r2} - 2)$-dimensional subspace, due to the two zero forcing equations shown in (4.65). Clearly if $\mathbf{a}$ in (4.68), which represents the direction with the largest power of the channel, has less energy projected on the $(n_{r2} - 2)$-dimensional subspace, the JAR performance will be poorer. Under LoS channels, the majority of the channel energy concentrates on a single path, or a single direction. Once this direction has a small projection on the subspace, the performance will be poor. In contrast, under NLoS channels, the channel energy evenly disperses on multiple paths. Only when all of these paths have a small projection on the subspace, the performance will be poor. In other words, the probability of a poor performance is lower under NLoS channels than that under LoS channels. Hence, on average, the JAR gap between SI-ZF-MRT and the upper bound is greater under LoS channels than that under NLoS channels.

iii) LB-MMSE is sensitive to ω and d. From Fig. 4.6, we observe that the performance of the LB-MMSE fluctuates as ω changes, and the fluctuation is different for different d. From Fig. 4.7, we observe that the performance of LB-MMSE has a U-shape as d increases, but behaves stable when d is large. To understand these, we need to go back to (4.62) and (4.63). From these two equations, we can see that the transmit AWVs are decided to maximize the SINR rather than minimize SI based on the local information. Taking (4.62) for illustration, since usually P_{11} is large, when $\mathbf{H}_{11}$ has a low rank[5], the eigenvector of $\mathbf{H}_{12}^{\mathrm{H}}\mathbf{H}_{12}$ has a high probability to locate within the null space of $\mathbf{H}_{11}$. In such a case, a high signal power can be achieved while little SI locates within the signal subspace; thus good performance is achieved. However, when $\mathbf{H}_{11}$ has a high or even full rank, SI will almost unavoidably locate within the signal subspace and affects the received SINR, and thus the performance will be poor. When d is small, the energy dispersion of $\mathbf{H}_{11}$ is sensitive to ω and d according to the SI channel model, and thus the JAR performance is also sensitive to ω and d. However, when d is large, the SI channel almost reduces to a directional channel with rank 1, and thus SI has a low probability to locate within the signal subspace. In such a case, LB-MMSE can stably achieve a near-optimal performance.

iv) Steering beamforming is also sensitive to ω, d, and SI. This is because steering beamforming does not even consider SI in the beamforming design. Meanwhile, from Fig. 4.7 it is found that steering beamforming becomes improved as d increases. In Fig. 4.7 (right), the improving speed of steering beamforming is faster than that in Fig. 4.7, because SI is reduced as d increases. This phenomenon suggests that when the near-field SI channel gradually reduces to a directional channel, the conventional beamforming schemes that simply steer toward each other may also achieve good performance, because usually the communication channel and SI channel have different steering angles. However, in practical FD communications where d is generally small, the SI channel does not have the feature of directivity; thus steering beamforming is much poorer than the other candidates, and the performance of steering beamforming does not show monotonicity with ω, as shown in Fig. 4.6. Thus, steering beamforming may not be a good choice for FD communications, where SI must be taken into account.

Then, we compare the JAR performances of the discussed schemes with separate arrays and the same array. Fig. 4.8 shows the comparison results with respect to SI under LoS and NLoS channels, respectively, where relevant parameters are listed in the captions. From these two figures, we observe that the schemes with separate arrays basically achieve better performance than those with the same array. This advantage is also due to that the different channel parameters of $\mathbf{H}_{12}$ and $\mathbf{H}_{21}$ when using separate arrays provide larger degrees of spatial freedom for beamforming than in the case of using the same array, where the channel parameters are the same. Moreover, both ZF-Max-Power and SI-ZF-MRT are insensitive to the increase of SI, thanks to the operation of ZF SI, while the performance of steering beamforming becomes poorer as the increase of SI, due to no operation of ZF SI. Interestingly, the JAR of LB-MMSE with separate arrays slowly decreases as the increase of SI whereas that with the same array changes little, which shows that LB-MMSE with the same array is more robust against the SI. To explain this, let us retrospect (4.58). In the

---

[5]This statement is just for illustration. In practice, $\mathbf{H}_{11}$ is generally with full rank except when $\omega = 0$ or π rad. However, when most energy of $\mathbf{H}_{11}$ locates at a low-dimensional subspace, the situation will be similar to the statement that $\mathbf{H}_{11}$ has a low rank.

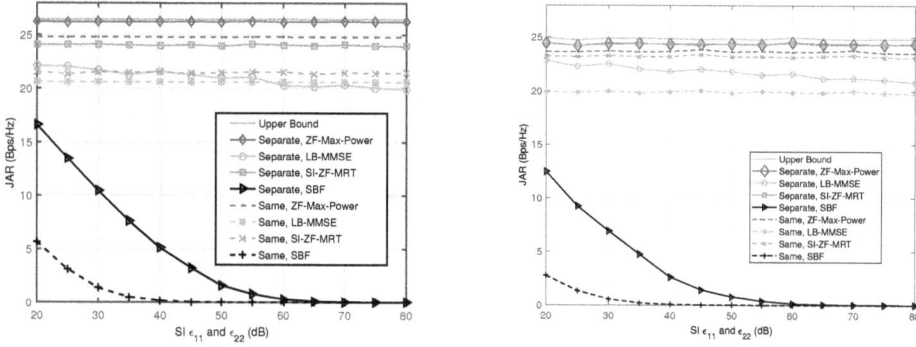

Figure 4.8 JAR comparison between different array settings (separate arrays versus the same array) under LoS (left) and NLoS (right) channels with varying SI. $P_{12} = P_{21} = 10$ dB. For the case of separate arrays, $\omega = 0.6\pi$ rad, $d/\lambda = 1$.

inequality (b) of (4.58),

$$\frac{\mathbf{w}_{t2}^H \mathbf{H}_{21}^H \mathbf{H}_{21} \mathbf{w}_{t2}}{\mathbf{w}_{t1}^H \left(\mathbf{I} + P_{11} \mathbf{H}_{11}^H \mathbf{H}_{11} \right) \mathbf{w}_{t1}},$$
$$\frac{\mathbf{w}_{t1}^H \mathbf{H}_{12}^H \mathbf{H}_{12} \mathbf{w}_{t1}}{\mathbf{w}_{t2}^H \left(\mathbf{I} + P_{22} \mathbf{H}_{22}^H \mathbf{H}_{22} \right) \mathbf{w}_{t2}},$$

$$(4.76)$$

can be roughly seen as the receive SINRs at Node #1 and Node #2 without considering the receive AWVs, respectively. However, in order to obtain closed-form expressions of the transmission AWVs, the denominators (or numerators) of these two components are exchanged and optimized, respectively, as shown in (4.60) and (4.61). This means that the optimizations in (4.60) and (4.61) are not to directly optimize the receive SINRs at Node #1 and Node #2. Hence, in general LB-MMSE is not so robust against the SI. However, in the case with the same array, the link from Node #1 to Node #2 is symmetric; thus the denominators (or numerators) of the two components in the inequality (b) of (4.58) can be seen equal or at least proportional to each other. In such a case, (4.60) and (4.61) are, in fact, to optimize the receive SINRs at Node #1 and Node #2 without considering the receive AWVs. Hence, LB-MMSE with the same array is relatively more robust against the SI.

Finally, we evaluate the effects of channel estimation errors on the presented schemes. For the SI channel, there exists Gaussian error, while for the communication channel, it is possible to miss some MPCs during the beam search process. Fig. 4.9 shows the effects of these estimation errors on the presented schemes with separate arrays (the results are similar with the same array) under LoS and NLoS channels, respectively, where the parameters are specified in the captions. From these two figures, we can observe that both ZF-Max-Power and LB-MMSE are relatively robust against the channel estimation error. Even only one MPC is acquired, they can achieve promising performance, especially under LoS channels. However, SI-ZF-MRT is not robust against the estimation error of the communication channel, i.e., if only one MPC is acquired, the performance of SI-ZF-MRT becomes rather poor. This is

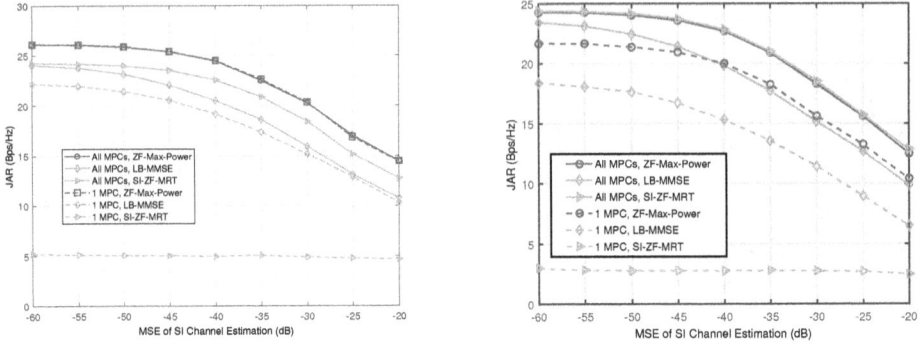

Figure 4.9 Effects of channel estimation errors on the presented schemes with separate arrays under LOS (left) and NLOS (right) channels. $P_{12} = P_{21} = 10$ dB, $P_{11} = P_{22} = 40$ dB, $\omega = \pi$ rad, $d/\lambda = 1$.

because the full channel information is involved in the SI ZF operation according to (4.65).

It is noteworthy that circuit imperfections are not taken into consideration in our system model, i.e., in (4.11) and (4.12). In a practical FD system, there are always Tx/Rx hardware and implementation imperfections, including low noise amplifier (LNA) noise figure, phase noise, in phase-quadrature (IQ) mismatch, nonlinear distortion of power amplifier (PA), etc. These imperfections may be more severe in high-frequency communication systems than in low-frequency systems, because of the higher carrier frequency and larger bandwidth. Since SI is strong, the Tx imperfections, which are carried by the transmitted signals s_1 and s_2 in (4.11) and (4.12), will also arrive at the Rx. When beamforming cancellation is adopted to force the SI to zero, the performance will be affected little by the Tx circuit imperfections provided that the AWV control is perfect, because all the SI, including the imperfections, can be filtered out by beamforming. However, in practice the AWV control may have error. In such a case the SI cannot be completely filtered out by beamforming, and the residual SI, which contains Tx imperfections, will degrade the system performance. Although baseband cancellation can be used to deal with the residual SI after beamforming cancellation, it basically cannot effectively cancel the residual Tx imperfections.

To further illustrate the effect of the circuit imperfections on the system performance, we model all the typical Tx imperfections as a zero-mean Gaussian distributed error vector magnitude (EVM) noise[44], and its average power can be measured in dB with respect to the transmit power. On the other hand, we also need to consider AWV control error, which can also be modeled as a zero-mean Gaussian variable, and its average power can be measured in dB with respect to the ℓ_2-norm of its corresponding weight. By exploiting this model, we can evaluate the effects of AWV error and EVM error on the JAR performance of ZF-Max-Power[6] as shown in Fig. 4.10,

---

[6]The effects are similar to the performances of the other schemes.

where relevant parameters are listed in the caption. From this figure, we can find that when AWV control is perfect or the AWV error is small enough, the SI as well as the EVM noise can be mitigated successfully by beamforming. However, when there is significant AWV error, the system performance deteriorates as the AWV error becomes greater. In such a case, baseband cancellation is needed to cancel the residual SI. From the figure, we observe that with baseband cancellation, the performance is greatly improved and does not depend on the AWV error. On the other hand, baseband cancellation can hardly mitigate the residual EVM noise, because it is difficult to estimate relevant parameters of the EVM noise[44]. Hence, we can observe that even when baseband cancellation is adopted, if EVM noise exists, the performance still deteriorates as the AWV error becomes greater.

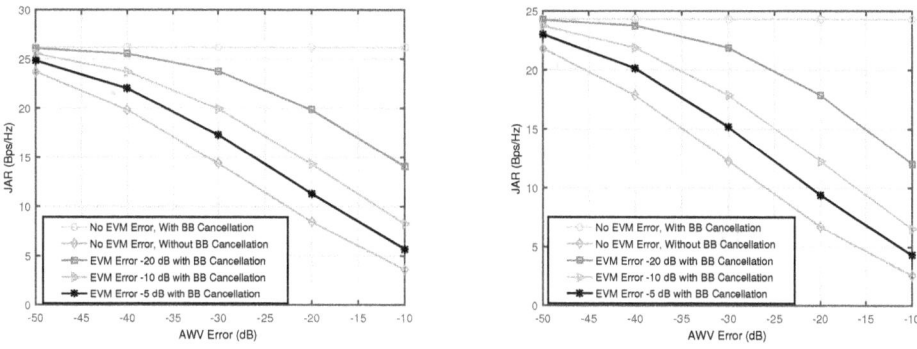

Figure 4.10 Effects of AWV error and EVM error on the JAR performance of ZF-Max-Power under LoS (left) and NLoS (right) channels. $P_{12} = P_{21} = 10$ dB, $P_{11} = P_{22} = 50$ dB, $\omega = 0.8\pi$ rad, $d/\lambda = 1$.

4.2.4 FD Communication Considering CM Constraint

Subject to the CM constraint, in regular communication analog beamforming/combining is generally realized by steering the Tx AWV and Rx AWV to the AoD and AoA of the strongest MPC, such that high array gain can be achieved. However, this method may not be viable in FD communication, where strong SI may exist. In other words, the conventional Tx/Rx AWV settings may not be able to effectively reduce SI, although they can achieve a higher array gain. As a result, the overall SINR may not be high, as pointed out in[20]. Hence, in FD communication SI driven beamforming/combining must be considered.

4.2.4.1 CM Beamforming Cancellation

The problem of CM beamforming cancellation is challenging because the objective function, e.g., the achievable sum rate (ASR) of the two nodes in an FD link, is usually complicated and non-concave due to the SI. Moreover, the CM constraints of the Tx/Rx AWVs are also non-concave/convex.

While it is difficult to find an optimal solution for CM beamforming cancellation, it is interesting and with significance to find sub-optimal solutions with low computational complexity. Here we discuss some preliminary solutions. Intuitively, as there is only phase uncertainty for each element of the Tx/Rx AWVs under the CM constraints, we may search from 0 to 2π for each element. Theoretically, provided small enough step size, an optimal solution can be found. However, as the numbers of antennas are large in general, the exhaustive search method is basically infeasible due to the exhibitively high computational complexity. Hence, we present the following two candidate methods:

- **Method 1:** To restrict the Tx/Rx AWVs to the steering vector space, i.e., we assume that the Tx/Rx AWVs are all steering vectors and have uncertainty only on the steering angles. Under this assumption, we can search over the steering angle spaces of the Tx/Rx AWVs and find the best steering angles for them to optimize the objective. This method is viable just because the SI channel also has spatial sparsity, as shown in Section 4.2.2.

- **Method 2:** To first find AWVs by solving the CM beamforming cancellation problem without considering the CM constraints, and then find AWVs that satisfy the CM constraints and have the minimal Euclidean distances from the already found AWVs.

Fig. 4.11 (left) shows the ASR comparison between different schemes in an FD communication system. The upper bound is obtained by assuming that SI is zero and there are no CM constraints. The "Beam Steering" method is to simply set Tx/Rx AWVs to steer along a strong MPC without considering the SI, which is a typical beamforming method used for regular communication. "Angle Search" refers to Method 1. To use Method 2, we adopt ZF-Matched Filter (ZF-MF) to obtain the beamforming vectors, and then add the CM constraints according to Method 2. ZF-MF is a linear beamforming algorithm, which performs matched filter (MF) at the Tx first and then performs ZF at the Rx.

From Fig. 4.11 (left), we can observe that a conventional beamforming approach (Beam Steering) which does not take SI into account cannot achieve a satisfactory performance. In addition, although the simple Angle Search and ZF-MF with CM methods are viable, there is a significant performance gap from the upper bound, which means that these two methods may work better in low SI cases; when SI is strong (as shown in the figure), these two methods need to work together with other SI cancellation methods, like antenna cancellation or baseband cancellation. Meanwhile, more efficient methods are in demand to deal with the SI under the CM constraint. In fact, it is just the CM constraint that limits the performance of Method 2. Without the CM constraint, ZF-MF can achieve a performance close to the upper bound.

On the other hand, we've mentioned in Section 4.2.2 that the SI channel also has spatial sparsity, which can be exploited to design methods to mitigate SI, e.g., the Angle Search method. As we can expect, the Tx/Rx array positions may significantly affect the performance of this category of methods. The right-hand side figure of

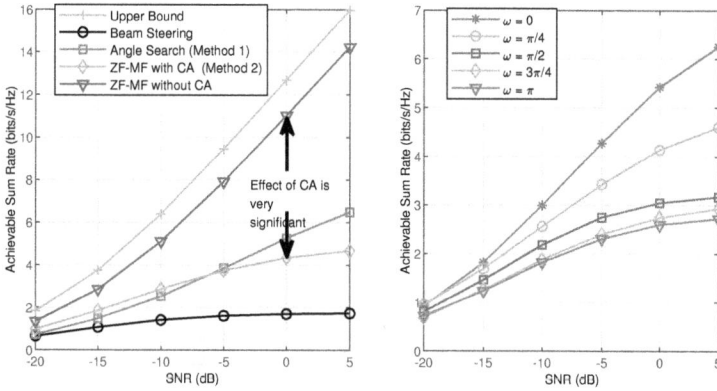

Figure 4.11 **Left:** Achievable sum rate comparison between different beamforming schemes in an FD communication system, where for each scheme the numbers of Tx and Rx antennas are 16 and 8, respectively, $\omega = \pi$ rad and $d/\lambda = 5$. **Right:** Achievable sum rate comparison of Angle Search with varying ω, where the number of Tx and Rx antennas is 16 at both nodes, $d/\lambda = 1$. In both figures, the signal/noise powers are the same at the two nodes, and the SI is 25 dB w.r.t. the signal power.

Fig. 4.11 shows the ASR comparison of Angle Search with varying ω. It is clear that ω can dramatically affect the performance, and thus may be considered as another DoF to optimize the ASR. It is noteworthy that different methods may favor different ω.

4.2.4.2 Beamforming Cancellation with Double RF Chains

We have seen that the CM constraints greatly limit the performance of beamforming cancellation. If we can find an approach to remove or avoid the CM constraints, beamforming cancellation can become very promising even without the need to work together with other SI cancellation methods[45]. Fortunately, the work in[46] provides a great opportunity to realize this. It is shown in[46] that an arbitrary vector $\mathbf{v}$ can be expressed as a linear combination of two CM vectors, i.e., $\mathbf{v} = \beta_1\mathbf{v}_1 + \beta_2\mathbf{v}_2$, where β_1 and β_2 are coefficients, $\mathbf{v}_1$ and $\mathbf{v}_2$ are CM vectors with carefully designed phase for each element[7]. This result implies that by doubling the number of Tx/Rx RF chains, although CM constraint holds on each RF chain, the equivalent beamforming vector $\mathbf{v}$ does not exhibit the CM constraint any more, and thus ZF-MF without CM in Fig. 4.11 (left) can be realized.

Fig. 4.12 (left) shows the structure with double Tx/Rx RF chains in an FD node with separate Tx/Rx arrays. The Tx/Rx arrays are driven by the two independent Tx/Rx RF chains with CM AWVs. An additional Tx/Rx RF chain increases an additional DoF to control the AWV, and thus the CM constraint does not appear

---

[7]In fact, this result can be extended to multi-stream transmission, i.e., an equivalent fully-digital N-stream precoding can be realized provided $2N$ RF chains with CM constraints in communications[46].

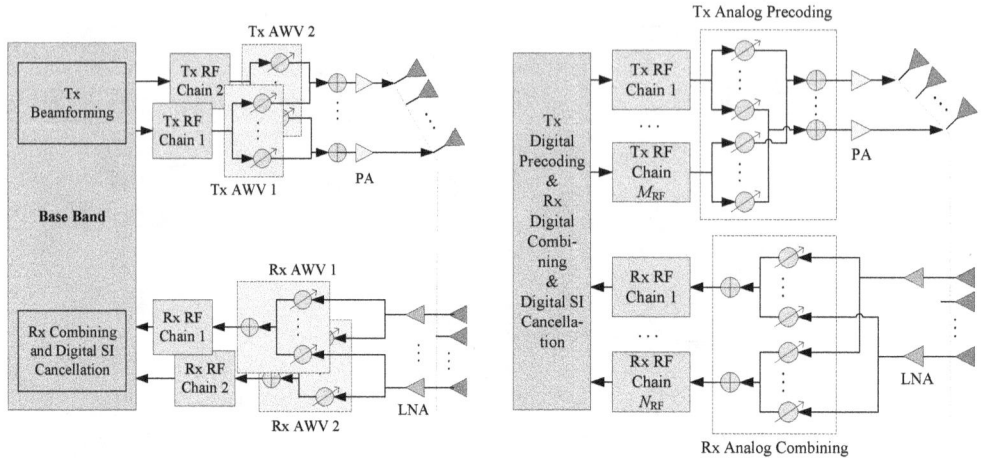

Figure 4.12 **Left**: FD communication with double Tx/Rx RF chains. **Right**: An FD hybrid precoding structure for the BS.

or is circumvented in the beamforming cancellation, according to the result achieved in[46]. On the other hand, doubling the Tx/Rx RF chains may also increase the cost and power consumptions of an FD node. Fortunately, as shown in the figure, the two Tx/Rx RF chains can share the same up/down-converter, PAs, and LNAs; hence the cost and power consumptions may not increase much.

4.2.5 Multi-User Scenario

4.2.5.1 Benefit of FD Transmission

In the previous section, we mainly discuss point-to-point FD transmission, which can be used for wireless backhaul link where bi-directional high data rate is required. In this section, we discuss the benefit of FD communication in a multi-user scenario, where a base station (BS) serves multiple users. Apparently, with FD transmission, the multi-user capacity can be greatly improved, because the achievable rate for each user can be almost doubled. In practice, however, this benefit may be offset since in some cases, one user does not need to transmit and receive simultaneously. For instance, when we make a call, we basically do not speak much while we are listening. Fortunately, with multiple users, the BS can still sufficiently exploit the FD benefit most of the time, by receiving from some users while transmitting to some other users at the same time.

Fig. 4.13 (left) shows a typical FD multi-user scenario, where an FD BS serves multiple users simultaneously. It is noteworthy that within these users, there are FD users, like user 2, as well as HD users, like user 1 and user 3. An FD user has the antenna setting shown in Fig. 4.1, while an HD user can only perform conventional beamforming, either Tx or Rx, but cannot perform FD Tx/Rx beamforming. In a conventional multi-user scenario, a BS usually exploits a hybrid precoding structure[47] to serve multiple users with spatial division multiple access (SDMA). Thus, in an FD multi-user scenario, an FD BS needs to exploit an FD hybrid precoding structure

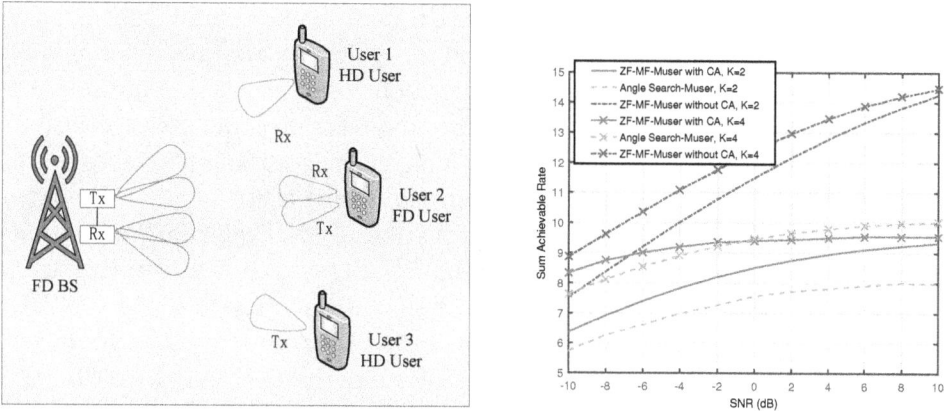

Figure 4.13 **Left**: An FD multi-user scenario, where an FD BS serves multiple users, including FD users and HD users. **Right**: Achievable sum rate comparison under a multi-user FD scenario, where one BS serves K FD users. For all the users and the BS, the number of Tx and Rx antennas is 16. $\omega = \pi/8$ rad, $d/\lambda = 5$, and the SI is 20 dB w.r.t. the signal power.

shown in the right-hand side figure of Fig. 4.12, where M_{RF} Tx RF chains share the same Tx array and N_{RF} Rx RF chains share the same Rx array. With this structure, the BS can transmit data to M_{RF} parallel users and meanwhile receives data from N_{RF} parallel users by using both SDMA and FD transmission.

As shown in the left-hand side figure of Fig. 4.13, at the same time and the same frequency band, the BS can serve these three users simultaneously. In particular, the BS transmits data to users 1 and 2, and distinguishes them by SDMA, meanwhile the BS receives data from users 2 and 3, and also distinguishes them by SDMA. From this scenario, we can find that even if there is no FD users in practice, the FD BS can still exploit the FD benefit by transmitting data to some users (like user 1) while at the same time receiving data from some other users (like user 3).

4.2.5.2 Interference Mitigation

The FD multi-user communication does not achieve the FD benefit at no cost. Instead, the FD BS must face more complicated interference management. It is known that in conventional multi-user systems, hybrid precoding must be exploited to mitigate multi-user interference (MUI) and meanwhile achieve array gains for multiple users[47]. In an FD multi-user system, the BS must deal with both MUI and SI and meanwhile achieve array gains, which is called FD hybrid precoding in this section.

Firstly, let us see the SI at the FD BS. Unlike the point-to-point case discussed in the previous section and shown in Fig. 4.1, where the SI only contains one single Tx data stream, under the multi-user scenario, the SI may include multiple independent Tx data streams for multiple downlink users, which means that the SI becomes multi-dimensional. Appropriate Tx/Rx AWVs must be designed to handle the multi-dimensional SI in addition to the MUI. On the other hand, subject to the particular antenna structure in the FD BS, the precoding matrix (also the combining matrix) has to be decomposed into the product of a digital matrix and an analog matrix, which correspond to digital precoding and analog precoding, respectively[47]. The analog matrix has the CM constraints, which challenges the interference mitigation. In fact, in a conventional multi-user scenario, where there is only MUI but no SI, the CM constraints already make the mitigation of MUI difficult. For instance, in[47] an approach was proposed to first perform BS-user analog beamforming one by one without considering MUI in analog precoding, and then try to mitigate MUI in digital precoding where there is no CM constraint. This approach can achieve good performance, because the BS-user analog beamforming basically induces little MUI due to the spatial sparsity of the high-frequency band channel. For this reason, this approach may also be used for the FD hybrid precoding here. In particular, we may first perform BS-user analog beamforming by exploiting Method 1 or Method 2 in the previous subsection, and then design digital precoding and combining matrices to mitigate MUI and the residual SI. In fact, considering that the SI channel also has the property of spatial sparsity, as introduced in Section 4.2.2, the spatially sparse precoding[13], which formulates the hybrid precoding problem as a sparse reconstruction problem and solves it with the principle of basis pursuit, can also be exploited to solve the FD hybrid precoding problem.

In addition, in the case that the system complexity is affordable, the method to double the number of Tx/Rx RF chains to remove the CM constraints in Section 4.2.4 can also be applied in the FD BS. In particular, the FD BS may be equipped with $2N_{RF}$ Rx RF chains and $2M_{RF}$ Tx RF chains to serve N_{RF} uplink users and M_{RF} downlink users simultaneously. In such a case, the CM constraints will not appear in the beamforming cancellation according to the results in[46], and we can perform fully digital precoding/combining to mitigate SI and MUI, which is much more flexible than the hybrid precoding/combining. In the case when the system complexity is limited that the BS can only be equipped with M_{RF} Tx RF chains (and N_{RF} Rx RF chains), there is a tradeoff between using these RF chains to serve M_{RF} downlink users with CM constraints and serving $M_{RF}/2$ downlink users but with the CM constraints circumvented. Channel conditions would determine which one is better.

The right-hand side figure of Fig. 4.13 shows a preliminary comparison of ASR under a multi-user FD scenario, where ZF-MF-Muser refers to the scheme that uses ZF-MF (cf. Section 4.2.4) between the BS and each user for SI cancellation, while Angle Search-Muser refers to the scheme that uses Angle Search (cf. Section 4.2.4) between the BS and each user for SI cancellation. The MUI is simply mitigated by using ZF filtering in the digital beamforming after SI cancellation. From this figure, we can see that due to the CM constraint, the SI and MUI may not be completely cancelled, as there are performance floors for both ZF-MF-Muser and Angle Search-Muser at the high SNR regime. Hence, more efficient methods are in demand to deal with the SI and MUI under the CM constraint. When the CM constraint is not considered, the performance can be greatly improved, which indicates that the method of doubling the number of Tx/Rx RF chains is also effective for the multi-user scenario.

4.3 BEAMFORMING FOR FULL-DUPLEX RELAY

One of the drawbacks of high-frequency communications is that obstacles on the ground may prevent the establishment of LoS links, which leads to severely attenuated received signal powers even if beamforming is applied. To address this issue, in this section, an FD relay is employed to increase the communication capacity of networks. Specifically, an FD-unmanned aerial vehicle (UAV) relay is deployed between a source node (SN) and a destination node (DN) to establish an LoS link, where large antenna arrays are employed for beamforming to enable directional beams facilitating high channel gains. Although physically separated antenna panels and directional antennas are usually used for transceivers, the small sidelobes of the radiation pattern, which are inevitable, may result in significant SI for FD relays The authors of[48] have shown that, in addition to 70–80 dB physical isolation realized by increasing the distance between a Tx antenna panel and an adjacent Rx antenna panel, 35–50 dB isolation via SI reduction[8] is needed to enable successful reception of signals in in-band FD wireless backhaul links. For this reason, we establish the corresponding optimization problem to maximize the reachable rate between SN and DN, and use Tx and Rx beamforming to reduce the SI at the FD relay.

4.3.1 System Model

We consider an end-to-end transmission scenario, where an SN serves a remote DN[9]. The SN and the DN are equipped with UPAs employing $N_{\text{S}}^{\text{tot}} = M_{\text{S}} \times N_{\text{S}}$ and $N_{\text{D}}^{\text{tot}} = M_{\text{D}} \times N_{\text{D}}$ antennas, respectively, to overcome the high path loss in the high-frequency band. Due to obstacles, such as ground buildings, the channel from the SN to the DN may be blocked. Thus, an FD relay, equipped with an $N_{\text{t}}^{\text{tot}} = M_{\text{t}} \times N_{\text{t}}$ Tx-UPA and an $N_{\text{r}}^{\text{tot}} = M_{\text{r}} \times N_{\text{r}}$ Rx-UPA, is deployed between the SN and the DN to improve system performance.

4.3.1.1 Signal Model

In the considered system, the SN transmits signal s_1 to the relay with power P_{S}, and concurrently, the relay transmits signal s_2 to the DN with power P_{V}, where $\mathbb{E}(|s_i|^2) = 1$ for $i = 1, 2$. Thus, the received signal at the relay is given by

$$\bar{y}_{\text{V}} = \mathbf{w}_{\text{r}}^{\text{H}} \mathbf{H}_{\text{S2R}} \mathbf{w}_{\text{S}} \sqrt{P_{\text{S}}} s_1 + \mathbf{w}_{\text{r}}^{\text{H}} \mathbf{H}_{\text{SI}} \mathbf{w}_{\text{t}} \sqrt{P_{\text{V}}} s_2 + \mathbf{n}_1, \qquad (4.77)$$

---

[8]SI reduction methods for FD terminals are usually partitioned into three classes: propagation-domain, analog-circuit-domain, and digital-domain techniques. Tx and Rx beamforming at the FD relay can be categorized as propagation-domain and analog-circuit-domain approaches, respectively.

[9]FD relays can be used to increase the end-to-end data rate between two nodes with poor link quality in beyond 5G (B5G) networks. Exemplary application scenarios include BS-to-user equipment (UE) communication, backhaul links[49], device-to-device communications[50], and communication between two terrestrial mobile BSs in emergency situations[51].

where $\mathbf{H}_{\text{S2R}} \in \mathbb{C}^{N_r^{\text{tot}} \times N_S^{\text{tot}}}$ is the channel matrix between the SN and the FD relay. $\mathbf{H}_{\text{SI}} \in \mathbb{C}^{N_r^{\text{tot}} \times N_t^{\text{tot}}}$ is the SI channel matrix between the Tx-UPA and the Rx-UPA at the relay. $\mathbf{n}_1$ denotes the white Gaussian noise at the relay having zero mean and power σ_1^2. $\mathbf{w}_S \in \mathbb{C}^{N_S^{\text{tot}} \times 1}$, $\mathbf{w}_r \in \mathbb{C}^{N_r^{\text{tot}} \times 1}$, and $\mathbf{w}_t \in \mathbb{C}^{N_t^{\text{tot}} \times 1}$ represent the SN beamforming vector, the Rx beamforming vector at the relay, and the Tx beamforming vector at the relay, respectively.

The received signal at the DN is given by

$$\bar{y}_D = \mathbf{w}_D^H \mathbf{H}_{\text{S2D}} \mathbf{w}_S \sqrt{P_S} s_1 + \mathbf{w}_D^H \mathbf{H}_{\text{R2D}} \mathbf{w}_t \sqrt{P_V} s_2 + \mathbf{n}_2, \tag{4.78}$$

where $\mathbf{H}_{\text{R2D}} \in \mathbb{C}^{N_D^{\text{tot}} \times N_t^{\text{tot}}}$ is the channel matrix between the relay and the DN. $\mathbf{H}_{\text{S2D}} \in \mathbb{C}^{N_D^{\text{tot}} \times N_S^{\text{tot}}}$ is the channel matrix between the SN and the DN. $\mathbf{w}_D \in \mathbb{C}^{N_D^{\text{tot}} \times 1}$ denotes the DN beamforming vector. $\mathbf{n}_2$ denotes the white Gaussian noise at the DN having zero mean and power σ_2^2.

As mentioned in Chapter 2, there are two main strategies for beamforming, i.e., digital beamforming and analog beamforming. Digital beamforming has large DoFs with high hardware cost. In contrast, analog beamforming is more energy efficient. In addition, for FD communication, analog-circuit-domain SI cancellation is usually performed before digital sampling to avoid saturation due to strong SI[3, 52]. For these reasons, analog beamforming is adopted for the considered FD relay, The employed analog beamforming vectors impose a CM constraint[53, 54, 55], i.e.,

$$|[\mathbf{w}_\tau]_n| = \frac{1}{\sqrt{N_\tau^{\text{tot}}}}, \ 1 \le n \le N_\tau^{\text{tot}}, \tau = \{\text{S}, \text{r}, \text{t}, \text{D}\}. \tag{4.79}$$

Then, we can obtain the achievable rates of the source node-to-relay (S2R) and relay-to-destination node (R2D) links as follows

$$R_{\text{S2R}} = \log_2 \left(1 + \frac{|\mathbf{w}_r^H \mathbf{H}_{\text{S2R}} \mathbf{w}_S|^2 P_S}{|\mathbf{w}_r^H \mathbf{H}_{\text{SI}} \mathbf{w}_t|^2 P_V + \sigma_1^2} \right), \tag{4.80}$$

$$R_{\text{R2D}} = \log_2 \left(1 + \frac{|\mathbf{w}_D^H \mathbf{H}_{\text{R2D}} \mathbf{w}_t|^2 P_V}{|\mathbf{w}_D^H \mathbf{H}_{\text{S2D}} \mathbf{w}_S|^2 P_S + \sigma_2^2} \right). \tag{4.81}$$

Since the source node-to-destination node (S2D) link has a small channel gain due to the assumed blockage, the signal received via the S2D link is treated as interference at DN. Note that the achievable rates in (4.80) and (4.81) hold for coherent detection. Therefore, the FD-UAV relay and DN need to know the effective channel gains $\mathbf{w}_r^H \mathbf{H}_{\text{S2R}} \mathbf{w}_S$ and $\mathbf{w}_D^H \mathbf{H}_{\text{R2D}} \mathbf{w}_t$, respectively. The achievable rate between the SN and the DN is the minimum of the rates of the S2R and R2D links, i.e.,

$$R_{\text{S2D}} = \min\{R_{\text{S2R}}, R_{\text{R2D}}\}. \tag{4.82}$$

4.3.1.2 Channel Model

Due to the directivity and sparsity of the far-field high-frequency band channel, the channel matrices of the S2R and R2D links can be expressed as a superposition of MPCs, where different paths have different AoDs and AoAs. Hence, the channel matrices of the S2R, R2D, and S2D links are modeled as follows[53, 54, 55]

$$
\mathbf{H}_{\mathrm{S2R}} = \chi_{\mathrm{S2R}} \beta_{\mathrm{S2R}}^{(0)} \mathbf{a}_{\mathrm{r}}(\theta_{\mathrm{r}}^{(0)}, \phi_{\mathrm{r}}^{(0)}) \mathbf{a}_{\mathrm{S}}^{\mathrm{H}}(\theta_{\mathrm{S}}^{(0)}, \phi_{\mathrm{S}}^{(0)})
$$
$$
+ \sum_{\ell=1}^{L_{\mathrm{S2R}}} \beta_{\mathrm{S2R}}^{(\ell)} \mathbf{a}_{\mathrm{r}}(\theta_{\mathrm{r}}^{(\ell)}, \phi_{\mathrm{r}}^{(\ell)}) \mathbf{a}_{\mathrm{S}}^{\mathrm{H}}(\theta_{\mathrm{S}}^{(\ell)}, \phi_{\mathrm{S}}^{(\ell)}),
\tag{4.83}
$$

$$
\mathbf{H}_{\mathrm{R2D}} = \chi_{\mathrm{R2D}} \beta_{\mathrm{d2D}}^{(0)} \mathbf{a}_{\mathrm{D}}(\theta_{\mathrm{D}}^{(0)}, \phi_{\mathrm{D}}^{(0)}) \mathbf{a}_{\mathrm{t}}^{\mathrm{H}}(\theta_{\mathrm{t}}^{(0)}, \phi_{\mathrm{t}}^{(0)})
$$
$$
+ \sum_{\ell=1}^{L_{\mathrm{R2D}}} \beta_{\mathrm{R2D}}^{(\ell)} \mathbf{a}_{\mathrm{D}}(\theta_{\mathrm{D}}^{(\ell)}, \phi_{\mathrm{D}}^{(\ell)}) \mathbf{a}_{\mathrm{t}}^{\mathrm{H}}(\theta_{\mathrm{t}}^{(\ell)}, \phi_{\mathrm{t}}^{(\ell)}),
\tag{4.84}
$$

$$
\mathbf{H}_{\mathrm{S2D}} = \sum_{\ell=1}^{L_{\mathrm{S2D}}} \beta_{\mathrm{S2D}}^{(\ell)} \mathbf{a}_{\mathrm{D}}(\theta_{\widetilde{\mathrm{D}}}^{(\ell)}, \phi_{\widetilde{\mathrm{D}}}^{(\ell)}) \mathbf{a}_{\mathrm{S}}^{\mathrm{H}}(\theta_{\widetilde{\mathrm{S}}}^{(\ell)}, \phi_{\widetilde{\mathrm{S}}}^{(\ell)}),
\tag{4.85}
$$

where index $\ell = 0$ represents the LoS component and indices $\ell \geq 1$ represent the NLoS components. L_{S2R}, L_{R2D}, and L_{S2D} are the total numbers of NLoS components for the S2R, R2D, and S2D channels, respectively. Random variables χ_{S2R} and χ_{R2D} are equal to 1 if the LoS path exists and equal to 0 otherwise. Furthermore, the LoS path from the SN to the DN is assumed to be blocked, which is the main motivation for deploying an FD relay. $\beta_{\mathrm{S2R}}^{(\ell)}$, $\beta_{\mathrm{R2D}}^{(\ell)}$, and $\beta_{\mathrm{S2D}}^{(\ell)}$ are the complex coefficients of the S2R, R2D, and S2D paths, respectively. $\theta_{\mathrm{S}}^{(\ell)}$, $\phi_{\mathrm{S}}^{(\ell)}$, $\theta_{\mathrm{r}}^{(\ell)}$, and $\phi_{\mathrm{r}}^{(\ell)}$ represent the elevation AoD, azimuth AoD, elevation AoA, and azimuth AoA of the S2R path, respectively. $\theta_{\mathrm{t}}^{(\ell)}$, $\phi_{\mathrm{t}}^{(\ell)}$, $\theta_{\mathrm{D}}^{(\ell)}$, and $\phi_{\mathrm{D}}^{(\ell)}$ represent the elevation AoD, azimuth AoD, elevation AoA, and azimuth AoA of the R2D path, respectively. $\theta_{\widetilde{\mathrm{B}}}^{(\ell)}$, $\phi_{\widetilde{\mathrm{B}}}^{(\ell)}$, $\theta_{\widetilde{\mathrm{U}}}^{(\ell)}$, and $\phi_{\widetilde{\mathrm{U}}}^{(\ell)}$ represent the elevation AoD, azimuth AoD, elevation AoA, and azimuth AoA of the S2D path, respectively. $\mathbf{a}_{\mathrm{S}}(\cdot)$, $\mathbf{a}_{\mathrm{r}}(\cdot)$, $\mathbf{a}_{\mathrm{t}}(\cdot)$, and $\mathbf{a}_{\mathrm{D}}(\cdot)$ are the steering vectors of the UPA at the SN, the Rx-UPA at the FD-UAV relay, the Tx-UPA at the FD relay, and the UPA at the DN, respectively. The steering vectors are given as follows[56]

$$
\mathbf{a}_{\tau}(\theta_{\tau}, \phi_{\tau}) = [1, \cdots, e^{j2\pi \frac{d}{\lambda} \cos \theta_{\tau}[(m-1)\cos \phi_{\tau} + (n-1)\sin \phi_{\tau}]},
$$
$$
\cdots, e^{j2\pi \frac{d}{\lambda} \cos \theta_{\tau}[(M_{\tau}^{\mathrm{tot}}-1)\cos \phi_{\tau} + (N_{\tau}^{\mathrm{tot}}-1)\sin \phi_{\tau}]}]^{\mathrm{T}},
\tag{4.86}
$$

where d is the spacing between adjacent antennas, λ is the carrier wavelength, $0 \leq m \leq M_{\tau}^{\mathrm{tot}}-1$, $0 \leq n \leq N_{\tau}^{\mathrm{tot}}-1$, and $\tau = \{\mathrm{S}, \mathrm{r}, \mathrm{t}, \mathrm{D}\}$. Particularly, for half-wavelength spacing arrays, we have $d = \lambda/2$.

For the LoS path of the SI channel at the FD relay, the far-field range condition, $R \geq 2D^2/\lambda$, where R is the distance between the Tx antenna and the Rx antenna and D is the diameter of the antenna aperture, does not hold in general. Thus, the SI channel has to be modeled using the near-field model as follows[57, 58, 59]

$$
[\mathbf{H}_{\mathrm{SI}}]_{m,n} = \beta_{\mathrm{SI}}^{(m,n)} \exp\left(-j2\pi \frac{r_{m,n}}{\lambda}\right),
\tag{4.87}
$$

where $\beta_{\mathrm{SI}}^{(m,n)}$ are the complex coefficients of the SI channel, and $r_{m,n}$ is the distance between the m-th Tx array element and the n-th Rx array element. Note that for the SI channel, NLoS paths may also exist, due to reflectors around the FD relay. Since the propagation distances of the NLoS paths are much longer than that of the LoS path, which leads to a higher attenuation, we focus on the LoS component of the SI channel[57,58,59]. Although the SI channel model is more complicated compared to the far-field channel model, the FD relay is expected to be able to acquire the corresponding channel state information (CSI), as the SI channel is only slowly varying[57]. In this section, we assume that for a given fixed position of the FD relay, instantaneous CSI is available at the SN, FD relay, and DN via channel estimation.

Next, we provide the models for the parameters of the channel matrices in (4.83)–(4.85), (4.87). We establish a coordinate system with the origin at the SN, and the three axes x, y, and z, are separately aligned with the directions of east, north, and vertical (upward), respectively. Without loss of generality, we assume the SN and the DN both have zero altitude, and the UPAs are parallel to the plane spanned by the x and y axes. Then, the coordinates of the DN are $(x_D, y_D, 0)$, and the coordinates of the FD relay are (x_V, y_V, h_V).

According to basic geometry, we obtain the parameters of the S2R link, including the distance and the AoDs and AoAs of the LoS path, as follows

$$
\begin{cases}
d_{S2R} = \sqrt{x_V^2 + y_V^2 + h_V^2}, \\
\theta_S^{(0)} = \theta_r^{(0)} = \arctan \dfrac{h_V}{\sqrt{x_V^2 + y_V^2}}, \\
\phi_S^{(0)} = \phi_r^{(0)} = \arctan \dfrac{y_V}{x_V}.
\end{cases}
\tag{4.88}
$$

Similarly, we obtain the parameters of the R2D link as

$$
\begin{cases}
d_{R2D} = \sqrt{(x_V - x_D)^2 + (y_V - y_D)^2 + h_V^2}, \\
\theta_t^{(0)} = \theta_D^{(0)} = \arctan \dfrac{h_V}{\sqrt{(x_V - x_D)^2 + (y_V - y_D)^2}}, \\
\phi_t^{(0)} = \phi_D^{(0)} = \arctan \dfrac{y_V - y_D}{x_V - x_D}.
\end{cases}
\tag{4.89}
$$

For the S2R, R2D, and S2D links, which are characterized by far-field channels, the AoDs and AoAs of the NLoS paths are assumed to be uniformly distributed. Considering the propagation conditions at high frequencies, the complex coefficients of the LoS and NLoS paths are modeled as[60]

$$
\beta_{S2R}^{(0)} = \frac{c}{4\pi f_c} d_{S2R}^{-\alpha_{LoS}/2}, \quad \beta_{R2D}^{(0)} = \frac{c}{4\pi f_c} d_{R2D}^{-\alpha_{LoS}/2},
\tag{4.90}
$$

$$
\begin{cases}
\beta_{S2R}^{(\ell)} = \dfrac{c}{4\pi f_c} d_{S2R}^{-\alpha_{NLoS}/2} X_1, & \text{for } \ell \geq 1, \\
\beta_{R2D}^{(\ell)} = \dfrac{c}{4\pi f_c} d_{R2D}^{-\alpha_{NLoS}/2} X_2, & \text{for } \ell \geq 1, \\
\beta_{S2d}^{(\ell)} = \dfrac{c}{4\pi f_c} d_{S2d}^{-\alpha_{NLoS}/2} X_3, & \text{for } \ell \geq 1,
\end{cases}
\tag{4.91}
$$

where c is the speed of light, f_c is the carrier frequency, and $d_{\text{S2D}} = \sqrt{x_{\text{D}}^2 + y_{\text{D}}^2}$ is the distance of the S2D link. α_{LoS} and α_{NLoS} are the large-scale path loss exponents for the LoS and NLoS links, respectively. X_i, $i = 1, 2, 3$, are the gains for the NLoS paths, which are assumed to be circular symmetric complex Gaussian random variables with zero mean and standard deviation σ_f, i.e., Rayleigh fading is assumed[61]. For the SI channel, the complex coefficient is given by[57, 58, 59]

$$\beta_{\text{SI}}^{(m,n)} = \frac{c}{4\pi f_c} r_{m,n}^{-\alpha_{\text{LoS}}/2}. \tag{4.92}$$

Besides, due to obstacles on the ground, the probabilities that an LoS path exists for S2R and R2D links are modeled as logistic functions of the elevation angles[62], i.e.,

$$\hat{P}_{\text{S2R}}^{\text{LoS}} = \frac{1}{1 + a \exp\left(-b\left(\frac{180}{\pi}\theta_r^{(0)} - a\right)\right)}, \tag{4.93}$$

$$\hat{P}_{\text{R2D}}^{\text{LoS}} = \frac{1}{1 + a \exp\left(-b\left(\frac{180}{\pi}\theta_t^{(0)} - a\right)\right)}, \tag{4.94}$$

where a and b are positive modeling parameters whose values depend on the propagation environment. Random variables χ_{S2d} and χ_{d2D} in (4.83) and (4.84) are generated based on the LoS probabilities in (4.93) and (4.94), respectively. Hereto, the statistical channel models for S2R, R2D, and S2D links have been provided. For the communication scenario considered in this section, the instantaneous channel responses are generated according to these statistical models.

4.3.2 Beamforming Design

In this subsection, we design the beamforming vectors for the given coordinates of the FD relay. At first, we assume that both the S2R and the R2D links have LOS paths, i.e.,

Definition 4.3.1. (Ideal Beamforming) *For ideal beamforming vectors* $\mathbf{w}_\tau$, $\tau = \{\text{S}, \text{r}, \text{t}, \text{D}\}$, *assuming an LoS environment, the FD-UAV relay system achieves the full array gains for the S2R and R2D links, respectively, while the SI and the interference caused by the S2D link are completely eliminated in the beamforming domain, i.e.,*

$$\begin{cases} \left|\mathbf{w}_r^H \mathbf{H}_{\text{S2R}} \mathbf{w}_\text{S}\right|^2 = \left|\beta_{\text{S2R}}^{(0)}\right|^2 N_\text{S}^{\text{tot}} N_r^{\text{tot}}, \\ \left|\mathbf{w}_\text{D}^H \mathbf{H}_{\text{R2D}} \mathbf{w}_\text{t}\right|^2 = \left|\beta_{\text{R2D}}^{(0)}\right|^2 N_\text{t}^{\text{tot}} N_\text{D}^{\text{tot}}, \\ \left|\mathbf{w}_r^H \mathbf{H}_{\text{SI}} \mathbf{w}_\text{t}\right|^2 = \left|\mathbf{w}_\text{D}^H \mathbf{H}_{\text{S2D}} \mathbf{w}_\text{S}\right|^2 = 0. \end{cases} \tag{4.95}$$

Substituting (4.90) and (4.95) into (4.80) and (4.81), for a pure LoS environment, we obtain upper bounds for the achievable rates of the S2R and R2D links as follows

$$\bar{R}_{\text{S2R}} = \log_2 \left(1 + \frac{c^2}{16\pi^2 f_c^2} \frac{N_{\text{S}}^{\text{tot}} N_{\text{r}}^{\text{tot}} P_{\text{S}}^{\text{tot}}}{d_{\text{S2R}}^{\alpha_{\text{LoS}}} \sigma_1^2} \right), \tag{4.96}$$

$$\bar{R}_{\text{R2D}} = \log_2 \left(1 + \frac{c^2}{16\pi^2 f_c^2} \frac{N_{\text{t}}^{\text{tot}} N_{\text{D}}^{\text{tot}} P_{\text{V}}^{\text{tot}}}{d_{\text{R2D}}^{\alpha_{\text{LoS}}} \sigma_2^2} \right). \tag{4.97}$$

Note that the upper bounds given by (4.96) and (4.97) are valid for a pure LoS environment without NLoS paths. When the NLoS paths are also considered, we obtain upper bounds for the achievable rates of the S2R and R2D links as follows

$$\bar{\bar{R}}_{\text{S2R}} = \log_2 \left(1 + \sum_{\ell=0}^{L_{\text{S2R}}} \left| \beta_{\text{S2R}}^{(\ell)} \right|^2 \frac{N_{\text{S}}^{\text{tot}} N_{\text{r}}^{\text{tot}} P_{\text{S}}^{\text{tot}}}{\sigma_1^2} \right), \tag{4.98}$$

$$\bar{\bar{R}}_{\text{R2D}} = \log_2 \left(1 + \sum_{\ell=0}^{L_{\text{R2D}}} \left| \beta_{\text{R2D}}^{(\ell)} \right|^2 \frac{N_{\text{t}}^{\text{tot}} N_{\text{D}}^{\text{tot}} P_{\text{V}}^{\text{tot}}}{\sigma_2^2} \right). \tag{4.99}$$

We refer to the achievable rates in (4.96) and (4.97) as *approximate upper bounds*, and to the achievable rates in (4.98) and (4.99) as *strict upper bounds*.

It is assumed that full CSI is available at the SN, the DN, and the FD relay, where both the LoS and NLoS components are considered for the S2R and the R2D links. Due to the non-convex CM constraints and the coupled variables, it is challenging to jointly optimize the beamforming vectors at the SN and DN. To address this issue, we present an alternating interference suppression (AIS) algorithm, which employs alternating optimization to design the beamforming vector at the SN, the beamforming vector at the DN, and the Tx/Rx beamforming vector at the FD relay. First, we initialize the beamforming vectors with the normalized steering vectors corresponding to the LoS paths for the S2R and R2D channels, i.e.,

$$\mathbf{w}_\tau^{(0)} = \frac{1}{\sqrt{N_\tau^{\text{tot}}}} \mathbf{a}_\tau(\theta_\tau^{(0)}, \phi_\tau^{(0)}), \tau = \{\text{S}, \text{r}, \text{t}, \text{D}\}. \tag{4.100}$$

Then, we start an iterative process. Given an SN beamforming vector, a DN beamforming vector, and a Tx beamforming vector, such that the received signal power of the R2D link and the interference from the S2D link are fixed, motivated by (4.81), we optimize the Rx beamforming vector to maximize the received signal power of the S2R link, while suppressing the SI. Specifically, in the k-th iteration, we solve the following problem:

$$\underset{\mathbf{w}_{\text{r}}}{\text{Maximize}} \quad \left| \mathbf{w}_{\text{r}}^{\text{H}} \mathbf{H}_{\text{S2R}} \mathbf{w}_{\text{S}}^{(k-1)} \right| \tag{4.101a}$$

$$\text{Subject to} \quad \left| \mathbf{w}_{\text{r}}^{\text{H}} \mathbf{H}_{\text{SI}} \mathbf{w}_{\text{t}}^{(k-1)} \right| \leq \eta_1^{(k)}, \tag{4.101b}$$

$$\left| [\mathbf{w}_{\text{r}}]_n \right| \leq \frac{1}{\sqrt{N_{\text{r}}^{\text{tot}}}}, \quad 1 \leq n \leq N_{\text{r}}^{\text{tot}}, \tag{4.101c}$$

where $\mathbf{w}_{\mathrm{S}}^{(k-1)}$ and $\mathbf{w}_{\mathrm{t}}^{(k-1)}$ are the fixed SN beamforming vector and Tx beamforming vector obtained in the $(k-1)$-th iteration, respectively, and $\eta_1^{(k)}$ is the interference suppression factor. The suppression factor successively decreases in each iteration. Besides, the CM constraint on the beamforming vector is relaxed to a convex constraint in Problem (4.101). We will show later that this relaxation has little influence on the performance.

Similarly, given the Rx beamforming vector obtained in Problem (4.101), i.e., $\mathbf{w}_{\mathrm{r}}^{(k)}$, and the DN beamforming vector $\mathbf{w}_{\mathrm{D}}^{(k-1)}$, such that the received signal power of the S2R link and the interference from the S2D link are fixed, motivated by (4.80), (4.81), we optimize the Tx beamforming vector to maximize the received signal power of the R2D link, while suppressing the SI. Specifically, we solve the following problem:

$$\underset{\mathbf{w}_{\mathrm{t}}}{\text{Maximize}} \quad \left| \mathbf{w}_{\mathrm{D}}^{(k-1)\mathrm{H}} \mathbf{H}_{\mathrm{R2D}} \mathbf{w}_{\mathrm{t}} \right| \tag{4.102a}$$

$$\text{Subject to} \quad \left| \mathbf{w}_{\mathrm{r}}^{(k)\mathrm{H}} \mathbf{H}_{\mathrm{SI}} \mathbf{w}_{\mathrm{t}} \right| \le \eta_2^{(k)}, \tag{4.102b}$$

$$\left| [\mathbf{w}_{\mathrm{t}}]_n \right| \le \frac{1}{\sqrt{N_{\mathrm{t}}^{\mathrm{tot}}}}, \quad 1 \le n \le N_{\mathrm{t}}^{\mathrm{tot}}, \tag{4.102c}$$

where $\eta_2^{(k)}$ is the interference suppression factor.

After obtaining the Rx beamforming vector $\mathbf{w}_{\mathrm{r}}^{(k)}$ and the Tx beamforming vector $\mathbf{w}_{\mathrm{t}}^{(k)}$ in the k-th iteration, we optimize the SN beamforming vector and DN beamforming vector in a similar manner. Specifically, given the fixed DN beamforming vector $\mathbf{w}_{\mathrm{D}}^{(k-1)}$, we optimize the SN beamforming vector to maximize the received signal power of the S2R link, while suppressing the interference caused by the S2D link, i.e.,

$$\underset{\mathbf{w}_{\mathrm{S}}}{\text{Maximize}} \quad \left| \mathbf{w}_{\mathrm{r}}^{(k)\mathrm{H}} \mathbf{H}_{\mathrm{S2R}} \mathbf{w}_{\mathrm{S}} \right| \tag{4.103a}$$

$$\text{Subject to} \quad \left| \mathbf{w}_{\mathrm{D}}^{(k-1)\mathrm{H}} \mathbf{H}_{\mathrm{S2D}} \mathbf{w}_{\mathrm{S}} \right| \le \eta_3^{(k)}, \tag{4.103b}$$

$$\left| [\mathbf{w}_{\mathrm{S}}]_n \right| \le \frac{1}{\sqrt{N_{\mathrm{S}}^{\mathrm{tot}}}}, \quad 1 \le n \le N_{\mathrm{S}}^{\mathrm{tot}}, \tag{4.103c}$$

Finally, we optimize the DN beamforming vector to maximize the received signal power of the R2D link, while suppressing the interference caused by the S2D link, i.e.,

$$\underset{\mathbf{w}_{\mathrm{D}}}{\text{Maximize}} \quad \left| \mathbf{w}_{\mathrm{D}}^{\mathrm{H}} \mathbf{H}_{\mathrm{R2D}} \mathbf{w}_{\mathrm{t}}^{(k)} \right| \tag{4.104a}$$

$$\text{Subject to} \quad \left| \mathbf{w}_{\mathrm{D}}^{\mathrm{H}} \mathbf{H}_{\mathrm{S2D}} \mathbf{w}_{\mathrm{S}}^{(k)} \right| \le \eta_4^{(k)}, \tag{4.104b}$$

$$\left| [\mathbf{w}_{\mathrm{D}}]_n \right| \le \frac{1}{\sqrt{N_{\mathrm{D}}^{\mathrm{tot}}}}, \quad 1 \le n \le N_{\mathrm{D}}^{\mathrm{tot}}, \tag{4.104c}$$

To ensure that the interferences from the SI channel and the S2D channel are reduced in each iteration, we set

$$\eta_i^{(k)} = \eta + \mu_i^{(k)}, \; for \; i = \{1, 2, 3, 4\}, \tag{4.105}$$

where η is a nonnegative lower bound for the interference suppression factor. One possible choice is

$$\begin{cases} \mu_1^{(k)} = \dfrac{\mu_2^{(k-1)}}{\kappa}, & \mu_2^{(k)} = \dfrac{\mu_1^{(k)}}{\kappa}, \\ \mu_3^{(k)} = \dfrac{\mu_4^{(k-1)}}{\kappa}, & \mu_4^{(k)} = \dfrac{\mu_3^{(k)}}{\kappa}, \end{cases} \tag{4.106}$$

where κ is defined as the step size for the reduction of the interference suppression factor. The iterative process can be stopped when the increase of the achievable rate is no larger than a threshold ϵ_r.

Problems (4.101), (4.102), (4.103), and (4.104) have a similar form. Thus, we only develop the solution of Problem (4.101) in detail, and the other problems can be solved in the same manner. For Problem (4.101), a convex objective function is maximized, which makes it a non-convex problem[63]. Fortunately, a phase rotation of the beamforming vectors does not impact the optimality of this problem. If $\mathbf{w}_r^\star$ is an optimal solution, then $\mathbf{w}_r^\star e^{j\pi\omega}$ is also an optimal solution. Exploiting this property, we can always find an optimal solution, where the argument of the magnitude operator $|\cdot|$ in the objective function of Problem (4.101) is a real number. Then, Problem (4.101) becomes equivalent to

$$\underset{\mathbf{w}_r}{\text{Maximize}} \quad \mathfrak{R}\left(\mathbf{w}_r^H \mathbf{H}_{\text{S2R}} \mathbf{w}_S^{(k-1)}\right) \tag{4.107a}$$

$$\text{Subject to} \quad \left| \mathbf{w}_r^H \mathbf{H}_{\text{SI}} \mathbf{w}_t^{(k-1)} \right| \leq \eta_1^{(k)}, \tag{4.107b}$$

$$\left| [\mathbf{w}_r]_n \right| \leq \frac{1}{\sqrt{N_r^{\text{tot}}}}, \; 1 \leq n \leq N_r^{\text{tot}}, \tag{4.107c}$$

where $\mathfrak{R}(\cdot)$ denotes the real part of a complex number. Problem (4.107) is a convex problem and can be solved by utilizing standard optimization tools, such as CVX[63].

After obtaining the optimal solution of Problems (4.101), (4.102), (4.103), and (4.104), which we denote by $\mathbf{w}_r^\circ$, $\mathbf{w}_t^\circ$, $\mathbf{w}_S^\circ$, and $\mathbf{w}_D^\circ$, respectively, we normalize the modulus of the beamforming vectors' elements to satisfy the CM constraint, i.e.,

$$\left[\mathbf{w}_\tau^{(k)}\right]_n = \frac{1}{\sqrt{N_\tau^{\text{tot}}}} \frac{[\mathbf{w}_\tau^\circ]_n}{|[\mathbf{w}_\tau^\circ]_n|}, \; 1 \leq n \leq N_\tau^{\text{tot}}, \tau = \{\text{S, r, t, D}\}. \tag{4.108}$$

During the alternating optimization of the Tx beamforming vector and Rx beamforming vector in Problems (4.101) and (4.102), respectively, the SI at the FD relay decreases successively, because the interference suppression factor decreases in each iteration. Similarly, the interference from the S2D link decreases successively, benefiting from the alternating optimization of the SN beamforming vector and DN beamforming vector in Problems (4.103) and (4.104), respectively. Meanwhile, the beam gains of the target signals are maximized. With the AIS algorithm, the interference suppression factor finally converges to its lower bound η, and thus the powers of the SI and the interference from the S2D link are no larger than $\eta^2 P_{\mathrm{V}}^{\mathrm{tot}}$ and $\eta^2 P_{\mathrm{S}}^{\mathrm{tot}}$, respectively. To maximize the achievable rate, the interference powers should be restricted to be smaller than the noise powers, i.e.,

$$\begin{cases} \eta^2 P_{\mathrm{V}}^{\mathrm{tot}} < \sigma_1^2, \\ \eta^2 P_{\mathrm{S}}^{\mathrm{tot}} < \sigma_2^2. \end{cases} \tag{4.109}$$

Hence, a small η is preferable to minimize the influence of the SI. However, a too small value of η leads to smaller gains of the target signals because of the stricter interference constraints in (4.101), (4.102), (4.103), and (4.104). In fact, there is a tradeoff between the powers of the interferences and the powers of the target signals.

Now, the influence of the relaxation and normalization of the beamforming vectors remains to be analyzed. To this end, we provide the following theorem.

Theorem 4.3.1. *There always exists an optimal solution of Problem (4.101), where at most one element of the optimal beamforming vector does not satisfy the CM constraint.*

Proof. For notational simplicity, we employ the definitions $\mathbf{h}_{\mathrm{S2d}} = \mathbf{H}_{\mathrm{S2d}} \mathbf{w}_{\mathrm{S}}^{(k-1)}$ and $\mathbf{h}_{\mathrm{SI}} = \mathbf{H}_{\mathrm{SI}} \mathbf{w}_{\mathrm{t}}^{(k-1)}$ in (4.101). Note that Problems (4.101), (4.102), (4.103), and (4.104) have a similar form, Theorem 4.3.1 holds for all four problems. We only present the proof for Problem (4.101). A similar proof can be provided for the other problems.

Let $\mathbf{w}_{\mathrm{r}}^{\circ}$ denote the optimal solution of Problem (4.101), which satisfies

$$\begin{cases} \mathbf{w}_{\mathrm{r}}^{\circ\mathrm{H}} \mathbf{h}_{\mathrm{S2d}} = l_1 e^{j\omega_1}, \\ \mathbf{w}_{\mathrm{r}}^{\circ\mathrm{H}} \mathbf{h}_{\mathrm{SI}} = l_2 e^{j\omega_2}, \end{cases} \tag{4.110}$$

where l_1 and ω_1 denote the modulus and phase of $\mathbf{w}_{\mathrm{r}}^{\circ\mathrm{H}} \mathbf{h}_{\mathrm{S2d}}$, respectively. l_2 and ω_2 denote the modulus and phase of $\mathbf{w}_{\mathrm{r}}^{\circ\mathrm{H}} \mathbf{h}_{\mathrm{SI}}$, respectively. According to the formulation of Problem (4.101), we know that $l_2 \leq \eta_1^{(k)}$ and l_1 is the maximum of the objective function.

Note that $N_{\mathrm{r}}^{\mathrm{tot}} \geq 2$ is an implicit precondition for beamforming at the FD relay. Assume that $\mathbf{w}_{\mathrm{r}}^{\circ}$ has two elements which do not satisfy the CM constraint, i.e.,

$$\begin{cases} |[\mathbf{w}_{\mathrm{r}}^{\circ}]_{\pi_1}| < \dfrac{1}{\sqrt{N_{\mathrm{r}}^{\mathrm{tot}}}}, \\ |[\mathbf{w}_{\mathrm{r}}^{\circ}]_{\pi_2}| < \dfrac{1}{\sqrt{N_{\mathrm{r}}^{\mathrm{tot}}}}, \end{cases} \tag{4.111}$$

where $\{\pi_n\} \subseteq \{1, 2, \cdots, N_{\mathrm{r}}^{\mathrm{tot}}\}$ is the sequence of the beamforming vector's indices. Furthermore, we keep $[\mathbf{w}_{\mathrm{r}}]_{\pi_n} = [\mathbf{w}_{\mathrm{r}}^{\circ}]_{\pi_n}$ fixed for $n = 3, 4, \cdots, N_{\mathrm{r}}^{\mathrm{tot}}$, and construct a new solution by adjusting $[\mathbf{w}_{\mathrm{r}}]_{\pi_1}$ and $[\mathbf{w}_{\mathrm{r}}]_{\pi_2}$, which can be obtained by solving the following problem:

$$\begin{array}{ll}
\underset{[\mathbf{w}_r]_{\pi_1}, [\mathbf{w}_r]_{\pi_2}}{\text{Maximize}} & \left|\mathbf{w}_r^H \mathbf{h}_{S2d}\right| \\
\text{Subject to} & \mathbf{w}_r^H \mathbf{h}_{SI} = l_2 e^{j\omega_2}, \\
& \left|[\mathbf{w}_r]_{\pi_1}\right| \leq \dfrac{1}{\sqrt{N_r^{\text{tot}}}}, \\
& \left|[\mathbf{w}_r]_{\pi_2}\right| \leq \dfrac{1}{\sqrt{N_r^{\text{tot}}}}.
\end{array} \tag{4.112}$$

Based on the assumption that $\mathbf{w}_r^\circ$ is the optimal solution of Problem (4.101), we know that $\mathbf{w}_r^\circ$ is also the optimal solution of Problem (4.112), because the feasible region of Problem (4.112) is a subset of that of Problem (4.101).

Next, we provide the following two lemmas to illustrate a key property of the solution, for

$$\begin{aligned}
\frac{[\mathbf{h}_{S2d}]_{\pi_1}}{[\mathbf{h}_{S2d}]_{\pi_2}} &\neq \frac{[\mathbf{h}_{SI}]_{\pi_1}}{[\mathbf{h}_{SI}]_{\pi_2}}, \\
\frac{[\mathbf{h}_{S2d}]_{\pi_1}}{[\mathbf{h}_{S2d}]_{\pi_2}} &= \frac{[\mathbf{h}_{SI}]_{\pi_1}}{[\mathbf{h}_{SI}]_{\pi_2}},
\end{aligned} \tag{4.113}$$

respectively.

Lemma 4.3.1. *If*

$$\frac{[\mathbf{h}_{S2d}]_{\pi_1}}{[\mathbf{h}_{S2d}]_{\pi_2}} \neq \frac{[\mathbf{h}_{SI}]_{\pi_1}}{[\mathbf{h}_{SI}]_{\pi_2}}, \tag{4.114}$$

the assumption

$$\begin{aligned}
\left|[\mathbf{w}_r^\circ]_{\pi_1}\right| &< \frac{1}{\sqrt{N_r^{\text{tot}}}}, \\
\left|[\mathbf{w}_r^\circ]_{\pi_2}\right| &< \frac{1}{\sqrt{N_r^{\text{tot}}}},
\end{aligned} \tag{4.115}$$

cannot hold.

Proof. If

$$\frac{[\mathbf{h}_{S2d}]_{\pi_1}}{[\mathbf{h}_{S2d}]_{\pi_2}} \neq \frac{[\mathbf{h}_{SI}]_{\pi_1}}{[\mathbf{h}_{SI}]_{\pi_2}}, \tag{4.116}$$

holds, according to the first constraint in Problem (4.112), we can express $[\mathbf{w}_r]_{\pi_2}$ as a function of $[\mathbf{w}_r]_{\pi_1}$, i.e.,

$$[\mathbf{w}_r]_{\pi_2}^* = \frac{l_2 e^{j\omega_2} - \sum_{n=3}^{N_r^{\text{tot}}} [\mathbf{w}_r^\circ]_{\pi_n}^* [\mathbf{h}_{SI}]_{\pi_n}}{[\mathbf{h}_{SI}]_{\pi_2}} - [\mathbf{w}_r]_{\pi_1}^* \frac{[\mathbf{h}_{SI}]_{\pi_1}}{[\mathbf{h}_{SI}]_{\pi_2}} \triangleq f_1\left([\mathbf{w}_r]_{\pi_1}\right). \tag{4.117}$$

Substituting (4.117) into the objective function of Problem (4.112), we obtain

$$\mathbf{w}_r^H \mathbf{h}_{S2d}$$

$$= [\mathbf{w}_r]_{\pi_1}^* [\mathbf{h}_{S2d}]_{\pi_1} + [\mathbf{w}_r]_{\pi_2}^* [\mathbf{h}_{S2d}]_{\pi_2} + \sum_{n=3}^{N_r^{tot}} [\mathbf{w}_r^\circ]_{\pi_n}^* [\mathbf{h}_{S2d}]_{\pi_n}$$

$$= [\mathbf{w}_r]_{\pi_1}^* \underbrace{\left([\mathbf{h}_{S2d}]_{\pi_1} - [\mathbf{h}_{S2d}]_{\pi_2} \frac{[\mathbf{h}_{SI}]_{\pi_1}}{[\mathbf{h}_{SI}]_{\pi_2}} \right)}_{=\hat{k}} \tag{4.118}$$

$$+ [\mathbf{h}_{S2d}]_{\pi_2} \underbrace{\frac{l_2 e^{j\omega_2} - \sum\limits_{n=3}^{N_r^{tot}} [\mathbf{w}_r^\circ]_{\pi_n}^* [\mathbf{h}_{SI}]_{\pi_n}}{[\mathbf{h}_{SI}]_{\pi_2}} \sum_{n=3}^{N_r^{tot}} [\mathbf{w}_r^\circ]_{\pi_n}^* [\mathbf{h}_{S2d}]_{\pi_n}}_{=\hat{b}}$$

$$\triangleq \hat{k}[\mathbf{w}_r]_{\pi_1}^* + \hat{b} \triangleq f_2\left([\mathbf{w}_r]_{\pi_1}\right).$$

Note that

$$\frac{[\mathbf{h}_{S2d}]_{\pi_1}}{[\mathbf{h}_{S2d}]_{\pi_2}} \neq \frac{[\mathbf{h}_{SI}]_{\pi_1}}{[\mathbf{h}_{SI}]_{\pi_2}}, \tag{4.119}$$

holds in Lemma 4.3.1. Thus, we have $\hat{k} \neq 0$ in (4.118). Because of the assumption

$$\left| [\mathbf{w}_r^\circ]_{\pi_1} \right| < \frac{1}{\sqrt{N_r^{tot}}},$$
$$\left| [\mathbf{w}_r^\circ]_{\pi_2} \right| < \frac{1}{\sqrt{N_r^{tot}}}, \tag{4.120}$$

we can always find a real number δ, which is positive and small enough to satisfy

$$\begin{cases} \left| [\mathbf{w}_r^\circ]_{\pi_1} \pm \delta \right| < \dfrac{1}{\sqrt{N_r^{tot}}}, \\[2mm] \left| f_1\left([\mathbf{w}_r^\circ]_{\pi_1} \pm \delta\right) \right| < \dfrac{1}{\sqrt{N_r^{tot}}}. \end{cases} \tag{4.121}$$

This means that $([\mathbf{w}_r^\circ]_{\pi_1} + \delta)$ and $([\mathbf{w}_r^\circ]_{\pi_1} - \delta)$ are both located in the feasible region of Problem (4.112). Since $[\mathbf{w}_r^\circ]_{\pi_1}$ is the optimal solution of Problem (4.112), the objective function at $[\mathbf{w}_r^\circ]_{\pi_1} + \delta$ and $[\mathbf{w}_r^\circ]_{\pi_1} - \delta$ is no larger than at $[\mathbf{w}_r^\circ]_{\pi_1}$, i.e.,

$$\begin{cases} \left| f_2\left([\mathbf{w}_r^\circ]_{\pi_1} + \delta\right) \right|^2 \leq \left| f_2\left([\mathbf{w}_r^\circ]_{\pi_1}\right) \right|^2, \\[2mm] \left| f_2\left([\mathbf{w}_r^\circ]_{\pi_1} - \delta\right) \right|^2 \leq \left| f_2\left([\mathbf{w}_r^\circ]_{\pi_1}\right) \right|^2, \end{cases} \tag{4.122}$$

According to the definition in (4.118), we obtain

$$\begin{cases} \left| \hat{k}[\mathbf{w}_r^\circ]_{\pi_1}^* + \hat{b} + \hat{k}\delta \right|^2 \leq \left| \hat{k}[\mathbf{w}_r^\circ]_{\pi_1}^* + \hat{b} \right|^2 \\[2mm] \left| \hat{k}[\mathbf{w}_r^\circ]_{\pi_1}^* + \hat{b} - \hat{k}\delta \right|^2 \leq \left| \hat{k}[\mathbf{w}_r^\circ]_{\pi_1}^* + \hat{b} \right|^2 \end{cases} \Rightarrow$$

$$\begin{cases} \Re\left(\left(\hat{k}[\mathbf{w}_r^\circ]_{\pi_1}^* + \hat{b} \right)^* \hat{k}\delta \right) + \left| \hat{k}\delta \right|^2 \leq 0 \\[2mm] -\Re\left(\left(\hat{k}[\mathbf{w}_r^\circ]_{\pi_1}^* + \hat{b} \right)^* \hat{k}\delta \right) + \left| \hat{k}\delta \right|^2 \leq 0 \end{cases} \Rightarrow 2\left| \hat{k}\delta \right|^2 \leq 0, \tag{4.123}$$

which contradicts the fact that $\hat{k} \neq 0$ and $\delta > 0$. Thus, we can conclude that the assumption that $\mathbf{w}_r^\circ$ has two elements that do not satisfy the CM constraint cannot hold when

$$\frac{[\mathbf{h}_{S2d}]_{\pi_1}}{[\mathbf{h}_{S2d}]_{\pi_2}} \neq \frac{[\mathbf{h}_{SI}]_{\pi_1}}{[\mathbf{h}_{SI}]_{\pi_2}}. \tag{4.124}$$

In other words, if there are any two elements that do not satisfy the CM constraint, they always have

$$\frac{[\mathbf{h}_{S2d}]_{\pi_1}}{[\mathbf{h}_{S2d}]_{\pi_2}} = \frac{[\mathbf{h}_{SI}]_{\pi_1}}{[\mathbf{h}_{SI}]_{\pi_2}}. \tag{4.125}$$

$\square$

Lemma 4.3.2. *If*

$$\frac{[\mathbf{h}_{S2d}]_{\pi_1}}{[\mathbf{h}_{S2d}]_{\pi_2}} = \frac{[\mathbf{h}_{SI}]_{\pi_1}}{[\mathbf{h}_{SI}]_{\pi_2}}, \tag{4.126}$$

holds, there always exists another optimal solution of Problem (4.112), where at least one of $[\mathbf{w}_r]_{\pi_1}$ and $[\mathbf{w}_r]_{\pi_2}$ satisfies the CM constraint.

Proof. Based on

$$\frac{[\mathbf{h}_{S2d}]_{\pi_1}}{[\mathbf{h}_{S2d}]_{\pi_2}} = \frac{[\mathbf{h}_{SI}]_{\pi_1}}{[\mathbf{h}_{SI}]_{\pi_2}}, \tag{4.127}$$

we obtain

$$\frac{[\mathbf{h}_{S2d}]_{\pi_1}}{[\mathbf{h}_{SI}]_{\pi_1}} = \frac{[\mathbf{h}_{S2d}]_{\pi_2}}{[\mathbf{h}_{SI}]_{\pi_2}} = \frac{[\mathbf{w}_r]_{\pi_1}^* [\mathbf{h}_{S2d}]_{\pi_1} + [\mathbf{w}_r]_{\pi_2}^* [\mathbf{h}_{S2d}]_{\pi_2}}{[\mathbf{w}_r]_{\pi_1}^* [\mathbf{h}_{SI}]_{\pi_1} + [\mathbf{w}_r]_{\pi_2}^* [\mathbf{h}_{SI}]_{\pi_2}} \triangleq \chi. \tag{4.128}$$

This indicates that

$$\begin{aligned} &[\mathbf{w}_r]_{\pi_1}^* [\mathbf{h}_{S2d}]_{\pi_1} + [\mathbf{w}_r]_{\pi_2}^* [\mathbf{h}_{S2d}]_{\pi_2}, \\ &[\mathbf{w}_r]_{\pi_1}^* [\mathbf{h}_{SI}]_{\pi_1} + [\mathbf{w}_r]_{\pi_2}^* [\mathbf{h}_{SI}]_{\pi_2}, \end{aligned} \tag{4.129}$$

always has the same ratio regardless of the values of $[\mathbf{w}_r]_{\pi_1}$ and $[\mathbf{w}_r]_{\pi_2}$. We call this property the *constant-ratio property*.

Since

$$\begin{aligned} |[\mathbf{w}_r^\circ]_{\pi_1}| &< \frac{1}{\sqrt{N_r^{\text{tot}}}}, \\ |[\mathbf{w}_r^\circ]_{\pi_2}| &< \frac{1}{\sqrt{N_r^{\text{tot}}}}, \end{aligned} \tag{4.130}$$

it is easy to see that

$$0 \leq \left| [\mathbf{w}_r^\circ]_{\pi_1}^* [\mathbf{h}_{S2d}]_{\pi_1} + [\mathbf{w}_r^\circ]_{\pi_2}^* [\mathbf{h}_{S2d}]_{\pi_2} \right| < \frac{1}{\sqrt{N_r^{\text{tot}}}} \left(|[\mathbf{h}_{S2d}]_{\pi_1}| + |[\mathbf{h}_{S2d}]_{\pi_2}| \right), \tag{4.131}$$

and

$$0 \leq \left| [\mathbf{w}_r^\circ]_{\pi_1}^* [\mathbf{h}_{SI}]_{\pi_1} + [\mathbf{w}_r^\circ]_{\pi_2}^* [\mathbf{h}_{SI}]_{\pi_2} \right| < \frac{1}{\sqrt{N_r^{\text{tot}}}} \left(|[\mathbf{h}_{SI}]_{\pi_1}| + |[\mathbf{h}_{SI}]_{\pi_2}| \right). \tag{4.132}$$

Next, we will consider two cases shown in Fig. 4.14. We define

$$
\begin{cases}
\bar{a} = \left| [\mathbf{w}_r^\circ]_{\pi_1}^* \, [\mathbf{h}_{S2d}]_{\pi_1} + [\mathbf{w}_r^\circ]_{\pi_2}^* \, [\mathbf{h}_{S2d}]_{\pi_2} \right|, \\[2mm]
\bar{b} = \dfrac{1}{\sqrt{N_r^{\text{tot}}}} \left| [\mathbf{h}_{S2d}]_{\pi_1} \right|, \\[2mm]
\bar{c} = \dfrac{1}{\sqrt{N_r^{\text{tot}}}} \left| [\mathbf{h}_{S2d}]_{\pi_2} \right|.
\end{cases}
\tag{4.133}
$$

The corresponding angles in Fig. 4.14 are defined as follows

$$
\begin{cases}
u = \angle \left([\mathbf{w}_r^\circ]_{\pi_1}^* \, [\mathbf{h}_{S2d}]_{\pi_1} + [\mathbf{w}_r^\circ]_{\pi_2}^* \, [\mathbf{h}_{S2d}]_{\pi_2} \right), \\[2mm]
v_1 = \arccos \dfrac{\bar{a}^2 + \bar{b}^2 - \bar{c}^2}{2\bar{a}\bar{b}}, \\[2mm]
v_2 = \arccos \dfrac{\bar{a}^2 + \bar{c}^2 - \bar{b}^2}{2\bar{a}\bar{c}}.
\end{cases}
\tag{4.134}
$$

Case 1: $\bar{a} \geq \left| \bar{b} - \bar{c} \right|$.

In this case, according to the constant-ratio property, it is easy to verify that

$$
\left| [\mathbf{w}_r^\circ]_{\pi_1}^* \, [\mathbf{h}_{SI}]_{\pi_1} + [\mathbf{w}_r^\circ]_{\pi_2}^* \, [\mathbf{h}_{SI}]_{\pi_2} \right| \geq \frac{1}{\sqrt{N_r^{\text{tot}}}} \left(\left| [\mathbf{h}_{SI}]_{\pi_1} \right| + \left| [\mathbf{h}_{SI}]_{\pi_2} \right| \right).
\tag{4.135}
$$

According to the triangle inequality, we can always find other $[\mathbf{w}_r]_{\pi_1}$ and $[\mathbf{w}_r]_{\pi_2}$ which satisfy the CM constraint. The basic idea is to adjust the phases of the two complex elements, and keep

$$
[\mathbf{w}_r]_{\pi_1}^* \, [\mathbf{h}_{S2d}]_{\pi_1} + [\mathbf{w}_r]_{\pi_2}^* \, [\mathbf{h}_{S2d}]_{\pi_2} = \bar{a} e^{ju},
\tag{4.136}
$$

unchanged in Fig. 4.14. The new solutions are generated as follows

$$
\begin{cases}
[\mathbf{w}_r^\diamond]_{\pi_1} = \dfrac{1}{\sqrt{N_r^{\text{tot}}}} e^{-j(u - v_1 - \vartheta_1)}, \\[2mm]
[\mathbf{w}_r^\diamond]_{\pi_2} = \dfrac{1}{\sqrt{N_r^{\text{tot}}}} e^{-j(u + v_2 - \vartheta_2)},
\end{cases}
\tag{4.137}
$$

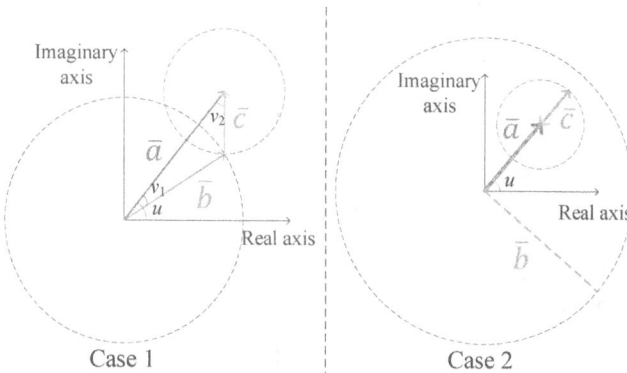

Figure 4.14 Illustration of the adjustment for the beamforming vector's elements.

where $\vartheta_1 = \angle([\mathbf{h}_{\mathrm{S2d}}]_{\pi_1})$ and $\vartheta_2 = \angle([\mathbf{h}_{\mathrm{S2d}}]_{\pi_2})$. Then, it is easy to verify that $[\mathbf{w}_{\mathrm{r}}^\circ]_{\pi_1}$ and $[\mathbf{w}_{\mathrm{r}}^\circ]_{\pi_2}$ in (4.137) satisfy

$$\begin{cases} [\mathbf{w}_{\mathrm{r}}^\circ]_{\pi_1}^* [\mathbf{h}_{\mathrm{S2d}}]_{\pi_1} + [\mathbf{w}_{\mathrm{r}}^\circ]_{\pi_2}^* [\mathbf{h}_{\mathrm{S2d}}]_{\pi_2} = [\mathbf{w}_{\mathrm{r}}^\circ]_{\pi_1}^* [\mathbf{h}_{\mathrm{S2d}}]_{\pi_1} + [\mathbf{w}_{\mathrm{r}}^\circ]_{\pi_2}^* [\mathbf{h}_{\mathrm{S2d}}]_{\pi_2}, \\ [\mathbf{w}_{\mathrm{r}}^\circ]_{\pi_1}^* [\mathbf{h}_{\mathrm{SI}}]_{\pi_1} + [\mathbf{w}_{\mathrm{r}}^\circ]_{\pi_2}^* [\mathbf{h}_{\mathrm{SI}}]_{\pi_2} = [\mathbf{w}_{\mathrm{r}}^\circ]_{\pi_1}^* [\mathbf{h}_{\mathrm{SI}}]_{\pi_1} + [\mathbf{w}_{\mathrm{r}}^\circ]_{\pi_2}^* [\mathbf{h}_{\mathrm{SI}}]_{\pi_2}, \end{cases} \tag{4.138}$$

which means that the designed $[\mathbf{w}_{\mathrm{r}}^\circ]_{\pi_1}$ and $[\mathbf{w}_{\mathrm{r}}^\circ]_{\pi_2}$ in (4.137) are also optimal solutions of Problem (4.112) for which all elements satisfy the CM constraint.

Case 2: $\bar{a} > |\bar{b} - \bar{c}|$.

In this case, according to the constant-ratio property, it is easy to verify that

$$\left| [\mathbf{w}_{\mathrm{r}}^\circ]_{\pi_1}^* [\mathbf{h}_{\mathrm{SI}}]_{\pi_1} + [\mathbf{w}_{\mathrm{r}}^\circ]_{\pi_2}^* [\mathbf{h}_{\mathrm{SI}}]_{\pi_2} \right| < \frac{1}{\sqrt{N_{\mathrm{r}}^{\mathrm{tot}}}} \left(\left| [\mathbf{h}_{\mathrm{SI}}]_{\pi_1} \right| + \left| [\mathbf{h}_{\mathrm{SI}}]_{\pi_2} \right| \right), \tag{4.139}$$

holds. This indicates that $[\mathbf{w}_{\mathrm{r}}]_{\pi_1}$ and $[\mathbf{w}_{\mathrm{r}}]_{\pi_2}$ cannot be adjusted such that both satisfy the CM constraint because the triangle inequality is not satisfied, i.e., the difference between the lengths of two sides is less than the length of the third side. However, we can adjust them such that one element satisfies the CM constraint. The basic idea is to enlarge the shorter side to satisfy the CM constraint, and then adjust the longer side to keep

$$[\mathbf{w}_{\mathrm{r}}]_{\pi_1}^* + [\mathbf{w}_{\mathrm{r}}]_{\pi_2}^* [\mathbf{h}_{\mathrm{S2d}}]_{\pi_2} = \bar{a} e^{ju}, \tag{4.140}$$

unchanged in Fig. 4.14.

Without loss of generality, we assume $\bar{b} \geq \bar{c}$ as shown in Fig. 4.14.[10] Then, we can generate a new solution as follows

$$\begin{cases} [\mathbf{w}_{\mathrm{r}}^\circ]_{\pi_1} = \left(\frac{1}{\sqrt{N_{\mathrm{r}}^{\mathrm{tot}}}} + \frac{a}{\left| [\mathbf{h}_{\mathrm{S2d}}]_{\pi_1} \right|} \right) e^{-j(u-\vartheta_1)}, \\ [\mathbf{w}_{\mathrm{r}}^\circ]_{\pi_2} = \frac{1}{\sqrt{N_{\mathrm{r}}^{\mathrm{tot}}}} e^{-j(u-\vartheta_2+\pi)}. \end{cases} \tag{4.141}$$

It is easy to verify that $[\mathbf{w}_{\mathrm{r}}^\circ]_{\pi_1}$ and $[\mathbf{w}_{\mathrm{r}}^\circ]_{\pi_2}$ in (4.141) satisfy (4.138), which means that they are also an optimal solution of Problem (4.112) for which only one element does not satisfy the CM constraint. Thus, we can conclude that if

$$\frac{[\mathbf{h}_{\mathrm{S2d}}]_{\pi_1}}{[\mathbf{h}_{\mathrm{S2d}}]_{\pi_2}} = \frac{[\mathbf{h}_{\mathrm{SI}}]_{\pi_1}}{[\mathbf{h}_{\mathrm{SI}}]_{\pi_2}}, \tag{4.142}$$

holds, we can always construct an optimal solution of Problem (4.112), where at most one element does not satisfy the CM constraint. □

Based on Lemma 4.3.1, we know that for any two elements of the beamforming vector which do not satisfy the CM constraint,

$$\frac{[\mathbf{h}_{\mathrm{S2d}}]_{\pi_1}}{[\mathbf{h}_{\mathrm{S2d}}]_{\pi_2}} \neq \frac{[\mathbf{h}_{\mathrm{SI}}]_{\pi_1}}{[\mathbf{h}_{\mathrm{SI}}]_{\pi_2}}, \tag{4.143}$$

---

[10]When $\bar{b} < \bar{c}$, we can construct new optimal solutions in a similar manner.

cannot hold. In other words, these elements always satisfy the *constant-ratio property* in Lemma 4.3.2. Then, for any two elements that do not satisfy the CM constraint, we can always construct a new solution based on Lemma 4.3.2, where at most one element does not satisfies the CM constraint. Note that if there are three or more elements that do not satisfy the CM constraint, this construction can be repeated until only one or zero elements do not satisfy the CM constraint. Thus, we can conclude that there always exists an optimal solution of Problem (4.101), for which at most one element of the optimal beamforming vector does not satisfy the CM constraint.

□

Theorem 4.3.1 suggests that the relaxation and normalization of the beamforming vectors in (4.108) have little influence on the rate performance because they impact at most one of their elements. In particular, when the number of antennas is large, the impact of a single element's normalization on the effective channel gain is small.

We summarize the solution of beamforming for FD relay systems in Algorithm 4.2. In line 1, we assume that the FD relay is in the optimal position. Then, in lines 2-22, we successively decrease the interferences by alternately solving Problems (4.101), (4.102), (4.103), and (4.104). The algorithm terminates if the improvement in the achievable rate from one iteration to the next falls below a threshold ϵ_r. The convergence of Algorithm 4.2 will be studied via simulations in the following content.

4.3.3 Performance Evaluation

In this subsection, we provide simulation results to evaluate the performance of the presented beamforming scheme for FD relay systems.

4.3.3.1 Simulation Setup

We adopt the channel models in (4.83), (4.84), (4.85), and (4.87), where the probabilities that an LoS path exists for the S2R and R2D channels are given by (4.93) and (4.94), respectively. The number of NLoS components for the S2R, R2D, and S2D channels are assumed to be identical, i.e.,

$$L_{\text{S2d}} = L_{\text{d2D}} = L_{\text{S2d}} = L. \tag{4.144}$$

Half-wavelength spacing UPAs are used at all nodes, and the Tx-UPA and Rx-UPA at the FD relay are parallel to each other with a distance of 10λ ($\approx$ 8 cm). For the presented AIS algorithm, the lower bound for the SI suppression factor is set to

$$\eta = \min \left\{ \frac{\sigma_1}{10\sqrt{P_{\text{S}}^{\text{tot}}}}, \frac{\sigma_2}{10\sqrt{P_{\text{V}}^{\text{tot}}}} \right\}, \tag{4.145}$$

such that the interference power is in the same range as the noise power. Each simulation point is averaged over 10^3 node distributions and channel realizations, where the DN is randomly distributed in a disk of radius 500 m, with the SN at its center.

Algorithm 4.2: Beamforming for FD relay systems.

Input: M_S, N_S, M_D, N_D, M_t, N_t, M_r, N_r, x_D, y_D,
  h_{min}, h_{max} P_S^{tot}, P_V^{tot}, σ_1, σ_2, f_c α_{LoS}, α_{NLoS},
  σ_f, a, b, ϵ_x, ϵ_y, ϵ_h, η, κ, ϵ_r.
Output: x_V°, y_V°, h_V°, $\mathbf{w}_S^\star$, $\mathbf{w}_D^\star$, $\mathbf{w}_r^\star$, $\mathbf{w}_t^\star$, $P_S^\star$, $P_V^\star$.
1: Set $(x_V^\circ, y_V^\circ, h_V^\circ) = (x_V^\star, y_V^\star, h_V^\star)$.
2: Estimate channel matrices $\mathbf{H}_{S2d}$, $\mathbf{H}_{d2D}$, $\mathbf{H}_{S2d}$, and $\mathbf{H}_{SI}$.
3: Initialize $k = 0$.
4: Initialize $\mathbf{w}_S^{(0)}$, $\mathbf{w}_D^{(0)}$, $\mathbf{w}_r^{(0)}$ and $\mathbf{w}_t^{(0)}$ according to (4.100).
5: Initialize $\mu_2^{(0)} = \left| \mathbf{w}_r^{(0)H} \mathbf{H}_{SI} \mathbf{w}_t^{(0)} \right|$.
6: Calculate $R_{S2d}^{(0)}$ according to (4.80) and define $R_{S2d}^{(-1)} = -\infty$.
7: **while** $R_{S2d}^{(k)} - R_{S2d}^{(k-1)} > \epsilon_r$ **do**
8:    $k = k + 1$.
9:    Update the suppression factor $\mu_i^{(k)} = \frac{\mu_{i+1}^{(k-1)}}{\kappa}$ and $\eta_i^{(k)} = \eta + \mu_i^{(k)}$ for $i = 1, 3$.
10:    Update the suppression factor $\mu_i^{(k)} = \frac{\mu_{i-1}^{(k)}}{\kappa}$ and $\eta_i^{(k)} = \eta + \mu_i^{(k)}$ $i = 2, 4$.
11:    Solve Problem (4.101) to obtain $\mathbf{w}_r^\circ$.
12:    Normalize $\mathbf{w}_r^\circ$ according to (4.108) and obtain $\mathbf{w}_r^{(k)}$.
13:    Solve Problem (4.102) to obtain $\mathbf{w}_t^\circ$.
14:    Normalize $\mathbf{w}_t^\circ$ according to (4.108) and obtain $\mathbf{w}_t^{(k)}$.
15:    Solve Problem (4.103) to obtain $\mathbf{w}_S^\circ$.
16:    Normalize $\mathbf{w}_S^\circ$ according to (4.108) and obtain $\mathbf{w}_S^{(k)}$.
17:    Solve Problem (4.104) to obtain $\mathbf{w}_D^\circ$.
18:    Normalize $\mathbf{w}_D^\circ$ according to (4.108) and obtain $\mathbf{w}_D^{(k)}$.
19:    Calculate $R_{S2d}^{(k)}$ according to (4.80).
20: **end while**
21: $\mathbf{w}_r^\star = \mathbf{w}_r^{(k)}$, $\mathbf{w}_t^\star = \mathbf{w}_t^{(k)}$, $P_S^\star = P_S^{(k)}$, and $P_V^\star = P_V^{(k)}$.
22: **return** $x_V^\star, y_V^\star, h_V^\star, \mathbf{w}_S^\star, \mathbf{w}_D^\star, \mathbf{w}_r^\star, \mathbf{w}_t^\star, P_S^\star, P_V^\star$.

Two upper bounds for the achievable rate for FD relay systems are considered. The presented approximate upper bound is obtained as the minimum of (4.96) and (4.97), while the presented strict upper bound is the minimum of (4.98) and (4.99).

4.3.3.2 Simulation Results

First, in Fig. 4.15, we evaluate the convergence of the presented AIS beamforming method (Algorithm 4.2) for different step sizes for the reduction of the interference suppression factor (i.e., κ in Algorithm 4.2). Identical sizes are adopted for the UPA at the SN, the UPA at the DN, and the Tx and Rx UPAs at the FD relay, i.e., 4×4 or 8×8. As can be observed, the presented ASIS beamforming method converges very fast to a value close to the performance upper bound, and the approximate upper bound is very close to the strict upper bound. These results confirm the assumption of a pure LoS environment in the previous content because the LoS path has much higher power compared to the NLoS paths.

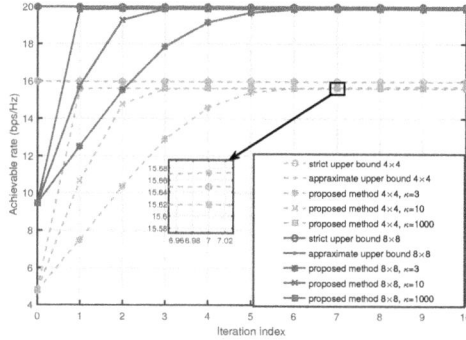

Figure 4.15 Evaluation of the convergence of the presented Algorithm 4.2 for different values of κ.

When the antenna array size is 4×4 at the FD relay, after convergence, the performance gap between the presented method and the upper bound is no more than 0.3 bps/Hz, and this gap reduces to 0.1 bps/Hz when the antenna array size is 8×8. For larger numbers of antennas, there are more DoFs for minimization of the SI. Thus, the performance gap between the presented method and the upper bound becomes smaller. The results in Fig. 4.15 demonstrate that the presented method can achieve a near-upper-bound performance in terms of the achievable rate. In addition, the speed of convergence of the presented AIS algorithm depends on the step size for the reduction of the suppression factor. For larger κ, the AIS algorithm converges faster. However, if κ is chosen too large, for example, $\kappa \to +\infty$, the SI decreases too fast in the first iteration for designing $\mathbf{w}_r^{(k)}$. As such, the effective channel gain of the S2R link may be much smaller than that of the R2D link, which negatively affects the achievable rate of the DN. Thus, to achieve a favorable tradeoff between the achievable rate and computational complexity, we set $\kappa = 10$ for the following simulations.

Fig. 4.16 shows the convergence performance of the presented AIS algorithm for different maximum transmit powers of the FD relay. For all considered cases, the presented algorithm converges to a near-upper-bound achievable rate within a few iterations, where all curves reach steady state after 4 iterations. Particularly, as the maximum transmit power at the FD relay increases, the number of iterations required for convergence increases. The reason is that a higher transmit power of the delay causes more SI, and thus more iterations are required to successively reduce the SI.

To shed more light on the properties of Algorithm 4.2, in Fig. 4.17, we show the change of the channel gains and transmit powers during the iterations. In particular, we show the normalized channel gains, which are the ratios of the effective channel

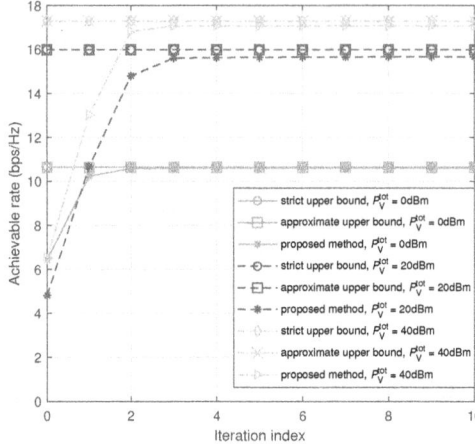

Figure 4.16 Evaluation of the convergence of the presented Algorithm 4.2 for different maximum transmit powers of the FD relay.

gains and the noise power in (4.80) and (4.81), i.e.,

$$
\begin{cases}
\left| \mathbf{w}_r^H \mathbf{H}_{S2d} \mathbf{w}_S \right|^2 / \sigma_1^2, \\[2mm]
\left| \mathbf{w}_r^H \mathbf{H}_{SI} \mathbf{w}_t \right|^2 / \sigma_1^2, \\[2mm]
\left| \mathbf{w}_D^H \mathbf{H}_{d2D} \mathbf{w}_t \right|^2 / \sigma_1^2, \\[2mm]
\left| \mathbf{w}_D^H \mathbf{H}_{S2d} \mathbf{w}_S \right|^2 / \sigma_1^2.
\end{cases}
\tag{4.146}
$$

As can be observed, the channel gain of the SI channel decreases fast and converges to the lower bound η^2/σ_1^2, since the SI suppression factor is reduced in each iteration in (4.101) and (4.102). The channel gain of the S2D channel is always lower than that of the SI channel because of the long transmission distance and the blockage of the LoS link between SN and DN. Besides, the channel gains of the S2R and R2D links remain almost unchanged during the iterations, which confirms the rational behind the presented AIS beamforming algorithm. This is also the reason why the achievable rate of the presented scheme can approach the performance upper bound. For the variation of transmit powers, during the first iteration, the transmit power of the FD relay is very low, while the SN transmits with the maximal power. This is because the S2R link suffers from high SI for the initially chosen beamforming vectors, and thus the FD relay reduces the transmit power to decrease the SI. After several iterations, the effective channel gain of the SI channel becomes lower, and thus the FD relay can increase its transmit power to improve the achievable rate of the R2D link.

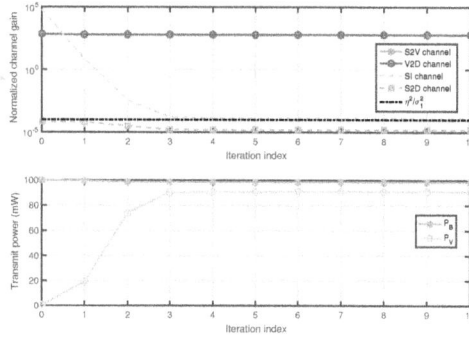

Figure 4.17 Normalized channel gains and transmit powers versus iteration index.

4.4 SUMMARY

This chapter introduced the opportunities, challenges, and possible solutions for array beamforming enabled FD transmission, where beamforming provides a new DoF for SI cancellation and FD enables improvement of the spectrum efficiency for array beamforming communication systems. We first introduced antenna configurations for FD transmission, and compared two typical settings, i.e., with separate Tx/Rx arrays or with the same array. It was shown that the configuration with separate Tx/Rx arrays provides flexibility to mitigate SI with signal processing methods, while it may increase some cost and space versus that with the same array. As SI is still significant in FD communication even with separate Tx/Rx antenna arrays, we need to mitigate it as much as possible, where beamforming design plays an important role.

Without considering the CM constraint of communication, we introduced a JTR-BF problem in the presence of significant SI. As the problem of finding the optimal beamforming vectors to maximize the JAR is non-convex, several sub-optimal solutions were presented. Firstly, ZF-Max-Power, which restricts the original problem by ZF SI and alternatively optimizes the desired power, was presented, and its convergence is proven. It was shown that the computational complexity of ZF-Max-Power is lower than that of Max-SINR by one order of magnitude. Next, two closed-form solutions, namely LB-MMSE and SI-ZF-MRT were introduced, by jointly using MMSE, ZF and MRT criteria. Performance evaluations showed that ZF-Max-Power approaches an upper bound on the JAR, and it needed only 2 iterations on average to achieve the convergence with random initial points. LB-MMSE and SI-ZF-MRT achieved sub-optimal performances. In addition, we found that ZF-Max-Power and SI-ZF-MRT are robust against the geometry of Tx/Rx antenna arrays due to the operation of ZF SI, while LB-MMSE is not. ZF-Max-Power and LB-MMSE are robust against channel estimation error, while SI-ZF-MRT is not. Furthermore, these schemes basically achieved better JAR performance with a separate-array setting than those sharing the same array. The results demonstrated the feasibility of FD communications and the effectiveness of beamforming cancellation.

With the structural constraints of analog beamforming taken into account, we presented several candidate solutions. Results showed that the CM constraints lead to a significant performance loss, and thus more efficient algorithms are required. Moreover, the CM constraint could be circumvented in the beamforming cancellation by doubling the number of Tx/Rx RF chains, and in such a case beamforming cancellation could achieve much better performance. Lastly, we considered an FD multi-user scenario, and showed that even if there are no FD users in an FD cellular system, the FD benefit could still be exploited in the FD BS. However, interference cancellation is more stringent in this scenario, because there are both SI and MUI. We also presented two candidate solutions for this scenario, where the hybrid beamforming problem is simplified by temporarily neglecting the MUI or the CM constraints. In addition, the approach of doubling the number of RF chains may also be workable here provided that the complexity is affordable.

Finally, we presented to employ an FD relay to improve the achievable rate of communication system, where the source node, destination node, and FD relay are all equipped with UPAs and use directional beams to overcome the high path loss of signals. Analog beamforming was utilized to mitigate the SI at the FD relay. We developed an iterative algorithm for optimization of the beamforming vectors. In each iteration, the beamforming vectors were optimized for maximization of the beam gains of the target signals and successive reduction of the interference. Simulation results demonstrated that the presented beamforming method for the FD relay system can closely approach a performance upper bound in terms of the achievable rate.

Bibliography

[1] Su-Khiong Yong, Pengfei Xia, and Alberto Valdes-Garcia. *60GHz Technology for Gbps WLAN and WPAN: from Theory to Practice*. Wiley, West Sussex, UK, 2011.

[2] Dinesh Bharadia, Emily McMilin, and Sachin Katti. Full duplex radios. In *Special Interest Group on Data Communication (SIGCOMM)*, Hong Kong, China, 2013. ACM.

[3] Ashutosh Sabharwal, Philip Schniter, Dongning Guo, Daniel W. Bliss, Sampath Rangarajan, and Risto Wichman. In-band full-duplex wireless: Challenges and opportunities. *IEEE J. Select. Areas Commun.*, 32(9):1637–1652, Sep. 2014.

[4] Tissana Kijsanayotin and James F. Buckwalter. Millimeter-wave dual-band, bidirectional amplifier and active circulator in a CMOS SOI process. *IEEE Trans. Microwave Theory Tech.*, 62(12):3028–3040, 2014.

[5] Zhenyu Xiao. Suboptimal spatial diversity scheme for 60 GHz millimeter-wave WLAN. *IEEE Commun. Lett.*, 17(9):1790–1793, Sep. 2013.

[6] Zhenyu Xiao, Lin Bai, and Jinho Choi. Iterative joint beamforming training with constant-amplitude phased arrays in millimeter-wave communications. *IEEE Commun. Lett.*, 18(5):829–832, May 2014.

[7] Melissa Duarte, Chris Dick, and Ashutosh Sabharwal. Experiment-driven characterization of full-duplex wireless systems. *IEEE Trans. Wireless Commun.*, 11(12):4296–4307, Dec. 2011.

[8] Melissa Duarte, Ashutosh Sabharwal, Vaneet Aggarwal, Rittwik Jana, Kadangode K. Ramakrishnan, Christopher W. Rice, and N.K. Shankaranarayanan. Design and characterization of a full-duplex multiantenna system for WiFi networks. *IEEE Trans. Veh. Technol.*, 63(3):1160–1177, Mar. 2014.

[9] Mayank Jain, Jung Il Choi, Taemin Kim, Dinesh Bharadia, Siddharth Seth, Kannan Srinivasan, Philip Levis, Sachin Katti, and Prasun Sinha. Practical, real-time, full duplex wireless. In *Proceedings of the 17th Annual International Conference on Mobile Computing and Networking*, pages 301–312, Las Vegas, Nevada, USA, 2011. ACM.

[10] Jung II Choi, Mayank Jain, Kannan Srinivasan, Philip Levis, and Sachin Katti. Achieving single channel, full duplex wireless communication. In *Proceedings of the 16th Annual International Conference on Mobile Computing and Networking (MobiCom 2010)*, pages 301–312, New York, NY, USA, Sep. 2010. IEEE.

[11] Alexander A. Maltsev, Roman Maslennikov, Alexey Sevastyanov, Artyom Lomayev, Alexey Khoryaev, Alexei Davydov, and Vladimir Ssorin. Characteristics of indoor millimeter-wave channel at 60 GHz in application to perspective WLAN system. In *European Conference on Antennas and Propagation (EuCAP)*, pages 1–5, Barcelona, 2010. IEEE.

[12] Minyoung Park and Helen K. Pan. A spatial diversity technique for IEEE 802.11 ad WLAN in 60 GHz band. *IEEE Commun. Lett.*, 16(8):1260–1262, Aug. 2012.

[13] Ahmed Alkhateeb, Jianhua Mo, Nuria González-Prelcic, and Robert W. Heath. MIMO precoding and combining solutions for millimeter-wave systems. *IEEE Commun. Mag.*, 52(12):122–131, Dec. 2014.

[14] Sooyoung Hur, Taejoon Kim, David J. Love, James V. Krogmeier, Timothy A. Thomas, and Amitava Ghosh. Millimeter wave beamforming for wireless backhaul and access in small cell networks. *IEEE Trans. Commun.*, 61(10):4391–4403, Oct. 2013.

[15] Jimmy Nsenga, Wim Van Thillo, François Horlin, Valery Ramon, André Bourdoux, and Rudy Lauwereins. Joint transmit and receive analog beamforming in 60 GHz MIMO multipath channels. In *IEEE International Conference on Communications (ICC)*, pages 1–5, Dresden, Germany, Jun. 2009. IEEE.

[16] Zhenyu Xiao, Pengfei Xia, and Xiang-Gen Xia. Enabling UAV cellular with millimeter-wave communication: potentials and approaches. *IEEE Commun. Mag.*, 54(5):66–73, 2016.

[17] Sundeep Rangan, Theodore S. Rappaport, and Elza Erkip. Millimeter-wave cellular wireless networks: Potentials and challenges. *Proceedings of the IEEE*, 102(3):366–385, 2014.

[18] Omar El Ayach, Sridhar Rajagopal, Shadi Abu-Surra, Zhouyue Pi, and Robert W. Heath. Spatially sparse precoding in millimeter wave MIMO systems. *IEEE Trans. Wireless Commun.*, 13(3):1499–1513, Mar. 2014.

[19] Alan J. Fenn. Evaluation of adaptive phased array antenna, far-field nulling performance in the near-field region. *IEEE Trans. Antennas Propagat.*, 38(2):173–185, Feb. 1990.

[20] Liangbin Li, Kaushik Josiam, and Rakesh Taori. Feasibility study on full-duplex wireless millimeter-wave systems. In *2014 IEEE International Conference on Acoustics, Speech and Signal Processing (ICASSP)*, pages 2769–2773, May 2014.

[21] Yan-Ping Liao, Hua Han, and Qiang Guo. Design of robust near-field multi-beam forming based on improved LCMV algorithm. *Journal of Information Hiding and Multimedia Signal Processing*, 6(4):783–791, Jul. 2015.

[22] Jing-ran Lin, Qi-cong Peng, and Huai-zong Shao. Near-field robust adaptive beamforming based on worst-case performance optimization. *International Journal of Electrical and Computer Engineering*, 1(6), 2006.

[23] Yang Tang, Branka Vucetic, and Yonghui Li. An iterative singular vectors estimation scheme for beamforming transmission and detection in MIMO systems. *IEEE Commun. Lett.*, 9(6):505–507, Jun. 2005.

[24] Bin Li, Zheng Zhou, Haijun Zhang, and Arumugam Nallanathan. Efficient beamforming training for 60-GHz millimeter-wave communications: a novel numerical optimization framework. *IEEE Trans. Veh. Technol.*, 63(2):703–717, Feb. 2014.

[25] Bin Li, Zheng Zhou, Weixia Zou, Xuebin Sun, and Guanglong Du. On the efficient beam-forming training for 60GHz wireless personal area networks. *IEEE Trans. Wireless Commun.*, 12(2):504–515, Feb. 2013.

[26] Byungjin Chun and Yong Hoon Lee. A spatial self-interference nullification method for full duplex amplify-and-forward MIMO relays. In *IEEE Wireless Commun. and Networking Conference (WCNC)*, pages 1–6, Sydney, Australia, Apr. 2010. IEEE.

[27] Taneli Riihonen, Stefan Werner, and Risto Wichman. Mitigation of loopback self-interference in full-duplex MIMO relays. *IEEE Trans. Signal Processing*, 59(12):5983–5993, Dec. 2011.

[28] Yi Liu, Xiang-Gen Xia, and Hailin Zhang. Distributed space-time coding for full-duplex asynchronous cooperative communications. *IEEE Trans. Wireless Commun.*, 11(7):2680–2688, Jul. 2012.

[29] Yi Liu, Xiang-Gen Xia, and Hailin Zhang. Distributed linear convolutional space-time coding for two-relay full-duplex asynchronous cooperative networks. *IEEE Trans. Wireless Commun.*, 12(12):6406–6417, Dec. 2013.

[30] Trevor Snow, Caleb Fulton, and William Johnson Chappell. Transmit–receive duplexing using digital beamforming system to cancel self-interference. *IEEE Trans. Microwave Theory Tech.*, 59(12):3494–3503, Dec. 2011.

[31] Krishna Gomadam, Viveck R. Cadambe, and Syed Ali Jafar. A distributed numerical approach to interference alignment and applications to wireless inter-ference networks. *IEEE Trans. Inform. Theory*, 57(6):3309–3322, 2011.

[32] Ignacio Santamaria, Oscar Gonzalez, Robert W. Heath, and Steven W. Peters. Maximum sum-rate interference alignment algorithms for MIMO channels. In *IEEE Global Telecommunications Conference (GLOBECOM 2010)*, pages 1–6. IEEE, 2010.

[33] Steven W. Peters and Robert W. Heath. Cooperative algorithms for MIMO interference channels. *Vehicular Technology, IEEE Transactions on*, 60(1):206–218, Jan. 2011.

[34] David A. Schmidt, Changxin Shi, Randall A. Berry, Michael L. Honig, and Wolf-gang Utschick. Comparison of distributed beamforming algorithms for MIMO interference networks. *IEEE Trans. Signal Processing*, 61(13):3476–3489, Jul. 2013.

[35] Moon-Sik Lee, Ji-Yong Park, Vladimir Katkovnik, Tatsuo Itoh, and Yong-Hoon Kim. Adaptive robust DOA estimation for a 60-GHz antenna-array system. *IEEE Trans. Veh. Technol.*, 56(5):3231–3237, Sep. 2007.

[36] Y. Ming Tsang, Ada S.Y. Poon, and Sateesh Addepalli. Coding the beams: Improving beamforming training in mmwave communication system. In *IEEE Global Telecommunications Conference (GLOBECOM 2011)*, pages 1–6, Hous-ton, TX, USA, Dec. 2011. IEEE.

[37] Zhenyu Xiao, Xiang-Gen Xia, Depeng Jin, and Ning Ge. Iterative eigenvalue de-composition and multipath-grouping Tx/Rx joint beamformings for millimeter-wave communications. *IEEE Trans. Wireless Commun.*, 14(3):1595–1607, Mar. 2015.

[38] Ahmed Alkhateeb, Omar El Ayach, Geert Leus, and Robert W. Heath. Channel estimation and hybrid precoding for millimeter wave cellular systems. *IEEE J. Sel. Top. Sign. Proces.*, 8(5):831–846, Oct. 2014.

[39] Yuexing Peng, Yonghui Li, and Peng Wang. An enhanced channel estimation method for millimeter wave systems with massive antenna arrays. *IEEE Com-mun. Lett.*, 19(9):1592–1595, Sep. 2015.

[40] Javier Vía, Ignacio Santamaría, Victor Elvira, and Ralf Eickhoff. A general criterion for analog Tx-Rx beamforming under OFDM transmissions. *IEEE Trans. Signal Processing*, 58(4):2155–2167, Apr. 2010.

[41] Matthew Kokshoorn, Peng Wang, Yonghui Li, and Branka Vucetic. Fast channel estimation for millimetre wave wireless systems using overlapped beam patterns. In *IEEE International Conference on Communications (ICC)*, pages 1304–1309, London, UK, Jun. 2015. IEEE.

[42] Dong-Woo Kang, Jeong-Geun Kim, Byung-Wook Min, and Gabriel M. Rebeiz. Single and four-element-ka-band transmit/receive phased-array silicon RFICs with 5-bit amplitude and phase control. *IEEE Trans. Microwave Theory Tech.*, 57(12):3534–3543, Dec. 2009.

[43] Qiang Li, Mingyi Hong, Hoi-To Wai, Ya-Feng Liu, Wing-Kin Ma, and Zhi-Quan Luo. Transmit solutions for MIMO wiretap channels using alternating optimization. *IEEE J. Select. Areas Commun.*, 31(9):1714–1727, Sep. 2013.

[44] Wei Li, Jorma Lilleberg, and Kari Rikkinen. On rate region analysis of half- and full-duplex OFDM communication links. *IEEE J. Select. Areas Commun.*, 32(9):1688–1698, 2014.

[45] Xiao Liu, Zhenyu Xiao, Lin Bai, Jinho Choi, Pengfei Xia, and Xiang-Gen Xia. Beamforming based full-duplex for millimeter-wave communication. *Sensors*, 16(7):1–22, Jul. 2016.

[46] Foad Sohrabi and Wei Yu. Hybrid digital and analog beamforming design for large-scale MIMO systems. In *IEEE International Conference on Acoustics, Speech and Signal Processing (ICASSP)*, pages 2929–2933, South Brisbane, QLD, 2015. IEEE.

[47] Ahmed Alkhateeb, Geert Leus, and Robert W. Heath. Limited feedback hybrid precoding for multi-user millimeter wave systems. *IEEE Trans. Wireless Commun.*, 14(11):6481–6494, 2015.

[48] Sridhar Rajagopal, Rakesh Taori, and Shadi Abu-Surra. Self-interference mitigation for in-band mmWave wireless backhaul. In *Proc. IEEE Consumer Commun. Netw. Conf.*, pages 551–556, Jan. 2014.

[49] Margarita Gapeyenko, Vitaly Petrov, Dmitri Moltchanov, Sergey Andreev, Nageen Himayat, and Yevgeni Koucheryavy. Flexible and reliable UAV-assisted backhaul operation in 5G mmWave cellular networks. *IEEE J. Select. Areas Commun.*, 36(11):2486–2496, Nov. 2018.

[50] Haichao Wang, Jinlong Wang, Guoru Ding, Jin Chen, Yuzhou Li, and Zhu Han. Spectrum sharing planning for full-duplex UAV relaying systems with underlaid D2D communications. *IEEE J. Select. Areas Commun.*, 36(9):1986–1999, Sep. 2018.

[51] Xianbin Cao, Peng Yang, Mohamed Alzenad, Xing Xi, Dapeng Wu, and Halim Yanikomeroglu. Airborne communication networks: A survey. *IEEE J. Select. Areas Commun.*, 36(9):1907–1926, Sep. 2018.

[52] Gang Liu, F. Richard Yu, Hong Ji, Victor C. M. Leung, and Xi Li. In-band full-duplex relaying: A survey, research issues and challenges. *IEEE Commun. Surveys Tuts.*, 17(2):500–524, 2nd Quart. 2015.

[53] Xinyu Gao, Linglong Dai, Shuangfeng Han, Chih-Lin I, and Robert W. Heath. Energy-efficient hybrid analog and digital precoding for mmWave MIMO systems with large antenna arrays. *IEEE J. Select. Areas Commun.*, 34(4):998–1009, Apr. 2016.

[54] Zhenyu Xiao, Lipeng Zhu, Jinho Choi, Pengfei Xia, and Xiang-Gen Xia. Joint power allocation and beamforming for non-orthogonal multiple access (NOMA) in 5G millimeter wave communications. *IEEE Trans. Wireless Commun.*, 17(5):2961–2974, May 2018.

[55] Lipeng Zhu, Jun Zhang, Zhenyu Xiao, Xianbin Cao, Dapeng Oliver Wu, and Xiang-Gen Xia. Millimeter-wave NOMA with user grouping, power allocation and hybrid beamforming. *IEEE Trans. Wireless Commun.*, 18(11):5065–5079, Nov. 2019.

[56] Constantine A. Balanis. *Antenna theory: analysis and design.* Hoboken, NJ, USA: Wiley, 2016.

[57] Zhenyu Xiao, Pengfei Xia, and Xiang-Gen Xia. Full-duplex millimeter-wave communication. *IEEE Wireless Commun. Mag.*, 16, Dec. 2017.

[58] K. Satyanarayana, Mohammed El-Hajjar, Ping-Heng Kuo, Alain Mourad, and Lajos Hanzo. Hybrid beamforming design for full-duplex millimeter wave communication. *IEEE Trans. Veh. Technol.*, 68(2):1394–1404, Feb. 2019.

[59] Yi Zhang, Ming Xiao, Shuai Han, Mikael Skoglund, and Weixiao Meng. On precoding and energy efficiency of full-duplex millimeter-wave relays. *IEEE Trans. Wireless Commun.*, 18(3):1943–1956, Mar. 2019.

[60] Theodore S. Rappaport, George R. MacCartney, Mathew K. Samimi, and Shu Sun. Wideband millimeter-wave propagation measurements and channel models for future wireless communication system design. *IEEE Trans. Commun.*, 63(9):3029–3056, Sep. 2015.

[61] David Tse and Pramod Viswanath. *Fundamentals of Wireless Communication.* Cambridge Univ. Press, New York, USA, 2005.

[62] Akram Al-Hourani, Sithamparanathan Kandeepan, and Simon Lardner. Optimal LAP altitude for maximum coverage. *IEEE Wireless Commun. Lett.*, 3(6):569–572, Dec. 2014.

[63] Stephen Boyd and Lieven Vandenberghe. *Convex Optimization*. Cambridge, U.K.: Cambridge Univ. Press, 2004.

CHAPTER 5

Array Beamforming Enabled 2-User NOMA

5.1 INTRODUCTION

Since the resource is limited in practical communication scenarios, the reuse of the resource can be used to improve effective capacity. On one hand, the full-duplex (FD) technology is able to improve spectrum efficiency through the reuse of carrier frequency, as introduced in Chapter 4, where effective capacity can be improved through reuse of spectrum. On the other hand, since the maximum number of users that can be served within one time/frequency/code resource block (RB) is very limited, it is able to improve the effective capacity through the reuse of RBs. Multiple access technologies have always been one of the key technologies for wireless communication systems, supporting the huge number of mobile devices in future networks. Thus, sufficient effective capacity is required to support the requirements on increasing data rate. In particular, for communication systems with large antenna array, the number of radio frequency (RF) chains is usually much smaller than that of antennas due to the hardware cost[1,2,3,4,5,6]. As a result, the maximum number of users that can be served within one RB is no larger than the number of RF chains. Then the conventional orthogonal multiple access (OMA) schemes, e.g., time division multiple access (TDMA), code division multiple access (CDMA), and orthogonal frequency-division multiple access (OFDMA), may face great difficulties to support the huge number of mobile devices for future networks[7].

In contrast, the non-OMA (NOMA) strategy can support multiple users in the same (time/frequency/code) RB by using superposition coding in the power domain[8,9,10,11]. By exploiting corresponding successive interference cancellation (SIC) in the power domain at Rxs, multiple users can be distinguished from each other, thus both the number of users and the spectrum efficiency can be increased. And then the effective capacity can be improved.

The basic idea of NOMA is as follows. We consider the downlink transmission from a BS to two users with different channel conditions. Let h_1 and h_2 denote the channel gains of the two users. For OMA strategies, the two users require two orthogonal time/frequency/code RB to ensure independent transmission, and the

DOI: 10.1201/9781003366362-5

achievable rates are, respectively, given by

$$
\begin{cases}
R_1^{\text{OMA}} = \dfrac{1}{2}\log_2(1 + \dfrac{h_1^2 P}{\sigma^2}), \\[2mm]
R_2^{\text{OMA}} = \dfrac{1}{2}\log_2(1 + \dfrac{h_2^2 P}{\sigma^2}),
\end{cases}
$$

where P denotes the total transmit power at the BS, and σ^2 is the noise power. Factor $\frac{1}{2}$ is because the RBs are allocated to two users. As can be observed, the two users cannot achieve their maximum achievable rate under OMA strategies due to the multiplexing loss. In contrast, NOMA enables the two users share the same time/frequency/code RB. Without loss of generality, we assume user 1 has a better channel condition compared to user 2, i.e., $h_1 \geq h_2$. Under the NOMA mechanism, the BS transmits the signals for the two users in the same RB with different power levels. The user with higher channel gain, i.e., user 1, employs SIC technologies to decode the signal for user 2 first, and then decode the signal for itself. While the user with lower channel gain, i.e., user 2, can directly decode the signal for itself by regarding the signal for user 1 as noise. Thus, the achievable rates for the two users are, respectively, given by

$$
\begin{cases}
R_1^{\text{NOMA}} = \log_2(1 + \dfrac{h_1^2 P_1}{\sigma^2}), \\[3mm]
R_2^{\text{NOMA}} = \log_2(1 + \dfrac{h_2^2 P_2}{h_2^2 P_1 + \sigma^2}),
\end{cases}
$$

where P_1 and P_2 with $P_1 + P_2 = P$ represent the allocated power to users 1 and 2, respectively. It is easy to verify

$$
R_1^{\text{NOMA}} + R_2^{\text{NOMA}} \geq R_1^{\text{OMA}} + R_2^{\text{OMA}}, \tag{5.1}
$$

which indicates that NOMA provides a higher spectrum efficiency compared to OMA strategies. The achievable rates of the two users can be adaptively adjusted according to their channel conditions to avoid a waste of the RB. In addition, we can observe that the performance gain of the achievable rate for NOMA highly depends on the difference of the channel gains for the two users. Especially for $h_1 = h_2$,

$$
R_1^{\text{NOMA}} + R_2^{\text{NOMA}} = R_1^{\text{OMA}} + R_2^{\text{OMA}}, \tag{5.2}
$$

always holds, which indicates that the superiority of NOMA does not exist anymore. It becomes more interesting when introducing array beamforming to NOMA systems. The effective channel gain of each user is determined by both the channel gain and the beamforming gain. In other words, beamforming provides a new degree of freedom (DoF) to enlarge the channel difference between the two users. Thus, more performance gain can be acquired for array beamforming enabled NOMA communication systems.

Despite these promising opportunities, beamforming enabled NOMA has some unique challenges. First, the beamforming design is much more difficult for NOMA systems. Since the multiple users may be located in different directions, a wide beam is required to cover all the users, but this may reduce the beam gain and offset the benefit of NOMA. Therefore, we consider multi-beam forming with different beam gains, where the key challenge is the constant modulus (CM) constraint due to phase shifters, which is non-convex and high-dimension in general.

Second, in addition to power allocation, beamforming constitutes a new DoF. Specifically, the effective channel gains of users can be artificially changed by beamforming. Power allocation intertwines with beamforming, because the achievable rates of the users depend on them both. As a result, we usually need to consider a joint power allocation and beamforming problem which is non-convex and may not be converted to a convex problem with simple manipulations. In this chapter, we introduce a solution to the problem, i.e., to decompose the original problem into two sub-problems: One is the power and beam gain allocation problem, the other is the beamforming problem under the CM constraint. We first solve the power and beam gain allocation problem to obtain the optimal solution of the power and beam gain of the two users. Then substitute them into the beamforming problem. As for the beamforming sub-problem, there also exists a challenge which is the CM constraint due to exploiting only phase shifters to control the antenna weights[12, 13, 14, 15]. In such a case, some relaxation may be induced to ease the problem. For instance, the CM constraint may be relaxed to minimize the maximal absolute weight of the antenna weights. Meanwhile, the beam gain requirements are relaxed from equality to inequality. By searching the optimal phases of the two beam gains, the problem can be transformed into several standard convex optimization problems.

Third, the challenge of user pairing intertwined with beamforming also exists. It is difficult to find the optimal user pairing for NOMA systems due to its enumeration feature[16]. What makes this issue even more difficult is beamforming also affects user pairing. In particular, the channel gains of the users are the main factors to affect user pairing in single-beam NOMA. However, in multi-beam NOMA, the relative angles between the users affect the beam gains and then indirectly affect user pairing, i.e., when the relative angle between the two users is small, there is no need to form two beams, and only one beam is able to cover the two users with high beam gain.

As shown in Fig. 5.1, the left sub-figure shows the situation with single-beam forming, while the right sub-figure shows the situation with multi-beam forming. In single-beam NOMA, as shown in the left sub-figure, we need to select two users as a NOMA group. User 3 can be paired with either user 1 or user 2. However, the relative angle between user 2 and user 3 is large, so a wide beam, i.e., Beam 2, needs to be formed to cover the group. In contrast, a narrower beam, i.e., Beam 1, needs to be formed to cover the group of users 3 and 1, because the relative angle between them is smaller. As a result, the achievable beam gain when pairing users 3 and 1 is higher than pairing users 3 and 2. The situation is different when using multi-beam NOMA, as shown in the right sub-figure of Fig. 5.1. With multi-beam forming, two narrow beams are formed to cover the two NOMA users. In such a case, it is almost the same to pair users 3 and 2 as to pair users 3 and 1, provided that the channel gains of users 1 and 2 are similar, because the beam gains to cover users 3 and 1 with Beams 3 and 1 are the same as those to cover users 3 and 2 with Beams 3 and 2. However, when the relative angle between user 3 and user 1 is very small, e.g., smaller than $2/N$ in the cosine angle domain, the BS may not form two different beams to cover them because the smallest beam width is $2/N$[13]. Instead, the BS may form only one narrow beam to cover both users. In such a case, both users can achieve a higher beam gain, and there is no need to form two beams.

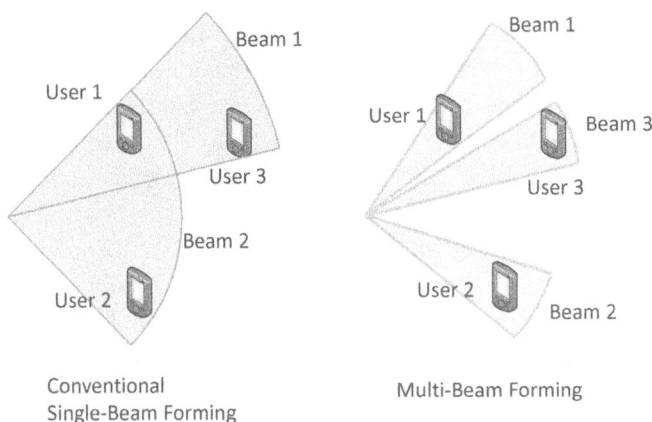

Figure 5.1 Beamforming affects user pairing in NOMA.

Last, the integration of NOMA and SDMA has new challenges. In conventional communication systems using SDMA, one RF chain can only form one beam, and thus one RF chain can only serve one user in general. By integrating NOMA, multiple users can be served in the same RB with multi-beam forming[17, 18]. However, the number of NOMA users is limited in general, because the achievable rates of the users with weak channel conditions decrease with the number of NOMA users due to multi-user interference (MUI). In such a case, a potential method to increase the total number of users is to use hybrid SDMA and NOMA. As shown in Fig. 5.2, for a fully connected phase-shifter based hybrid beamforming structure, a higher DoF for beamforming design results in a higher array gain and lower interference. The hybrid structure has M RF chains, and thus by using SDMA, the structure can support at most M users. If NOMA is exploited with each RF chain and SDMA between different RF chains, one RF chain can serve K NOMA users. Hence, the hybrid SDMA and NOMA strategy can support at most $M \times K$ users. When the number of users is small and the MUI from different NOMA groups can be ignored, the system can be configured as multiple independent analog-beamforming NOMA. Power allocation is performed among only the users of the same group. However, when the number of users is not small and the MUI from other NOMA groups needs to be considered, the design is more complicated. All the RF chains jointly serve all the users and power allocation is performed among all the users. This mode can be regarded as an overloaded SDMA system, where the number of users is larger than that of the RF chains.

For hybrid SDMA and NOMA systems, there are two ways to improve the sum-rate performance. One way is to increase the beam gains of the target signals, and the other way is to decrease the beam gains of the interference signals. With the larger number of RF chains and antennas, higher beam gains and lower interference can be obtained. An alternative way is to first design the analog beamforming to maximize array gain. Then, the digital beamforming can be designed by using some classical algorithms to minimize the inter-beam interference, such as ZF, MMSE. Especially for uplink NOMA with hybrid beamforming, parallel interference cancellation (PIC) can also be utilized at Rx (BS) to further decrease the MUI. In addition to suppressing, the MUI in the beamforming domain, user pairing/grouping can also be elaborated, due to the directivity of the high-frequency band channels. For instance, the users with high channel correlation can be assigned to different groups to minimize the inter-group interference, while the users with low channel correlation can be assigned to the same group to make full use of the array gain[19].

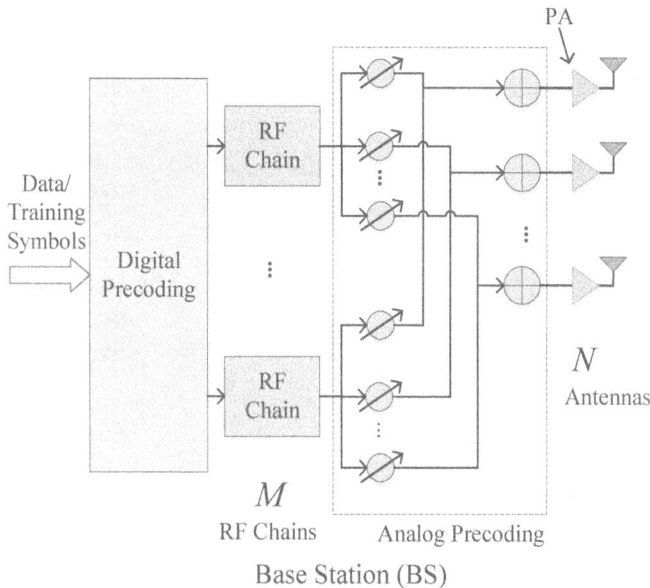

Figure 5.2 System structure of hybrid analog/digital beamforming.

In this chapter, we first give an overview of the unique challenges of beamforming enabled NOMA and some possible solutions are presented. Then, we formulate multi-beam forming problems for user coverage in NOMA systems and introduce some relaxation methods to handle the CM constraint. We introduce three approaches for multi-beam forming, i.e., sub-array technique, optimization technique, and intelligent search technique. In the third section, we consider the downlink scenario. The system model, problem decomposition, solution of the power, and beam gain allocation problem and solution of the relaxed beamforming problem are introduced. Moreover, we discuss the optimal decoding order under specific conditions. In the forth section, we consider the uplink scenario. The basic idea of the uplink transmission problem is almost the same with the downlink scenario. Nevertheless, as the maximal power has been set for the two users and the power cannot be further increased. We introduce a new solution to relax and solve the beamforming problem. Furthermore, it has been proved that in uplink transmission, the optimal decoding order can be identically determined as that in downlink transmission.

5.2 MULTI-BEAM FORMING

In conventional beamforming technologies, subject to the hardware constraints, usually only a single beam is formed with each chain[13, 20, 21]. If NOMA is used, the BS needs to serve multiple users in one time slot, and a narrow beam may not cover all the users. Instead, a wide beam may be required to cover all the served users in that time slot, and the beam width depends on the relative angles between these users. This may reduce the beam gain and in turn offset the benefit of NOMA, because the beam gain is roughly inversely proportional to the beam width, as shown in Fig. 5.3. The NOMA BS with analog beamforming needs to serve two users. When serving users 2 and 3, the BS only needs to form a narrow beam because the angle gap between users 2 and 3 is small. In such a case, the beam gain will be high. However, when serving users 1 and 4, the BS has to form a much wider beam because the angle gap between users 1 and 4 is much larger. As a result, the beam gain of the BS will be much lower, which will degrade the performance of NOMA.

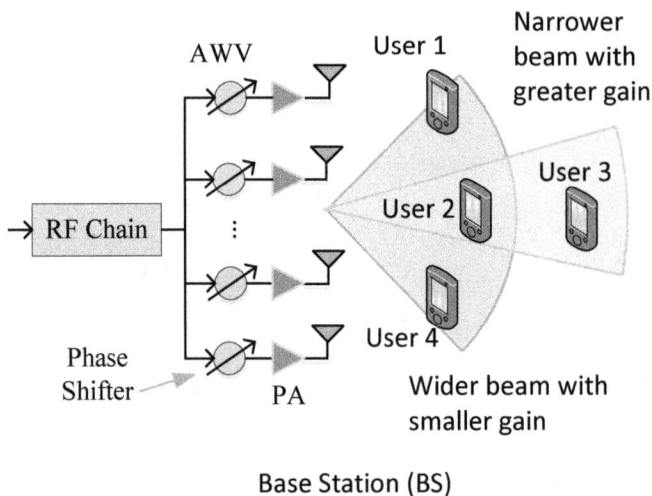

Figure 5.3 Millimeter-wave-NOMA with analog beamforming at the BS.

As single-beam forming does not behave efficiently and robustly enough, multi-beam forming with a single RF chain is considered, meaning that the BS can form multiple narrow beams to steer toward multiple NOMA users simultaneously. As shown in Fig. 5.4, the BS with multi-beam forming forms two narrow beams to cover two NOMA users (user 1 and user 2) in the same time slot. Intuitively, since multi-beam forming covers a narrower range than single-beam forming, a higher beam gain can be achieved, and the beam gain will not be reduced even when the angles of the two users become larger. More importantly, with multi-beam forming, the beam gains for different NOMA users can be different. This feature is very important for NOMA, because in addition to the DoF in the power domain, it provides another DoF, i.e., beamforming, to improve the performance of NOMA.

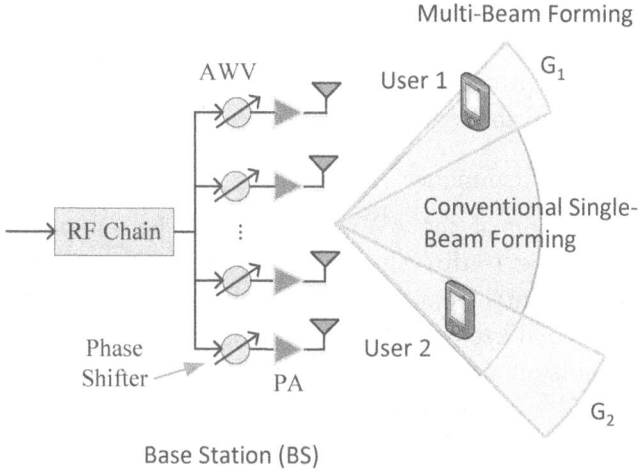

Figure 5.4 Comparison between single-beam forming and multi-beam forming for NOMA.

Compared with single-beam forming, multi-beam forming has distinctive advantages, but the antenna weight vector (AWV) design is more challenging as mentioned above. To achieve multi-beam forming, there are three promising techniques, i.e., sub-array technique, optimization technique, and intelligent search technique.

1) Sub-Array Technique

As we need to form multiple beams, a natural method is to divide a large antenna array, with a given number of antennas, into multiple sub-arrays, and let them steer to different directions, i.e., set the AWVs of these sub-arrays to steering vectors associated with different directions. Thanks to the small sidelobe of the large antenna array, the interference between different sub-arrays is small. Thus, the beam gain is roughly linear to the number of antennas for each sub-array.

2) Optimization Technique

Since the number of antennas is large in general, the dimension (the number of variables) of the optimization problem will be large. The key is how to deal with the CM constraint and the beam gain constraints, which are all non-convex. For instance, if we want to design $\mathbf{w}$ to form two different beams with gains c_1 and c_2, respectively, it is natural to minimize $\|\mathbf{w}\|_2^2$ (or α) subject to the CM constraint $|\mathbf{w}| = \alpha\mathbf{1}$ and gain constraint $|\mathbf{w}^H\mathbf{a}(\theta_k)| = c_k, k = 1, 2$, where $\mathbf{a}(\theta_k)$ is a given steering vector toward the direction θ_k. However, with these equality constraints, the problem is generally difficult to solve, not only because they are non-convex constraints, but also because the equality constraints are usually too strict to find an appropriate AWV. In such a case, some relaxation is induced to ease the problem. For instance, we may relax the CM constraints to minimize the maximal absolute weight of the antenna weights, i.e., to minimize α subject to $|\mathbf{w}| \prec \alpha \mathbf{1}$ where $\prec$ is componentwise inequality. Meanwhile, we may relax the beam gain requirements from equality to inequality, i.e., $|\mathbf{w}^H \mathbf{a}(\theta_k)| \geq c_k$. Then, by searching the optimal phases of the two beam gains $\mathbf{w}^H \mathbf{a}(\theta_k)$, the problem can be transformed into several standard convex optimization problem[21,22].

3) Intelligent Search Technique

Since the dimension of the AWV is very high for antenna array, it is computation prohibitive to directly search the optimal solution. Alternatively, we can introduce the intelligent search methods, e.g., particle swarm optimization (PSO) and artificial bee colony (ABC) algorithm[19,23]. Due to the interference between the multiple users, the formulation of the achievable rate is not convex in general. Besides, the CM constraint on the AWV is also highly non-convex. As a result, PSO algorithm may converge to a sub-optimal solution very fast and the globally optimal solution is missed. To improve the search capacity, some improvements can be conducted for PSO. As shown in Fig. 5.5, first, the search space defined by the CM constraint can be relaxed as a convex set, i.e., $\{|[\mathbf{w}]_n| \leq \frac{1}{\sqrt{N}}\}$. Then, we can define the boundaries of the relaxed search space, the inner boundary $\{|[\mathbf{w}]_n| = \frac{t}{T}\frac{1}{\sqrt{N}}\}$, and the outer boundary $\{|[\mathbf{w}]_n| = \frac{1}{\sqrt{N}}\}$, where $t = 1, 2, \cdots, T$ is the iteration index and T is the total number of the iterations. As we can see, the inner boundary is dynamically increasing from zero to the outer boundary during the iteration, while the outer boundary is fixed and indeed equivalent to the CM constraint. For each iteration, the positions and velocities of the particles (AWVs) are updated according to the PSO principle. If one particle moves out of the boundaries, then it should be adjusted to the closest boundary. With this operation, the particles can move around the whole search space and finally converge to the outer boundary to satisfy the CM constraint. Since the particles in the presented boundary-compressed PSO (BC-PSO) algorithm have a more opportunities to perceive global information, it outperforms the classical PSO algorithm in terms of multi-beam forming[19,23].

The presented three multi-beam forming techniques can achieve an increasing beamforming performance, with the expense of higher computational complexity. Sub-array technique has the lowest computational complexity, but it results in the beam-gain loss due to the beam broadening. Optimization technique can achieve a tradeoff between the computational complexity and the beam-gain performance. However, it requires an elaborate optimization design for different system models. Intelligent search technique has the near-optimal beamforming performance, while the computational complexity may be high when the dimension of the AWV is large.

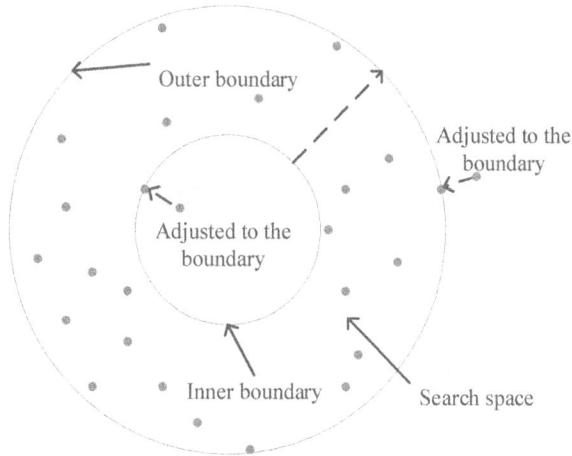

Figure 5.5 Illustration of the principle for BC-PSO.

5.3 DOWNLINK TRANSMISSION

In this section, without loss of generality, we consider a downlink 2-user NOMA problem, i.e., maximization of the sum rate of a 2-user NOMA, as shown in Fig. 5.6. In this problem, we need to find the beamforming vector (BFV) to steer toward the two users are simultaneously subject to an analog beamforming structure, while allocating appropriate power to them.

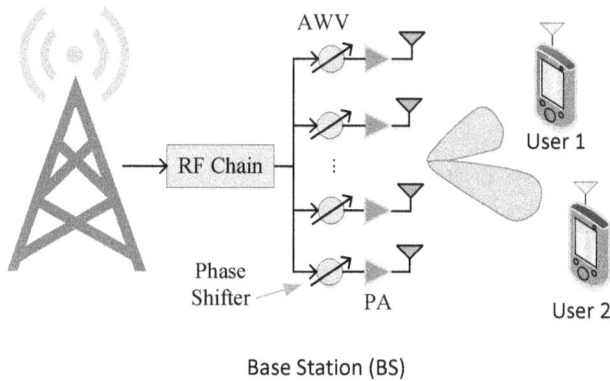

Figure 5.6 Illustration of a millimeter-wave mobile cell, where one BS with N antennas serves multiple users with a single antenna.

In[20], combining NOMA in millimeter-wave communications, random steering single-beam forming was proposed, which can work only in a special case that the NOMA users are close to each other. In[24], the new concept of beamspace multiple-input multiple-output (MIMO) NOMA with a lens-array hybrid beamforming structure was firstly presented to use multi-beam forming to serve multiple NOMA users with arbitrary locations, thus the limit that the number of supported users cannot be larger than the number of RF chains can be broken. However, the power allocation problem is studied under fixed beam pattern when lens array is considered. In[25], sub-channel assignment and power allocation are optimized to maximize the energy efficiency for a downlink NOMA network, while beamforming is not included. In[26], the optimization of sum rate in the downlink MIMO-NOMA system is studied using a full digital beamforming structure. Different from[20, 24, 25, 26], we consider joint power allocation and beamforming to maximize the sum rate of a 2-user NOMA system using an analog beamforming structure with a phased array.

5.3.1 Problem Formulation

As mentioned above, the considered problem, i.e., the power allocation and beam-forming problem in downlink transmission, is non-convex and may not be converted to a convex problem with simple manipulations, and to solve the problem directly by using the existing optimization tools is infeasible. On the other hand, to directly search the optimal solution is computationally prohibitive because the number of variables is large in general. Hence, we present a sub-optimal solution to this problem. The basic idea is to decompose the original joint beamforming and power allocation problem into two sub-problems: One is a power and beam gain allocation problem, and the other is a beamforming problem under the CM constraint. The two sub-problems are relatively easy to solve compared to the original problem. Thus, a sub-optimal solution is obtained.

1) System Model

As shown in Fig. 5.6, the BS equips with an N-element antenna array to serve two users with a single antenna. Noted that in this case, the users also use an antenna array, and Rx beamforming can be done first[27, 28]. Then the reception processing at each user can be seen equivalent to a single-antenna Rx, and the presented solution in this section can be used. At the BS, each antenna branch has a phase shifter and a PA to drive the antenna. Generally, all the PAs have the same scaling factor. Thus, the beamforming vector, i.e., AWV, has CM elements. When the BS transmits a signal s_k to user $k, k = 1, 2$, where $\mathbb{E}(|s_k|^2) = 1$, with transmit power p_k. The total transmit power is restricted to P. With 2-user NOMA, s_1 and s_2 are superimposed as

$$s = \sqrt{p_1}s_1 + \sqrt{p_2}s_2. \tag{5.3}$$

The received signals at user 1 and user 2 are

$$\begin{cases} y_1 = \mathbf{h}_1^{\mathrm{H}}\mathbf{w}(\sqrt{p_1}s_1 + \sqrt{p_2}s_2) + n_1, \\ y_2 = \mathbf{h}_2^{\mathrm{H}}\mathbf{w}(\sqrt{p_1}s_1 + \sqrt{p_2}s_2) + n_2, \end{cases} \tag{5.4}$$

where $\mathbf{h}_k$ is the channel response vector between user k and the BS, $\mathbf{w}$ denotes a CM beamforming vector with $|[\mathbf{w}]_n| = \frac{1}{\sqrt{N}}$ for $n = 1, 2, ..., N$, and n_k denotes the Gaussian white noise at user k with power σ^2.

As the multipath is mainly caused by reflection, the number of MPCs is small in general. And the high-frequency band channel has directionality and appears spatial sparsity in the angle domain, different MPCs have different AoDs. Without loss of generality, we adopt the channel model assuming a ULA with a half-wavelength antenna space. Then, we use the channel model (1.131) introduced in Chapter 1, which can be expressed as

$$\bar{\mathbf{h}}_k = \sum_{\ell=1}^{L_k} \lambda_{k,\ell} \mathbf{a}(N, \theta_{k,\ell}), \tag{5.5}$$

where $\lambda_{k,\ell}$, $\theta_{k,\ell}$ are the complex coefficient and cos(AoD) of the ℓ-th MPC of the channel vector for user k, respectively, L_k is the total number of MPCs for user k, $\mathbf{a}(\cdot)$ is the steering vector of a ULA defined in (1.70), i.e.,

$$\mathbf{a}(N, \theta) = [e^{j\pi 0\theta}, e^{j\pi 1\theta}, e^{j\pi 2\theta}, \cdots, e^{j\pi(N-1)\theta}], \tag{5.6}$$

which depends on the array geometry. Let $\theta_{k,\ell}$ denote the real AoD of the ℓ-th MPC for user k, then we have $\theta_{k,\ell} = \cos(\Omega_{k,\ell})$. Therefore, $\theta_{k,\ell}$ is within the range $[-1, 1]$. For convenience and without loss of generality, in the rest of this section, $\theta_{k,\ell}$ is also called AoD.

For each user, the BS would perform beamforming toward the angle direction along the AoD of the strongest MPC to achieve a high array gain. In general, if there is no blockage between the BS and a user, the LoS component will be adopted for beamforming. However, if the above conditions are not met, i.e., there is no LoS component, the strongest NLoS path would be selected for beamforming. Briefly speaking, we can always obtain one effective channel and thus the channel model can be written as

$$\mathbf{h}_k = \lambda_k \mathbf{a}(N, \theta_k), \tag{5.7}$$

where $\lambda_k = \lambda_{k,m_k}$ and $\theta_k = \theta_{k,m_k}$. Here m_k denotes the index of the strongest MPC for user i. Without loss of generality, we assume $|\lambda_1| \geq |\lambda_2|$, which means that the channel gain of user 1 is better.

2) Decoding Order

Next, we will introduce the decoding order in multi-beam NOMA. In contrast to the conventional NOMA with single-beam NOMA, the decoding order of multi-beam NOMA depends on both channel gain and beamforming gain. We consider two cases of the decoding order.

Case 1: s_1 is decoded first. In such a case, user 1 directly decodes s_1 by treating the signal component of s_2 as noise. In comparison, user 2 first decodes s_1 and subtracts the signal component s_1 from the received signal y_2; then it decodes s_2. Therefore, user 2 can decode s_2 without the interference from s_1.

With this decoding method, the achievable rates of user $k, k = 1, 2$, denoted by R_k, are represented as

$$R_1^{(1)} = \log_2(1 + \frac{\left|\mathbf{h}_1^H\mathbf{w}\right|^2 p_1}{\left|\mathbf{h}_1^H\mathbf{w}\right|^2 p_2 + \sigma^2}),$$

$$R_2^{(1)} = \log_2(1 + \frac{\left|\mathbf{h}_2^H\mathbf{w}\right|^2 p_2}{\sigma^2}). \tag{5.8}$$

Note that there is an implicit assumption for this result, i.e., the observed signal-to-interference-plus-noise ratio (SINR) at user 2 for decoding s_1 should be no less than that at user 1, otherwise s_1 cannot be correctly decoded at user 1, and the interference at user 2, i.e., the signal component of s_1, cannot be completely removed. We call this implicit assumption is an *implicit SINR constraint*, and it is expressed as

$$\frac{\left|\mathbf{h}_2^H\mathbf{w}\right|^2 p_1}{\left|\mathbf{h}_2^H\mathbf{w}\right|^2 p_2 + \sigma^2} \geq \frac{\left|\mathbf{h}_1^H\mathbf{w}\right|^2 p_1}{\left|\mathbf{h}_1^H\mathbf{w}\right|^2 p_2 + \sigma^2}, \tag{5.9}$$

which is equivalent to

$$\left|\mathbf{h}_2^H\mathbf{w}\right|^2 \geq \left|\mathbf{h}_1^H\mathbf{w}\right|^2. \tag{5.10}$$

Case 2: s_2 is decoded first. Similarly, with this decoding method, the achievable rates of user $k, k = 1, 2$, denoted by R_k, are represented as

$$R_1^{(2)} = \log_2(1 + \frac{\left|\mathbf{h}_1^H\mathbf{w}\right|^2 p_1}{\sigma^2}),$$

$$R_2^{(2)} = \log_2(1 + \frac{\left|\mathbf{h}_2^H\mathbf{w}\right|^2 p_2}{\left|\mathbf{h}_2^H\mathbf{w}\right|^2 p_1 + \sigma^2}). \tag{5.11}$$

The implicit SINR constraint is expressed as

$$\frac{\left|\mathbf{h}_1^H\mathbf{w}\right|^2 p_2}{\left|\mathbf{h}_1^H\mathbf{w}\right|^2 p_1 + \sigma^2} \geq \frac{\left|\mathbf{h}_2^H\mathbf{w}\right|^2 p_2}{\left|\mathbf{h}_2^H\mathbf{w}\right|^2 p_1 + \sigma^2}, \tag{5.12}$$

which is equivalent to

$$\left|\mathbf{h}_1^H\mathbf{w}\right|^2 \geq \left|\mathbf{h}_2^H\mathbf{w}\right|^2. \tag{5.13}$$

It is noted that since we assume that the channel gain of user 1 is better, usually Case 2 can achieve a better sum rate than Case 1. The optimal sum rate of Case 2 is better than that of Case 1 if $r_1 = r_2$, where r_1 denotes the minimal rate constraint for user i. However, if $r_1 \neq r_2$, it is possible that Case 1 is better than Case 2. In such a case, we can solve the problems under both cases, and select the case with a higher achievable sum rate as the optimal decoding order.

3) Problem Formulation

The problem is how to maximize the achievable sum rate of the two users when there are minimal rate constraints for the two users. And as a result of minimal rate constraints, the power allocation intertwines with the beamforming design, which makes the problem complicated under the system setup.

We choose Case 2 as the decoding order, then the problem is formulated by

$$\underset{p_1, p_2, \mathbf{w}}{\text{Maximize}} \quad R_1 + R_2 \tag{5.14a}$$

$$\text{Subject to} \quad R_1 \geq r_1, \tag{5.14b}$$

$$R_2 \geq r_2, \tag{5.14c}$$

$$p_1 + p_2 = P, \tag{5.14d}$$

$$|[\mathbf{w}]_n| = \frac{1}{\sqrt{N}}, \quad n = 1, 2, ..., N, \tag{5.14e}$$

$$\left|\mathbf{h}_1^H \mathbf{w}\right|^2 \geq \left|\mathbf{h}_2^H \mathbf{w}\right|^2, \tag{5.14f}$$

where $|[\mathbf{w}]_n| = \frac{1}{\sqrt{N}}$ is the CM constraint due to using the phase shifters in each antenna branch at the BS, $R_k = R_k^{(2)}$ as defined in (5.11). It is noteworthy that the decoding order can also be Case 1, and another problem can be formulated accordingly. The two problems with different decoding orders are similar to each other, which means that if an approach can be used to solve one of them, it can also be used to solve the other.

5.3.2 Solution of the Problem

1) Problem Decomposition

Clearly, directly solving Problem (5.14) by using existing optimization tools is infeasible because the problem is non-convex and may not be converted to a convex problem with simple manipulations. Hence, we introduce another solution. Since power allocation intertwines with beamforming under the CM constraint; we first try to decompose them. Let

$$c_1 = \left|\mathbf{h}_1^H \mathbf{w}\right|^2, \tag{5.15}$$

and

$$c_2 = \left|\mathbf{h}_2^H \mathbf{w}\right|^2, \tag{5.16}$$

denote the beam gains for user 1 and user 2, respectively. We have the following lemma.

Lemma 5.3.1. *With the ideal beamforming, the beam gains satisfy*

$$\frac{c_1}{|\lambda_1|^2} + \frac{c_2}{|\lambda_2|^2} = N, \tag{5.17}$$

where N is the number of antennas.

Proof. With the ideal beamforming, there is no side lobe, i.e., the beam gains along the directions of the two users are significant, while the beam gains along the other directions are zeros. Moreover, the beam pattern for each user is flat with a beam width of $2/N$, which is the same as that of an arbitrary steering vector shown in (5.6) [14, 29]. Under such an ideal condition, the average power of an ideal beamforming vector $\mathbf{w}$ in the beam domain is

$$
\begin{aligned}
\frac{1}{2}\int_{-1}^{1}\left|\mathbf{a}(N,\theta)^{\mathrm{H}}\mathbf{w}\right|^{2}d\theta &= \frac{1}{N}\left(\left|\mathbf{a}(N,\theta_1)^{\mathrm{H}}\mathbf{w}\right|^{2} + \left|\mathbf{a}(N,\theta_2)^{\mathrm{H}}\mathbf{w}\right|^{2}\right) \\
&= \frac{1}{N}\left(\frac{c_1}{|\lambda_1|^{2}} + \frac{c_2}{|\lambda_2|^{2}}\right).
\end{aligned}
\tag{5.18}
$$

On the other hand, for an arbitrary AWV $\mathbf{w}$, we have

$$
\begin{aligned}
&\frac{1}{2}\int_{-1}^{1}|\mathbf{a}(N,\theta)^{\mathrm{H}}\mathbf{w}|^{2}d\theta \\
&= \frac{1}{2}\int_{-1}^{1}\sum_{m=1}^{N}[\mathbf{w}]_m e^{-j\pi(m-1)\theta}\sum_{n=1}^{N}[\mathbf{w}]_n^{*}e^{j\pi(n-1)\theta}d\theta \\
&= \frac{1}{2}\int_{-1}^{1}\sum_{m=1}^{N}\sum_{n=1}^{N}[\mathbf{w}]_m[\mathbf{w}]_n^{*}e^{-j\pi(m-1)\theta}e^{j\pi(n-1)\theta}d\theta \\
&= \sum_{m=1}^{N}[\mathbf{w}]_m[\mathbf{w}]_m^{*} + \frac{1}{2}\sum_{m=1}^{N}\sum_{n=1,\,n\neq m}^{N}[\mathbf{w}]_m[\mathbf{w}]_n^{*}\int_{-1}^{1}e^{j\pi(n-m)\theta}d\theta \\
&= \|\mathbf{w}\|^{2}.
\end{aligned}
\tag{5.19}
$$

Under the CM constraint, we have $\|\mathbf{w}\|^{2}=1$. Thus

$$
\frac{1}{2}\int_{-1}^{1}\left|\mathbf{a}(N,\theta)^{\mathrm{H}}\mathbf{w}\right|^{2}d\theta = \frac{1}{N}\left(\frac{c_1}{|\lambda_1|^{2}} + \frac{c_2}{|\lambda_2|^{2}}\right) = 1 \Rightarrow \frac{c_1}{|\lambda_1|^{2}} + \frac{c_2}{|\lambda_2|^{2}} = N. \tag{5.20}
$$

□

Based on Lemma 5.3.1, Problem (5.14) can be re-described as

$$
\begin{aligned}
&\underset{p_1,p_2,c_1,c_2}{\text{Maximize}} && R_1 + R_2 && \text{(5.21a)} \\
&\text{Subject to} && R_1 \geq r_1, && \text{(5.21b)} \\
& && R_2 \geq r_2, && \text{(5.21c)} \\
& && p_1 + p_2 = P, && \text{(5.21d)} \\
& && \frac{c_1}{|\lambda_1|^{2}} + \frac{c_2}{|\lambda_2|^{2}} = N, && \text{(5.21e)} \\
& && c_1 \geq c_2, && \text{(5.21f)}
\end{aligned}
$$

where $|\mathbf{h}_i^{\mathrm{H}}\mathbf{w}|^2$ is replaced by the beam gain $c_k, k = 1,2$. The CM constraint is not involved in the above Problem (5.21), but will be considered in the beamforming sub-problem.

It is worthy to note that we consider Case 2 because when $r_1 = r_2$, Case 2 is the optimal decoding order in the downlink model. The proof is as follows.

Proof. With Case 1, the achievable rates of user $k, k = 1, 2$, are represented as

$$\begin{cases} R_1^{(1)} = \log_2(1 + \frac{c_1 p_1}{c_1 p_2 + \sigma^2}), \\ R_2^{(1)} = \log_2(1 + \frac{c_2 p_2}{\sigma^2}), \end{cases} \tag{5.22}$$

where $c_1 \le c_2$. The other constraints are the same as those of Case 2, as shown in (5.21).

With Case 2, the achievable rates of user $i, i = 1, 2$, are represented as

$$\begin{cases} R_1^{(2)} = \log_2(1 + \frac{c_1 p_1}{\sigma^2}), \\ R_2^{(2)} = \log_2(1 + \frac{c_2 p_2}{c_2 p_1 + \sigma^2}), \end{cases} \tag{5.23}$$

where $c_1 \ge c_2$.

Assume that the optimal solution for Case 1 is $\{c_1^{(1)}, c_2^{(1)}, p_1^{(1)}, p_2^{(1)}\}$. We have

$$c_1^{(1)} \le c_2^{(1)}, \tag{5.24}$$

and

$$\frac{c_1^{(1)}}{|\lambda_1|^2} + \frac{c_2^{(1)}}{|\lambda_2|^2} = N. \tag{5.25}$$

Hence, we can obtain

$$\begin{aligned} & \left(N - \frac{c_2^{(1)}}{|\lambda_2|^2} \right) |\lambda_1|^2 \le c_2^{(1)} \\ \Rightarrow & N \le \frac{c_2^{(1)}}{|\lambda_2|^2} + \frac{c_2^{(1)}}{|\lambda_1|^2} \Rightarrow N \le \frac{|\lambda_1|^2 + |\lambda_2|^2}{|\lambda_1|^2 |\lambda_2|^2} c_2^{(1)} \\ \Rightarrow & N \left(|\lambda_1|^2 - |\lambda_2|^2 \right) \le \frac{|\lambda_1|^4 - |\lambda_2|^4}{|\lambda_1|^2 |\lambda_2|^2} c_2^{(1)} \\ \Rightarrow & \left(N - \frac{c_2^{(1)}}{|\lambda_2|^2} \right) |\lambda_1|^2 \le \left(N - \frac{c_2^{(1)}}{|\lambda_1|^2} \right) |\lambda_2|^2. \end{aligned} \tag{5.26}$$

For Case 2, we need to find a set of parameters in the feasible region of Problem (5.21), which can obtain higher sum rate than Case 1. For this reason, we choose

$$\{c_1^{(2)} = c_2^{(1)}, \ c_2^{(2)} = (N - \frac{c_1^{(2)}}{|\lambda_1|^2}) |\lambda_2|^2, \ p_1^{(2)} = p_2^{(1)}, \ p_2^{(2)} = p_1^{(1)}\}. \tag{5.27}$$

Thus, we have

$$c_2^{(2)} = (N - \frac{c_1^{(2)}}{|\lambda_1|^2}) |\lambda_2|^2 = (N - \frac{c_2^{(1)}}{|\lambda_1|^2}) |\lambda_2|^2 \ge (N - \frac{c_2^{(1)}}{|\lambda_2|^2}) |\lambda_1|^2 = c_1^{(1)}. \tag{5.28}$$

It is worthy to note that the order of beam gains is

$$c_1^{(1)} \le c_2^{(2)} \le \bar{c} \le c_1^{(2)} = c_2^{(1)}, \tag{5.29}$$

where $\bar{c}$ is the equilibrium point we define above. Furthermore, we have

$$\begin{cases} R_1^{(2)} = \log_2(1 + \dfrac{c_1^{(2)} p_1^{(2)}}{\sigma^2}) = \log_2(1 + \dfrac{c_2^{(1)} p_2^{(1)}}{\sigma^2}) = R_2^{(1)}, \\[3mm] R_2^{(2)} = \log_2(1 + \dfrac{c_2^{(2)} p_2^{(2)}}{c_2^{(2)} p_1^{(2)} + \sigma^2}) \ge \log_2(1 + \dfrac{c_1^{(1)} p_1^{(1)}}{c_1^{(1)} p_2^{(1)} + \sigma^2}) = R_1^{(1)}. \end{cases} \tag{5.30}$$

So

$$R_1^{(2)} + R_2^{(2)} \ge R_1^{(1)} + R_2^{(1)}, \tag{5.31}$$

i.e., decoding user 2 first is optimal if $r_1 = r_2$. □

In order to optimize the sum rate, the BS tends to allocate more beam gain to the user with a higher channel gain to further intensify the channel differences.

Next is the beamforming problem. As analyzed above, we formulate the original beamforming problem as

$$\mathbf{w} \in \mathbb{C}^N \tag{5.32a}$$

$$\text{Subject to} \quad \left| \mathbf{h}_1^H \mathbf{w} \right|^2 = c_1, \tag{5.32b}$$

$$\left| \mathbf{h}_2^H \mathbf{w} \right|^2 = c_2, \tag{5.32c}$$

$$\left| [\mathbf{w}]_n \right| = \frac{1}{\sqrt{N}}, \quad n = 1, 2, ..., N. \tag{5.32d}$$

Since an appropriate weighing vector $\mathbf{w}$ may not be found in most cases under the CM constraints. Therefore, we relax the equality constraints and formulate the beamforming problem as

$$\underset{\mathbf{w}}{\text{Minimize}} \quad \alpha \tag{5.33a}$$

$$\text{Subject to} \quad [\mathbf{w}]_n [\mathbf{w}]_n^* \le \alpha; \ n = 1, 2, \cdots, N, \tag{5.33b}$$

$$\left| \mathbf{h}_1^H \mathbf{w} \right| \ge \sqrt{c_1}, \tag{5.33c}$$

$$\left| \mathbf{h}_2^H \mathbf{w} \right| \ge \sqrt{c_2}. \tag{5.33d}$$

Finally, considering that the normalized $\mathbf{w}$ may not satisfy the strict gain constraint

$$|\mathbf{h}_k^H \mathbf{w}|^2 = c_k, k = 1, 2, \tag{5.34}$$

we need to substitute it into the original problem, i.e., (5.14), to reset the user powers. Then we obtain the final solution.

With the above manipulations, Problem (5.14) is decomposed into Problems (5.21) and (5.33), which are independent power and beam gain allocation and beamforming sub-problems. Problem (5.21) is a relaxation of the original problem, since the CM beamforming is bypassed, which enlarges the feasible region. Problem (5.33) addresses the beamforming issue, but it is also a relaxation of the CM beamforming problem. Although the original problem is hard to solve, the two sub-problems are relatively easy to solve. Next, we will first solve Problem (5.21), and obtain the optimal solution $\{c_1^\star, c_2^\star, p_1^\star, p_2^\star\}$. The optimal values of c_1 and c_2 are used as gain constraints in Problem (5.33). Then we solve Problem (5.33) and obtain an appropriate $\mathbf{w}$. Afterwards, we normalize $\mathbf{w}$ to satisfy the CM constraint. Finally, substituting $\mathbf{w}$ into problem (5.14), we obtain the final solution of power allocation $\{p_1, p_2\}$.

2) Solution of the Power and Beam Gain Allocation Sub-Problem

As previously mentioned, when $r_1 = r_2$, Case 2 is the optimal decoding order in the downlink model, thus the optimal decoding is confirmed. So the expression of achievable rates are exclusive, i.e.,

$$R_1 = R_1^{(2)}, R_2 = R_2^{(2)}. \tag{5.35}$$

The implicit SINR constraint

$$\frac{c_1 p_2}{c_1 p_1 + \sigma^2} \geq \frac{c_2 p_2}{c_2 p_1 + \sigma^2}, \tag{5.36}$$

is equivalent to $c_1 \geq c_2$. In addition, there are two equality constraints in Problem (5.21), namely

$$p_1 + p_2 = P, \tag{5.37}$$

and

$$\frac{c_1}{|\lambda_1|^2} + \frac{c_2}{|\lambda_2|^2} = N. \tag{5.38}$$

Thus, we have

$$p_2 = P - p_1, \tag{5.39}$$

and

$$c_2 = (N - \frac{c_1}{|\lambda_1|^2}) |\lambda_2|^2. \tag{5.40}$$

Substituting them into Problem (5.21), we obtain

$$\text{Maximize}_{p_1, c_1} \quad f(c_1, p_1) = R_1 + R_2 \tag{5.41a}$$

$$\text{Subject to} \quad R_1 \geq r_1, \tag{5.41b}$$

$$R_2 \geq r_2, \tag{5.41c}$$

$$c_1 \geq c_2, \tag{5.41d}$$

where R_1 and R_2 become

$$\begin{cases} R_1 = \log_2(1 + \frac{c_1 p_1}{\sigma^2}), \\ R_2 = \log_2 \left(1 + \frac{(|\lambda_2|^2 N - \frac{|\lambda_2|^2}{|\lambda_1|^2} c_1)(P - p_1)}{(|\lambda_2|^2 N - \frac{|\lambda_2|^2}{|\lambda_1|^2} c_1) p_1 + \sigma^2} \right). \end{cases} \tag{5.42}$$

The objective function $f(c_1, p_1)$ with two variables c_1 and p_1 is given by

$$f(c_1, p_1) = \log_2(1 + \frac{c_1 p_1}{\sigma^2}) + \log_2\left(1 + \frac{(|\lambda_2|^2 N - \frac{|\lambda_2|^2}{|\lambda_1|^2}c_1)(P - p_1)}{(|\lambda_2|^2 N - \frac{|\lambda_2|^2}{|\lambda_1|^2}c_1)p_1 + \sigma^2}\right). \quad (5.43)$$

It is known that the maximum point of a continuous function in a bounded closed set is either an extreme point or located on the boundary. Thus, to solve Problem (5.41), we first obtain the extreme points of the objective function. If the three inequality constraints are satisfied at any one of these extreme points, it may be an optimal solution. Otherwise, the optimal solution should be within the boundary defined by the three inequality constraints. Hence, we start by obtaining the stationary points of the objective function. Note that a stationary point may not be necessarily an extreme point; it may be a saddle point.

Let the gradient of $f(c_1, p_1)$ be zero, i.e.,

$$(\frac{\partial f}{\partial c_1}, \frac{\partial f}{\partial p_1}) = (0, 0). \quad (5.44)$$

Then, we obtain one stationary point (c_{1m}, p_{1m}):

$$\begin{cases} c_{1m} = \dfrac{|\lambda_1|^2 |\lambda_2|^2 N}{|\lambda_1|^2 + |\lambda_2|^2}, \\ p_{1m} = \dfrac{|\lambda_2|^2 (|\lambda_1|^2 + |\lambda_2|^2)P\sigma^2}{|\lambda_1|^4 |\lambda_2|^2 NP + (|\lambda_1|^2 + |\lambda_2|^2)^2 \sigma^2}. \end{cases} \quad (5.45)$$

After some derivations, we can obtain

$$f(c_{1m}, p_{1m}) = \log_2\left(1 + \frac{|\lambda_1|^2 |\lambda_2|^2 NP}{(|\lambda_1|^2 + |\lambda_2|^2)\sigma^2}\right). \quad (5.46)$$

To tell whether this stationary point is an extreme point or just a saddle point, the values of the functions at some other points need to be examined. For simplicity, the intersection points of the boundaries, i.e., the maximum value and minimum value of c_1, p_1, namely

$$(0, 0), (N|\lambda_1|^2, P), (N|\lambda_1|^2, 0), (0, P). \quad (5.47)$$

Then we can obtain the values

$$f(0, 0) = \log_2(1 + \frac{|\lambda_2|^2 NP}{\sigma^2}), \quad (5.48)$$

$$f(N|\lambda_1|^2, P) = \log_2(1 + \frac{|\lambda_1|^2 NP}{\sigma^2}), \quad (5.49)$$

and the values of the objective function at points $(N|\lambda_1|^2, 0)$ and $(0, P)$ are, respectively,

$$f(N|\lambda_1|^2, 0) = 0, \quad (5.50)$$

$$f(0, P) = 0. \quad (5.51)$$

Clearly, $f(N|\lambda_1|^2, 0)$ and $f(0, P)$ are smaller than $f(c_{1m}, p_{1m})$, while $f(0,0)$ and $f(N|\lambda_1|^2, P)$ are greater than $f(c_{1m}, p_{1m})$. Hence, the point (c_{1m}, p_{1m}) is just a saddle point instead of an optimum of the objective function. Hence, the function does not have an extreme point. The optimal solution of Problem (5.41) locates within the boundary of the feasible region defined by one of the inequality constraints. As there are three inequality constraints in Problem (5.41), the feasible region is enclosed by three boundaries corresponding to the three inequality constraints, and the optimal solution may locate within any one of them. Thus, there are three cases:

Case 1: If the optimal solution is within the boundary $c_1 = c_2$, we have $c_1 = c_{1m}$ as shown in (5.45), and in such a case the objective function $f(c_1, p_1)$ does not depend on p_1, i.e., $f(c_1, p_1) = f(c_{1m}, p_{1m})$ as shown in (5.46), provided that (c_1, p_1) satisfies the two rate constraints.

Case 2: If the optimal solution is within the boundary $R_1 = r_1$, we have

$$p_1 = \frac{(2^{r_1} - 1)\sigma^2}{c_1}. \tag{5.52}$$

Substituting it into the objective function $f(c_1, p_1)$, we find that the objective function now has only one variable c_1. By letting the derivative of the objective function with respect to c_1 equal zero, and solving the equation, we can obtain two roots, a positive one and a negative one. Clearly the negative one does not satisfy the condition that the beam gain c_1 is positive. Thus, the obtained positive root is the optimal value of c_1. Then we achieve the following optimal solution:

$$\begin{cases} c_{1,1} = \dfrac{|\lambda_1|^2 \left[(2^{1+r_1} - 2)|\lambda_2|^2 NP - 2\sqrt{G}\right]}{2P\left[(2^{r_1} - 1)|\lambda_2|^2 - |\lambda_1|^2\right]}, \\[4mm] p_{1,1} = \dfrac{(2^{r_1} - 1)\sigma^2}{c_{1,1}}, \end{cases} \tag{5.53}$$

where

$$\begin{aligned} G = &(2^{r_1} - 1)|\lambda_1|^2 |\lambda_2|^2 N^2 P^2 + \\ &\left[(2^{r_1} - 1)|\lambda_1|^2 + (2^{1+r_1} - 2^{2r_1} - 1)|\lambda_2|^2\right]NP\sigma^2. \end{aligned} \tag{5.54}$$

It is noteworthy that Case 2 implicitly requires $c_{1,1} \geq c_{1m}$, such that

$$f(c_{1,1}, p_{1,1}) \geq f(c_{1m}, p_{1a}) = f(c_{1m}, p_{1m}), \tag{5.55}$$

where (c_{1m}, p_{1a}) is the intersection point of boundary $c_1 = c_2$ and boundary $R_1 = r_1$. Otherwise, if $c_{1,1} < c_{1m}$, in the region of $c_1 \geq c_{1m}$ within boundary $R_1 = r_1$, $f(c_1, p_1)$ is monotonically descending as c_1. In such a case,

$$f(c_{1m}, p_{1m}) = f(c_{1m}, p_{1a}) \geq f(c_1, p_1), \tag{5.56}$$

which means that the optimal solution locates within boundary $c_1 = c_2$ instead of boundary $R_2 = r_2$.

Case 3: If the optimal solution is within the boundary $R_2 = r_2$, analogously, we can obtain the optimal solution:

$$\begin{cases} c_{1,2} = |\lambda_1|^2 \left(N - \dfrac{\sqrt{(2^{r_2} - 1)NP\sigma^2}}{|\lambda_2|\,P} \right), \\[4mm] p_{1,2} = \dfrac{(N\,|\lambda_1|^2 - c_1)\,|\lambda_2|^2\,P - (2^{r_2} - 1)\,|\lambda_1|^2\,\sigma^2}{2^{r_2}\,|\lambda_2|^2\,(N\,|\lambda_1|^2 - c_1)}. \end{cases} \tag{5.57}$$

Similarly, Case 3 implicitly requires $c_{1,2} \geq c_{1m}$; otherwise the optimal solution locates within boundary $c_1 = c_2$ instead of boundary $R_2 = r_2$.

Hereto, the power and beam gain allocation sub-problem, i.e., the optimal solution of Problem (5.21) and $\{c_1^\star, c_2^\star, p_1^\star, p_2^\star\}$ under the assumption of ideal beamforming, has been solved. However, $\{c_1^\star, c_2^\star, p_1^\star, p_2^\star\}$ may not be an optimal solution of the original problem because a beamforming vector with beam gains $\{c_1^\star, c_2^\star\}$ may not be found under the CM constraint. Hence, the optimal achievable sum rate of Problem (5.21) is an upper bound of that of the original problem.

3) Solution of the Beamforming Sub-Problem

Define

$$\begin{cases} b_1 = \dfrac{c_1^\star}{|\lambda_1|^2}, \\[4mm] b_2 = \dfrac{c_2^\star}{|\lambda_2|^2}. \end{cases} \tag{5.58}$$

As mentioned in (5.33), the beamforming sub-problem can be rewritten as

$$\underset{\mathbf{w}}{\text{Minimize}} \quad \alpha \tag{5.59a}$$

$$\text{Subject to} \quad [\mathbf{w}]_n[\mathbf{w}]_n^* \leq \alpha; \ n = 1, 2, \cdots, N, \tag{5.59b}$$

$$\left| \mathbf{a}_1^H \mathbf{w} \right| \geq \sqrt{b_1}, \tag{5.59c}$$

$$\left| \mathbf{a}_2^H \mathbf{w} \right| \geq \sqrt{b_2}, \tag{5.59d}$$

where $\mathbf{a}_k \triangleq \mathbf{a}(N, \theta_k)$ for $k = 1, 2$.

It is clear that an arbitrary phase rotation can be added to the vector $\mathbf{w}$ in Problem (5.59) without affecting the beam gains. Thus, if $\mathbf{w}$ is optimal, so is $\mathbf{w}e^{j\phi}$, where ϕ is an arbitrary phase within $[0, 2\pi)$. Without loss of generality, we may then choose ϕ so that $\mathbf{h}_1^H \mathbf{w}$ is real. Problem (5.59) is tantamount to

$$\underset{\mathbf{w}}{\text{Minimize}} \quad \alpha \tag{5.60a}$$

$$\text{Subject to} \quad [\mathbf{w}]_n[\mathbf{w}]_n^* \leq \alpha; \ n = 1, 2, \cdots, N, \tag{5.60b}$$

$$\Re(\mathbf{a}_1^H \mathbf{w}) \geq \sqrt{b_1}, \tag{5.60c}$$

$$\left| \mathbf{a}_2^H \mathbf{w} \right| \geq \sqrt{b_2}. \tag{5.60d}$$

Substituting the expression of $\mathbf{a}_k$, we can rewrite the above problem as

$$\underset{\mathbf{w}}{\text{Minimize}} \quad \underset{i}{\text{Max}} \left\{ [\mathbf{w}]_n [\mathbf{w}]_n^* \right\} \tag{5.61a}$$

$$\text{Subject to} \quad \Re \left(\sum_{n=1}^{N} [\mathbf{w}]_n e^{j\psi_n} \right) \geq \sqrt{b_1}, \tag{5.61b}$$

$$\left| \sum_{n=1}^{N} [\mathbf{w}]_n e^{j\eta_n} \right| \geq \sqrt{b_2}, \tag{5.61c}$$

where

$$\begin{cases} \psi_n = (n-1)\pi\theta_1, \\ \eta_n = (n-1)\pi\theta_2. \end{cases} \tag{5.62}$$

Problem (5.61) is still non-convex because of the absolute value operation. Thus, we split it into a serial of convex optimization problems and obtain:

$$\underset{\mathbf{w}}{\text{Minimize}} \quad \underset{i}{\text{Max}} \, [\mathbf{w}]_n [\mathbf{w}]_n^* \tag{5.63a}$$

$$\text{Subject to} \quad \Re \left(\sum_{n=1}^{N} [\mathbf{w}]_n e^{j\psi_n} \right) \geq \sqrt{b_1}, \tag{5.63b}$$

$$\Re \left(\left(\sum_{n=1}^{N} [\mathbf{w}]_n e^{j\eta_n} \right) e^{j\frac{m}{M_p}2\pi} \right) \geq \sqrt{b_2}, \tag{5.63c}$$

where M_p is the number of total candidate phases, $m = 1, 2, \cdots, M_p$. Each of these M_p problems can be efficiently solved by using standard convex optimization tools. We select the solution with the minimal objective among the M_p optimal solutions as the optimal solution $\mathbf{w}_0^\star$, and then we normalize it such that the vector has unit power: $\mathbf{w}_1^\star = \mathbf{w}_0^\star / \|\mathbf{w}_0^\star\|$.

However, there is no guarantee that the obtained $\mathbf{w}_1^\star$ by solving (5.59) satisfies the CM constraint. Hence, we still need CM normalization, i.e., to normalize $\mathbf{w}_1^\star$ to $\mathbf{w}^\star$, so as to satisfy the CM constraint with the phases of its elements the same. The CM normalization is given by

$$[\mathbf{w}^\star]_n = \frac{[\mathbf{w}_1^\star]_n}{\sqrt{N}|[\mathbf{w}_1^\star]_n|}, \quad n = 1, 2, ..., N. \tag{5.64}$$

If the modulus of the elements of $\mathbf{w}_1^\star$ are different with each other, the CM normalization would result in significant influence, and the finally achieved beam gains, i.e., c_k, would be far away from $c_k^\star$. In such a case, the sum rate performance would not be satisfactory. Hence, we need to evaluate the impact of the CM normalization on performance. To this end, we give the following theorem.

Theorem 5.3.1. *If $\mathbf{w}^\star$ is the optimal solution of Problem (5.63), then, at least $N-1$ elements of $\mathbf{w}^\star$ have the same modulus, and the modulus of the remaining element cannot be larger than that of the $N-1$ elements.*

Proof. Let $\mathbf{w}_{\text{opt}}$ represent the optimal solution of Problem (5.63), and assume

$$
\begin{cases}
\sum_{i=1}^{N} [\mathbf{w}_{\text{opt}}]_i e^{j\theta_i} = d_1, \\
\sum_{i=1}^{N} [\mathbf{w}_{\text{opt}}]_i e^{j\eta_i} = d_2 e^{-j\frac{m}{M}2\pi},
\end{cases}
\tag{5.65}
$$

where d_1 and d_2 are positive real values.

Lemma 5.3.2. *Given d_1, d_2, Problem (5.63) is equivalent to*

$$
\begin{aligned}
\underset{\mathbf{w}}{\text{Minimize}} \quad & \underset{i}{\text{Max}} \left\{ [\mathbf{w}]_i [\mathbf{w}]_i^* \right\} \\
\text{Subject to} \quad & \sum_{i=1}^{N} [\mathbf{w}]_i e^{j\theta_i} = d_1, \\
& \sum_{i=1}^{N} [\mathbf{w}]_i e^{j\eta_i} = d_2 e^{-j\frac{m}{M}2\pi}.
\end{aligned}
\tag{5.66}
$$

Proof. According to the definitions of d_1 and d_2, the optimal solution of Problem (5.63), i.e., $\mathbf{w}_{\text{opt}}$, is a feasible solution of Problem (5.66).

On the other hand, since $d_1 \geq \sqrt{b_1}$ and $d_2 \geq \sqrt{b_2}$, the optimal solution of Problem (5.66) must be a feasible solution of Problem (5.63).

In summary, Problem (5.66) is equivalent to Problem (5.63). $\qquad\square$

Lemma 5.3.3. *There are at least $N-1$ elements of $\mathbf{w}_{\text{opt}}$ which have the same modulus, where $\mathbf{w}_{\text{opt}}$ is the optimal solution of Problem (5.66), and the modulus of the remaining element cannot be larger than that of the $N-1$ elements.*

Proof. We first sort the absolute values of the weights in $\mathbf{w}_{\text{opt}}$ as

$$
|[\mathbf{w}_{\text{opt}}]_{\pi_1}| \leq |[\mathbf{w}_{\text{opt}}]_{\pi_2}| \leq |[\mathbf{w}_{\text{opt}}]_{\pi_3}| \leq \cdots \leq |[\mathbf{w}_{\text{opt}}]_{\pi_N}|.
\tag{5.67}
$$

Then Lemma 5.3.3 is equivalent to that the inequalities after $|[\mathbf{w}_{\text{opt}}]_{\pi_2}|$ are all equalities. We prove it by using the contradiction method, i.e., we assume that there is at least one strictly less-than sign after $|[\mathbf{w}_{\text{opt}}]_{\pi_2}|$, and then we prove that $\mathbf{w}_{\text{opt}}$ is not optimal.

Since the two constraints of Problem (5.66) are two linear equations in the space $\mathbb{C}^N$, we have $N-2$ DoFs to adjust the elements of vector $\mathbf{w}$. To make full use of the available DoFs, $[\mathbf{w}_{\text{opt}}]_{\pi_1}$ and $[\mathbf{w}_{\text{opt}}]_{\pi_2}$ can always be expressed as linear combinations of non-homogeneous linear equations consisting of the two constraints, which means that the feasible region of Problem (5.66) has a dimension of $N-2$. Without loss of generality, let

$$
\mathbf{w}_0 \triangleq [[\mathbf{w}_{\text{opt}}]_{\pi_3}, [\mathbf{w}_{\text{opt}}]_{\pi_4}, ..., [\mathbf{w}_{\text{opt}}]_{\pi_N}]^{\text{T}},
\tag{5.68}
$$

and $[\mathbf{w}_{\text{opt}}]_{\pi_1}$ and $[\mathbf{w}_{\text{opt}}]_{\pi_2}$ can be expressed as

$$
\begin{cases}
[\mathbf{w}_{\text{opt}}]_{\pi_1} = \mathbf{f}_1^{\text{H}} \mathbf{w}_0 + \beta_1, \\
[\mathbf{w}_{\text{opt}}]_{\pi_2} = \mathbf{f}_2^{\text{H}} \mathbf{w}_0 + \beta_2,
\end{cases}
\tag{5.69}
$$

where $\mathbf{f}_1$ and $\mathbf{f}_2$ are the combination coefficient vectors. Parameters β_1 and β_2 are constant terms because the constraints of Problem (5.66) have the form of non-homogeneous equations and will be specified later. In the following, we will construct $\bar{\mathbf{w}}$, a better solution than $\mathbf{w}_{\mathrm{opt}}$.

We consider a point in the feasible region $\mathbb{C}^{N-2}$ close to $\mathbf{w}_0$, i.e., $\bar{\mathbf{w}}_0 = \frac{1}{1+\delta}\mathbf{w}_0$, where δ is a small positive variable that is very close to zero. Let $w_1 = \mathbf{f}_1^{\mathrm{H}}\bar{\mathbf{w}}_0 + \beta_1$ and $w_2 = \mathbf{f}_2^{\mathrm{H}}\bar{\mathbf{w}}_0 + \beta_2$. Then β_1 and β_2 along with f_1 and f_2 can be determined so that $\bar{\mathbf{w}} = [w_1, w_2, \bar{\mathbf{w}}_0^{\mathrm{T}}]^{\mathrm{T}}$ is a feasible point of Problem (5.66). We have

$$\begin{cases} |w_1| = |\mathbf{f}_1^{\mathrm{H}}\bar{\mathbf{w}}_0 + \beta_1|, \\ |w_2| = |\mathbf{f}_2^{\mathrm{H}}\bar{\mathbf{w}}_0 + \beta_2|. \end{cases} \tag{5.70}$$

When there is at least one strictly less-than sign after $|[\mathbf{w}_{\mathrm{opt}}]_{\pi_2}|$ in (5.67), we have

$$|[\mathbf{w}_{\mathrm{opt}}]_{\pi_1}| \leq |[\mathbf{w}_{\mathrm{opt}}]_{\pi_2}| < |[\mathbf{w}_{\mathrm{opt}}]_{\pi_N}|. \tag{5.71}$$

Suppose

$$|[\mathbf{w}_{\mathrm{opt}}]_{\pi_N}| - |[\mathbf{w}_{\mathrm{opt}}]_{\pi_1}| = \varepsilon_1, \tag{5.72}$$

where $\varepsilon_1 > 0$. In a sequel,

$$\begin{aligned} |[\bar{\mathbf{w}}]_{\pi_N}| - |w_1| &= \frac{1}{1+\delta}|[\mathbf{w}_{\mathrm{opt}}]_{\pi_N}| - \frac{1}{1+\delta}|[\mathbf{w}_{\mathrm{opt}}]_{\pi_1}| + \frac{1}{1+\delta}|[\mathbf{w}_{\mathrm{opt}}]_{\pi_1}| - |w_1| \\ &= \frac{\varepsilon_1}{1+\delta} + \frac{1}{1+\delta}|\mathbf{f}_1^{\mathrm{H}}\mathbf{w}_0 + \beta_1| - |\mathbf{f}_1^{\mathrm{H}}\bar{\mathbf{w}}_0 + \beta_1| \\ &= \frac{\varepsilon_1}{1+\delta} + |\frac{1}{1+\delta}\mathbf{f}_1^{\mathrm{H}}\mathbf{w}_0 + \frac{1}{1+\delta}\beta_1| - |\frac{1}{1+\delta}\mathbf{f}_1^{\mathrm{H}}\mathbf{w}_0 + \beta_1| \\ &\geq \frac{\varepsilon_1}{1+\delta} - |(\frac{1}{1+\delta}\mathbf{f}_1^{\mathrm{H}}\mathbf{w}_0 + \frac{1}{1+\delta}\beta_1) - (\frac{1}{1+\delta}\mathbf{f}_1^{\mathrm{H}}\mathbf{w}_0 + \beta_1)| \\ &= \frac{\varepsilon_1 - \delta|\beta_1|}{1+\delta}. \end{aligned} \tag{5.73}$$

Hence, there exists a sufficiently small

$$\delta_1 = \frac{\varepsilon_1}{1+|\beta_1|}, \tag{5.74}$$

such that

$$\frac{\varepsilon_1 - \delta_1|\beta_1|}{1+\delta} > 0, \text{ i.e. } |w_1| < |[\bar{\mathbf{w}}]_{\pi_N}|. \tag{5.75}$$

Similarly, supposing

$$|[\mathbf{w}_{\mathrm{opt}}]_{\pi_N}| - |[\mathbf{w}_{\mathrm{opt}}]_{\pi_2}| = \varepsilon_2, \tag{5.76}$$

we can conclude that there exists a sufficiently small

$$\delta_2 = \frac{\varepsilon_1}{1+|\beta_2|}, \tag{5.77}$$

such that

$$|w_2| < |[\bar{\mathbf{w}}]_{\pi_N}|. \tag{5.78}$$

Let $\delta = \min\{\delta_1, \delta_2\}$, so there is always

$$\begin{cases} |w_1| < |[\bar{\mathbf{w}}]_{\pi_N}|, \\ |w_2| < |[\bar{\mathbf{w}}]_{\pi_N}|. \end{cases} \tag{5.79}$$

In other words, $[\bar{\mathbf{w}}]_{\pi_N}$ is the largest-modulus element of $\bar{\mathbf{w}}$. Therefore,

$$\underset{i}{\text{Max}} \{[\bar{\mathbf{w}}]_i [\bar{\mathbf{w}}]_i^*\} = |[\bar{\mathbf{w}}]_{\pi_N}|^2 < |[\mathbf{w}_{\text{opt}}]_{\pi_N}|^2, \tag{5.80}$$

which means that $\bar{\mathbf{w}}$ is a better solution of Problem (5.66) than $\mathbf{w}_{\text{opt}}$. This is contradictory against that $\mathbf{w}_{\text{opt}}$ is the optimal solution of Problem (5.66); so the assumption that there is at least one strictly less-than sign after $|[\mathbf{w}_{\text{opt}}]_{\pi_2}|$ in (5.67) does not hold. Hence, the inequalities after $|[\mathbf{w}_{\text{opt}}]_{\pi_2}|$ in (5.67) are all equalities, i.e., Lemma 5.3.3 holds. □

Combining Lemma 5.3.2 and Lemma 5.3.3, we can easily conclude that Theorem 5.3.1 holds. □

According to Theorem 5.3.1, since $\mathbf{w}_0^\star$ is the optimal solution of Problem (5.63), and $\mathbf{w}_1^\star$ is the l_2-norm normalization of $\mathbf{w}_0^\star$, at least $N-1$ elements of $\mathbf{w}_1^\star$ have the same modulus, and the remaining one element has a modulus no larger than that of the $N-1$ elements, i.e.,

$$0 \leq |[\mathbf{w}_1^\star]_1| \leq \frac{1}{\sqrt{N}}, \tag{5.81}$$

and

$$\frac{1}{\sqrt{N}} \leq |[\mathbf{w}_1^\star]_2| = |[\mathbf{w}_1^\star]_3| = ... = |[\mathbf{w}_1^\star]_N| \leq \frac{1}{\sqrt{N-1}}. \tag{5.82}$$

Theorem 5.3.2. *Given an arbitrary CM vector* $\mathbf{b}$, *i.e.,* $|[\mathbf{b}]_i| = 1$ *for* $i = 1, 2, ..., N$, *we have*

$$||\mathbf{b}^H \mathbf{w}^\star| - |\mathbf{b}^H \mathbf{w}_1^\star|| < \frac{2}{\sqrt{N}}. \tag{5.83}$$

where $\mathbf{w}_1^\star$ *is the optimal solution of Problem (5.63) with unit power, and* $\mathbf{w}^\star$ *is the optimal solution after CM normalization, i.e., (5.64).*

Proof.

$$||\mathbf{b}^H \mathbf{w}^\star| - |\mathbf{b}^H \mathbf{w}_1^\star|| \leq |\mathbf{b}^H \mathbf{w}^\star - \mathbf{b}^H \mathbf{w}_1^\star|$$
$$\leq |[\mathbf{b}]_1^*([\mathbf{w}^\star]_1 - [\mathbf{w}_1^\star]_1)| + |[\mathbf{b}]_2^*([\mathbf{w}^\star]_2 - [\mathbf{w}_1^\star]_2)| + \cdots + |[\mathbf{b}]_N^*([\mathbf{w}^\star]_N - [\mathbf{w}_1^\star]_N)|$$
$$= |[\mathbf{w}^\star]_1 - [\mathbf{w}_1^\star]_1| + |[\mathbf{w}^\star]_2 - [\mathbf{w}_1^\star]_2| + \cdots + |[\mathbf{w}^\star]_N - [\mathbf{w}_1^\star]_N|. \tag{5.84}$$

According to the CM normalization in (5.64), the elements of $\mathbf{w}^\star$ have exactly the same phases as those of $\mathbf{w}_1^\star$; hence

$$|[\mathbf{w}^\star]_i - [\mathbf{w}_1^\star]_i| = ||[\mathbf{w}^\star]_i| - |[\mathbf{w}_1^\star]_i|| = |\frac{1}{\sqrt{N}} - |[\mathbf{w}_1^\star]_i||, \tag{5.85}$$

where $i = 1, 2, ..., N$.

Since $0 \leq |[\mathbf{w}_1^\star]_1| \leq \frac{1}{\sqrt{N}}$, we have $|[\mathbf{w}^\star]_1 - [\mathbf{w}_1^\star]_1| \leq \frac{1}{\sqrt{N}}$. Since $\frac{1}{\sqrt{N}} \leq |[\mathbf{w}_1^\star]_2| = |[\mathbf{w}_1^\star]_3| = ... = |[\mathbf{w}_1^\star]_N| \leq \frac{1}{\sqrt{N-1}}$, we have

$$\left|[\mathbf{w}^\star]_i - [\mathbf{w}_1^\star]_i\right| \leq \frac{1}{\sqrt{N-1}} - \frac{1}{\sqrt{N}}, \tag{5.86}$$

where $i = 2, 3, ..., N$.

Hence, we obtain

$$\left|\|\mathbf{b}^H\mathbf{w}^\star| - |\mathbf{b}^H\mathbf{w}_1^\star\|\right| \leq \frac{1}{\sqrt{N}} + (N-1)\left(\frac{1}{\sqrt{N-1}} - \frac{1}{\sqrt{N}}\right) < \frac{2}{\sqrt{N}}. \tag{5.87}$$

$\square$

According to Theorem 5.3.2, since the AWV for user k, i.e., $\mathbf{a}_k$, $k = 1, 2$, is a CM vector, $\left|\|\mathbf{a}_k^H\mathbf{w}^\star| - |\mathbf{a}_k^H\mathbf{w}_1^\star\|\right| < \frac{2}{\sqrt{N}}$, which means that the CM normalization has a limited influence on the desired beam gains, and the influence decreases when N increases.

4) Solution of the Original Problem

Substituting the obtained normalized $\mathbf{w}^\star$ above into the original problem, i.e., (5.14), we obtain

$$\underset{p_1}{\text{Maximize}} \quad \log_2(1 + \frac{c_1 p_1}{\sigma^2}) + \log_2(1 + \frac{c_2(P - p_1)}{c_2 p_1 + \sigma^2}) \tag{5.88a}$$

$$\text{Subject to} \quad \log_2(1 + \frac{c_1 p_1}{\sigma^2}) \geq r_1, \tag{5.88b}$$

$$\log_2(1 + \frac{c_2(P - p_1)}{c_2 p_1 + \sigma^2}) \geq r_2, \tag{5.88c}$$

where the beam gain $c_k = |\mathbf{h}_k^H\mathbf{w}^\star|^2$, $k = 1, 2$, are fixed values. Problem (5.88) is a single-variable optimization problem. The feasible region and the monotonicity of the objective function are distinct. We can easily obtain the final power allocation $\{p_1, p_2\}$ by solving the problem above.

Although the finally obtained solution $\{p_1, p_2, \mathbf{w}\}$ is sub-optimal, the achieved sum rate performance is close to the performance bound. This is because via solving Problem (5.21), we actually obtain an upper bound of the sum rate performance of the original problem. However, since it is difficult and sometimes not possible to obtain a CM beamformer which strictly satisfies the equality gain constraints, we formulate the beamforming sub-problem as (5.59). This sub-problem tries to let the power of each antenna be the same while satisfying the gain constraints, but does not guarantee the CM constraint. Thus, a CM normalization is required. Nevertheless, according to Theorems 5.3.1 and 5.3.2, the resulting performance loss due to the CM normalization is little.

In addition to the optimization method, some intuitive approaches can also be considered. For instance, to maximize the sum rate, most power or beam gain should be allocated to user 1, which has better channel quality, while only necessary power or beam gain should be allocated to user 2 to satisfy the rate constraint. Besides, for user 2, although its achievable rate increases with both P_2 and c_2, increasing P_2 is more efficient, because increasing c_2 also increases MUI, i.e., the signal of user 1 at user 2, while increasing P_2 reduces MUI on the contrary, since P_1 is reduced accordingly. Using these intuitive observations, we may set appropriate powers and beam gains for the two users, and then determine the beamforming vector.

5.3.3 Performance Evaluations

First, we will compare the ideal beam pattern with the designed beam pattern obtained by solving problem (5.59). We assume

$$
\begin{cases}
|\lambda_1| = 0.8, \\
|\lambda_1| = 0.5, \\
\theta_1 = -0.25, \\
\theta_2 = 0.4.
\end{cases} \tag{5.89}
$$

The desired beam gains are

$$
\begin{cases}
c_1^\star = N/2, \\
c_2^\star = (N - c_1^\star/|\lambda_1|^2)|\lambda_2|^2,
\end{cases} \tag{5.90}
$$

where N is the number of antennas at the BS. M_p in (5.63) is set to 20 in this simulation as well as the following simulations, which is large enough to obtain the best solution. As the CM normalization affects the shape of beam pattern, we compare the desired ideal beam pattern with the designed beam patterns before and after the CM normalization, i.e., the beam patterns computed with $\mathbf{w}_1^\star$ and $\mathbf{w}^\star$ in (5.64). Fig. 5.7 shows the comparison results with $N = 16, 32, 64$, and from this figure we can find that the beam gains are significant along the desired user directions, and both the beam patterns before and after the CM normalization are close to the ideal beam pattern along the user directions, which not only demonstrates that the CM normalization has little influence on the beam gains along the user directions, but also shows that the solution of the beamforming sub-problem is reasonable.

In addition to the beam pattern, we also compare the user beam gain with varying number of antennas in Fig. 5.8, where the parameter settings are the same as those in Fig. 5.7.

Fig. 5.8 shows that the user gains before and after the CM normalization are almost the same as each other, which demonstrates again that the CM normalization has little impact on the beamforming performance. Moreover, there is a small gap between the designed user gains and the ideal beam gains for both users (as well as the sum beam gain). This is because the designed beam pattern has side lobes which reduces the gains along the user directions. In comparison, an ideal beam pattern does not have side lobes. Fortunately, the gap increases slowly as N increases when $N \leq 40$, and almost does not increase when $N > 40$, which shows that the proposed beamforming method behaves robust against the number of antennas.

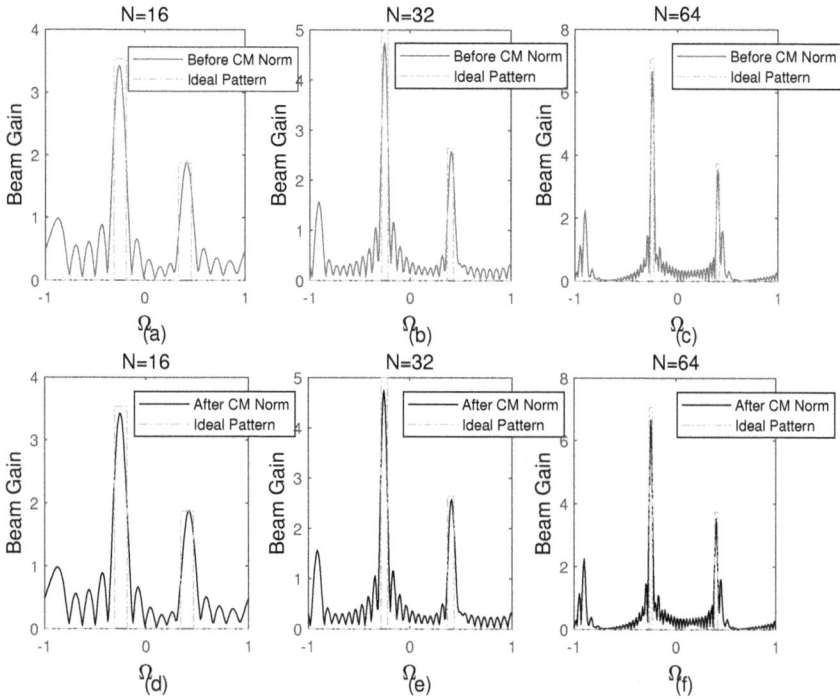

Figure 5.7 Comparison between the ideal beam pattern and the designed beam patterns before and after the CM normalization.

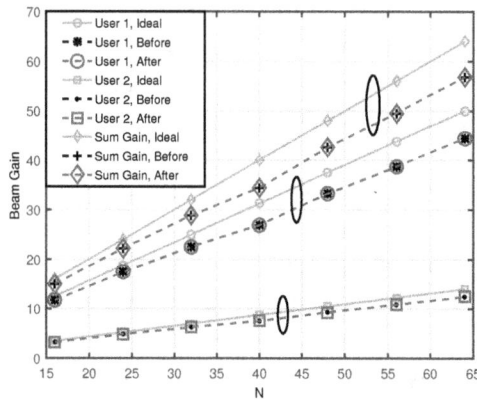

Figure 5.8 Comparison of user beam gains between the ideal beam gains and the designed beam gains before/after the CM normalization, where the sum gain refers to the summation of the beam gains of user 1 and user 2.

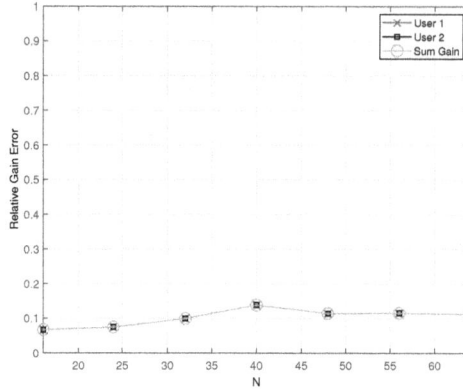

Figure 5.9 Relative gain errors versus the ideal beam gains of user 1, user 2 and the sum gain.

Fig. 5.9 shows the relative gain errors of user 1, user 2 and the sum gain versus the ideal/desired beam gains, where the parameter settings are the same as those in Fig. 5.8. Interestingly, from Fig. 5.9 we find that the relative beam gains of user 1 and user 2, as well as the sum beam gain, are almost the same as each other. Moreover, the relative gain errors are small, roughly around 0.1, and they increase slowly as N increases when $N \leq 40$, and almost does not increase when $N > 40$. This result not only demonstrates again that the proposed beamforming method behaves robust against the number of antennas, but also shows the rational of Lemma 5.3.1, i.e., the sum beam gain can be roughly seen a constant versus N.

The above evaluations show that the solution of the beamforming sub-problem is reasonably close to the ideal one.

Next, we evaluate the overall performance. Fig. 5.10 shows the comparison between the performance bound and the designed achievable rates with varying rate constraint. The performance bound refers to the achievable rate obtained by solving only the power and beam gain allocation sub-problem (5.21), i.e., with parameters $\{c_1^\star, c_2^\star, p_1^\star, p_2^\star\}$, where the beamforming is assumed ideal. The designed performance refers to the achievable rate obtained by solving the original Problem (5.14). Relevant parameter settings are $\sigma^2 = 1$ mW, $P = 100$ mW, $N = 32$, $|\lambda_1| = 0.8$, $|\lambda_1| = 0.5$, $\theta_1 = -0.25$, $\theta_2 = 0.4$. From Fig. 5.10, we can find that the designed achievable rates are close to the achievable rate bound for both user 1 and user 2, as well as the sum rate, which demonstrates that the proposed solution to the original problem is rational and effective, i.e., it can achieve close-to-bound performance. On the other hand, we can find that most power or beam gain is allocated to user 1, which has the better channel condition, so as to optimize the sum rate. Only necessary power or beam gain is allocated to user 2 to satisfy the rate constraint. That is why user 2 always achieves an achievable rate equal to the rate constraint.

Fig. 5.11 shows the comparison between the ideal values of parameters and the designed values with varying rate constraint. The ideal values $\{c_1, c_2, p_1, p_2\}$ refer

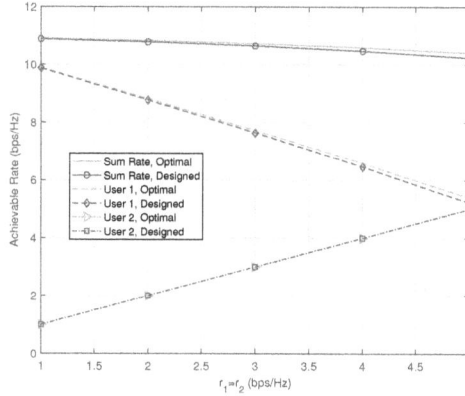

Figure 5.10 Comparison between the performance bound and the designed achievable rates with varying rate constraint.

to $\{c_1^\star,\ c_2^\star,\ p_1^\star,\ p_2^\star\}$, i.e., the optimal solution of the power and beam gain allocation sub-problem (5.21), while the designed $\{c_1,\ c_2\}$ refer to $\{|\mathbf{h}_1^H\mathbf{w}^\star|^2,\ |\mathbf{h}_2^H\mathbf{w}^\star|^2\}$, i.e., the beam gains achieved by the final solution to the original Problem (5.14). The parameter settings are the same as those in Fig. 5.10. As shown in Fig. 5.11, the designed beam gains are close to the ideal gains. The gap between the designed gain and the ideal gain for user 1 is due to the fact that there are side lobes for the designed beam pattern, while for the ideal beam pattern there is no side lobe. More importantly, for user 1, which has a better channel condition, the beam gain is much higher than user 2, while the allocated power is smaller than user 2, which illustrates that necessary power should be allocated to the user with a worse channel condition to satisfy the rate constraint, and the rest power is allocated to the better one to maximize the sum rate. Also, we can observe that as the rate constraint increases,

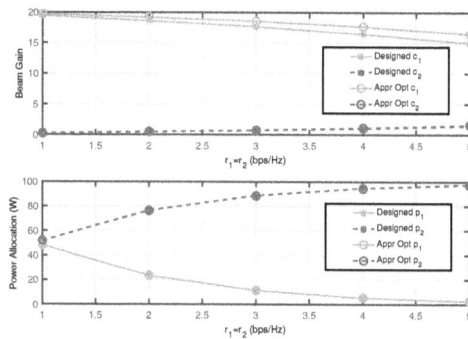

Figure 5.11 Comparison between the ideal values of the parameters and the designed values with varying rate constraint.

the beam gain and power of user 1 decrease, while those of user 2 increase, but the varying speed of beam gain is much slower than that of power for both users.

Fig. 5.12 shows the comparison between the performance bound and the designed achievable rates with varying total power to noise ratio. Relevant parameter settings are the same as Fig. 5.10 and $r_1 = r_2 = 3$ bps/Hz. This figure shows the similar results as Fig. 5.10, i.e., the designed achievable rates are close to the performance bounds for both user 1 and user 2, as well as the sum rate, and most power or beam gain is allocated to user 1 to optimize the sum rate, while only necessary power or beam gain is allocated to user 2 to satisfy the rate constraint.

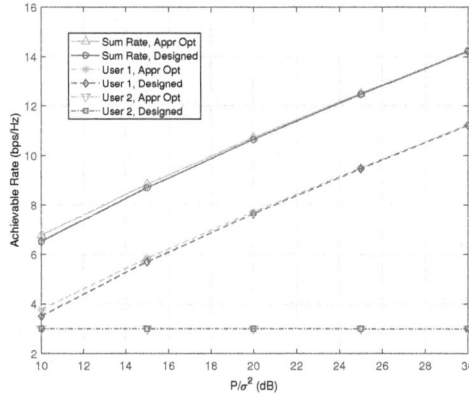

Figure 5.12 Comparison between the performance bound and the designed achievable rates with varying total power to noise ratio.

Fig. 5.13 shows the comparison between the ideal values of parameters and the designed values with varying total power to noise ratio. The parameter settings are the same as those in Fig. 5.12, and $\sigma^2 = 1$ mW here. Except for the same results mentioned before, the figure also shows that as P/σ^2 increases, the beam gain and power of user 1 increase, while the beam gain of user 2 decreases on the contrary.

Fig. 5.11 and Fig. 5.13 show that for user 2, the beam gain is small in general, and varies slowly as the rate constraint and the total power to noise ratio increases. This is because, as shown in (5.21), when increasing the beam gain the interference from user 1 also increases, but when increasing the power, the interference does not increases. Hence, for user 2 the beamforming gain is small in general. Most of the beam gain is allocated to user 1, because for user 1 the interference from user 2 can be decoded and removed by using SIC.

Last but not least, we compare the performance of multi-beam NOMA with that of TDMA. Figs. 5.14 (right) and 5.14 (left) show the comparison results of sum rate between theoretical multi-beam NOMA whose performance is under the effective channel model (5.7), practical multi-beam NOMA and TDMA whose performances are under the original channel model (5.5), where $N = 32$ and $L_1 = L_2 = L = 4$. User 1 has a better channel condition than user 2, i.e., the average power ratio of them is $(1/0.3)^2$. For Fig. 5.14 (right), $\sigma^2 = 1$ mW and $P = 100$ mW, while

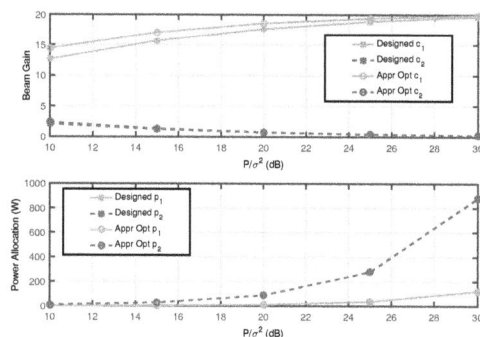

Figure 5.13 Comparison between the ideal values of the parameters and the designed values with varying total power to noise ratio.

for Fig. 5.14 (left) $r_1 = r_2 = 3$ bps/Hz. Both LoS and NLoS channel models are considered. For LoS channel, the first path is the LoS path, which has a constant power, i.e., $|\lambda_1| = 1$ (0 dB), while the coefficients of the other 3 NLoS paths, i.e., $\{\lambda_i\}_{i=2,3,4}$, obey the complex Gaussian distribution with zero mean, and each of them has an average power of $-10/-15$ dB. For the NLoS channel, the 4 paths are all NLoS paths with zero-mean complex Gaussian distributed coefficients, and each of them has an average power of $1/\sqrt{L}$. Each point in Figs. 5.14 (right) and 5.14 (left) is the average performance based on 10^3 channel realizations. With each channel realization, the optimal parameters are obtained by the proposed solution, and the theoretical/practical performances are obtained by computing the sum rates with the effective/original channel. The performance of TDMA is obtained based on the assumption that the beam gains of user 1 and user 2 are equal, i.e., $N/2$. From these two figures we can observe that the theoretical performance is very close to the practical performance, which demonstrates the rational of the proposed method. Moreover, the performance of multi-beam NOMA is significantly better than that of TDMA under both the LoS and NLoS channels.

5.4 UPLINK TRANSMISSION

Similar to the downlink transmission, we consider joint power control and multi-beam forming to maximize the sum rate of a 2-user NOMA system. The same with the downlink transmission, the problem is non-convex and the number of variables is large due to the large number of antennas. Hence, we also present to decompose the original problem into two sub-problems, which are relatively easy to solve. One sub-problem is a power control and beam gain allocation problem. For this sub-problem, we are able to find an optimal solution. While the other is a multi-directional beamforming sub-problem under the CM constraint. As directly solving this sub-problem is challenging, we are able to convert it into a standard convex optimization problem.

However, in addition to the difference in the signal model, the uplink CM beamforming turns to have more strict constraints than the downlink CM beamforming,

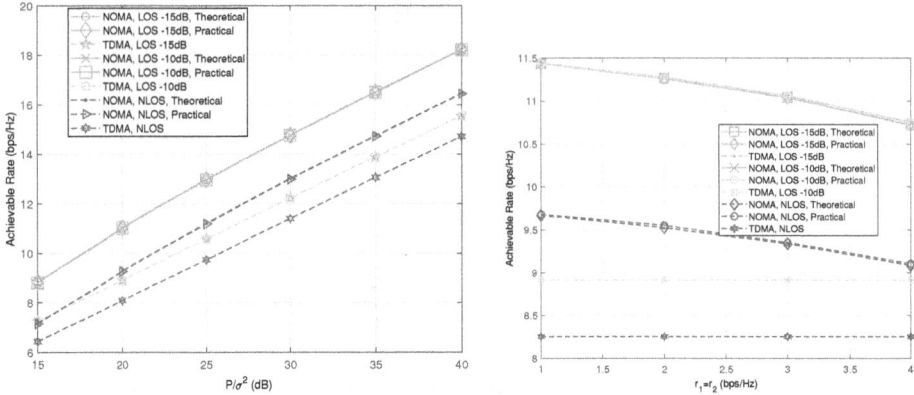

Figure 5.14 **Left**: Comparison of sum rate between theoretical NOMA, practical NOMA and TDMA with varying total power to noise ratio. **Right**: Comparison of sum rate between theoretical NOMA, practical NOMA and TDMA with varying rate constraint.

and thus the beamforming method proposed in downlink transmission may not work, which will be introduced in Section 5.4.2.

5.4.1 System Model and Problem Formulation

1) System Model

In fact, in NOMA systems, the MUI will increase with the number of users served within one time/frequency/code RB, which degrades the average rate of each user and increases the average decoding delay. Hence, the number of NOMA users is not large in general. Without loss of generality, we consider an uplink scenario with two users as shown in Fig. 5.15, where a BS equipped with an N-element antenna array serves two users with a single antenna.

At the BS, each antenna has a phase shifter and a low noise amplifier (LNA) to drive the antenna. As is the same reason in downlink transmission, all the LNAs have the same scaling factor. The AWV has CM elements, User k, $k = 1$, 2, transmits a signal s_k to the BS, where $\mathbb{E}(|s_k|^2) = 1$, with transmit power p_k. The total transmit power of each user is restricted to P. With 2-user NOMA, signals s_1 and s_2 are superimposed at the BS as

$$y = \mathbf{h}_1^{\mathrm{H}} \mathbf{w} \sqrt{p_1} s_1 + \mathbf{h}_2^{\mathrm{H}} \mathbf{w} \sqrt{p_2} s_2 + \mathbf{n}^{\mathrm{H}} \mathbf{w}, \tag{5.91}$$

where $\mathbf{h}_i$ is the channel response vectors between user i and the BS, $\mathbf{w}$ denotes a CM beamforming vector with $|[\mathbf{w}]_n| = \frac{1}{\sqrt{N}}$ for $n = 1, 2, ..., N$, and $\mathbf{n}$ is an N-dimension vector that denotes the Gaussian white noises of N antennas at the BS with power σ^2.

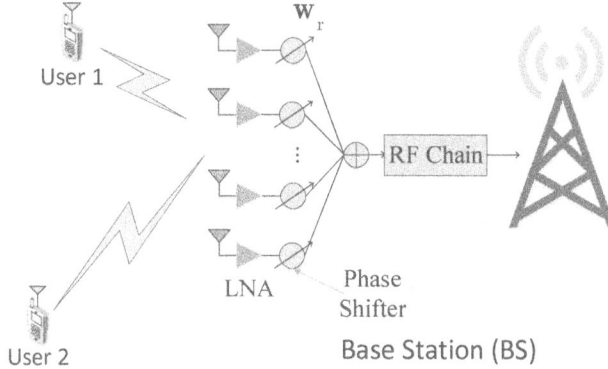

Figure 5.15 Illustration of a mobile cell, where one BS with N antennas serves multiple users with one single antenna.

As the channel is the same with the downlink transmission, the channel can be expressed as

$$\bar{\mathbf{h}}_k = \sum_{\ell=1}^{L_k} \lambda_{k,\ell} \mathbf{a}(N, \theta_{k,\ell}), \tag{5.92}$$

where $\theta_{k,l}$ is the cos(AoA) of the ℓ-th MPC of the channel vector for user k which in different with that in the downlink transmission. $\mathbf{a}(\cdot)$ is the steering vector function defined in (1.70).

Similarly with the downlink, the simplified channel model can be obtained

$$\mathbf{h}_k = \lambda_k \mathbf{a}(N, \theta_k), \tag{5.93}$$

where $\lambda_k = \lambda_{k,m_k}$ and $\theta_k = \theta_{k,m_k}$. Here m_k denotes the index of the strongest MPC for user k. Without loss of generality, we assume $|\lambda_1| \geq |\lambda_2|$, which means that the channel gain of user 1 is better.

2) Decoding Order

In multi-beam NOMA, the decoding order depends on both channel gain and beamforming gain. There are also two cases for the 2-user uplink NOMA system.

Case 1: s_1 is decoded first. Then s_2 is decoded after subtracting the signal component of s_1. With this decoding method, the achievable rates of user k, $k = 1, 2$, denoted by R_k are represented as

$$\begin{cases} R_1^{(1)} = \log_2(1 + \dfrac{|\mathbf{h}_1^{\mathrm{H}}\mathbf{w}|^2 p_1}{|\mathbf{h}_2^{\mathrm{H}}\mathbf{w}|^2 p_2 + \sigma^2}), \\ R_2^{(1)} = \log_2(1 + \dfrac{|\mathbf{h}_2^{\mathrm{H}}\mathbf{w}|^2 p_2}{\sigma^2}). \end{cases} \tag{5.94}$$

Case 2: s_2 is decoded first. Then s_1 is decoded after subtracting the signal component of s_2. With this decoding method, the achievable rates of user k, $k = 1, 2$, denoted by R_k are represented as

$$\begin{cases} R_1^{(2)} = \log_2(1 + \dfrac{|\mathbf{h}_1^{\mathrm{H}}\mathbf{w}|^2 p_1}{\sigma^2}), \\[4mm] R_2^{(2)} = \log_2(1 + \dfrac{|\mathbf{h}_2^{\mathrm{H}}\mathbf{w}|^2 p_2}{|\mathbf{h}_1^{\mathrm{H}}\mathbf{w}|^2 p_1 + \sigma^2}). \end{cases} \tag{5.95}$$

The expressions of the achievable sum rate of under different decoding orders are identical, which can be calculated directly as

$$R_1 + R_2 = \log_2(1 + \frac{|\mathbf{h}_1^{\mathrm{H}}\mathbf{w}|^2 p_1 + |\mathbf{h}_2^{\mathrm{H}}\mathbf{w}|^2 p_2}{\sigma^2}). \tag{5.96}$$

Noted that in the uplink transmission, the achievable sum rate of Case 1 is better than that of Case 2.

3) Problem Formulation

It is the same with downlink transmission, the problem is how to maximize the achievable sum rate of the two users provided that the channel is known a priori. Also, with the minimal rate constraints for the two users, the power control intertwines with the beamforming design. The problem is formulated by

$$\underset{p_1, p_2, \mathbf{w}}{\text{Maximize}} \quad R_1 + R_2 \tag{5.97a}$$

$$\text{Subject to} \quad R_1 \geq r_1, \tag{5.97b}$$

$$R_2 \geq r_2, \tag{5.97c}$$

$$0 \leq p_1, p_2 \leq P, \tag{5.97d}$$

$$|[\mathbf{w}]_n| = \frac{1}{\sqrt{N}}, \quad n = 1, 2, ..., N, \tag{5.97e}$$

where $|[\mathbf{w}]_n| = \frac{1}{\sqrt{N}}$ is the CM constraint due to using the phase shifters in each antenna branch at the BS.

5.4.2 Solution of the Problem

1) Problem Decomposition

Lemma 5.3.1 also applies to uplink transmission. Based on Lemma 5.3.1, we can rewrite Problem (5.97) with the beamforming gains. Since the achievable rate expressions are different for different decoding orders, the problems are also different for different cases:

Case 1:

$$\underset{p_1,p_2,c_1,c_2}{\text{Maximize}} \quad \log_2\left(1 + \frac{c_1 p_1 + c_2 p_2}{\sigma^2}\right) \tag{5.98a}$$

$$\text{Subject to} \quad \log_2\left(1 + \frac{c_1 p_1}{c_2 p_2 + \sigma^2}\right) \geq r_1, \tag{5.98b}$$

$$\log_2\left(1 + \frac{c_2 p_2}{\sigma^2}\right) \geq r_2, \tag{5.98c}$$

$$0 \leq p_1, p_2 \leq P, \tag{5.98d}$$

$$\frac{c_1}{|\lambda_1|^2} + \frac{c_2}{|\lambda_2|^2} = N. \tag{5.98e}$$

Case 2:

$$\underset{p_1,p_2,c_1,c_2}{\text{Maximize}} \quad \log_2\left(1 + \frac{c_1 p_1 + c_2 p_2}{\sigma^2}\right) \tag{5.99a}$$

$$\text{Subject to} \quad \log_2\left(1 + \frac{c_1 p_1}{\sigma^2}\right) \geq r_1, \tag{5.99b}$$

$$\log_2\left(1 + \frac{c_2 p_2}{c_1 p_1 + \sigma^2}\right) \geq r_2, \tag{5.99c}$$

$$0 \leq p_1, p_2 \leq P, \tag{5.99d}$$

$$\frac{c_1}{|\lambda_1|^2} + \frac{c_2}{|\lambda_2|^2} = N. \tag{5.99e}$$

The CM constraint is not involved in Problem (5.98) or Problem (5.99), but will be considered in the beamforming sub-problem. It is worthy to note that the objective functions are the same under different decoding orders, which are different with the downlink scenario. For this reason, the optimal decoding order can be uniquely determined. The proof will be introduced later.

Next, we consider the beamforming problem. Aforesaid in Section 5.2, the beamforming problem in uplink transmission can be formulated as

$$\mathbf{w} \in \mathbb{C}^N \tag{5.100a}$$

$$\text{Subject to} \quad \left|\mathbf{h}_1^H \mathbf{w}\right|^2 = c_1, \tag{5.100b}$$

$$\left|\mathbf{h}_2^H \mathbf{w}\right|^2 = c_2, \tag{5.100c}$$

$$|[\mathbf{w}]_n| = \frac{1}{\sqrt{N}}, \ n = 1, 2, ..., N. \tag{5.100d}$$

This is an optimization problem which can form two narrow beams steering to two different users. Thus, we are able to form the narrow beams with a particular beam pattern, which can be synthesized by the optimization approach.

With above manipulations, Problem (5.97) is decomposed into Problem (5.98) and Problem (5.100a), which are independent sub-problems. Next, we will first solve Problem (5.98), and obtain the optimal solution $\{c_1^\star, c_2^\star, p_1^\star, p_2^\star\}$ of (5.98). Then $c_2^\star$ is used as the gain constraints in Problem (5.100a). We solve Problem (5.100a) and obtain an appropriate $\mathbf{w}^\circ$. Although the obtained solution

$\{p_1^\star, p_2^\star, \mathbf{w}^\circ\}$ is not globally optimal, the achieved sum rate performance is close to the upper bound.

2) Solution of the Power Control and Beam Gain Allocation Sub-Problem

As the optimal sum rate of Case 1 is better than that of Case 2, we just show the solution of Problem (5.98) in detail. We first figure out the optimal $\{p_1^\star, p_2^\star\}$ and then the optimal beam gains $\{c_1^\star, c_2^\star\}$.

Lemma 5.4.1. *With the ideal beamforming, the optimal transmit power is*

$$\begin{cases} p_1^\star = P, \\ p_2^\star = P. \end{cases} \tag{5.101}$$

Proof. Suppose the optimal solution of Problem (5.98) is $p_1 = p_1^\star$, $p_2 = p_2^\star$, $c_1 = c_1^\star$, $c_2 = c_2^\star$. With the optimal solution, the optimal user rates are $R_1 = R_1^\star$ and $R_2 = R_2^\star$, respectively.

Assume $p_1^\star < P$. We consider the parameter settings $p_1 = P > p_1^\star$, $p_2 = p_2^\star$, $c_1 = c_1^\star$, $c_2 = c_2^\star$. Then we have

$$\begin{cases} R_1 = \log_2(1 + \dfrac{c_1^\star P}{c_2^\star p_2^\star + \sigma^2}) > R_1^\star \geq r_1, \\ R_2 = \log_2(1 + \dfrac{c_2^\star p_2^\star}{\sigma^2}) = R_2^\star \geq r_2, \\ R_1 + R_2 > R_1^\star + R_2^\star, \end{cases} \tag{5.102}$$

which means that the rate constraints in Problem (5.98) are all satisfied while the value of the objective function becomes larger. Hence, the assumption of $p_1^\star < P$ does not hold. We have $p_1^\star = P$.

Next, we consider the parameter settings $\{p_1, p_2, c_1, c_2\}$, where their values are

$$\begin{cases} p_2 = P > p_2^\star, \\ c_2 = \dfrac{c_2^\star p_2^\star}{p_2} = \dfrac{c_2^\star p_2^\star}{P} < c_2^\star, \\ p_1 = p_1^\star, \\ c_1 = |\lambda_1|^2 \left(N - \dfrac{c_2}{|\lambda_2|^2}\right) > c_1^\star. \end{cases} \tag{5.103}$$

The intention of the setting above is to improve p_2 and keep $c_2 p_2 = c_2^\star p_2^\star$ unchanged. Then, the achievable rates are

$$\begin{cases} R_1 = \log_2(1 + \dfrac{c_1 p_1}{c_2 p_2 + \sigma^2}) = \log_2(1 + \dfrac{c_1 p_1^\star}{c_2^\star p_2^\star + \sigma^2}) > R_1^\star \geq r_1, \\ R_2 = \log_2(1 + \dfrac{c_2 p_2}{\sigma^2}) = \log_2(1 + \dfrac{c_2^\star p_2^\star}{\sigma^2}) = R_2^\star \geq r_2, \\ R_1 + R_2 > R_1^\star + R_2^\star, \end{cases} \tag{5.104}$$

which means that the rate constraints in Problem (5.98) are all satisfied while the value of the objective function becomes larger. Hence, the assumption of $p_2^\star < P$ does not hold. We have $p_2^\star = P$.

With the above analyses, the value of the objective function can always increase in the feasible domain when increasing p_1 or p_2. Hereto, the optimal values of p_1 and p_2 are

$$\begin{cases} p_1^\star = P, \\ p_2^\star = P. \end{cases} \tag{5.105}$$

$\square$

According to Lemma 5.3.1 and Lemma 5.4.1, we have

$$\begin{cases} p_1 = P, \\ p_2 = P, \\ c_1 = |\lambda_1|^2 \left(N - \dfrac{c_2}{|\lambda_2|^2}\right). \end{cases} \tag{5.106}$$

Substituting them into Problem (5.98), there is only one independent variable c_2 now. Hence, we can transform the problem as

$$\underset{c_2}{\text{Maximize}} \quad \log_2\left(1 + \frac{(|\lambda_1|^2 N - (\frac{|\lambda_1|^2}{|\lambda_2|^2} - 1)c_2)P}{\sigma^2}\right) \tag{5.107a}$$

$$\text{Subject to} \quad \log_2\left(1 + \frac{|\lambda_1|^2 \left(N - \frac{c_2}{|\lambda_2|^2}\right)P}{c_2 P + \sigma^2}\right) \geq r_1, \tag{5.107b}$$

$$\log_2\left(1 + \frac{c_2 P}{\sigma^2}\right) \geq r_2. \tag{5.107c}$$

As $|\lambda_1| \geq |\lambda_2|$, we have

$$-\left(\frac{|\lambda_1|^2}{|\lambda_2|^2} - 1\right) \leq 0. \tag{5.108}$$

The objective function is monotonically decreasing for c_2, so the infimum of c_2 is optimal. Furthermore, R_1 is decreasing for c_2 and R_2 is increasing for c_2. The lower-bound of c_2 depends on the second constraint $R_2 \geq r_2$ of Problem (5.107).

$$\log_2\left(1 + \frac{c_2 P}{\sigma^2}\right) \geq r_2 \Rightarrow c_2 \geq \frac{(2^{r_2} - 1)\sigma^2}{P}. \tag{5.109}$$

Hereto, we have solved the power control and beam gain allocation sub-problem in Case 1, i.e., Problem (5.98). As Problem (5.98) and Problem (5.99) are similar to each other, which means Problem (5.99) can also be solved by the above method. As mentioned before, the optimal decoding order can be uniquely determined, and we have the following theorem.

Theorem 5.4.1. *The maxima of the objective function in the power control and beam gain allocation sub-problem under Case 1 is larger than that under Case 2.*

Proof. Lemma 5.3.1 and Lemma 5.4.1 are still applicable in Case 2. We have

$$
\begin{cases}
p_1 = P, \\
p_2 = P, \\
c_1 = |\lambda_1|^2 \left(N - \dfrac{c_2}{|\lambda_2|^2} \right).
\end{cases}
\tag{5.110}
$$

Substituting them into Problem (5.99), there is only one independent variable c_2 now. Hence, we can transform the problem as

$$
\begin{aligned}
\underset{c_2}{\text{Maximize}} \quad & \log_2\left(1 + \frac{|\lambda_1|^2 N - (\frac{|\lambda_1|^2}{|\lambda_2|^2} - 1)c_2)P}{\sigma^2}\right), \\
\text{Subject to} \quad & \log_2\left(1 + \frac{|\lambda_1|^2 (N - \frac{c_2}{|\lambda_2|^2})P}{\sigma^2}\right) \geq r_1, \\
& \log_2\left(1 + \frac{c_2 P}{|\lambda_1|^2 (N - \frac{c_2}{|\lambda_2|^2})P + \sigma^2}\right) \geq r_2.
\end{aligned}
\tag{5.111}
$$

Similar to Case 1, the objective function in (5.111) is monotonically decreasing for c_2, so the infimum of c_2 is optimal. Furthermore, R_1 is decreasing for c_2 and R_2 is increasing for c_2. The lower-bound of c_2 depends on the second constraint $R_2 \geq r_2$ of Problem (5.111).

$$
\begin{aligned}
& \log_2\left(1 + \frac{c_2 P}{|\lambda_1|^2 (N - \frac{c_2}{|\lambda_2|^2})P + \sigma^2}\right) \geq r_2 \\
\Leftrightarrow\ & c_2 \geq \frac{(|\lambda_1|^2 NP + \sigma^2)(2^{r_2} - 1)}{(1 + \frac{|\lambda_1|^2}{|\lambda_2|^2}(2^{r_2} - 1))P}.
\end{aligned}
\tag{5.112}
$$

The lower-bound of c_2 can be obtained when $R_2 = r_2$ in both cases. Denote them as $c_2^{(1)}$ and $c_2^{(2)}$, respectively, and we have

$$
\log_2\left(1 + \frac{c_2^{(1)} P}{\sigma^2}\right) = r_2 = \log_2\left(1 + \frac{c_2^{(2)} P}{|\lambda_1|^2 (N - \frac{c_2^{(2)}}{|\lambda_2|^2})P + \sigma^2}\right)
$$
$$
\leq \log_2\left(1 + \frac{c_2^{(2)} P}{\sigma^2}\right) \Leftrightarrow c_2^{(1)} \leq c_2^{(2)}.
\tag{5.113}
$$

Aforesaid, the objective functions in Problems (5.107) and (5.111) are identical, which is monotonically decreasing for the variable c_2. Then we can conclude that the optimal solution of Case 1 is better than Case 2, because $c_2^{(1)} \leq c_2^{(2)}$. □

Theorem 5.4.1 shows that the optimal order is to decode the signal of user 1, i.e., the one with higher channel gain. And the optimal values of $|\mathbf{h}_2^{\mathrm{H}} \mathbf{w}|^2$ and $|\mathbf{h}_1^{\mathrm{H}} \mathbf{w}|^2$ are

$$
\begin{cases}
c_2^\star = \dfrac{(2^{r_2} - 1)\sigma^2}{P}, \\
c_1^\star = |\lambda_1|^2 \left(N - \dfrac{c_2^\star}{|\lambda_2|^2} \right).
\end{cases}
\tag{5.114}
$$

As the optimal solution $\{c_1^\star, c_2^\star, p_1^\star, p_2^\star\}$ is obtained under the assumption of the ideal beamforming, i.e., assuming Lemma 5.3.1 holds, however, similar to the downlink transmission, $\{c_1^\star, c_2^\star, p_1^\star, p_2^\star\}$ may not be an optimal solution of the original problem, i.e., Problem (5.97), because a BFV with beam gains $\{c_1^\star, c_2^\star\}$ may not be found under the CM constraint. Hence, the optimal achievable sum rate of Problem (5.98) is an upper bound of that of the original problem.

3) Solution of the Beamforming Sub-Problem

Next, we solve the multi-beam forming problem, i.e., Problem (5.100a), to design an appropriate $\mathbf{w}$ to realize the user beam gains $c_1^\star$ and $c_2^\star$. However, as we have mentioned before, the BFV with beam gains $\{c_1^\star, c_2^\star\}$ may not be found because of the sidelobe in beam pattern. Proper relaxation should be adopted to obtain the appropriate $\mathbf{w}$ in Problem (5.100a). As mentioned in Section 5.2, the downlink beamforming method does not apply to uplink beamforming because once the finally obtained beam gains (after beamforming) are less than the desired values, the power cannot be further increased to meet the original rate constraints. Hence, there is a new beamforming method to relax Problem (5.100a) where the CM constraint is strictly satisfied and the power re-adjustment is evitable.

The key to relax the problem is to deal with the two equality constraints

$$
\begin{cases}
|\mathbf{h}_1^H \mathbf{w}|^2 = c_1^\star, \\
|\mathbf{h}_2^H \mathbf{w}|^2 = c_2^\star.
\end{cases}
\tag{5.115}
$$

Since we have proven that the optimal beam gain of user 2 in Problem (5.98) is the lower bound, we relax the corresponding equality to inequality

$$
\left|\mathbf{h}_2^H \mathbf{w}\right|^2 \geq c_2^\star.
\tag{5.116}
$$

As we have proven that the optimal power control of Problem (5.98) is $p_1^\star = p_2^\star = P$, the objective function is equal to

$$
\log_2\left(1 + \frac{(|\mathbf{h}_1^H \mathbf{w}|^2 + |\mathbf{h}_2^H \mathbf{w}|^2)\,P}{\sigma^2}\right),
\tag{5.117}
$$

which is equivalent to maximize

$$
|\mathbf{h}_1^H \mathbf{w}|^2 + |\mathbf{h}_2^H \mathbf{w}|^2.
\tag{5.118}
$$

We have proven that the objective function is monotonically increasing for c_1 and monotonically decreasing for c_2 under Lemma 5.3.1. Hence, we should increase c_1 (c_2 decreases accordingly) to maximize the objective function. Hence, maximizing

$$
|\mathbf{h}_1^H \mathbf{w}|^2 + |\mathbf{h}_2^H \mathbf{w}|^2,
\tag{5.119}
$$

could be replaced by maximizing $|\mathbf{h}_1^H \mathbf{w}|^2$. Problem (5.100a) can be relaxed as

$$\text{Maximize} \quad \left|\mathbf{h}_1^H \mathbf{w}\right|^2 \tag{5.120a}$$

$$\text{Subject to} \quad \left|\mathbf{h}_2^H \mathbf{w}\right|^2 \geq c_2^\star, \tag{5.120b}$$

$$|[\mathbf{w}]_n| = \frac{1}{\sqrt{N}}, \; n = 1, 2, ..., N. \tag{5.120c}$$

Define

$$g = \sqrt{\frac{c_2^\star}{|\lambda_2|^2}}, \tag{5.121}$$

the problem above can be rewritten as

$$\text{Maximize} \quad \left|\mathbf{a}_1^H \mathbf{w}\right| \tag{5.122a}$$

$$\text{Subject to} \quad \left|\mathbf{a}_2^H \mathbf{w}\right| \geq g, \tag{5.122b}$$

$$|[\mathbf{w}]_n| = \frac{1}{\sqrt{N}}, \; n = 1, 2, ..., N, \tag{5.122c}$$

where

$$\mathbf{a}_k \triangleq \mathbf{a}(N, \theta_k), k = 1, 2. \tag{5.123}$$

Problem (5.122) is also non-convex. The problem is still difficult to solve due to the equality constraints. Therefore, we relax the equality constraints $|[\mathbf{w}]_n| = \frac{1}{\sqrt{N}}$ with inequality constraints $|[\mathbf{w}]_n| \leq \frac{1}{\sqrt{N}}$, which is convex. We reformulate the beamforming problem as

$$\text{Maximize} \quad \left|\mathbf{a}_1^H \mathbf{w}\right| \tag{5.124a}$$

$$\text{Subject to} \quad \left|\mathbf{a}_2^H \mathbf{w}\right| \geq g, \tag{5.124b}$$

$$|[\mathbf{w}]_n| \leq \frac{1}{\sqrt{N}}, \; n = 1, 2, ..., N. \tag{5.124c}$$

Then, we have the following theorem.

Theorem 5.4.2. *If $\mathbf{w}_0$ is the optimal solution of Problem (5.124), then,* $|[\mathbf{w}_0]_n| = \frac{1}{\sqrt{N}}, \; n = 1, 2, ..., N.$

Proof. Let $\mathbf{w}_0$ represent the optimal solution of Problem (5.124), and

$$\begin{cases} \mathbf{a}_1^H \mathbf{w}_0 = d_1 e^{\theta_1 j}, \\ \mathbf{a}_2^H \mathbf{w}_0 = d_2 e^{\theta_2 j}, \end{cases} \tag{5.125}$$

where d_i and θ_i denote the modulus and phase of $\mathbf{a}_i^H \mathbf{w}_0$, respectively. We will show $|[\mathbf{w}_0]_i| = \frac{1}{\sqrt{N}}$ for $i = 1, 2, ..., N$. Hence, we will only prove that

$|[\mathbf{w}_0]_1| = \frac{1}{\sqrt{N}}$ in detail, while $|[\mathbf{w}_0]_i| = \frac{1}{\sqrt{N}}$ for $i = 2, 3, ..., N$ can be proven similarly. As the modulus of $[\mathbf{a}_i]_1, i = 1, 2$ is 1, we have

$$|[\mathbf{a}_i]_1[\mathbf{w}_0]_1| = |[\mathbf{w}_0]_1| \triangleq l. \tag{5.126}$$

Denote

$$\begin{cases} [\mathbf{a}_1^H]_1[\mathbf{w}_0]_1 = le^{\mu_1 j}, \\ [\mathbf{a}_2^H]_1[\mathbf{w}_0]_1 = le^{\mu_2 j}, \end{cases} \tag{5.127}$$

and

$$\begin{cases} \displaystyle\sum_{k=2}^{N}[\mathbf{a}_1^H]_k[\mathbf{w}_0]_k = b_1 e^{\nu_1 j}, \\ \displaystyle\sum_{k=2}^{N}[\mathbf{a}_2^H]_k[\mathbf{w}_0]_k = b_2 e^{\nu_2 j}. \end{cases} \tag{5.128}$$

Obviously,

$$le^{\mu_i j} + b_i e^{\nu_i j} = d_i e^{\theta_i j}. \tag{5.129}$$

Note that the phase difference between $[\mathbf{a}_1^H]_1[\mathbf{w}_0]_1$ and $[\mathbf{a}_2^H]_1[\mathbf{w}_0]_1$, i.e., $\mu_2 - \mu_1$, does not dependent on $[\mathbf{w}_0]_1$. Next, we will show that the optimal $[\mathbf{w}_0]_1$ must be on the constraint boundary $|[\mathbf{w}_0]_1| = \frac{1}{\sqrt{N}}$.

For the constraints in Problem (5.124). For fixed $[\mathbf{w}_0]_k$, $k = 2, 3, \cdots, N$, the constraints for $[\mathbf{w}_0]_1$ are

$$\begin{cases} |\mathbf{a}_2^H \mathbf{w}_0| = |le^{\mu_2 j} + b_2 e^{\nu_2 j}| \ge g, \\ |[\mathbf{w}_0]_1| = l \le \dfrac{1}{\sqrt{N}}. \end{cases} \tag{5.130}$$

Consider the above variables in the polar coordinate system, where the constraints (5.130) denote a feasible region in the 2-dimensional plane, where

$$|le^{\mu_2 j} + b_2 e^{\nu_2 j}| \ge g, \tag{5.131}$$

is the outside part of a circle and $l \le \frac{1}{\sqrt{N}}$ is the inside part of a circle. Hence, the feasible region of (5.130), denoted by S_1, is a closed set with two boundaries. Noted that S_1 is not empty because there is at least one point, $[\mathbf{w}_0]_1$. One boundary is the equation

$$|le^{\mu_2 j} + b_2 e^{\nu_2 j}| = g. \tag{5.132}$$

We define it as the inner boundary of S_1. The other is the equation $l = \frac{1}{\sqrt{N}}$. We define it as the outer boundary of S_1. The shape of S_1 depends on the relative position relation between the two circles, i.e., included, intersecting, internally tangent, externally tangent and separate, which are shown in Fig. 5.16, where S_2 is defined below.

It is assumed that the objective function of Problem (5.124) is the maximum at the point $[\mathbf{w}_0]_1$, which is described by

$$\begin{aligned} |\mathbf{a}_1^H \mathbf{w}_0| &= |le^{\mu_1 j} + b_1 e^{\nu_1 j}| = d_1 \\ \Leftrightarrow &|le^{\mu_1 j + (\mu_2 - \mu_1)j} + b_1 e^{\nu_1 j + (\mu_2 - \mu_1)j}| = d_1 \\ \Leftrightarrow &|le^{\mu_2 j} + b_1 e^{\nu_1 j + (\mu_2 - \mu_1)j}| = d_1, \end{aligned} \tag{5.133}$$

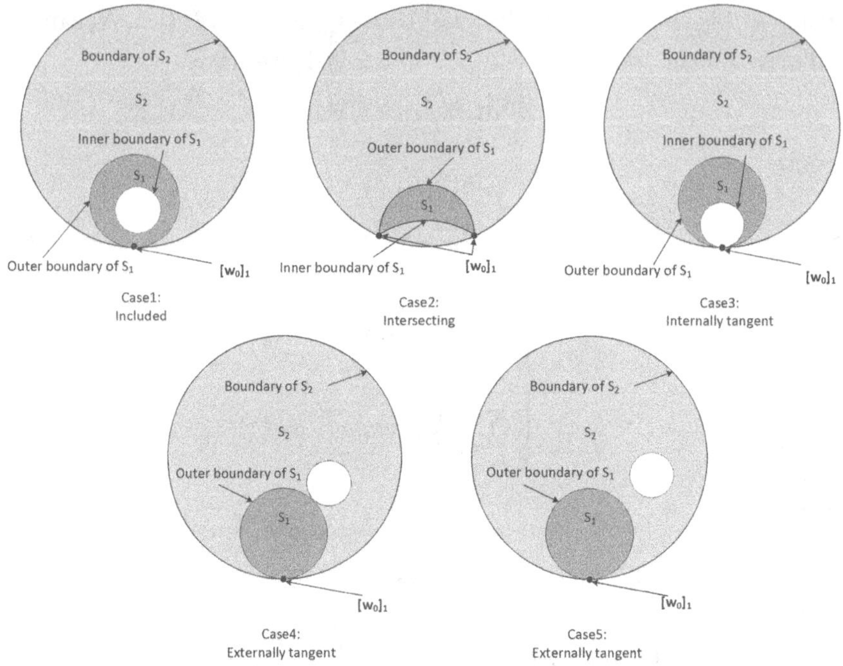

Figure 5.16 Illustration of the relative position relation between S_1 and S_2. On one hand, $S_1 \subseteq S_2$; On the other hand, $[\mathbf{w}_0]_1$ is the interchapter between S_1 and the boundary of S_2. Thus, no matter what the shape of S_1 is, $[\mathbf{w}_0]_1$ must be located in the outer boundary of S_1.

where b_1, ν_1 and $\mu_2 - \mu_1$ are constants. In other words, d_1 is the maximum distance from the point

$$-b_1 e^{\nu_1 j + (\mu_2 - \mu_1) j}, \tag{5.134}$$

to the region S_1. If we draw a circle centered at the point

$$-b_1 e^{\nu_1 j + (\mu_2 - \mu_1) j}, \tag{5.135}$$

with the radius of d_1, then S_1 is certainly located inside of this circle. Otherwise, the point outside this circle is optimal, which is contradictory to the assumption. The inside part of this circle is described by

$$|le^{\mu_2 j} + b_1 e^{\nu_1 j + (\mu_2 - \mu_1) j}| \le d_1, \tag{5.136}$$

and denoted by S_2 (see also Fig. 5.16). In particular, we define the equation

$$|le^{\mu_2 j} + b_1 e^{\nu_1 j + (\mu_2 - \mu_1) j}| = d_1, \tag{5.137}$$

as the boundary of S_2. Then we have $S_1 \subseteq S_2$. It can be seen that $\mathbf{w}_0$ is located in the outer boundary of S_1 in Fig. 5.16, no matter what the shape of S_1 is. Thus, $|[\mathbf{w}_0]_1| = \frac{1}{\sqrt{N}}$. $\qquad\square$

According to Theorem 5.4.1, Problem (5.122) is equivalent to Problem (5.124). It is clear that an arbitrary phase rotation can be added to the vector $\mathbf{w}$ in Problem (5.124) without affecting the beam gains. Thus, if $\mathbf{w}^\star$ is optimal, so is $\mathbf{w}^\star e^{j\phi}$, where ϕ is an arbitrary phase within $[0, 2\pi)$. Without loss of generality, we may then choose ϕ so that $\mathbf{a}_1^H \mathbf{w}$ is real and non-negative. Problem (5.124) is tantamount to

$$\underset{\mathbf{w}}{\text{Maximize}} \quad \mathbf{a}_1^H \mathbf{w} \tag{5.138a}$$

$$\text{Subject to} \quad \left|\mathbf{a}_2^H \mathbf{w}\right| \geq g, \tag{5.138b}$$

$$\left|[\mathbf{w}]_k\right| \leq \frac{1}{\sqrt{N}}, \ k = 1, 2, ..., N. \tag{5.138c}$$

Problem (5.138) is still not convex because of the absolute value operation in the first constraint. Thus, we can split it into a serial of convex optimization problems, i.e., we assume different phases for $\mathbf{a}_2^H \mathbf{w}$ and obtain M convex problems

$$\underset{\mathbf{w}}{\text{Maximize}} \quad \mathbf{a}_1^H \mathbf{w} \tag{5.139a}$$

$$\text{Subject to} \quad \Re(\mathbf{a}_2^H \mathbf{w} e^{2\pi j \frac{m}{M_p}}) \geq g, \tag{5.139b}$$

$$\left|[\mathbf{w}]_n\right| \leq \frac{1}{\sqrt{N}}, \ n = 1, 2, ..., N. \tag{5.139c}$$

where M_p is the number of total candidate phases, $m = 1, 2, \cdots, M_p$. The phase of $\mathbf{a}_2^H \mathbf{w}$, from 0 to 2π, is divided into M_p fragments. The search of the optimal phase is more accurate as M_p increases, but meanwhile the computational complexity is also increased. There is a tradeoff between the accuracy of the optimal solution and the computational complexity. Each of these M_p problems can be efficiently solved by using standard convex optimization tools. We select the solution with the maximal objective among the M_p optimal solutions as the final solution $\mathbf{w}^\star$.

Hereto, we have obtained a sub-optimal solution of the original problem (5.97), i.e., $\{p_1^\star, p_2^\star, \mathbf{w}^\star\}$. Although we have some relaxations in the sub-problems, the solution $\{p_1^\star, p_2^\star, \mathbf{w}^\star\}$ obtained from the sub-problems (5.98) and (5.120) is located in the feasible region of Problem (5.97). And we have the following theorem.

Theorem 5.4.3. *If the feasible region of Problem (5.97) is not empty, then $\{p_1^\star, p_2^\star, \mathbf{w}^\star\}$ is a solution of Problem (5.97).*

Proof. It is obvious that $p_i^\star = P$, $i = 1, 2$, satisfies the power constraint for user i. And the CM constraint for the BFV $\mathbf{w}^\circ$ is also considered in Problem (5.120). Thus we just need to verify that

$$\begin{cases} R_1^\star \geq r_1, \\ R_2^\star \geq r_2, \end{cases} \tag{5.140}$$

where $R_i^\star$, $i = 1, 2$ is the achievable rate of user i under the proposed solution $\{p_1^\star, p_2^\star, \mathbf{w}^\circ\}$.

On one hand, we have

$$R_2^\star = \log_2(1 + \frac{|\mathbf{h}_2^H \mathbf{w}^\circ|^2 p_2^\star}{\sigma^2}) \geq \log_2(1 + \frac{c_2^\star P}{\sigma^2}) = r_2. \qquad (5.141)$$

On the other hand, in Problem (5.107), the optimal solution is located on the boundary of $R_2 = r_2$, which means only necessary beam gain is allocated to user 2 to satisfy the minimum rate constraint and the rest of beam gain is all allocated to user 1. Similar to (5.120), we try to maximize the beam gain of user 1 while the beam gain of user 2 just ensures the minimum gain to satisfy the rate constraint. Thus the combination of Problem (5.107) and (5.120) is equivalent to

$$\begin{aligned}
\underset{p_1,p_2,\mathbf{w}}{\text{Maximize}} \quad & R_1 \\
\text{Subject to} \quad & R_2 \geq r_2, \\
& 0 \leq p_1, p_2 \leq P, \\
& |[\mathbf{w}]_k| = \frac{1}{\sqrt{N}}, \quad k = 1, 2, ..., N.
\end{aligned} \qquad (5.142)$$

Assume that $R_1^\star < r_1$, which means that under the constraints of Problem (5.142), the maximum value of R_1 is smaller than r_1. In other words, the constraint $R_1 \geq r_1$ in Problem (5.97) cannot be feasible. The feasible region of Problem (5.97) is empty. However, we have assumed that the feasible region of Problem (5.97) is not empty in Theorem 5.4.3, which is contradictory. Thus, we must have $R_1^\star \geq r_1$. □

4) Consideration of the Finite Resolution Analog Beamforming

In the problem formulation and solution, we have assumed that the phase of beamforming is continuous, i.e., the phase shifters have infinite resolution. To reduce the hardware cost, finite resolution analog beamforming (FRAB) is usually adopted in practice[30,31]. Hence, we consider the implementation of FRAB in this part.

With FRAB, all the elements of the beamforming $\mathbf{w}$ should be drawn from the following vector:

$$\overline{\mathbf{w}} = \left[1, e^{j\frac{2\pi}{N_f}}, \cdots, e^{j\frac{(N_f-1)2\pi}{N_f}}\right], \qquad (5.143)$$

where N_f is the number of supported shifts.

Hence, after obtaining the beamforming $\mathbf{w}^\star$ with the continuous phase model, the k-th element of $\overline{\mathbf{w}}$ is chosen to replace the i-th element of $\mathbf{w}$ based on the following criterion[30]:

$$k_k = \underset{1 \leq n \leq N_f}{\arg\min} |[\mathbf{w}^\star]_k - [\overline{\mathbf{w}}_n]|, \qquad (5.144)$$

where $\mathbf{w}^\star$ is the solution of (5.120). The intention of (5.144) is to choose the closest one among the available discrete phases for each element of $\mathbf{w}^\star$. The simulation results show that FRAB leads to little performance loss with a normal phase resolution.

5.4.3 Performance Evaluations

In this subsection, we evaluate the performance of the proposed joint power control and beamforming method. First, we start from the performance evaluation of the beamforming phase.

Fig. 5.17 compares the ideal beam pattern with the designed beam pattern. The designed beam pattern is obtained by solving Problem (5.122), where N_f is set to 20. The ideal beam pattern is obtained by the definition of ideal beamforming below Lemma 5.3.1. We assume $|\lambda_1| = 0.9$, $|\lambda_2| = 0.4$, $\theta_1 = -0.7$, $\theta_2 = 0.5$. The desired beam gains are

$$\begin{cases} c_1^\star = 2N/3, \\ c_2^\star = (N - c_1^\star/|\lambda_1|^2)|\lambda_2|^2, \end{cases} \tag{5.145}$$

where N is the number of antennas at the BS. M_p in (5.139) is set to 20 in this simulation as well as the following simulations, which is large enough to obtain the best solution. Fig. 5.17 shows the comparison results with $N = 8$, 16, 32, 64, and from this figure we can find that the beam gains are significant along the desired user directions, and the beam pattern designed is close to the ideal beam pattern along the user directions, which demonstrates that the solution of the beamforming sub-problem is reasonable. The designed beam pattern almost coincides with the designed beam pattern of FRAB, especially along the user directions. This result demonstrates that the proposed beamforming method works well with FRAB.

In addition to the beam pattern comparison, we also compare the user beam gains with different numbers of antennas in Fig. 5.18, where the parameter settings are the same as those in Fig. 5.17. From Fig. 5.18, we can observe that the designed gain

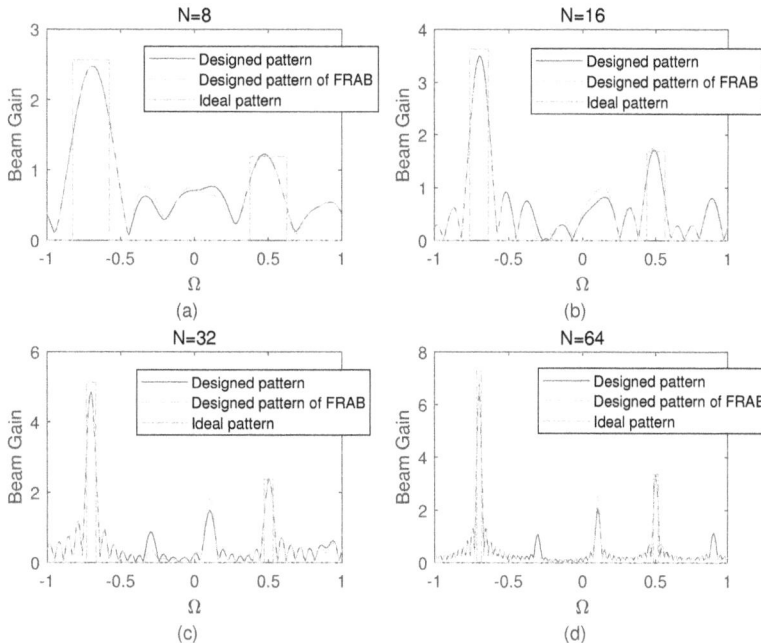

Figure 5.17 Comparison between the ideal beam pattern and the designed beam pattern.

of user 2 is equal to the ideal, and there is a small gap between the designed user gain and the ideal beam gain for user 1 (as well as the sum beam gain). The small gap is because the designed beam pattern has side lobes, which reduces the gains along the user 1 directions. In comparison, an ideal beam pattern does not have side lobes. Fortunately, the gap only increases slowly as N increases when $N \leq 40$, and almost does not increase when $N > 40$, which shows that the proposed beamforming method behaves robust against the number of antennas.

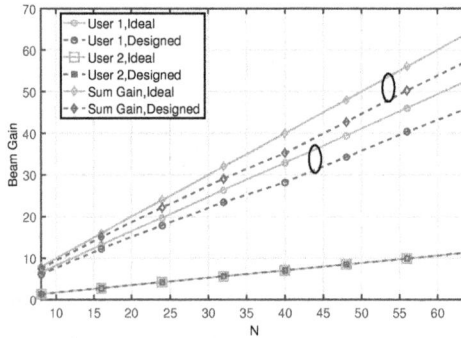

Figure 5.18 Comparison of user beam gains between the ideal beam gain and the designed beam gain, where the sum gain refers to the summation of the beam gains of user 1 and user 2.

Fig. 5.19 shows the average relative gain errors of user 1, user 2 and the sum gain versus the ideal/desired beam gains. The parameter settings are $|\lambda_1| = 0.9, |\lambda_2| = 0.4$. The AoAs of users θ_1 and θ_2 randomly range in $[-1, 1]$ with uniform distribution, and there is a constraint

$$2/N < |\theta_1 - \theta_2| < 2 - 2/N, \tag{5.146}$$

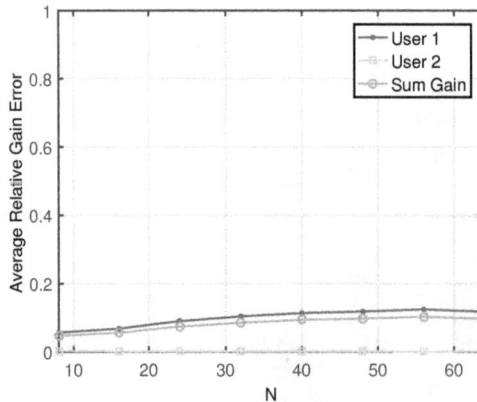

Figure 5.19 Average relative gain errors versus the ideal beam gains of user 1, user 2 and the sum gain.

because the width of beam gains we designed is $2/N$ in general. The desired beam gains are

$$\begin{cases} c_1^\star = 2N/3, \\ c_2^\star = (N - c_1^\star/|\lambda_1|^2)|\lambda_2|^2, \end{cases} \tag{5.147}$$

where N is the number of antennas at the BS. Each point in Fig. 5.19 is the average performance based on 10^3 beamforming realizations. We find that the relative gain error of user 2 is near zero, which shows that the beamforming setting is almost ideal for user 2. The relative gain error of user 1 is roughly around 0.1, and the relative error of sum gain is no more than 0.1, and they increase slowly as N increases when $N \leq 56$, and almost does not increase when $N > 40$. This result not only demonstrates again that the proposed beamforming method behaves robust against the number of antennas, but also shows the rational of Lemma 5.3.1, i.e., the sum beam gain can be roughly seen as a constant versus N.

The above evaluations show that the solution of the beamforming sub-problem is reasonably close to the ideal one. Next, we evaluate the overall performance.

Fig. 5.20 (left) shows the comparison between the performance bound and the designed achievable rates with varying rate constraint while Fig. 5.20 (right) shows that with varying maximal power to noise ratio. The performance bound refers to the achievable rate obtained by solving only the power control and beam gain allocation sub-problem, i.e., with parameters $\{c_1^\star, c_2^\star, p_1^\star, p_2^\star\}$, where the beamforming is assumed ideal. The designed performance refers to the achievable rate obtained by solving both the power control and beam gain allocation and beamforming sub-problems, i.e., (5.105) and solution of Problem (5.139). Relevant parameter settings in Fig. 5.20 (left) are $\sigma^2 = 1$ mW, $P = 100$ mW, $N = 32$, $|\lambda_1| = 0.9$, $|\lambda_2| = 0.2$, $\theta_1 = -0.7$, $\theta_2 = 0.5$. In Fig. 5.20 (right), the parameters, i.e., $|\lambda_1|$, $|\lambda_2|$, θ_1 and θ_2, are the same with Fig. 5.20 (left) with $r_1 = r_2 = 3$ bps/Hz. From Fig. 5.20 (left) and Fig. 5.20 (right), we can find that the designed achievable rates are close to the ideal achievable rates for both user 1 and user 2, as well as the sum rate, which demonstrates that the

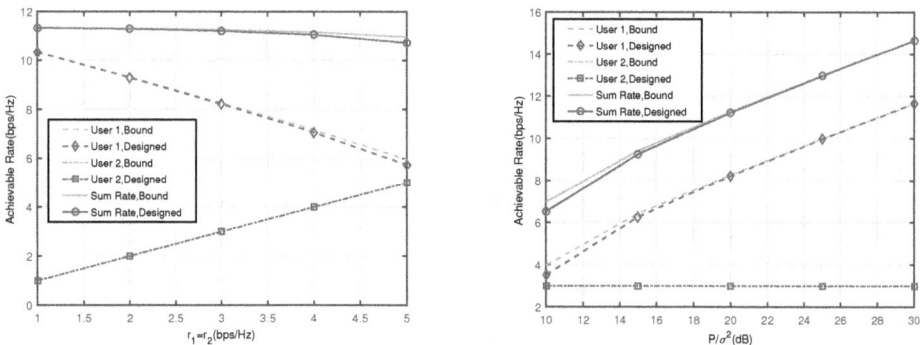

Figure 5.20 **Left**: Comparison between the performance bound and the designed achievable rates with varying rate constraint. **Right**: Comparison between the performance bound and the designed achievable rates with varying maximal power to noise ratio.

proposed solution to the original problem is rational and effective, i.e., it can achieve near-optimal performance. On the other hand, we can find that most beam gain is allocated to user 1, which has the better channel condition, so as to optimize the sum rate. Only necessary beam gain is allocated to user 2 to satisfy the rate constraint. That is why user 2 always achieves an achievable rate equal to the rate constraint.

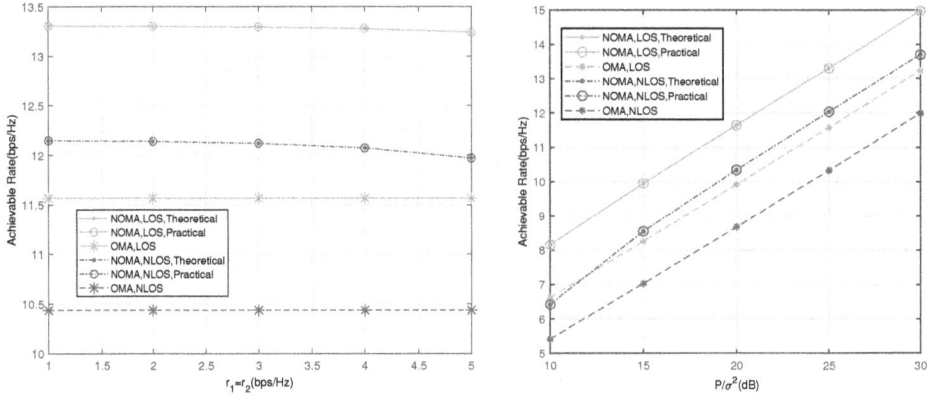

Figure 5.21 **Left**: Comparison of sum rate between theoretical NOMA, practical NOMA and OMA with varying rate constraint. **Right**: Comparison of sum rate between theoretical NOMA, practical NOMA and OMA with varying maximal power to noise ratio.

Figs. 5.21(left) and 5.21(right) show the comparison results of sum rate between theoretical NOMA, practical NOMA and OMA with varying rate constraint and varying maximal power to noise ratio, respectively, where $N = 32$ and $L_1 = L_2 = L = 4$. user 1 has a better channel condition than user 2, i.e., the average power ratio of them is $(1/0.3)^2$. For Fig. 5.21(left), $\frac{P}{\sigma^2} = 25$ dB, while for Fig. 5.21(right) $r_1 = r_2 = 2$ bps/Hz. Both LoS and NLoS channel models are considered. For LoS channel, the first path is the LoS path, which has a constant power, i.e., $|\lambda_1| = 1$ (0 dB), while the coefficients of the other 3 NLoS paths, i.e., $\{\lambda_i\}_{i=2,3,4}$, obey the complex Gaussian distribution with zero mean, and each of them has an average power of -15 dB. For the NLoS channel, the four paths are all NLoS paths with zero-mean complex Gaussian distributed coefficients, and each of them has an average power of $1/\sqrt{L}$. Each point in Figs. 5.21(left) and 5.21(right) is the average performance based on 10^3 channel realizations. With each channel realization, the optimal parameters are obtained by the proposed solution, and the theoretical/practical performances are obtained by computing the sum rates with the effective/original channel in (5.92) and (5.93). The achievable rates of NOMA users are computed by the solution of Problem (5.97). The performance of OMA is obtained based on the assumption that the beams gains of user 1 and user 2 are equal, i.e., $N/2$, and the instantaneous signal power for each user is $2P$. Thus, the achievable rate of OMA user is

$$R_i^{\text{OMA}} = \frac{1}{2} \log_2 \left(1 + \frac{|\lambda_i|^2 NP}{\sigma^2} \right), \quad i = 1, 2. \tag{5.148}$$

From these two figures we can observe that the theoretical performance is very close to the practical performance, which demonstrates the rational of the proposed method. Moreover, the performance of NOMA is significantly better than that of OMA under both the LoS and NLoS channels.

5.5 SUMMARY

In this chapter, array beamforming enabled NOMA was introduced to improve the capacity of a BS. The first problem is to ensure the coverage of multiple users which are served by the same data stream but located at different directions. Thus, we formulated optimization problems of multi-beam forming, which are non-convex and high-dimensional in general. As the CM constraint for AWV is highly non-convex, some relaxations were induced to make the problem tractable. Next, we introduced the downlink scenario for multi-beam NOMA as well as the joint power allocation and beamforming problem. While in the fourth section, we considered the uplink scenario for array beamforming enabled NOMA. The basic idea of the two scenarios is the same, i.e., to decompose the original problem into two independent problems: One is a power and beam gain allocation problem, and the other is a multi-beam forming problem under the CM constraint. In the power and beam gain allocation problem, we found an optimal solution under the assumption of ideal beamforming which provides an upper bound of the original problem. Then, we substituted the optimal solution for the beamforming problem to design the AWV. By searching the optimal phases of the two beam gains, the problem could be transformed into several standard convex optimization problems. Extensive performance evaluations verified the rational of the proposed solution, and showed that the solution could achieve close-to-bound sum-rate performance, which is distinctively better than OMA strategies.

Bibliography

[1] Su-Khiong Yong, Pengfei Xia, and Alberto Valdes-Garcia. *60GHz Technology for Gbps WLAN and WPAN: from Theory to Practice.* John Wiley & Sons, 2011.

[2] Zhenyu Xiao, Pengfei Xia, and Xiang-Gen Xia. Full-duplex millimeter-wave communication. *IEEE Wireless Commun.*, 24(6):136–143, 2017.

[3] Junyi Wang, Zhou Lan, Chang-woo Pyo, Tuncer Baykas, Chin-sean Sum, Mohammad Azizur Rahman, Jing Gao, Ryuhei Funada, Fumihide Kojima, Hiroshi Harada, et al. Beam codebook based beamforming protocol for multi-gbps millimeter-wave WPAN systems. *IEEE J. Select. Areas Commun.*, 27(8):1390–1399, 2009.

[4] Ahmed Alkhateeb, Jianhua Mo, Nuria Gonzalez-Prelcic, and Robert W. Heath. MIMO precoding and combining solutions for millimeter-wave systems. *IEEE Commun. Mag.*, 52(12):122–131, 2014.

[5] Wonil Roh, Ji-Yun Seol, Jeongho Park, Byunghwan Lee, Jaekon Lee, Yungsoo Kim, Jaeweon Cho, Kyungwhoon Cheun, and Farshid Aryanfar. Millimeter-wave beamforming as an enabling technology for 5G cellular communications: Theoretical feasibility and prototype results. *IEEE Commun. Mag.*, 52(2):106–113, 2014.

[6] Shu Sun, Theodore S. Rappaport, Robert W. Heath, Andrew Nix, and Sundeep Rangan. MIMO for millimeter-wave wireless communications: Beamforming, spatial multiplexing, or both? *IEEE Commun. Mag.*, 52(12):110–121, 2014.

[7] Jeffrey G. Andrews, Stefano Buzzi, Wan Choi, Stephen V. Hanly, Angel Lozano, Anthony C. K. Soong, and Jianzhong Charlie Zhang. What will 5G be? *IEEE J. Select. Areas Commun.*, 32(6):1065–1082, 2014.

[8] Zhiqiang Wei, Lou Zhao, Jiajia Guo, Derrick Wing Kwan Ng, and Jinhong Yuan. A multi-beam NOMA framework for hybrid mmWave systems. In *2018 IEEE International Conference on Communications (ICC)*, pages 1–7. IEEE, 2018.

[9] Zhiguo Ding, Xianfu Lei, George K Karagiannidis, Robert Schober, Jinhong Yuan, and Vijay K. Bhargava. A survey on non-orthogonal multiple access for 5G networks: Research challenges and future trends. *IEEE J. Select. Areas Commun.*, 35(10):2181–2195, 2017.

[10] Linglong Dai, Bichai Wang, Zhiguo Ding, Zhaocheng Wang, Sheng Chen, and Lajos Hanzo. A survey of non-orthogonal multiple access for 5G. *IEEE Commun. Surveys & Tutorials*, 20(3):2294–2323, 2018.

[11] Qi Sun, Shuangfeng Han, I Chin-Lin, and Zhengang Pan. On the ergodic capacity of MIMO NOMA systems. *IEEE Wireless Commun. Lett.*, 4(4):405–408, 2015.

[12] Yong Niu, Chuhan Gao, Yong Li, Li Su, Depeng Jin, and Athanasios V. Vasilakos. Exploiting device-to-device communications in joint scheduling of access and backhaul for mmWave small cells. *IEEE J. Select. Areas Commun.*, 33(10):2052–2069, 2015.

[13] Zhenyu Xiao, Tong He, Pengfei Xia, and Xiang-Gen Xia. Hierarchical codebook design for beamforming training in millimeter-wave communication. *IEEE Trans. Wireless Commun.*, 15(5):3380–3392, 2016.

[14] Zhenyu Xiao, Pengfei Xia, and Xiang-Gen Xia. Codebook design for millimeter-wave channel estimation with hybrid precoding structure. *IEEE Trans. Wireless Commun.*, 16(1):141–153, 2016.

[15] Zhenyu Xiao, Hang Dong, Lin Bai, Pengfei Xia, and Xiang-Gen Xia. Enhanced channel estimation and codebook design for millimeter-wave communication. *IEEE Trans. Veh. Technol.*, 67(10):9393–9405, 2018.

[16] Zhiguo Ding, Pingzhi Fan, and H. Vincent Poor. Impact of user pairing on 5G nonorthogonal multiple-access downlink transmissions. *IEEE Trans. Veh. Technol.*, 65(8):6010–6023, 2015.

[17] Chen Sun, Xiqi Gao, Shi Jin, Michail Matthaiou, Zhi Ding, and Chengshan Xiao. Beam division multiple access transmission for massive mimo communications. *IEEE Trans. Commun.*, 63(6):2170–2184, 2015.

[18] Chen Sun, Xiqi Gao, and Zhi Ding. BDMA in multicell massive MIMO communications: Power allocation algorithms. *IEEE Trans. on Signal Processing*, 65(11):2962–2974, 2017.

[19] Lipeng Zhu, Jun Zhang, Zhenyu Xiao, Xianbin Cao, Dapeng Oliver Wu, and Xiang-Gen Xia. Millimeter-wave NOMA with user grouping, power allocation and hybrid beamforming. *IEEE Trans. Wireless Commun.*, 18(11):5065–5079, 2019.

[20] Zhiguo Ding, Pingzhi Fan, and H. Vincent Poor. Random beamforming in millimeter-wave NOMA networks. *IEEE Access*, 5:7667–7681, 2017.

[21] Zhenyu Xiao, Lipeng Zhu, Jinho Choi, Pengfei Xia, and Xiang-Gen Xia. Joint power allocation and beamforming for non-orthogonal multiple access (noma) in 5G millimeter wave communications. *IEEE Trans. Wireless Commun.*, 17(5):2961–2974, 2018.

[22] Lipeng Zhu, Jun Zhang, Zhenyu Xiao, Xianbin Cao, Dapeng Oliver Wu, and Xiang-Gen Xia. Joint power control and beamforming for uplink non-orthogonal multiple access in 5G millimeter-wave communications. *IEEE Trans. Wireless Commun.*, 17(9):6177–6189, 2018.

[23] Lipeng Zhu, Jun Zhang, Zhenyu Xiao, Xianbin Cao, Dapeng Oliver Wu, and Xiang-Gen Xia. Joint Tx-Rx beamforming and power allocation for 5G millimeter-wave non-orthogonal multiple access networks. *IEEE Trans. Commun.*, 67(7):5114–5125, 2019.

[24] Bichai Wang, Linglong Dai, Zhaocheng Wang, Ning Ge, and Shidong Zhou. Spectrum and energy-efficient beamspace MIMO-NOMA for millimeter-wave communications using lens antenna array. *IEEE J. Select. Areas Commun.*, 35(10):2370–2382, 2017.

[25] Fang Fang, Haijun Zhang, Julian Cheng, and Victor C.M. Leung. Energy-efficient resource allocation for downlink non-orthogonal multiple access network. *IEEE Trans. Commun.*, 64(9):3722–3732, 2016.

[26] Muhammad Fainan Hanif, Zhiguo Ding, Tharmalingam Ratnarajah, and George K. Karagiannidis. A minorization-maximization method for optimizing sum rate in the downlink of non-orthogonal multiple access systems. *IEEE Trans. Signal Processing*, 64(1):76–88, 2015.

[27] Xiangrong Wang, Elias Aboutanios, Matthew Trinkle, and Moeness G. Amin. Reconfigurable adaptive array beamforming by antenna selection. *IEEE Trans. Signal Processing*, 62(9):2385–2396, 2014.

[28] Moeness G. Amin, Xiangrong Wang, Yimin D. Zhang, Fauzia Ahmad, and Elias Aboutanios. Sparse arrays and sampling for interference mitigation and DOA estimation in GNSS. *Proceedings of the IEEE*, 104(6):1302–1317, 2016.

[29] David Tse and Pramod Viswanath. *Fundamentals of Wireless Communication.* Cambridge university press, 2005.

[30] Zhiguo Ding, Linglong Dai, Robert Schober, and H. Vincent Poor. NOMA meets finite resolution analog beamforming in massive MIMO and millimeter-wave networks. *IEEE Commun. Lett.*, 21(8):1879–1882, 2017.

[31] Ahmed Alkhateeb, Young-Han Nam, Jianzhong Zhang, and Robert W. Heath. Massive MIMO combining with switches. *IEEE Wireless Commun. Lett.*, 5(3):232–235, 2016.

CHAPTER 6

Array Beamforming Enabled Multi-User NOMA

6.1 INTRODUCTION

With the coming of 5G mobile communication, the urgent requirements of high spectrum efficiency, low latency, low cost and massive connectivity pose great challenges[1,2,3,4]. The the conventional orthogonal multiple access (OMA) techniques may not meet the insistent requirements of mobile Internet and Internet of Things (IoT) due to massive connectivity. Hence, non-orthogonal multiple access (NOMA) has become a promising candidate technology for 5G[1,2,3,5,6,7].

The basic idea of NOMA is the reuse of resource blocks (RB), i.e., to serve multiple users in an orthogonal frequency/time RB by transmitting signals of different users with different powers. In Chapter 5, we have introduced the two user NOMA for uplink and downlink transmission. Moreover, the corresponding power allocation and decoding order problem for two user NOMA have already been solved.

Following in Chapter 6, we generalize the scenario to a multiple user case. For a multi-user NOMA system, the joint optimization problem can be still transformed into a power allocation problem and an equivalent beamforming problem. Nevertheless, the optimization becomes more complicated due to the high dimensional variables. In addition, the dimension of the power allocation variables is much lower than that of the antenna weight vector (AWV), and the power allocation variables have linear or convex constraints in general, which is more tractable compared with the constant modulus (CM) constraint on the AWV. As can be seen from the two observations above, the multi-user NOMA is different from the two-user NOMA and brings new challenges.

In this chapter, we investigate the multi-user NOMA. We consider three different problems in this chapter. In the second section, we consider the user fairness for downlink NOMA networks. To improve the overall data rate, we maximize the minimal achievable rate among multiple users. In the third section, we consider joint Tx-Rx beamforming and power allocation in NOMA networks. We present a boundary-compresses particle swarm optimization (BC-PSO) algorithm to solve this problem and obtain a sub-optimal solution. In the fourth section, we consider the hybrid

DOI: 10.1201/9781003366362-6

analog-digital beamforming structure at the BS, where the user grouping and the joint optimization of hybrid analog-digital beamforming and power allocation are explored to improve the sum rate of the users.

6.2 USER FAIRNESS FOR NOMA

It is worthwhile to point out that although there are several literatures exploring the user fairness for NOMA and multiple-input multiple-output (MIMO) NOMA, the key feature of millimeter-wave channel, i.e., analog beamforming, was not considered[8,9,10,11,12]. Thus, the proposed approaches in[8,9,10,11,12] cannot be directly used to solve the joint beamforming and power allocation problem for user fairness in the millimeter-wave-NOMA system. In this section, power allocation and beamforming are jointly optimized. As the problem is non-convex and the dimension of the optimization variables is large, it is difficult to solve this problem with the existing optimization tools. To this end, we solve this problem with two stages and obtain a sub-optimal solution. In the first stage, we obtain closed-form optimal power allocation with an arbitrary fixed beamforming vector, which reduces the joint optimization problem into an equivalent beamforming problem. Then, in the second stage, we present an appropriate beamforming algorithm utilizing the spatial sparsity in the angle domain of the channel.

The rest of the section is organized as follows. In Section 6.2.1, we present the system model and formulate the problem. In Section 6.2.2, we propose the solution. In Section 6.2.3, simulation results are given to demonstrate the performance of the proposed solution.

6.2.1 System Model and Problem Formation

6.2.1.1 System Model

As shown in Fig. 6.1, in our downlink communications system, the base station (BS) is equipped with a single radio frequency (RF) chain and an N-antenna phased array [1]. K users with a single antenna are served by the same RB. Each antenna is driven by the power amplifier (PA) and phase shifter (PS). With only one RF chain at the BS, we only need to consider the analog beamforming, which has a low hardware complexity and it can also be used in the hybrid beamforming structure.

The BS transmits a signal s_k to user k, $k = 1, 2, \cdots, K$, with transmit power p_k, where $\mathbb{E}(|s_k|^2) = 1$. The total transmit power of the BS is P. The received signal for user k is

$$y_k = \mathbf{h}_k^{\mathrm{H}} \mathbf{w} \sum_{k=1}^{K} \sqrt{p_k} s_k + n_k, \tag{6.1}$$

---

[1]Note that the phased array with only one RF chain is easy to implement and has a low hardware cost. Thus, we consider the analog beamforming in this chapter, not the hybrid beamforming. Since the designs of analog beamforming and digital beamforming are usually separate[13,14,15,16], the proposed analog beamforming scheme can also be used in a hybrid beamforming system.

where $\mathbf{h}_k$ is the channel response vector between the BS and user k, $\mathbf{w}$ is the antenna weight vector (AWV), i.e., analog beamforming vector, and n_k denotes the Gaussian white noise at user k with power σ^2.

The conventional single phase shifter (SPS) implementation at the BS is shown in Fig. 6.1(a), where each antenna branch has a single PS. The elements of the AWV are complex numbers, whose modulus and phases are controlled by the PA and PS, respectively. In general, all the PAs have the same scaling factor to reduce the hardware complexity. Thus, the AWV of the SPS implementation has CM elements, which is denoted by

$$||[\mathbf{w}]_i| = \frac{1}{\sqrt{N}}, \ i = 1, 2, ..., N. \tag{6.2}$$

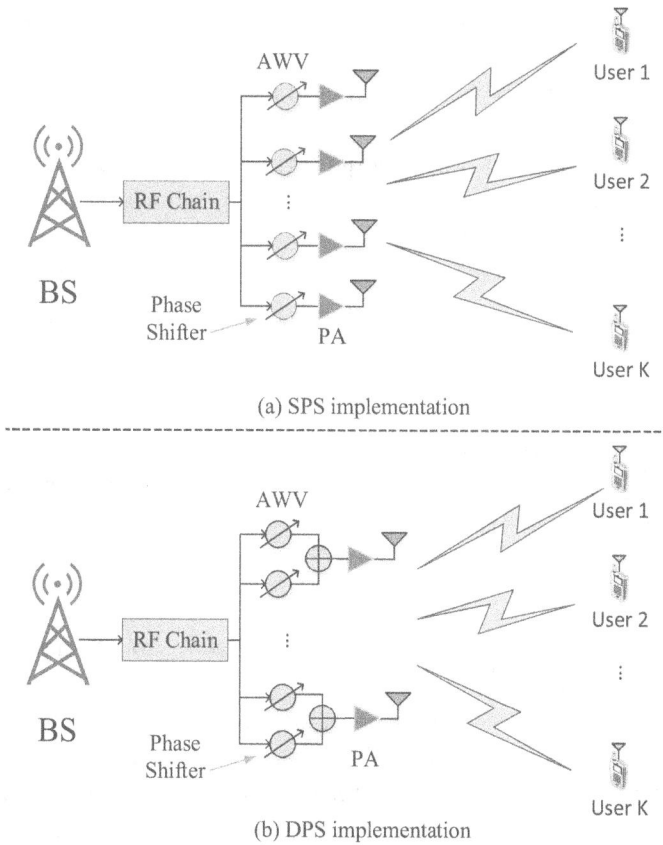

(a) SPS implementation

(b) DPS implementation

Figure 6.1 Illustration of a mobile cell, where one BS with N antennas serves multiple users with one single antenna.

Since the above constraint for the SPS implementation is non-convex, it results in high computational complexity of designing the AWV if the design space has a large dimension. Although some codebooks have been designed for the SPS implementation, there is still a tradeoff between the beamforming performance and the computational complexity[17, 18]. To address this problem, a new implementation named double phase shifter (DPS) was proposed in[19, 20], which is shown in Fig. 6.1(b). For the DPS implementation, each antenna is driven by the summation of the two independent PSs. Although the modulus of each PS is constant, the phases of two PSs can be adjusted to achieve different modulus in each antenna branch. Thus, the modulus constraint is relaxed to

$$|[\mathbf{w}]_i| \leq \frac{2}{\sqrt{N}}, \ i = 1, 2, ..., N. \tag{6.3}$$

By doubling the number of PSs, the new constraint becomes convex and therefore makes it more tractable to develop low-complexity design approaches. With this implementation, it is possible to achieve a better beamforming performance with lower computational complexity.

The channel between BS and user k is a millimeter-wave channel [2]. Subject to the limited scattering in millimeter-wave-band, multipath is mainly caused by reflection. As the number of the multipath components (MPCs) is small in general, the channel has directionality and appears spatial sparsity in the angle domain[17, 21, 22, 23, 24, 25]. Different MPCs have different angles of departure (AoDs). Without loss of generality, we adopt the directional channel model assuming a uniform linear array (ULA) with a half-wavelength antenna space.

Then, similar to the model in previous chapters, i.e., (1.131) and (5.5), the channel can be expressed as

$$\mathbf{h}_k = \sum_{\ell=1}^{L_k} \lambda_{k,\ell} \mathbf{a}(N, \theta_{k,\ell}), \tag{6.4}$$

where $\lambda_{k,\ell}$, $\theta_{k,\ell}$ are the complex coefficient and cos(AoD) of the ℓ-th MPC of the channel vector for user k, respectively. L_k is the total number of MPCs for user k, $\mathbf{a}(\cdot)$ is the steering vector of a ULA as shown in (1.70), which can be expressed as

$$\mathbf{a}(N, \theta) = [e^{j\pi 0\theta}, e^{j\pi 1\theta}, e^{j\pi 2\theta}, \cdots, e^{j\pi(N-1)\theta}]^{\mathrm{T}}, \tag{6.5}$$

Let $\Omega_{k,\ell}$ denote the real AoD of the ℓ-th MPC for user k, and we have $\theta_{k,\ell} = \cos(\Omega_{k,\ell})$. Therefore, $\theta_{k,\ell}$ is within the range $[-1, 1]$.

In general, the optimal decoding order of NOMA is the increasing order of the effective channel gains[3, 26], i.e., $|\mathbf{h}_k^{\mathrm{H}} \mathbf{w}|^2$. However, we cannot determine the order of the effective channel gains before beamforming design. For simplicity, we utilize the increasing order of uses' channel gains as the decoding order. We will illustrate the rational of selecting the increasing-channel-gain decoding order, and verify that it can achieve near optimal performance by simulations. Without loss of generality, we assume

$$\|\mathbf{h}_1\|_2 \geq \|\mathbf{h}_2\|_2 \geq \cdots \geq \|\mathbf{h}_K\|_2. \tag{6.6}$$

---

[2]In this chapter, we focus on the resource allocation for millimeter-wave-NOMA. We assume the CSI is known at BS here. The channel estimation with low complexity can be referred in[17] and[18].

Therefore, user k can decode s_n, $k+1 \leq n \leq K$, and then remove them from the received signal in a successive manner. The signals for user m, $1 \leq m \leq k-1$, are treated as noise. Thus, the achievable rate of user k is given by

$$R_k = \log_2(1 + \frac{\left|\mathbf{h}_k^{\mathrm{H}}\mathbf{w}\right|^2 p_k}{\left|\mathbf{h}_k^{\mathrm{H}}\mathbf{w}\right|^2 \sum_{m=1}^{k-1} p_m + \sigma^2}). \tag{6.7}$$

6.2.1.2 Problem Formulation

As aforementioned, both beamforming and power allocation have an important effect on the performance of the NOMA system. To improve the overall data rate and guarantee user fairness, we formulate a problem to maximize the minimal achievable rate (the max-min fairness) among the K users, where beamforming and power allocation are jointly optimized. The problem is formulated as

$$\underset{\{p_k\},\mathbf{w}}{\text{Max}} \ \underset{k}{\min}\{R_k\} \tag{6.8a}$$

$$\text{s.t.} \quad C_1: \ p_k \geq 0, \quad k = 1, 2, \cdots, K, \tag{6.8b}$$

$$C_2: \ \sum_{k=1}^{K} p_k \leq P, \tag{6.8c}$$

$$C_3: \ \|\mathbf{w}\|_2 \leq 1, \tag{6.8d}$$

$$C_4: \ |[\mathbf{w}]_i| = \frac{1}{\sqrt{N}} \text{ or } |[\mathbf{w}]_i| \leq \frac{2}{\sqrt{N}}, \ i = 1, 2, ..., N, \tag{6.8e}$$

where R_k denotes the achievable rate of user k as defined in (6.7) and $\min_k\{R_k\}$ is the minimal achievable rate among the K served users. The constraint C_1 indicates that the power allocation to each user should be non-negative. C_2 is the transmit power constraint, where P is the total transmit power. C_3 is the unit norm constraint on the AWV. As the modulus constraints on the AWV for different PS implementations are different, we distinguish them in the constraint C_4. It will be shown later, with the same computational complexity, the DPS implementation can achieve a better performance compared with the SPS implementation.

The above problem is challenging, not only due to the non-convex formulation, but also due to that the variables to be optimized are entangled with each other. It is computationally prohibitive to directly search the optimal solution, because the dimension of the optimization variables is $N + K$, which is large in general. Next, we will propose a sub-optimal solution with promising performance but low computational complexity.

6.2.2 Solution of the Problem

As the modulus constraints for SPS and DPS implementations are different, we first solve the problem without considering the constraint C_4. Thus, Problem (6.8) is simplified as

$$\underset{\{p_k\},\mathbf{w}}{\text{Max}} \ \min_k \{R_k\} \tag{6.9a}$$

$$\text{s.t.} \quad C_1: \ p_k \geq 0, \quad k = 1, 2, \cdots, K, \tag{6.9b}$$

$$C_2: \ \sum_{k=1}^{K} p_k \leq P, \tag{6.9c}$$

$$C_3: \ \|\mathbf{w}\|_2 \leq 1. \tag{6.9d}$$

We will solve Problem (6.9) first, and then particularly consider the modulus constraints.

Problem (6.9) is still difficult due to the non-convex formulation, so we propose a sub-optimal solution with two stages. In the first stage, we obtain the closed-form optimal power allocation with an arbitrary fixed AWV. Then, in the second stage, we propose an appropriate beamforming algorithm utilizing the angle-domain spatial sparsity of the millimeter-wave channel. It is worth pointing out that a closed-form solution of the optimal power allocation obtained in the first stage is a function of the AWV. Thus, we can substitute it to Problem (6.9) and solve the beamforming problem in the second stage without loss of optimality.

6.2.2.1 Optimal Power Allocation with an Arbitrary Fixed AWV

First, we introduce a variable to simplify Problem (6.9). Denote the minimal achievable rate among the K users as r. Then Problem (6.9) can be re-written as

$$\underset{\{p_k\},\mathbf{w},r}{\text{Max}} \ r \tag{6.10a}$$

$$\text{s.t.} \quad C_0: \ R_k \geq r, \quad k = 1, 2, \cdots, K, \tag{6.10b}$$

$$C_1: \ p_k \geq 0, \quad k = 1, 2, \cdots, K, \tag{6.10c}$$

$$C_2: \ \sum_{k=1}^{K} p_k \leq P, \tag{6.10d}$$

$$C_3: \ \|\mathbf{w}\|_2 \leq 1, \tag{6.10e}$$

where the constraints $C_0: \ R_k \geq r, \ k = 1, 2, \cdots, K$, are necessary and sufficient conditions of the fact that r is the minimal achievable rate among the served users. On one hand, as r is the minimal rate, the achievable rate of each user should be no less than r. On the other hand, there is at least one user, whose achievable rate R_{k_m} is equal to r; otherwise we can always improve r to minish the gap between R_{k_m} and r.

We give the following theorem to obtain the optimal solution of power allocation of Problem (6.10) with an arbitrary fixed AWV.

Theorem 6.2.1. *Given an arbitrary fixed* $\mathbf{w_0}$, *the optimal power allocation of Problem* (6.10) *is*

$$
\begin{cases}
p_1 = \eta \dfrac{\sigma^2}{\left|\mathbf{h}_1^{\mathrm{H}}\mathbf{w}_0\right|^2}, \\[2mm]
p_2 = \eta\left(p_1 + \dfrac{\sigma^2}{\left|\mathbf{h}_2^{\mathrm{H}}\mathbf{w}_0\right|^2}\right), \\[2mm]
\vdots \\[2mm]
p_K = \eta\left(\displaystyle\sum_{m=1}^{K-1} p_m + \dfrac{\sigma^2}{\left|\mathbf{h}_K^{\mathrm{H}}\mathbf{w}_0\right|^2}\right),
\end{cases}
\tag{6.11}
$$

where $\eta = 2^r - 1$, *and with the optimal power allocation,* $R_k = r$, $k = 1, 2, \cdots, K$.

Before proving Theorem 6.2.1, we give Lemma 6.2.1 for the summation of the optimal power allocation in (6.11), which is a function of η.

Lemma 6.2.1. *The summation of power allocation in* (6.11) *is*

$$
g(\eta) \triangleq \sum_{k=1}^{K} p_k = \sum_{k=1}^{K} \frac{\eta(1+\eta)^{K-k}\sigma^2}{\left|\mathbf{h}_K^{\mathrm{H}}\mathbf{w}_0\right|^2}.
\tag{6.12}
$$

Proof. We prove Lemma 6.2.1 with mathematical induction.

When $m = 1$, (6.12) is easy to verify

$$
p_1 = \eta \frac{\sigma^2}{\left|\mathbf{h}_1^{\mathrm{H}}\mathbf{w}_0\right|^2}.
\tag{6.13}
$$

When $m = n$, $n \geq 1$, assume that

$$
\sum_{k=1}^{n} p_k = \sum_{k=1}^{n} \frac{\eta(1+\eta)^{n-k}\sigma^2}{\left|\mathbf{h}_k^{\mathrm{H}}\mathbf{w}_0\right|^2}.
\tag{6.14}
$$

When $m = n + 1$, based on (6.14), we have

$$
\begin{aligned}
\sum_{k=1}^{n+1} p_k &= \sum_{k=1}^{n} p_k + \eta\left(\sum_{k=1}^{n} p_k + \frac{\sigma^2}{\left|\mathbf{h}_n^{\mathrm{H}}\mathbf{w}_0\right|^2}\right) \\
&= (1+\eta)\sum_{k=1}^{n} p_k + \eta\frac{\sigma^2}{\left|\mathbf{h}_n^{\mathrm{H}}\mathbf{w}_0\right|^2} \\
&= (1+\eta)\sum_{k=1}^{n} \frac{\eta(1+\eta)^{n-k}\sigma^2}{\left|\mathbf{h}_k^{\mathrm{H}}\mathbf{w}_0\right|^2} + \eta\frac{\sigma^2}{\left|\mathbf{h}_n^{\mathrm{H}}\mathbf{w}_0\right|^2} \\
&= \sum_{k=1}^{n+1} \frac{\eta(1+\eta)^{n+1-k}\sigma^2}{\left|\mathbf{h}_k^{\mathrm{H}}\mathbf{w}_0\right|^2}.
\end{aligned}
\tag{6.15}
$$

Finally, we can conclude that (6.12) is true. □

Based on Lemma 6.2.1, the proof of Theorem 6.2.1 is presented as follows.

Proof. The organization of the proof is as follows. First, given an optimal solution of Problem (6.10), we can always generate another optimal solution with the expression in (6.11), which satisfies the condition $R_k = r$ in Theorem 6.2.1, and the existence of the optimal solution under the condition $R_k = r$ is proved. Then, we prove the uniqueness of the optimal solution by using the contradiction, where we assume $R_k > r$ and derive a contradiction with the optimality.

Without loss of generality, we denote $\{p_k^\star, r^\star\}$ as one optimal solution of Problem (6.10) with fixed $\mathbf{w}_0$, where the achievable rate of user k is denoted by $R_k^\star$, and let $\eta^\star = 2^{r^\star} - 1$.

With $\eta^\star$, we can generate another solution $\{p_k^\circ, r^\star\}$ of Problem (6.10), where

$$
\begin{cases}
p_1^\circ = \eta^\star \dfrac{\sigma^2}{\left|\mathbf{h}_1^H \mathbf{w}_0\right|^2}, \\[2mm]
p_2^\circ = \eta^\star (p_1^\circ + \dfrac{\sigma^2}{\left|\mathbf{h}_2^H \mathbf{w}_0\right|^2}), \\[2mm]
\quad \vdots \\[2mm]
p_K^\circ = \eta^\star (\sum\limits_{m=1}^{K-1} p_m^\circ + \dfrac{\sigma^2}{\left|\mathbf{h}_K^H \mathbf{w}_0\right|^2}).
\end{cases}
\tag{6.16}
$$

The following lemma shows that this solution is also an optimal one.

Lemma 6.2.2. *The solution $\{p_k^\circ, r^\star\}$ is also an optimal solution of Problem (6.10), and the achievable rates under this parameter setting always satisfy $R_k^\circ = r^\star$, $1 \le k \le K$.*

Proof. First, we need to verify that the constraints C_0, C_1 and C_2 are all satisfied.

According to the expression of (6.16), it is obvious that $p_k^\circ \ge 0$, which means that the constraint C_1 is satisfied.

In addition, according to the assumption that $\{p_k^\star, r^\star\}$ is an optimal solution, we have

$$
r^\star \le R_k^\star
$$

$$
\Rightarrow \eta^\star \le \frac{\left|\mathbf{h}_k^H \mathbf{w}_0\right|^2 p_k^\star}{\left|\mathbf{h}_k^H \mathbf{w}_0\right|^2 \sum\limits_{m=1}^{k-1} p_m^\star + \sigma^2}
\tag{6.17}
$$

$$
\Rightarrow \eta^\star (\sum\limits_{m=1}^{k-1} p_m^\star + \frac{\sigma^2}{\left|\mathbf{h}_k^H \mathbf{w}_0\right|^2}) \le p_k^\star.
$$

Next, we use mathematical induction to prove that $p_k^\circ \le p_k^\star$, $k = 1, 2, \cdots, K$.

When $k = 1$, according to (6.17) we have

$$
p_1^\circ \le p_1^\star.
\tag{6.18}
$$

When $k = n$, $n \geq 1$, assume

$$p_1^\circ \leq p_1^\star, \cdots, p_n^\circ \leq p_n^\star. \tag{6.19}$$

According to (6.17) we have

$$p_{n+1}^\circ = \eta^\star \left(\sum_{m=1}^{n} p_m^\circ + \frac{\sigma^2}{\left| \mathbf{h}_{n+1}^{\mathrm{H}} \mathbf{w}_0 \right|^2} \right) \leq \eta^\star \left(\sum_{m=1}^{n} p_m^\star + \frac{\sigma^2}{\left| \mathbf{h}_{n+1}^{\mathrm{H}} \mathbf{w}_0 \right|^2} \right) \leq p_{n+1}^\star. \tag{6.20}$$

Thus, we can conclude that $p_k^\circ \leq p_k^\star$, $k = 1, 2, \cdots, K$, and we have

$$\sum_{k=1}^{K} p_k^\circ \leq \sum_{k=1}^{K} p_k^\star \leq P, \tag{6.21}$$

which means that the constraint C_2 is satisfied.

With the considered solution $(p_k^\circ, r^\star)$, we have

$$
\begin{aligned}
R_k^\circ &= \log_2 \left(1 + \frac{\left| \mathbf{h}_k^{\mathrm{H}} \mathbf{w}_0 \right|^2 p_k^\circ}{\left| \mathbf{h}_k^{\mathrm{H}} \mathbf{w}_0 \right|^2 \sum\limits_{m=1}^{k-1} p_m^\circ + \sigma^2} \right) \\
&= \log_2 \left(1 + \frac{p_k^\circ}{\sum\limits_{m=1}^{k-1} p_m^\circ + \frac{\sigma^2}{\left| \mathbf{h}_k^{\mathrm{H}} \mathbf{w}_0 \right|^2}} \right) \\
&\overset{(a)}{=} \log_2 (1 + \eta^\star) \\
&= r^\star,
\end{aligned}
\tag{6.22}
$$

where (a) is based on (6.16). The above equation means that the constraint C_0 is satisfied.

Since $\{p_k^\circ, r^\star\}$ can satisfy all the constraints, and $R_k^\circ = r^\star$, $1 \leq k \leq K$, it is also an optimal solution of Problem (6.10). □

As both $\{p_k^\circ, r^\star\}$ and $\{p_k^\star, r^\star\}$ are optimal solutions of Problem (6.10), we will prove that they are, in fact, the same. For this sake, we need to prove that $R_k^\star = r^\star$, $1 \leq k \leq K$. We assume that there exists one user whose achievable rate is strictly larger than $r^\star$, i.e., $R_{k_0}^\star > r^\star$, and we will prove that this assumption does not hold as follows.

As we have assumed that $R_{k_0}^\star > r^\star$, we have

$$R_{k_0}^\star > R_{k_0}^\circ = r^\star. \tag{6.23}$$

In addition, we have proven that $p_k^\circ \leq p_k^\star$ (see the proof in (6.18) and (6.20)). According to the expression of R_k in (6.102), it is straightforward to derive $p_{k_0}^\star > p_{k_0}^\circ$.

We define another solution $\{p_k^\Delta, r^\Delta\}$, where $r^\Delta = r^\star + \delta$, and

$$
\begin{cases}
p_1^\Delta = \eta^\Delta \dfrac{\sigma^2}{|\mathbf{h}_1^H \mathbf{w}_0|^2}, \\[2mm]
p_2^\Delta = \eta^\Delta (p_1^\Delta + \dfrac{\sigma^2}{|\mathbf{h}_2^H \mathbf{w}_0|^2}), \\[2mm]
\quad \vdots \\[2mm]
p_K^\Delta = \eta^\Delta (\sum\limits_{m=1}^{K-1} p_m^\Delta + \dfrac{\sigma^2}{|\mathbf{h}_K^H \mathbf{w}_0|^2}),
\end{cases}
\tag{6.24}
$$

where

$$
\begin{cases}
\eta^\Delta = 2^{r^\Delta} - 1, \\
\delta > 0.
\end{cases}
\tag{6.25}
$$

Thus, we have $\eta^\Delta > \eta^\star$.

Next, we prove that $\{p_k^\Delta, r^\Delta\}$ is within the feasible region of Problem (6.10). Similar to the proof in Lemma 6.3.1, we can prove that $\{p_k^\Delta \geq 0\}$ and $R_k^\Delta = r^\Delta > r^\star$, $1 \leq k \leq K$, which means that the constraints C_0 and C_1 are satisfied. According to Lemma 6.2.1, the summation of power allocation in (6.16) and (6.24) are $g(\eta^\star)$ and $g(\eta^\Delta)$, respectively. As we have proven that $p_{k_0}^\star > p_{k_0}^\circ$, we have $g(\eta^\star) < P$. Otherwise, if $g(\eta^\star) = P$, we have

$$
\sum_{k=1}^{K} p_k^\star > \sum_{k=1}^{K} p_k^\circ = g(\eta^\star) = P,
\tag{6.26}
$$

which is contradictory to Constraint C_2 in Problem (6.10). As $g(\eta)$ is an increasing function for η, we can always find a small positive δ, which satisfies

$$
g(\eta^\star + \delta) < P,
\tag{6.27}
$$

i.e.,

$$
g(\eta^\Delta) < P.
\tag{6.28}
$$

Thus, the constraint C_2 is satisfied with sufficiently small δ.

In brief, $\{p_k^\Delta, r^\Delta\}$ is within the feasible region of Problem (6.10) provided that δ is small enough. However, we have $R_k^\Delta = r^\Delta > r^\star$, $1 \leq k \leq K$, which means that the solution $\{p_k^\Delta, r^\Delta\}$ is better than $\{p_k^\star, r^\star\}$, which is contradictory to the fact that $\{p_k^\star, r^\star\}$ is an optimal solution. Thus, the assumption that there exists one user whose achievable rate is strictly larger than $r^\star$ does not hold. Equivalently, the achievable rates of users under the optimal power allocation satisfy $R_k^\star = r^\star = R_k^\circ$, $1 \leq k \leq K$. Solve the equation set above and we can obtain that $\{p_k^\star, r^\star\}$ is the same as $\{p_k^\circ, r^\star\}$, and the optimal power allocation of Problem (6.10) is given by (6.11). □

According to Theorem 6.2.1 and Lemma 6.2.1, Problem (6.10) can be equivalently written as

$$\underset{\mathbf{w},\eta}{\text{Max}} \quad \eta \tag{6.29a}$$

$$\text{s.t.} \quad \sum_{k=1}^{K} p_k = \sum_{k=1}^{K} \frac{\eta(1+\eta)^{K-k}\sigma^2}{|\mathbf{h}_k^{\text{H}}\mathbf{w}|^2} \leq P, \tag{6.29b}$$

$$\|\mathbf{w}\|_2 \leq 1, \tag{6.29c}$$

where $\eta = 2^r - 1$.

Hereto, the first stage to solve Problem (6.9) is finished, where the optimal power allocation is obtained, and thus the original problem with entangled power allocation and beamforming is reduced to a pure beamforming problem as shown in (6.29).

6.2.2.2 Beamforming Design with Optimal Power Allocation

The remaining task is to solve Problem (6.29) and obtain $\mathbf{w}$; then the closed-form expression of $\{p_k, \ k = 1, 2, \cdots, K\}$ can be obtained by (6.11). The main challenge is that the first constraint is non-convex, where $\mathbf{w}$ and η are entangled. As the dimension of $\mathbf{w}$, i.e., N, is large in general, it is computationally prohibitive to directly search the optimal solution. However, the introduced variable η is only 1-dimensional. We can search the maximal value of η in the range of $[0, \Gamma]$ with the bisection method, where Γ is the search upper bound. According to the definition of $\eta = 2^r - 1$, η, in fact, represents the minimal signal-to-interference-plus-noise ratio (SINR) among the K users. If we allocate all the beam gain and power to the user with the best channel condition, i.e., user 1, then user 1 can achieve the highest SINR $\Gamma = (\sum_{n=1}^{N} |[\mathbf{h}_1]_n|)^2 P/(N\sigma^2)$. Thus, we select Γ as the search upper bound. Given a fixed η, we judge whether an appropriate $\mathbf{w}$ can be found in the feasible region of Problem (6.29). Thus, we need to solve the following problem

$$\underset{\mathbf{w}}{\text{Min}} \quad f(\mathbf{w}) \triangleq \sum_{k=1}^{K} \frac{\eta(1+\eta)^{K-k}\sigma^2}{|\mathbf{h}_k^{\text{H}}\mathbf{w}|^2} \tag{6.30a}$$

$$\text{s.t.} \quad \|\mathbf{w}\|_2 \leq 1. \tag{6.30b}$$

Given η, if the minimal value of the objective function in Problem (6.30) is no larger than P, which means that a feasible solution can be found with the given η, we enlarge η and solve Problem (6.30) again. If the minimal value of the objective function in Problem (6.30) is larger than P, i.e., a feasible solution cannot be found with the given η, we reduce η and solve Problem (6.30) again. The stopping criterion of the bisection search is that η meets an accuracy requirement.

To solve Problem (6.30), some approximate manipulations are required to simplify the beamforming problem. Retrospecting the characteristic of the millimeter-wave channel, the channel response vectors of different users are approximatively orthogonal due to the spatial sparsity in the angle domain [3], which is

$$\frac{\mathbf{h}_m^{\mathrm{H}}}{\|\mathbf{h}_m^{\mathrm{H}}\|_2} \frac{\mathbf{h}_n}{\|\mathbf{h}_n\|_2} \approx \begin{cases} 1, & \text{if } m = n, \\ 0, & \text{if } m \neq n. \end{cases} \tag{6.31}$$

With this approximation, $\{\mathbf{h}_k/\|\mathbf{h}_k\|_2, \ k = 1, 2, \cdots, K\}$ can be considered as an orthonormal basis of a subspace in $\mathbb{C}^N$. We say the subspace expanded by $\{\mathbf{h}_k/\|\mathbf{h}_k\|_2, \ k = 1, 2, \cdots, K\}$ is a *channel space*. In Problem (6.30), most beam gains are inclined to focus on the users' directions. Thus, the AWV should be located in the channel space, which can be written as

$$\mathbf{w} = \sum_{k=1}^{K} \alpha_k \frac{\mathbf{h}_k}{\|\mathbf{h}_k\|_2}, \tag{6.32}$$

where $\{\alpha_k, \ k = 1, 2, \cdots, K\}$ are the coordinates of $\mathbf{w}$ in the channel space. Substituting (6.32) into Problem (6.30), we have

$$\operatorname*{Min}_{\{\alpha_k\}} \quad \sum_{k=1}^{K} \frac{\eta(1+\eta)^{K-k}\sigma^2}{\alpha_k^2 \|\mathbf{h}_k\|_2^2} \tag{6.33a}$$

$$\text{s.t.} \quad \sum_{k=1}^{K} \alpha_k^2 = 1. \tag{6.33b}$$

Note that the norm constraint $\|\mathbf{w}\|_2 \leq 1$ is replaced by $\|\mathbf{w}\|_2 = 1$ here, because the norm of optimal $\mathbf{w}$ is surely 1. Assuming that $\mathbf{w}^\star$ is optimal and $\|\mathbf{w}^\star\|_2 < 1$, we can always normalize the AWV to get a better solution of $\mathbf{w}^\star/\|\mathbf{w}^\star\|_2$.

To solve Problem (6.33), we define the Lagrange function as

$$L(\alpha, \lambda) = \sum_{k=1}^{K} \frac{\eta(1+\eta)^{K-k}\sigma^2}{\alpha_k^2 \|\mathbf{h}_k\|_2^2} + \lambda(\sum_{k=1}^{K} \alpha_k^2 - 1). \tag{6.34}$$

The Karush-Kuhn-Tucker (KKT) conditions can be obtained by the following equation[27],

$$\begin{cases} \dfrac{\partial L}{\partial \alpha_k} = 0, \ k = 1, 2, \cdots, K, \\[2mm] \dfrac{\partial L}{\partial \lambda} = 0. \end{cases} \tag{6.35}$$

---

[3]It is worthwhile noting that if N is large, the probability of two users located in the similar directions is small. If it happens, the proposed beamforming solution can also be used, consequently with some performance loss. It will be shown in the simulations that the achievable rate performance under this assumption is close to the upper bound.

From the KKT conditions, we can obtain the solution of Problem (6.33), which is given by

$$
\frac{\partial L}{\partial \alpha_k} = 0
$$

$$
\Rightarrow \frac{-2\eta(1+\eta)^{K-k}\sigma^2}{\alpha_k^3 \|\mathbf{h}_k\|_2^2} + 2\lambda\alpha_k = 0
$$

$$
\Rightarrow \alpha_k = \sqrt[4]{\frac{\eta(1+\eta)^{K-k}\sigma^2}{\lambda\|\mathbf{h}_k\|_2^2}} \tag{6.36}
$$

$$
\Rightarrow \alpha_k \propto \sqrt[4]{\frac{\eta(1+\eta)^{K-k}}{\|\mathbf{h}_k\|_2^2}}.
$$

Thus, the designed AWV in Problem (6.30) is given by

$$
\begin{cases}
\bar{\mathbf{w}} = \sum_{k=1}^{K} \sqrt[4]{\frac{\eta(1+\eta)^{K-k}}{\|\mathbf{h}_k\|_2^2}} \frac{\mathbf{h}_k}{\|\mathbf{h}_k\|_2}, \\
\mathbf{w} = \frac{\bar{\mathbf{w}}}{\|\bar{\mathbf{w}}\|_2}.
\end{cases} \tag{6.37}
$$

In summary, we give Algorithm 6.1 to solve Problem (6.29).

Algorithm 6.1: AWV design

Input: Channel response vectors: $\mathbf{h}_k$, $k = 1, 2, \cdots, K$;
   Total transmit power: P;
   Noise power: σ^2;
   The search accuracy ϵ.
Output: η and $\mathbf{w}$.
1: $\eta_{\min} = 0$, $\eta_{\max} = \Gamma$.
2: **while** $\eta_{\max} - \eta_{\min} > \epsilon$ **do**
3:    $\eta = (\eta_{\max} + \eta_{\min})/2$;
4:    Calculate $\mathbf{w}$ according to (6.37) and the objective function in
      Problem (6.30): $f(\mathbf{w})$.
5:    **if** $f(\mathbf{w}) > P$ **then**
6:       $\eta_{\max} = \eta$.
7:    **else**
8:       $\eta_{\min} = \eta$.
9:    **end if**
10: **end while**
11: **return** η and $\mathbf{w}$.

Hereto, we have solved Problem (6.9) and obtain the solution $\{p_k^\star, \mathbf{w}^\star\}$, where the AWV is obtained in Algorithm 6.1 and the power allocation is given in (6.11). The AWV is approximately optimal while the power allocation is optimal for the designed AWV. A remaining problem is to verify the rational of the decoding order. We will consider this problem next.

6.2.2.3 Decoding Order

When formulating Problem (6.8), we assume that the decoding order of signals is the increasing order of the channel gains. Next, we will verify that the order of the effective channel gains after beamforming design is the same with the channel-gain order. The effective channel gain for user k is

$$|\mathbf{h}_k^{\mathrm{H}}\mathbf{w}|^2 \propto |\mathbf{h}_k^{\mathrm{H}}\bar{\mathbf{w}}|^2 = \left| \sum_{m=1}^{K} \sqrt[4]{\frac{\eta(1+\eta)^{K-m}}{\|\mathbf{h}_m\|_2^2}} \frac{\mathbf{h}_k^{\mathrm{H}}\mathbf{h}_m}{\|\mathbf{h}_m\|_2} \right|^2$$

$$\overset{(a)}{=} \left| \sqrt[4]{\frac{\eta(1+\eta)^{K-k}}{\|\mathbf{h}_k\|_2^2}} \frac{\mathbf{h}_k^{\mathrm{H}}\mathbf{h}_k}{\|\mathbf{h}_k\|_2} \right|^2 \qquad (6.38)$$

$$= \sqrt{\eta(1+\eta)^{K-k}} \|\mathbf{h}_k\|_2,$$

where (a) is according to the orthogonal assumption of the channel response vectors. As $\eta = 2^r - 1 > 0$,

$$\sqrt{\eta(1+\eta)^{K-k}}, \qquad (6.39)$$

is decreasing for k. We have assumed that the order of the users' channel gains is

$$\|\mathbf{h}_1\|_2 \geq \|\mathbf{h}_2\|_2 \geq \cdots \geq \|\mathbf{h}_K\|_2. \qquad (6.40)$$

Thus, under the orthogonal assumption of the channel response vectors, the order of users' effective channel gains is

$$|\mathbf{h}_1^{\mathrm{H}}\mathbf{w}|^2 \geq |\mathbf{h}_2^{\mathrm{H}}\mathbf{w}|^2 \geq \cdots |\mathbf{h}_K^{\mathrm{H}}\mathbf{w}|^2. \qquad (6.41)$$

As shown in (6.41), the order of the effective channel gains is the same as that of channel gains. However, this property may not hold if we utilize other decoding orders, which indicates that the increasing-channel-gain decoding order is more reasonable. In the simulations, we will compare the performance of different decoding orders and find that the performance of increasing-channel-gain decoding order is very close to the performance of the optimal decoding order.

6.2.2.4 Consideration of Modulus Constraints

When solving Problem (6.9), the additional modulus constraints on the AWV were not considered. Next, we will consider the modulus constraints and solve the original problem, i.e., Problem (6.8). As we have shown in the system model, the modulus constraints on the elements of the AWV are (6.2) and (6.3) for SPS and DPS implementations, respectively. Some additional normalized operations on the designed AWV are required to satisfy the constraints. For the SPS implementation, the CM normalization is given by

$$[\mathbf{w}_S]_i = \frac{[\mathbf{w}]_i}{\sqrt{N}|[\mathbf{w}]_i|}, \quad i = 1, 2, \cdots, N. \qquad (6.42)$$

where $\mathbf{w}_S$ denotes the AWV for SPS implementation. For the DPS implementation, the modulus normalization is given by

$$[\mathbf{w}_D]_i = \begin{cases} [\mathbf{w}]_i, & \text{if } |[\mathbf{w}]_i| \leq \dfrac{2}{\sqrt{N}}, \\ \dfrac{2}{\sqrt{N}} \dfrac{[\mathbf{w}]_i}{|[\mathbf{w}]_i|}, & \text{if } |[\mathbf{w}]_i| > \dfrac{2}{\sqrt{N}}. \end{cases} \tag{6.43}$$

where $\mathbf{w}_D$ denotes the AWV for DPS implementation. Each element of $\mathbf{w}_D$ is the sum weight of the corresponding antenna branch, and it needs to be decomposed into two components, which can be expressed as

$$[\mathbf{w}_D]_i \triangleq a_i e^{j\omega_i} = \frac{1}{\sqrt{N}} e^{j(\omega_i + \varphi_i)} + \frac{1}{\sqrt{N}} e^{j(\omega_i - \varphi_i)}, \tag{6.44}$$

where $a_i \in [0, \frac{2}{\sqrt{N}}]$ and $\omega_i \in [0, 2\pi)$ are the modulus and the phase of $[\mathbf{w}_D]_i$, respectively, and $\varphi_i = \arccos(\frac{\sqrt{N}a_i}{2})$. Thus, the weights of the two PSs corresponding to $[\mathbf{w}_D]_i$ are

$$\begin{cases} [\tilde{\mathbf{w}}_D]_{2i-1} = \dfrac{1}{\sqrt{N}} e^{j(\omega_i + \varphi_i)}, \\ [\tilde{\mathbf{w}}_D]_{2i} = \dfrac{1}{\sqrt{N}} e^{j(\omega_i - \varphi_i)}. \end{cases} \tag{6.45}$$

6.2.2.5 Computational Complexity

As we obtained the CF optimal power allocation with an arbitrary fixed AWV, the computational complexity is mainly caused by the beamforming algorithm in the second stage. In Algorithm 6.1, the total search time for η is $T = \log_2(\frac{\Gamma}{\epsilon})$, where Γ is the search upper bound and ϵ is the search accuracy. Thus, the computational complexity of the proposed method is $\mathcal{O}(T)$, which does not increase with N or K. However, if we directly search the solution of Problem (6.8) and obtain the globally optimal solution, the total complexity is

$$\mathcal{O}((\frac{1}{\epsilon})^{N+K}), \tag{6.46}$$

which exponentially increases as N and K.

6.2.3 Performance Simulations

In this subsection, we provide simulation results to verify the performance of the proposed joint beamforming and power allocation method in the millimeter-wave-NOMA system. We adopt the channel model in (6.4) in the simulations, where the users are uniformly distributed from 10 m to 500 m away from the BS, and the mean square value of the complex coefficient is 1 at the node 100 m away from the BS. The number of MPCs for all the users are $L = 4$. Both LoS and NLoS channel models are considered. For the LoS channel, the average power of the NLoS paths is 15 dB weaker than that of the LoS path. For the NLoS channel, the coefficient of each path has an average power of $1/\sqrt{L}$. The search accuracy in Algorithm 6.1 is $\epsilon = 10^{-6}$.

In the simulations, the minimal achievable rates of "Ideal NOMA/OMA", "SPS-NOMA/SPS-OMA" and "DPS-NOMA/DPS-OMA" are based on the beamforming given in (6.37), (6.42) and (6.43), which are corresponding to the beamforming without CM constraint, with the SPS implementation and with the DPS implementation, respectively. The achievable rate of "Fully digital MIMO" is corresponding to the millimeter-wave MIMO with zero-forcing (ZF) precoding in[14]. The achievable rate of "NOMA upper bound" is corresponding to solving Problem (6.30) with PSO. The density of particles is sufficiently high, and thus the obtained minimal achievable rate can be treated as the upper bound.

We first show the power allocation and the effective channel gains in Figs. 6.2 (left) and 6.2 (right), respectively, where the LoS channel model is adopted [4]. Each point is an average result from 10^4 channel realizations. From Fig. 6.2 (left) we can find that most power is allocated to user 4, the user with the lowest channel gain. Less power is allocated to the users with higher channel gains, so as to reduce interference. Despite all of these, it can be observed from Fig. 6.2 (right) that the effective channel gain of user 4 is still the lowest. The user with a better channel gain has a higher effective channel gain with the proposed solution, which verifies the conclusion about the decoding order. It is noteworthy that the effective channel gains of user 1 and user 4 go increasing and decreasing, respectively, when P/σ^2 becomes higher, which is the result of joint power allocation and beamforming. It indicates that when the total power is high, power and beam gain should be jointly allocated to enlarge the difference of the effective channel gains to achieve a larger minimal user rate.

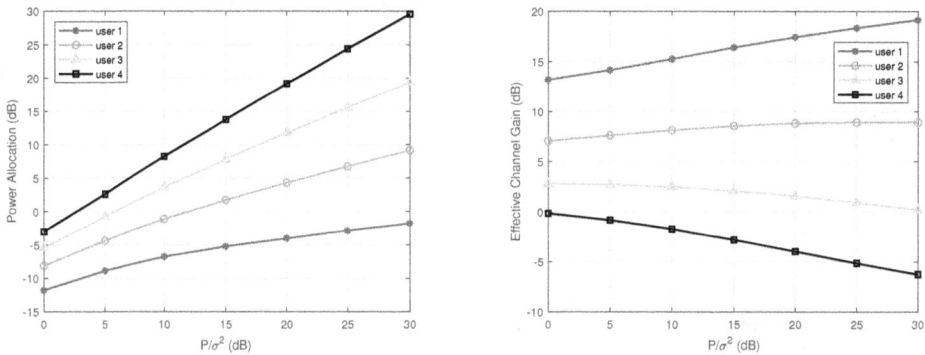

Figure 6.2 **Left**: Power allocation with varying total power to noise ratio, where $N = 32$ and $K = 4$. **Right**: Effective channel gains with varying total power to noise ratio, where $N = 32$ and $K = 4$.

Next, we compare the performance between the considered NOMA system and an OMA system. We give the following method to calculate the minimal achievable rates in a K-user OMA system, where TDMA is used without loss of generality.

---

[4]Similar results can be observed when the NLoS channel model is adopted; thus the results are not presented here for conciseness.

If all the time slots are allocated to user k, the achievable rate for user k is

$$\bar{R}_k = \log_2(1 + \frac{|\mathbf{h}_k^{\mathrm{H}}\mathbf{w}|^2 P}{\sigma^2}). \tag{6.47}$$

Assume that the time division is ideal, which means that the time slot can be allocated to the users with any proportion. To maximize the minimal achievable rate of the K users, more time should be allocated to the users with lower channel gains, such that the achievable rates of the K users are equal. Thus, the time allocation for user k is

$$\beta_k = \frac{1/\bar{R}_k}{\sum\limits_{m=1}^{K} 1/\bar{R}_m}. \tag{6.48}$$

Then the achievable rate of user k in the OMA system is

$$R_k^{\mathrm{OMA}} = \beta_k \bar{R}_k = \frac{1}{\sum\limits_{m=1}^{K} 1/\bar{R}_m}, \tag{6.49}$$

where all the users have the same achievable rate.

Figs. 6.3 (left) and 6.3 (right) show the comparison result of the minimal achievable rates between the NOMA, OMA and MIMO systems with varying transmit power to noise ratio and with varying number of users, respectively. Each point in the figures is the average performance of 10^4 LoS channel realizations. We can find that the minimal achievable rates of SPS-NOMA are lower than that of DPS-NOMA, which is very close to the minimal achievable rates of ideal NOMA, this is because the strict modulus normalization on the AWV for SPS results in significant performance loss, while the modulus normalization on the AWV for DPS is more relaxed and has little impact on the rate performance. In addition, the minimal achievable rates of the NOMA system is distinctly better than those of the OMA system for all the cases, and superiority is more significant when the total power to noise ratio is higher. In Fig. 6.3 (left) the rate gain between NOMA and OMA becomes large as the transmit power to noise increases. In Fig. 6.3 (right), the minimal achievable rates of both NOMA and OMA decreases as the number of users increases. This is mainly due to that the orthogonality of the channel vectors of the users becomes weakened, which deteriorates the beamforming performance and in turn the minimal achievable rate performance. Besides, the minimal achievable rates of the MIMO system are larger than that of the NOMA and OMA systems. The reason is that the number of RF chains in the proposed NOMA system is 1, while there are N RF chains in the MIMO system.

Since the circuit power consumptions for the SPS, the DPS, and the fully digital MIMO implementations are different, we provide the comparison of the energy efficiency in Figs. 6.4 (left) and 6.4 (right). The energy efficiency is defined as the radio between the minimum achievable rate and the average power consumption for each user[14], i.e.,

$$\mathrm{EE} = \frac{R_{\min}}{P + N_{\mathrm{RF}}P_{\mathrm{RF}} + N_{\mathrm{PS}}P_{\mathrm{PS}} + P_{\mathrm{BB}}(\mathrm{bps/Hz/W})}, \tag{6.50}$$

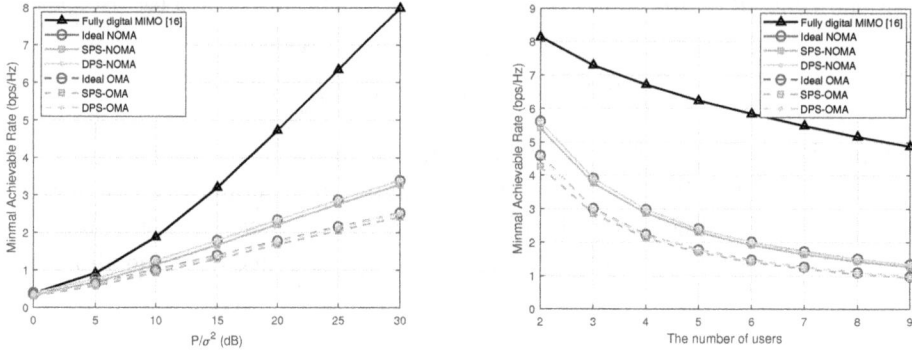

Figure 6.3 **Left**:Comparison of the minimal achievable rates between NOMA, OMA, and MIMO systems with varying total power to noise ratio, where $N = 32$ and $K = 4$. **Right**:Comparison of the minimal achievable rates between NOMA, OMA, and MIMO systems with varying number of users, where $N = 32$ and the average transmit power to noise for each user is 20 dB.

where $P = 30$ mW is the total transmit power, $P_{\text{RF}} = 300$ mW is the power consumed by each RF chain, $P_{\text{PS}} = 40$ mW is the power consumption of each PS, and $P_{\text{BB}} = 200$ mW is the baseband power consumption. N_{RF} is the number of the RF chains, which is equal to 1 for the SPS/DPS implementations and N for the fully digital MIMO implementation. N_{PS} is the number of the PSs, which is equal to N for the SPS implementation and $2N$ for the DPS implementation.

As shown in Fig. 6.4 (left), the NOMA system can achieve higher energy efficiency than both the OMA and MIMO systems. Since the number of the RF chains is equal to the number of antennas at the BS in the MIMO system, the high circuit power consumption of the RF chain (300 mW for each one) results in low energy efficiency. On the other hand, the energy efficiency for the SPS implementation is higher than that for the DPS implementation. The reason is that the number of the PSs for the DPS implementation is twice of that for the SPS implementation. Similar results can be obtained in Fig. 6.4 (right), where the NOMA system has a higher energy efficiency than both the OMA and MIMO systems in the most instances. However, as the number of the users increases, the energy efficiency of NOMA/OMA remains stable, while the energy efficiency of MIMO increases. The reason is that the number of the RF chains limits the performance of the proposed NOMA system. An alternative approach to increase the spectrum efficiency and the energy efficiency is using the hybrid beamforming structure in a NOMA system, which would be a good future work.

Fig. 6.5 shows the modulus of the elements of AWVs in (6.37), where $N = 32$, $K = 4$ and $P/\sigma^2 = 25$ dB. We show the 1st, 8th, 16th and 32th elements of 200 AWVs with different channel realizations. It can be seen that the modulus of the AWV's elements are mainly distributed around $1/\sqrt{N}$, and almost all of them have a modulus less than $2/\sqrt{N}$. The reason is as follows. As shown in (6.37), the solution of AWV is the weighted summation of the normalized channel response vectors of

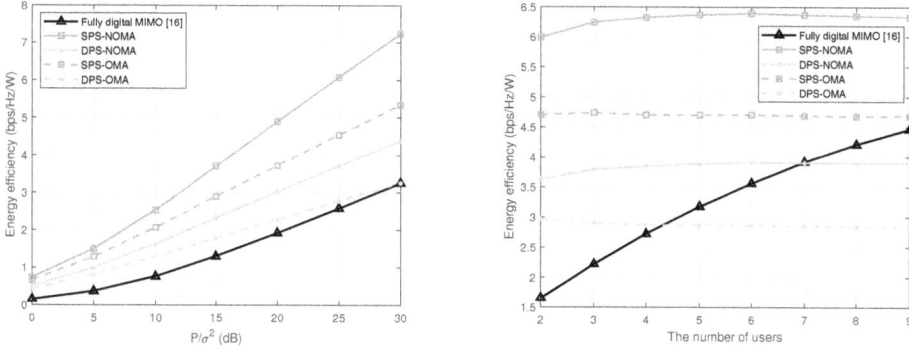

Figure 6.4 **Left**: Comparison of the energy efficiency between NOMA, OMA, and MIMO systems with varying total power to noise ratio, where $N = 32$ and $K = 4$. **Right**: Comparison of the energy efficiency between between NOMA, OMA, and MIMO systems with varying number of users, where $N = 32$ and the average transmit power to noise for each user is 20 dB.

the users. According to the channel model in (6.4), the channel response vector is the weighted summation of the steering vectors, whose elements have unit modulus. Since $N \gg K$ and $N \gg L_k$, it is impossible to exist one element of the AWV whose modulus is larger than the modulus of all the other elements. Thus, after the normalization in (6.37), the elements of the AWV have proportional modulus, which is distributed around $1/\sqrt{N}$. For this reason, the achievable performance of the DPS-NOMA is close to that of the ideal NOMA without modulus constraint, and the modulus normalization for the DPS implementation results in a limited performance loss.

In the second stage of the proposed solution, we have assumed that the channel response vectors are orthogonal and then found an appropriate AWV in (6.30). To evaluate the impact of this approximation, we compare the performance of the proposed solution with the upper-bound performance. Limited by the computational complexity, we provide the simulation results with a relatively small-scale antenna array, i.e., $N = 8, 16$. The comparison result is shown in Fig. 6.6, where each point is averaged from 10^3 LoS channel realizations. The minimal achievable rate of Ideal NOMA is based on the beamforming given in (6.37), which is corresponding to the beamforming without the CM constraint and the orthogonality assumption of the channel vectors between the NOMA users. As we can see, when $N = 8$, the performance gap between the proposed solution and the upper bound is no more than 0.25 bps/Hz. When $N = 16$, the performance gap is even smaller, i.e., no more than 0.2 bps/Hz. The reason is that the orthogonality of the channel vectors becomes better when N is larger. Thus, the approximation of the beamforming design in Problem (6.30) has limited impact on the system performance, and the proposed sub-optimal solution can achieve a near-upper-bound performance, especially when N is large.

Fig. 6.7 compares the minimal achievable rates of NOMA under the LoS and NLoS channel models with varying total power to noise ratio. The number of antennas is

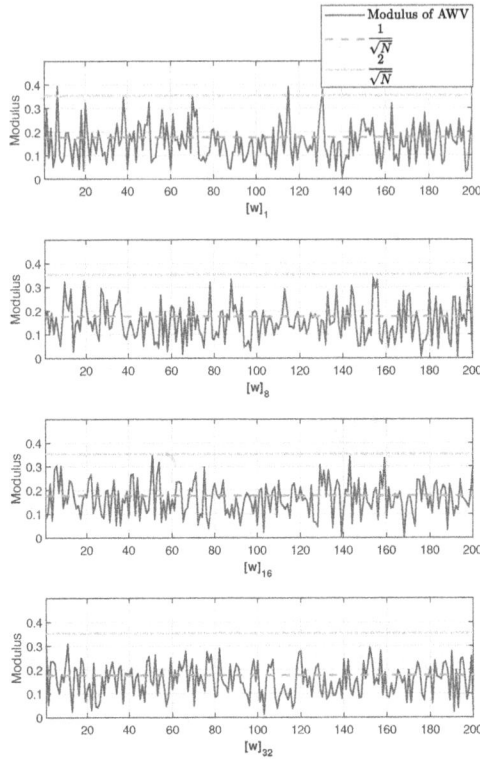

Figure 6.5 Modulus of the elements of the AWVs, where $N = 32$, $K = 4$ and $P/\sigma^2 = 25$ dB.

$N = 16, 64, 256$, respectively. The number of users is $K = 4$. Each point in Fig. 6.7 is the average performance of 10^4 channel realizations. It can be seen that the performance of DPS-NOMA with the LoS channel model is slightly better than that with the NLoS channel model, because the channel power is more centralized for the LoS channel. However, the performance gap between them is quite small, especially when N is large. The reason is that according to (6.38), the effective channel gain is linear to $\|\mathbf{h}_k\|_2$, the norm of the channel vector, rather than that of the power of the strongest path. Thus, the performance gap of DPS-NOMA with the LoS and NLoS channel models is small.

The simulations above are all based on the increasing-channel-gain decoding order. Next, we will show the impact of the decoding order on the NOMA system. Fig. 6.8 shows the performance comparison between different decoding orders with varying total power to noise ratio, where $N = 32$ and $K = 4$. There are 24 decoding orders in total for the 4 users. Each point in Fig. 6.8 is the average performance of 10^4 LoS channel realizations. The minimal achievable rates of the 24 decoding orders are all calculated. The order with the highest minimal achievable rate is chosen as the optimal order and the order with the lowest minimal achievable rate is chosen as the

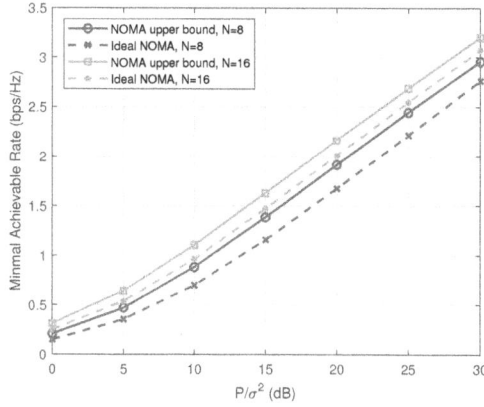

Figure 6.6 Comparison of the minimal achievable rates between the proposed solution and the upper bound with varying total power to noise ratio, where $K = 4$.

worst order. The increasing-channel-gain order is the one adopted in our solution, while the decreasing-channel-gain order is one for comparison. From the figure we can find that there is a significant performance gap between the optimal order and the worst order, which means that the decoding order has an important impact on the performance of NOMA. Moreover, the performance with the increasing-channel-gain order is almost the same as the optimal one, while the performance with the decreasing-channel-gain order is almost the same as the worst one. This result shows the rational of adopting the increasing-channel-gain order in our solution.

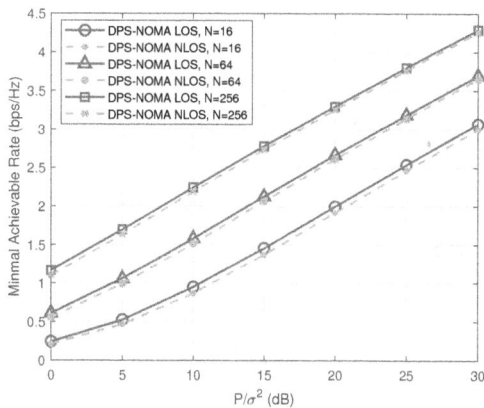

Figure 6.7 Performance comparison between LoS and NLoS channel models with varying total power to noise ratio, where $K = 4$.

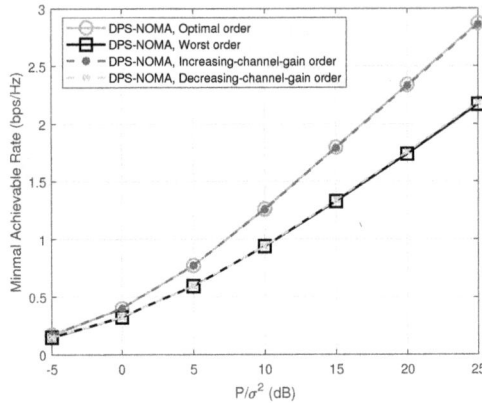

Figure 6.8 Comparison of the minimal achievable rates under different decoding orders with varying total power to noise ratio, where $N = 32$ and $K = 4$.

6.3 JOINT TX-RX BEAMFORMING FOR NOMA

In this section we solve a max-sum problem with joint Tx-Rx beamforming and power allocation. This problem is with a different type from that in[28]; thus particular study is necessary, and a different approach is expected to solve this different type problem. We formulate an optimization problem to maximize the achievable sum rate (ASR) of the multiple users, and meanwhile each user has a minimum rate constraint. As the formulated problem is non-convex and it cannot be directly solved by using the existing optimization tools, we propose a sub-optimal solution with three stages. In the first stage, the optimal power allocation with a closed form is obtained for an arbitrary fixed Tx-Rx beamforming. In the second stage, we obtain the optimal Rx beamforming with a closed form for arbitrary fixed Tx beamforming. In the third stage, by substituting the optimal solutions of the previous two stages into the original problem, a Tx beamforming problem is formulated. We propose a BC-PSO algorithm to solve this problem and obtain a sub-optimal solution.

The rest of this section is organized as follows. In Section 6.3.1, we present the system model and formulate the problem. In Section 6.3.2, we present the solution.

6.3.1 System Model and Problem Formulation

6.3.1.1 System Model

In this section, we consider a downlink communications system. As shown in Fig. 6.9, the BS serves K users simultaneously. The numbers of the antennas equipped at the BS and each user are N and M, respectively. Each antenna at the BS is driven by a PS and a PA, while an antenna at the users is driven by a PS and low noise amplifier (LNA). The number of the RF chains at the BS and the user sides is 1, which means that pure analog beamforming is utilized.

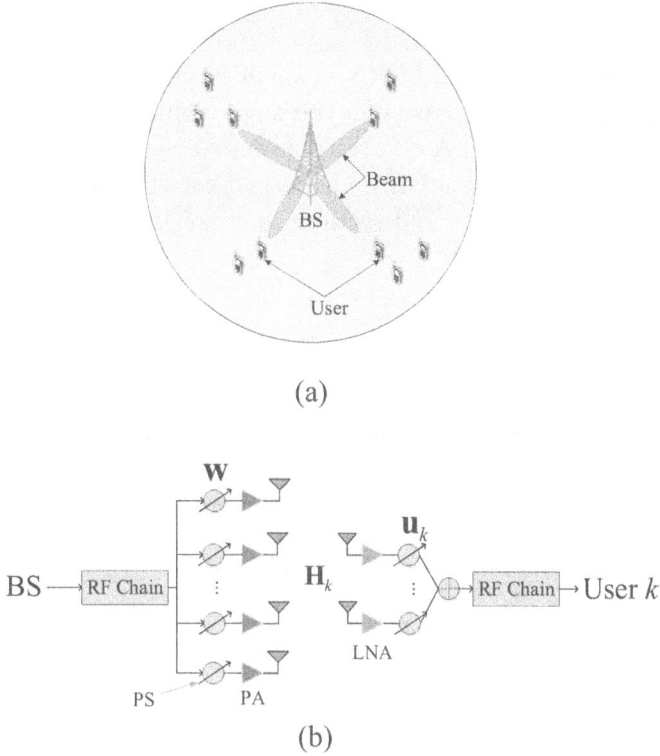

Figure 6.9 (a) Illustration of a mobile cell, where BS serves K users simultaneously. (b) Illustration of the channel architecture, where BS is equipped with a single RF chain and N antennas, while each user is equipped with a single RF chain and M antennas.

The BS transmits a signal s_k to user k, $k = 1, 2, \cdots, K$, with transmit power p_k, where $\mathbb{E}(|s_k|^2) = 1$. The total transmit power of the BS is P. Thus, the received signal for user k is

$$y_k = \mathbf{u}_k^{\mathrm{H}} \mathbf{H}_k \mathbf{w} \sum_{k=1}^{K} \sqrt{p_k} s_k + n_k, \qquad (6.51)$$

where $\mathbf{H}_k$ with dimension $M \times N$ is the channel response matrix between the BS and user k, and n_k denotes the Gaussian white noise at user k with power σ^2, $\mathbf{w}$ and $\mathbf{u}_k$ are the Tx beamforming vector of the BS and the Rx beamforming vector of user k, respectively. In general, the scaling factors of PA and LNA are constant. Thus, the Tx beamforming vector and Rx beamforming vector have CM constraints[28, 29, 30], i.e.,

$$\left| [\mathbf{w}]_n \right| = \frac{1}{\sqrt{N}}, \ 1 \le n \le N, \qquad (6.52)$$

$$\left| [\mathbf{u}_k]_m \right| = \frac{1}{\sqrt{M}}, \ 1 \le m \le M, \ 1 \le k \le K. \qquad (6.53)$$

The channel between the BS and user k is a millimeter-wave channel [5]. Subject to limited scattering in the millimeter-wave band, multipath is mainly caused by reflection. As the number of the MPCs is small in general, the channel has directionality and appears spatial sparsity in the angle domain[17,21,22,23,24,25]. Different MPCs have different AoDs and AoAs. Without loss of generality, we adopt the directional channel model assuming a ULA with a half-wavelength antenna space. Then, similar to the previous chapters and sections, a channel between the BS and user k can be expressed as[17,21,22,23,24,25]

$$\mathbf{H}_k = \sum_{\ell=1}^{L_k} \lambda_{k,\ell} \mathbf{a}_\mathrm{r}(\theta_{k,\ell}) \mathbf{a}_\mathrm{t}^\mathrm{H}(\psi_{k,\ell}), \tag{6.54}$$

where $\lambda_{k,\ell}$, $\theta_{k,\ell}$ and $\psi_{k,\ell}$ are the complex coefficient, cos(AoD) and cos(AoA) of the ℓ-th MPC of the channel vector for user k, respectively. L_k is the total number of MPCs for user k, $\mathbf{a}_\mathrm{t}(\cdot)$ and $\mathbf{a}_\mathrm{r}(\cdot)$ are steering vectors defined as

$$\mathbf{a}_\mathrm{t}(\theta) = [e^{j\pi 0\theta}, e^{j\pi 1\theta}, e^{j\pi 2\theta}, \cdots, e^{j\pi(N-1)\theta}]^\mathrm{T}, \tag{6.55}$$

$$\mathbf{a}_\mathrm{r}(\psi) = [e^{j\pi 0\psi}, e^{j\pi 1\psi}, e^{j\pi 2\psi}, \cdots, e^{j\pi(M-1)\psi}]^\mathrm{T}, \tag{6.56}$$

which depend on the array geometry.

6.3.1.2 Achievable Rate

As discussed in the previous section, the optimal decoding order for NOMA is the increasing order of the users' channel gains in general. However, for the NOMA with analog beamforming structure, the effective channel gains of the users are determined by both the channel gains and the beamforming gains. Thus, we need to sort the effective channel gains first, and then determine the decoding order. Without loss of generality, we assume that the order of the effective channel gains is

$$\left|\mathbf{u}_{\pi_1}^\mathrm{H} \mathbf{H}_{\pi_1} \mathbf{w}\right|^2 \geq \left|\mathbf{u}_{\pi_2}^\mathrm{H} \mathbf{H}_{\pi_2} \mathbf{w}\right|^2 \geq \cdots \geq \left|\mathbf{u}_{\pi_K}^\mathrm{H} \mathbf{H}_{\pi_K} \mathbf{w}\right|^2, \tag{6.57}$$

and thus the optimal decoding order is the increasing order of the effective channel gains[14,29,31]. Therefore, user π_k can decode s_{π_n}, $k+1 \leq n \leq K$, and then remove them from the received signal in a successive manner. The signals for user π_m, $1 \leq m \leq k-1$, are treated as interference. Thus, the achievable rate of user π_k is denoted by

$$R_{\pi_k} = \log_2 \left(1 + \frac{\left|\mathbf{u}_{\pi_k}^\mathrm{H} \mathbf{H}_{\pi_k} \mathbf{w}\right|^2 p_{\pi_k}}{\left|\mathbf{u}_{\pi_k}^\mathrm{H} \mathbf{H}_{\pi_k} \mathbf{w}\right|^2 \sum_{m=1}^{k-1} p_{\pi_m} + \sigma^2} \right). \tag{6.58}$$

---

[5]In this section, we assume the channel is known by the BS. The channel estimation with low complexity can be referred to[17] and[18].

The ASR of the proposed NOMA system is

$$R_{\text{sum}} = \sum_{k=1}^{K} R_k, \tag{6.59}$$

where $R_k \in \{R_{\pi_k}, k = 1, 2, ..., K\}$ depend on the decoding order.

6.3.1.3 Problem Formulation

To improve the overall data rate, we formulate a joint Tx-Rx beamforming and power allocation problem to maximize the ASR of the K users, where each user has a minimum rate constraint. The problem is formulated as

$$\underset{\{p_k\},\{\mathbf{u}_k\},\mathbf{w}}{\text{Maximize}} \quad R_{\text{sum}} \tag{6.60a}$$

$$\text{Subject to} \quad C_1 : R_k \geq r_k, \quad \forall k, \tag{6.60b}$$

$$C_2 : p_k \geq 0, \quad \forall k, \tag{6.60c}$$

$$C_3 : \sum_{k=1}^{K} p_k \leq P, \tag{6.60d}$$

$$C_4 : |[\mathbf{u}_k]_m| = \frac{1}{\sqrt{M}}, \quad \forall k, m, \tag{6.60e}$$

$$C_5 : |[\mathbf{w}]_n| = \frac{1}{\sqrt{N}}, \quad \forall n, \tag{6.60f}$$

where the constraint C_1 is the minimum rate constraint for each user. The constraint C_2 indicates that the power allocation to each user should be positive. The constraint C_3 is the total transmit power constraint, where the total power is no more than P. C_4 and C_5 are the CM constraints for the Rx beamforming vectors and Tx beamforming vector, respectively. In Problem (6.60), the formulation of the achievable rate is not convex/concave, and the CM constraints for the Rx beamforming vectors and Tx beamforming vector are also not convex/concave[28,29,30]. Thus, Problem (6.60) is not a convex/concave problem.

The total dimension of the variables in Problem (6.60) is $N + MK + K$, which is large in general. Direct search for the optimal solution results in heavy computational load, which is hard to accomplish in practice. To solve Problem (6.60), there are two main challenges. One is that the optimized variables are entangled with each other, which makes the formulation non-convex. The other is that the expression of R_{sum} depends on the decoding order. In general, the optimal decoding order is the increasing order of the users' effective channel gains. However, the order of effective channel gains varies with different Tx beamforming vectors and Rx beamforming vectors. In other words, given different Tx beamforming vectors and Rx beamforming vectors, the objective function in Problem (6.60), i.e., the ASR of the users, has different expressions. The two challenges make it infeasible to solve Problem (6.60) by using the existing optimization tools. Next, we will present a sub-optimal solution with promising performance but low computational complexity.

6.3.2 Solution of the Problem

As the optimized variables are entangled with each other in Problem (6.60), we propose a sub-optimal solution with three stages in this subsection. In the first stage, we obtain the optimal power allocation with a closed form for an arbitrary fixed Tx beamforming vector and arbitrary fixed Rx beamforming vectors. In the second stage, the optimal Rx beamforming vectors are obtained with a closed form for an arbitrary fixed Tx beamforming vector. Based on the two stages, the variables of power allocation and Rx beamforming vectors can be expressed as a function of the Tx beamforming vector. Substituting them into Problem (6.60), Problem (6.60) can be simplified as a Tx beamforming problem. Finally in the third stage, we propose the BC-PSO algorithm to solve the Tx beamforming problem and obtain a sub-optimal Tx beamforming vector.

6.3.2.1 Optimal Power Allocation with Arbitrary Fixed Beamforming Vectors

As we have analyzed before, an essential challenge to solve Problem (6.60) is the variation of the decoding order. However, given an arbitrary fixed Tx beamforming vector $\mathbf{w}$ and an arbitrary fixed Rx beamforming vectors $\mathbf{u}_k$, the order of the effective channel gains is fixed. For notational simplicity and without loss of generality, we assume

$$\left|\mathbf{u}_1^H\mathbf{H}_1\mathbf{w}\right|^2 \geq \left|\mathbf{u}_2^H\mathbf{H}_2\mathbf{w}\right|^2 \geq \cdots \geq \left|\mathbf{u}_K^H\mathbf{H}_K\mathbf{w}\right|^2, \tag{6.61}$$

in this part [6]. The original problem can be simplified as

$$\underset{\{p_k\}}{\text{Maximize}} \quad R_{\text{sum}} \tag{6.62a}$$

$$\text{Subject to} \quad C_1 : R_k \geq r_k, \quad \forall k, \tag{6.62b}$$

$$C_2 : p_k \geq 0, \quad \forall k, \tag{6.62c}$$

$$C_3 : \sum_{k=1}^{K} p_k \leq P, \tag{6.62d}$$

where the beamforming vectors are arbitrary and fixed. To solve Problem (6.62), we give the following Lemma first.

Lemma 6.3.1. *The optimal power allocation in Problem (6.62) must satisfy*

$$\sum_{k=1}^{K} p_k = P. \tag{6.63}$$

---

[6]Index k represents the user with the k-th highest effective channel gain. This simplification has no influence on the generality for solving the power allocation problem. In fact, given different Tx beamforming vectors and Rx beamforming vectors, the order of the effective channel gains may change, but we can always reorder them descendingly and define the user with the k-th highest effective channel gain as user k. This operation will be realized in Algorithm 6.1.

Proof. We prove Lemma 6.3.1 by using contradiction. Denote the optimal power allocation in Problem (6.62) as $\{p_k^\star\}$, and the achievable rate of user k under optimal power allocation is denoted by $R_k^\star$. Assume $\sum_{k=1}^{K} p_k^\star < P$.

Consider the following parameter settings

$$
\begin{cases}
p_k = p_k^\star, \ 1 \leq k \leq K - 1, \\
p_K = P - \sum_{k=1}^{K-1} p_k^\star.
\end{cases}
\tag{6.64}
$$

It is easy to verify that $R_k = R_k^\star$, $1 \leq k \leq K - 1$, and $R_K > R_K^\star$, which means that the parameter settings in (6.64) can satisfy the minimum rate constraint as well as improve the ASR in Problem (6.62). It is in contrast to the assumption that $\{p_k^\star\}$ is optimal. Thus, we have $\sum_{k=1}^{K} p_k^\star = P$. $\qquad\square$

According to Lemma 6.3.1, Problem (6.62) is equivalent to

$$
\underset{\{p_k\}}{\text{Maximize}} \quad R_{\text{sum}} \tag{6.65a}
$$

$$
\text{Subject to} \quad C_1 : R_k \geq r_k, \ \forall k, \tag{6.65b}
$$

$$
C_2 : p_k \geq 0, \ \forall k, \tag{6.65c}
$$

$$
C_3 : \sum_{k=1}^{K} p_k = P. \tag{6.65d}
$$

As the number of users is K, it is difficult to directly obtain the optimal power allocation for all the users. Thus, we commence from a simplified case, where only two adjacent users can adjust the transmit power while the other users have fixed transmit powers. The details are shown in the following Lemma.

Lemma 6.3.2. *For any k_0 ranging from 2 to K, if p_k, $k = 1, 2, \cdots, p_{k_0-2}, p_{k_0+1}, \cdots, p_K$, are all fixed, then R_{sum} in Problem (6.65) is decreasing with p_{k_0}.*

Proof. With fixed p_k, $k = 1, 2, \cdots, p_{k_0-2}, p_{k_0+1}, \cdots, p_K$, it is easy to verify that R_k, $k = 1, 2, \cdots, p_{k_0-2}, p_{k_0+1}, \cdots, p_K$, are constants. According to the constraint C_3 in Problem (6.65), we have

$$
p_{k_0-1} + p_{k_0} + \sum_{k \neq k_0-1, k_0} p_k = P
$$

$$
\Rightarrow p_{k_0-1} = (P - \sum_{k \neq k_0-1, k_0} p_k) - p_{k_0} \triangleq \tilde{P} - p_{k_0}.
\tag{6.66}
$$

Thus, there is only one independent variable p_{k_0} in Problem (6.65). The derivative of the objective function R_{sum} is shown in (6.67).

$$
\begin{aligned}
\frac{d\,R_{\text{sum}}}{d\,p_{k_0}} &= \frac{d\,(R_{k_0} + R_{k_0-1})}{d\,p_{k_0}} \\[4pt]
&= \frac{d\left[\log_2\left(1 + \dfrac{\left|\mathbf{u}_{k_0}^{\mathrm{H}}\mathbf{H}_{k_0}\mathbf{w}\right|^2 p_{k_0}}{\left|\mathbf{u}_{k_0}^{\mathrm{H}}\mathbf{H}_{k_0}\mathbf{w}\right|^2\left(\sum\limits_{m=1}^{k_0-2} p_m + \widetilde{P} - p_{k_0}\right) + \sigma^2}\right)\right]}{d\,p_{k_0}} \\[6pt]
&\quad + \frac{d\left[\log_2\left(1 + \dfrac{\left|\mathbf{u}_{k_0-1}^{\mathrm{H}}\mathbf{H}_{k_0-1}\mathbf{w}\right|^2(\widetilde{P} - p_{k_0})}{\left|\mathbf{u}_{k_0-1}^{\mathrm{H}}\mathbf{H}_{k_0-1}\mathbf{w}\right|^2\sum\limits_{m=1}^{k_0-2} p_m + \sigma^2}\right)\right]}{d\,p_{k_0}} \\[6pt]
&= \frac{1}{\ln 2}\frac{\left|\mathbf{u}_{k_0}^{\mathrm{H}}\mathbf{H}_{k_0}\mathbf{w}\right|^2}{\left|\mathbf{u}_{k_0}^{\mathrm{H}}\mathbf{H}_{k_0}\mathbf{w}\right|^2\sum\limits_{m=1}^{k_0-1} p_m + \sigma^2} - \frac{1}{\ln 2}\frac{\left|\mathbf{u}_{k_0-1}^{\mathrm{H}}\mathbf{H}_{k_0-1}\mathbf{w}\right|^2}{\left|\mathbf{u}_{k_0-1}^{\mathrm{H}}\mathbf{H}_{k_0-1}\mathbf{w}\right|^2\sum\limits_{m=1}^{k_0-1} p_m + \sigma^2} \\[6pt]
&= \frac{1}{\ln 2}\frac{\left(\left|\mathbf{u}_{k_0}^{\mathrm{H}}\mathbf{H}_{k_0}\mathbf{w}\right|^2 - \left|\mathbf{u}_{k_0-1}^{\mathrm{H}}\mathbf{H}_{k_0-1}\mathbf{w}\right|^2\right)\sigma^2}{\left(\left|\mathbf{u}_{k_0}^{\mathrm{H}}\mathbf{H}_{k_0}\mathbf{w}\right|^2\sum\limits_{m=1}^{k_0-1} p_m + \sigma^2\right)\left(\left|\mathbf{u}_{k_0-1}^{\mathrm{H}}\mathbf{H}_{k_0-1}\mathbf{w}\right|^2\sum\limits_{m=1}^{k_0-1} p_m + \sigma^2\right)}.
\end{aligned}
\tag{6.67}
$$

As we have assumed that

$$
\left|\mathbf{u}_1^{\mathrm{H}}\mathbf{H}_1\mathbf{w}\right|^2 \geq \left|\mathbf{u}_2^{\mathrm{H}}\mathbf{H}_2\mathbf{w}\right|^2 \geq \cdots \geq \left|\mathbf{u}_K^{\mathrm{H}}\mathbf{H}_K\mathbf{w}\right|^2,
\tag{6.68}
$$

thus

$$
\frac{dR_{\text{sum}}}{dp_{k_0}} \leq 0.
\tag{6.69}
$$

We can conclude that R_{sum} is decreasing with p_{k_0}. $\qquad\square$

Based on Lemma 6.3.2, we can find that the priority of power allocation in Problem (6.65) is $p_1 \succ p_2 \succ \cdots \succ p_K$, where $\succ$ denotes higher priority. In other words, the power allocated to the users with lower effective channel gains is only necessary to satisfy the minimum rate constraints, and all of the remaining power should be allocated to the user with the highest effective channel gain to maximize the ASR. We give the following theorem to illustrate this property and obtain the optimal power allocation.

Theorem 6.3.1. *The optimal solution in Problem (6.65) must satisfy $R_k = r_k$, $2 \le k \le K$, and the optimal power allocation is given by*

$$
\begin{cases}
p_K^\star = \dfrac{\eta_K}{\eta_K + 1}\left(P + \dfrac{\sigma^2}{\left|\mathbf{u}_K^H \mathbf{H}_K \mathbf{w}\right|^2}\right), \\[3mm]
p_{K-1}^\star = \dfrac{\eta_{K-1}}{\eta_{K-1} + 1}\left(P - p_K^\star + \dfrac{\sigma^2}{\left|\mathbf{u}_{K-1}^H \mathbf{H}_{K-1} \mathbf{w}\right|^2}\right), \\[3mm]
\vdots \\[3mm]
p_2^\star = \dfrac{\eta_2}{\eta_2 + 1}\left(P - \displaystyle\sum_{m=3}^{K} p_m^\star + \dfrac{\sigma^2}{\left|\mathbf{u}_2^H \mathbf{H}_2 \mathbf{w}\right|^2}\right), \\[3mm]
p_1^\star = P - \displaystyle\sum_{m=2}^{K} p_m^\star,
\end{cases}
\tag{6.70}
$$

where $\eta_k = 2^{r_k} - 1$.

Proof. Assume that the optimal power allocation of Problem (6.65) is $\{p_k^\star\}$, and the achievable rate of user k under optimal power allocation is $R_k^\star$. Also assume that there is one user whose achievable rate is lager than its minimum rate constraint, i.e., $R_{k_0}^\star > r_{k_0}$, where k_0 is ranging from 2 to K. Consider the parameter settings bellow,

$$
\begin{cases}
p_k = p_k^\star, \;\; k = 1, 2, \cdots, k_0 - 2, k_0 + 1, \cdots, K, \\
p_{k_0 - 1} = p_{k_0 - 1}^\star + \delta, \\
p_{k_0} = p_{k_0}^\star - \delta,
\end{cases}
\tag{6.71}
$$

where

$$
\delta = \dfrac{S + \left|\mathbf{u}_{k_0}^H \mathbf{H}_{k_0} \mathbf{w}\right|^2 p_{k_0}^\star - \sqrt{2^{r_{k_0}} S\left(S + \left|\mathbf{u}_{k_0}^H \mathbf{H}_{k_0} \mathbf{w}\right|^2 p_{k_0}^\star\right)}}{\left|\mathbf{u}_{k_0}^H \mathbf{H}_{k_0} \mathbf{w}\right|^2},
\tag{6.72}
$$

and

$$
S = \left|\mathbf{u}_{k_0}^H \mathbf{H}_{k_0} \mathbf{w}\right|^2 \sum_{m=1}^{k_0 - 1} p_m^\star + \sigma^2.
\tag{6.73}
$$

According to the assumption of $R_{k_0}^\star > r_{k_0}$, we have

$$
\begin{aligned}
& 1 + \dfrac{\left|\mathbf{u}_{k_0}^H \mathbf{H}_{k_0} \mathbf{w}\right|^2 p_{k_0}^\star}{\left|\mathbf{u}_{k_0}^H \mathbf{H}_{k_0} \mathbf{w}\right|^2 \sum\limits_{m=1}^{k_0 - 1} p_m^\star + \sigma^2} > 2^{r_{k_0}} \\
\Leftrightarrow & S + \left|\mathbf{u}_{k_0}^H \mathbf{H}_{k_0} \mathbf{w}\right|^2 p_{k_0}^\star > 2^{r_{k_0}} S \\
\Leftrightarrow & \left(S + \left|\mathbf{u}_{k_0}^H \mathbf{H}_{k_0} \mathbf{w}\right|^2 p_{k_0}^\star\right)^2 > 2^{r_{k_0}} S\left(S + \left|\mathbf{u}_{k_0}^H \mathbf{H}_{k_0} \mathbf{w}\right|^2 p_{k_0}^\star\right) \\
\Leftrightarrow & S + \left|\mathbf{u}_{k_0}^H \mathbf{H}_{k_0} \mathbf{w}\right|^2 p_{k_0}^\star > \sqrt{2^{r_{k_0}} S\left(S + \left|\mathbf{u}_{k_0}^H \mathbf{H}_{k_0} \mathbf{w}\right|^2 p_{k_0}^\star\right)} \\
\Leftrightarrow & \delta > 0.
\end{aligned}
\tag{6.74}
$$

Then, we calculate the achievable rates of the users. As we have $p_k = p_k^\star$, $k = 1, 2, \cdots, k_0 - 2, k_0 + 1, \cdots, K$, it is easy to verify that

$$R_k = R_k^\star \geq r_k, \ k = 1, 2, \cdots, k_0 - 2, k_0 + 1, \cdots, K. \tag{6.75}$$

According to $\delta > 0$, we have

$$
\begin{aligned}
R_{k_0-1} &= \log_2(1 + \frac{\left|\mathbf{u}_{k_0-1}^{\mathrm{H}}\mathbf{H}_{k_0-1}\mathbf{w}\right|^2 p_{k_0-1}}{\left|\mathbf{u}_{k_0-1}^{\mathrm{H}}\mathbf{H}_{k_0-1}\mathbf{w}\right|^2 \sum\limits_{m=1}^{k_0-2} p_m + \sigma^2}) \\
&= \log_2(1 + \frac{\left|\mathbf{u}_{k_0-1}^{\mathrm{H}}\mathbf{H}_{k_0-1}\mathbf{w}\right|^2 (p_{k_0-1}^\star + \delta)}{\left|\mathbf{u}_{k_0-1}^{\mathrm{H}}\mathbf{H}_{k_0-1}\mathbf{w}\right|^2 \sum\limits_{m=1}^{k_0-2} p_m^\star + \sigma^2}) \\
&> \log_2(1 + \frac{\left|\mathbf{u}_{k_0-1}^{\mathrm{H}}\mathbf{H}_{k_0-1}\mathbf{w}\right|^2 p_{k_0-1}^\star}{\left|\mathbf{u}_{k_0-1}^{\mathrm{H}}\mathbf{H}_{k_0-1}\mathbf{w}\right|^2 \sum\limits_{m=1}^{k_0-2} p_m^\star + \sigma^2}) \\
&= R_{k_0-1}^\star \geq r_{k_0-1},
\end{aligned}
\tag{6.76}
$$

and according to the expression of δ, we have

$$
\begin{aligned}
R_{k_0} &= \log_2(1 + \frac{\left|\mathbf{u}_{k_0}^{\mathrm{H}}\mathbf{H}_{k_0}\mathbf{w}\right|^2 p_{k_0}}{\left|\mathbf{u}_{k_0}^{\mathrm{H}}\mathbf{H}_{k_0}\mathbf{w}\right|^2 \sum\limits_{m=1}^{k_0-1} p_m + \sigma^2}) \\
&= \log_2(1 + \frac{\left|\mathbf{u}_{k_0}^{\mathrm{H}}\mathbf{H}_{k_0}\mathbf{w}\right|^2 (p_{k_0}^\star - \delta)}{\left|\mathbf{u}_{k_0}^{\mathrm{H}}\mathbf{H}_{k_0}\mathbf{w}\right|^2 \sum\limits_{m=1}^{k_0-1} p_m^\star + \sigma^2}) \\
&= \log_2(\frac{\sqrt{2^{r_{k_0}} S (S + \left|\mathbf{u}_{k_0}^{\mathrm{H}}\mathbf{H}_{k_0}\mathbf{w}\right|^2 p_{k_0}^\star)}}{S}) \\
&= \log_2 \sqrt{2^{r_{k_0}} (1 + \frac{\left|\mathbf{u}_{k_0}^{\mathrm{H}}\mathbf{H}_{k_0}\mathbf{w}\right|^2 p_{k_0}^\star}{\left|\mathbf{u}_{k_0}^{\mathrm{H}}\mathbf{H}_{k_0}\mathbf{w}\right|^2 \sum\limits_{m=1}^{k_0-1} p_m^\star + \sigma^2})} \\
&= \frac{R_{k_0}^\star + r_{k_0}}{2} > r_{k_0}.
\end{aligned}
\tag{6.77}
$$

Based on Lemma 6.3.2, when $p_k = p_k^\star$, $k = 1, 2, \cdots, k_0 - 2, k_0 + 1, \cdots, K$, R_{sum} is decreasing for p_{k_0}. Due to $p_{k_0} = p_{k_0}^\star - \delta < p_{k_0}^\star$, we have

$$R_{\mathrm{sum}} > R_{\mathrm{sum}}^\star, \tag{6.78}$$

which means that under the parameter settings of $\{p_k\}$, the minimum rate constraints for all the users are satisfied, and meanwhile the ASR increases. It is in contrast to the assumption that $\{p_k^\star\}$ is optimal. Thus, we have $R_k^\star = r_k$, $2 \leq k \leq K$.

Finally, solve the following equation set

$$\begin{cases} R_k = r_k, \ 2 \leq k \leq K, \\ \sum_{k=1}^{K} p_k = P. \end{cases} \tag{6.79}$$

We can obtain that the optimal power allocation of Problem (6.65) is given by (6.70).
□

Based on Theorem 6.3.1, we can find that although the users with lower effective channel gains are prior in the decoding order, the achievable rates of them have no gain compared with the minimum rate constraints. The power allocated to users 2-K is only necessary to satisfy the minimum rate constraint. The performance gain of NOMA depends mainly on the user with the highest effective channel gain, i.e., user 1.

6.3.2.2 Optimal Rx Beamforming Vectors with an Arbitrary Fixed Tx Beamforming Vector

In the first stage, we obtained the closed-form power allocation with arbitrary fixed beamforming vectors as shown in (6.70). In the second stage, we will handle the Rx beamforming. Given an arbitrary fixed Tx beamforming vector, Problem (6.60) is simplified as

$$\underset{\{p_k\},\{\mathbf{u}_k\}}{\text{Maximize}} \quad R_{\text{sum}} \tag{6.80a}$$

$$\text{Subject to} \quad C_1 \ : \ R_k \geq r_k, \quad \forall k, \tag{6.80b}$$

$$C_2 \ : \ p_k \geq 0, \quad \forall k, \tag{6.80c}$$

$$C_3 \ : \ \sum_{k=1}^{K} p_k \leq P, \tag{6.80d}$$

$$C_4 \ : \ |[\mathbf{u}_k]_m| = \frac{1}{\sqrt{M}}, \quad \forall k, m. \tag{6.80e}$$

To obtain the optimal Rx beamforming, we have the following theorem.

Theorem 6.3.2. *The optimal solution of the Rx beamforming vectors in Problem* (6.80) *is*

$$[\mathbf{u}_k^\star]_m = \frac{1}{\sqrt{M}} \frac{[\mathbf{H}_k \mathbf{w}]_m}{|[\mathbf{H}_k \mathbf{w}]_m|}, \quad \forall k, m. \tag{6.81}$$

Proof. As the Tx beamforming vector is fixed, $\mathbf{H}_k \mathbf{w}$, $1 \leq k \leq K$, are all constant vectors. Given an arbitrary decoding order of $\pi_K, \pi_{K-1}, \cdots, \pi_1$, we introduce intermediate variables b_{π_k}, where

$$b_{\pi_k} = \left| \mathbf{u}_{\pi_k}^{\text{H}} \mathbf{H}_{\pi_k} \mathbf{w} \right|^2, 1 \leq k \leq K. \tag{6.82}$$

Thus, the partial derivative of the achievable rate is

$$\frac{\partial R_{\pi_s}}{\partial b_{\pi_k}} \mid \{\pi_s = \pi_k\} = \frac{\partial \, \log_2 \left(1 + \frac{b_{\pi_k} p_{\pi_k}}{b_{\pi_k} \sum\limits_{m=1}^{k-1} p_{\pi_m} + \sigma^2} \right)}{\partial b_{\pi_k}}$$

$$= \frac{1}{\ln 2} \frac{p_{\pi_k} \sigma^2}{\left(b_{\pi_k} \sum\limits_{m=1}^{k-1} p_{\pi_m} + \sigma^2 \right) \left(b_{\pi_k} \sum\limits_{m=1}^{k} p_{\pi_m} + \sigma^2 \right)} \geq 0, \qquad (6.83)$$

$$\frac{\partial R_{\pi_s}}{\partial b_{\pi_k}} \mid \{\pi_s \neq \pi_k\} = 0. \qquad (6.84)$$

The achievable rate of user π_k is increasing with b_{π_k}, while the achievable rates of the other users are independent of b_{π_k}. Thus, to maximize the ASR, we can always adjust the Rx beamforming vector for each user to maximize b_{π_k}, $1 \leq k \leq K$. For user π_k, as $\mathbf{H}_{\pi_k} \mathbf{w}$ is a constant vector, we just need to let the phase of each element of $[\mathbf{u}_{\pi_k}]$ be the same as the phase of the corresponding element of $[\mathbf{H}_{\pi_k} \mathbf{w}]$, which is not influenced by the decoding order. Thus, under any decoding orders, the optimal solution of the Rx beamforming vectors is always given by (6.81). □

Based on Theorem 6.3.1 and Theorem 6.3.2, we can further obtain the ASR of the K users, which is given by

$$R_{\text{sum}} \triangleq R(\mathbf{w}) = \sum_{k=2}^{K} r_k + \log_2 \left(1 + \frac{|\mathbf{u}_1^{\star \mathrm{H}} \mathbf{H}_1 \mathbf{w}|^2 p_1^{\star}}{\sigma^2} \right), \qquad (6.85)$$

where $p_1^{\star}$ and $\mathbf{u}_1^{\star}$ are both functions of $\mathbf{w}$, whose definitions are given by (6.70) and (6.81), respectively. From (6.85), we can find that the value of R_{sum} is only determined by the Tx beamforming vector. Next, we will give the approach of Tx beamforming design in the third stage.

6.3.2.3 *Design of Tx Beamforming Vector with BC-PSO*

According to Theorem 6.3.1 and Theorem 6.3.2, Problem (6.60) can be transformed into a Tx beamforming problem, i.e.,

$$\underset{\mathbf{w}}{\text{Maximize}} \quad R(\mathbf{w}) \qquad (6.86\text{a})$$

$$\text{Subject to} \quad |[\mathbf{w}]_n| = \frac{1}{\sqrt{N}}, \quad 1 \leq n \leq N, \qquad (6.86\text{b})$$

where $R(\mathbf{w})$ is the ASR of K users shown in (6.85). Although the explicit expression of $R(\mathbf{w})$ can be obtained according to (6.70), (6.81) and (6.85), the highly non-convex formulation makes it complicated to solve Problem (6.86) directly. In addition, the dimension of the Tx beamforming vector, i.e., N, is large in general, so it is computationally prohibitive to directly search the optimal solution.

To solve this difficult problem, PSO is an alternative approach[32,33,34]. First, we give the basics of PSO.

6.3.2.4 Basics of PSO

In the N-dimensional search space $\mathcal{S}$, the I particles in the swarm are randomly initialized with position $\mathbf{x}$ and velocity $\mathbf{v}$. Each particle has a memory for its best found position $\mathbf{p}_{\text{best}}$ and the globally best position $\mathbf{g}_{\text{best}}$, where the goodness of a position is evaluated by the fitness function. For each iteration, the velocity and position of each particle are updated based on

$$[\mathbf{v}]_n = \omega[\mathbf{v}]_n + c_1 \times \text{rand} \times ([\mathbf{p}_{\text{best}}]_n - [\mathbf{x}]_n) + c_2 \times \text{rand} \times ([\mathbf{g}_{\text{best}}]_n - [\mathbf{x}]_n),$$
$$[\mathbf{x}]_n = [\mathbf{x}]_n + [\mathbf{v}]_n, \tag{6.87}$$

for $n = 1, 2, \cdots, N$. The parameter ω is the inertia weight of velocity. In general, ω is linearly decreasing to improve the convergence speed. The parameters c_1 and c_2 are the cognitive ratio and social ratio, respectively. The random number function rand returns to a number between 0.0 and 1.0 with uniform distribution. After calculating the fitness function for each particle, the locally and globally best positions, i.e., $\mathbf{p}_{\text{best}}$ and $\mathbf{g}_{\text{best}}$, are updated. In such a manner, the particles diffuse around the search space and may find the globally optimal solution.

However, the CM constraint in Problem (6.86) makes the search space highly non-convex. The particles may converge to a locally optimal solution with a high probability. Thus, directly using PSO in Problem (6.86) may not obtain a promising performance. To this end, we propose a modified approach, i.e., BC-PSO. In the proposed approach, the feasible region is relaxed to a convex set, i.e., $|[\mathbf{w}]_n| \leq \frac{1}{\sqrt{N}}$. The boundary-compressed approach is proposed to guarantee that the particles satisfy the CM constraint. The details of the BC-PSO algorithm is shown bellow.

6.3.2.5 Implementation of BC-PSO

Define the search space of Problem (6.86) as

$$\mathcal{S} = \{\mathbf{w} \,|\, |[\mathbf{w}]_n| \leq \frac{1}{\sqrt{N}}, 1 \leq n \leq N\}, \tag{6.88}$$

which has two boundaries. The outer boundary is defined as

$$\{|[\mathbf{w}]_n| = \frac{1}{\sqrt{N}}, 1 \leq n \leq N\}, \tag{6.89}$$

while the inner boundary is

$$\{|[\mathbf{w}]_n| = d_t, 1 \leq n \leq N\}. \tag{6.90}$$

d_t is a dynamic parameter, which is linear to the number of iterations. The initial value of d_t is 0, and it increases linearly for each iteration until $d_t = \frac{1}{\sqrt{N}}$. For each iteration, if the particle moves across the outer/inner boundary, then it is adjusted onto the boundary. With this implementation, the particles can move throughout the relaxed search space and converges to the outer boundary eventually.

On the other hand, the definitions of the fitness function for different particles are different. The reason is that the order of effective channel gains may change when the particles move, which results in the change of the ASR's expression. Thus, when implementing the BC-PSO algorithm here, we should reorder the effective channel gains first in each iteration, and then obtain the fitness function, i.e., $R(\mathbf{w})$, according to (6.85).

In summary, we give Algorithm 6.2 to solve Problem (6.60). Hereto, we solve the original problem. In the proposed solution, the power allocation and Rx beamforming are optimal, while the Tx beamforming is sub-optimal.

6.3.2.6 *Computational Complexity*

As we obtained the closed-form optimal power allocation and Rx beamforming vectors with an arbitrary fixed Tx beamforming vector, the computational complexity is mainly caused by Tx beamforming design in the third stage. In Algorithm 6.2, the total computational complexity is $\mathcal{O}(N)$, which linearly increases with N and does not increase with M or K. In contrast, if the direct search method is adopted, and the number of the candidate values for each variable in Problem (6.60) is G, the complexity of directly searching the globally optimal solution is $\mathcal{O}(G^{N+MK+K})$, which exponentially increases with N, M and K.

6.3.3 Performance Simulations

In this subsection, we provide the simulation results to verify the performance of the proposed joint Tx-Rx beamforming and power allocation approach in the NOMA system. We adopt the channel model shown in (6.54), where the users are uniformly distributed from 10 m to 500 m away from the BS, and the channel gain of the node 100 m away from the BS has an average power of 0 dB to noise power. The number of MPCs for all the users is $L = 4$. Both LoS and NLoS channel models are considered. For the LoS channel, the average power of the NLoS paths is 15 dB weaker than that of the LoS path. For the NLoS channel, the coefficient of each path has an average power of $1/\sqrt{L}$. For each channel realization in the simulations, the channel gains of the users are sorted by

$$\|\mathbf{H}_1\|_{\mathrm{F}} \geq \|\mathbf{H}_2\|_{\mathrm{F}} \geq \cdots \geq \|\mathbf{H}_K\|_{\mathrm{F}}. \tag{6.91}$$

The corresponding parameter settings in Algorithm 6.2 are $I = 800, T = 50, c_1 = c_2 = 1.4, \omega_{\max} = 0.9, \omega_{\min} = 0.4$.

First, we compare the performance between the considered NOMA system and an OMA system. The achievable rate of user k in an OMA system is

$$R_k^{\mathrm{OMA}} = \frac{1}{K} \log_2 \left(1 + \frac{|\mathbf{u}_k^{\star\mathrm{H}} \mathbf{H}_k \mathbf{w}^\star|^2 P}{\sigma^2} \right), \tag{6.92}$$

where the factor $1/K$ is due to the multiplexing loss in OMA. $\mathbf{u}_k^\star$ and $\mathbf{w}^\star$ are the Rx beamforming vector and Tx beamforming vector given in Algorithm 6.2, respectively.

Algorithm 6.2: Implementation of BC-PSO

Input: Number of antennas: M and N;
      Number of particle swarm: I;
      Maximum number of iterations: T;
      Scaling factors: c_1 and c_2;
      Range of inertia weight: $\omega_{\max}$ and $\omega_{\min}$.

Output: $p_k^\star$, $\mathbf{u}_k^\star$ and $\mathbf{w}^\star$

1: Initialize the position $\mathbf{x}_i = \mathbf{w}_i$ and velocity $\mathbf{v}_i$.
2: Find the globally best solution position $\mathbf{g}_{\text{best}}$.
3: **for** $t = 1 : T$ **do**
4:    $\omega = \omega_{\max} - \frac{t}{T}(\omega_{\max} - \omega_{\min})$.
5:    $d_t = \frac{t}{T\sqrt{N}}$.
6:    **for** $i = 1 : I$ **do**
7:      **for** $n = 1 : N$ **do**
8:        Update $[\mathbf{v}_i]_n$ and $[\mathbf{x}_i]_n$ based on (6.87).
9:        **if** $|[\mathbf{x}_i]_n| < d_t$ **then**
10:          $[\mathbf{x}_i]_n = \frac{d_t[\mathbf{x}_i]_n}{|[\mathbf{x}_i]_n|}$.
11:        **end if**
12:        **if** $|[\mathbf{x}_i]_n| > \frac{1}{\sqrt{N}}$ **then**
13:          $[\mathbf{x}_i]_n = \frac{[\mathbf{x}_i]_n}{\sqrt{N}|[\mathbf{x}_i]_n|}$.
14:        **end if**
15:        **if** $|[\mathbf{p}_{\text{best},i}]_n| < d_t$ **then**
16:          $[\mathbf{p}_{\text{best},i}]_n = \frac{d_t[\mathbf{p}_{\text{best},i}]_n}{|[\mathbf{p}_{\text{best},i}]_n|}$.
17:        **end if**
18:        Obtain the optimal Rx beamforming vectors $\mathbf{u}_k^\star$ according to (6.81).
19:        Reorder the effective channel gains of the users.
20:        Obtain the optimal power allocation $p_k^\star$ according to (6.70).
21:        Obtain the fitness function $R(\mathbf{w})$ according to (6.85).
22:      **end for**
23:      Update $\mathbf{p}_{\text{best},i}$.
24:    **end for**
25:    Update $\mathbf{g}_{\text{best}}$.
26: **end for**
27: $\mathbf{w}^\star = \mathbf{g}_{\text{best}}$.
28: **return** $p_k^\star$, $\mathbf{u}_k^\star$ and $\mathbf{w}^\star$.

Fig. 6.10 compares the ASRs between the proposed NOMA algorithm, the NOMA approach in[29] and OMA with varying total power to noise ratio. Each point in Fig. 6.10 is the average performance of 10^3 LoS channel realizations. Significantly, the performance of the proposed NOMA system is distinctly better than that of the OMA system, as well as better than that of the solution in[29]. Particularly when P/σ^2 is low, the superiority of the presented algorithm is more conspicuous compared with the

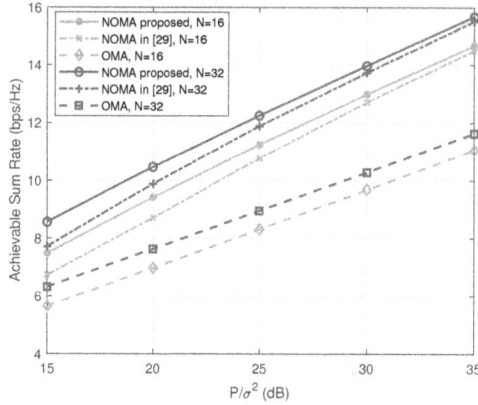

Figure 6.10 Comparison of the ASRs between the NOMA and OMA systems with varying total power to noise ratio, where $M = 1$, $N = 16, 32$, $K = 2$ and $r_k = 1.5$ bps/Hz.

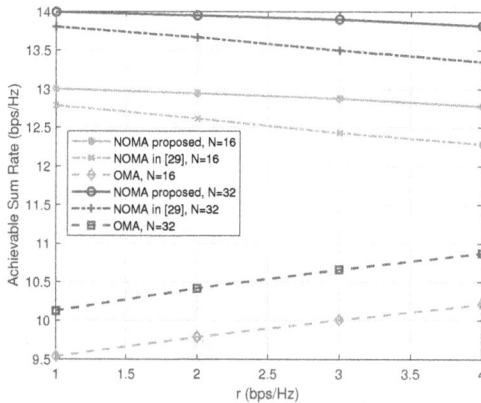

Figure 6.11 Comparison of the ASRs between the NOMA and OMA systems with varying minimum rate constraint, where $M = 1$, $N = 16, 32$, $K = 2$, $r_k = r$ and $P/\sigma^2 = 30$ dB.

approach in[29]. The reason is that given a designed beamforming vector, the solutions of power allocation in this section and[29] are both optimal. Thus, the performance gap is mainly caused by the beamforming design. Significantly, the proposed algorithm can always find a better beamforming vector than that of the approach in[29]. As shown in (6.58), the achievable rate is determined by the product of the effective channel gain and the transmit power. When the total transmit power becomes lower, the effective channel gain becomes the main portion to determine the ASR, so the superiority of the proposed NOMA algorithm is relatively conspicuous.

Fig. 6.11 shows the comparison result of the ASRs between the proposed NOMA algorithm, the mmNOMA approach in[29] and OMA with varying minimum rate constraint. Each point in Fig. 6.11 is the average performance of 10^3 LoS channel realizations. Similar to the result in Fig. 6.10, we can find that the proposed NOMA algorithm can achieve a higher ASR than that of NOMA in[29], as well as higher than the ASR of the OMA system. Particularly when r increases, the superiority of the proposed algorithm is more conspicuous compared with the approach in[29]. The results indicate that the presented beamforming design is better than that of the approach in[29], especially when the minimum rate constraint is large.

We show the power allocation and the effective channel gains in Figs. 6.12 (left) and 6.12 (right). Each point in Figs. 6.12 (left) and 6.12 (right) is an average result of 10^3 LoS channel realizations. From the two figures, we can find that the effective channel gain of user 1, the user with the highest channel gain, is distinctly larger than that of the other users. The user with a better channel gain has a higher effective channel gain with the proposed solution. In Fig. 6.12 (left), the effective channel gains of user 1 and user 3 go increasing and decreasing, respectively, when P/σ^2 becomes higher. It indicates that when the total transmit power is high, power and beam gains should be allocated jointly to enlarge the difference of the effective channel gains to obtain a higher ASR. In contrast, the power allocation and the effective channel gain of user 1 go decreasing while the power allocation and the effective channel gain of user 3 go increasing, when r becomes higher in Fig. 6.12 (right). It indicates that more power and beam gain should be allocated to the users with worse channel gains to satisfy the constraint, when the minimum rate constraint is high.

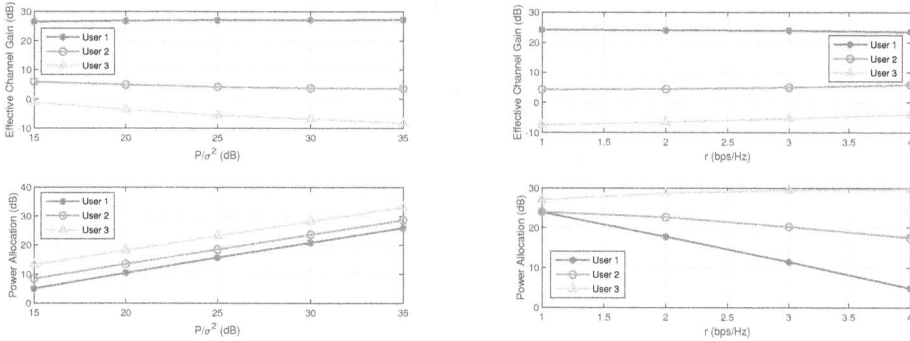

Figure 6.12 **Left**: The values of effective channel gains and power allocation for different users with varying total power to noise ratio, where $M = 4$, $N = 16$, $K = 3$ and $r_k = 1.5$ bps/Hz. **Right**: The values of effective channel gains and power allocation for different users with varying minimum rate constraint, where $M = 4$, $N = 16$, $K = 3$, $r_k = r$ and $P/\sigma^2 = 30$ dB.

Fig. 6.13 compares the ASRs between NOMA and OMA systems with varying number of users. For fairness, each user has an average transmit power to noise ratio of 30 dB. Each point in Fig. 6.13 is the average performance of 10^3 LoS channel realizations. It can be observed again that the NOMA can outperform the OMA, especially when N is large. It can be seen that the ASR of the NOMA users improves as the number of users increases, while the ASR of the OMA users is always around a

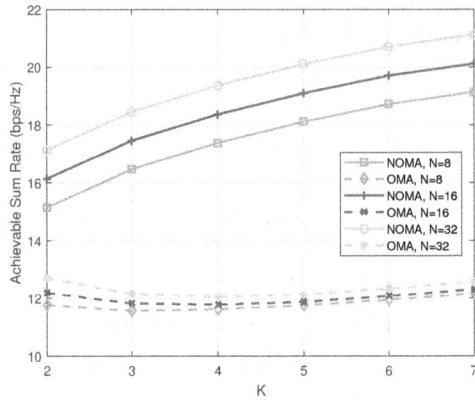

Figure 6.13 Comparison of the ASRs between the NOMA and OMA systems with varying number of the users, where $M = 4$, $N = 8, 16, 32$, $r_k = 1.5$ bps/Hz and $\frac{P}{K\sigma^2} = 30$ dB.

low value without obvious improvement. The results prove that NOMA can achieve a higher spectrum efficiency compared with OMA when the number of users increases.

Figs. 6.14 (left) and 6.14 (right) compare the ASRs of NOMA system between the LoS and NLoS channel models with varying total power to noise ratio and with varying minimum rate constraint, respectively. Each point in this two figures is the average performance of 10^3 channel realizations. It can be seen that the performance with the LoS channel model is distinctly better than that with the NLoS channel model, because the beam gain is more centralized for the LoS channel. Particularly, when P/σ^2 is small and r is large, the performance gap between the LoS channel model and NLoS channel model is larger.

In the third stage of the solution, we proposed the BC-PSO algorithm and obtained a sub-optimal solution. The convergence of the proposed algorithm is evaluated in Fig. 6.15. When $N = 8$, 16, 32, 64, the curve of the ASR tends to be stable after 7, 15, 30, 60 iterations, respectively. We can find that the number of iterations that the algorithm converges is roughly linear to N, which indicates that the proposed BC-PSO algorithm has a linear convergence rate against the number of antennas at the BS.

To evaluate the stability of the proposed approach, we compare the performance of the proposed BC-PSO and the classical PSO in Fig. 6.16, where PSO is corresponding to directly solving Problem (6.86) in the search space of

$$\mathcal{S} = \{\mathbf{w} \mid |[\mathbf{w}]_n| = \frac{1}{\sqrt{N}}, 1 \leq n \leq N\}, \tag{6.93}$$

while BC-PSO is corresponding to the proposed approach in Algorithm 6.2. With the same one channel realization, we solve Problem (6.86) with the PSO algorithm and the BC-PSO algorithm for 1000 times with different initializations. It can be seen that the ASRs with BC-PSO are distinctly higher than that with PSO. The curves for

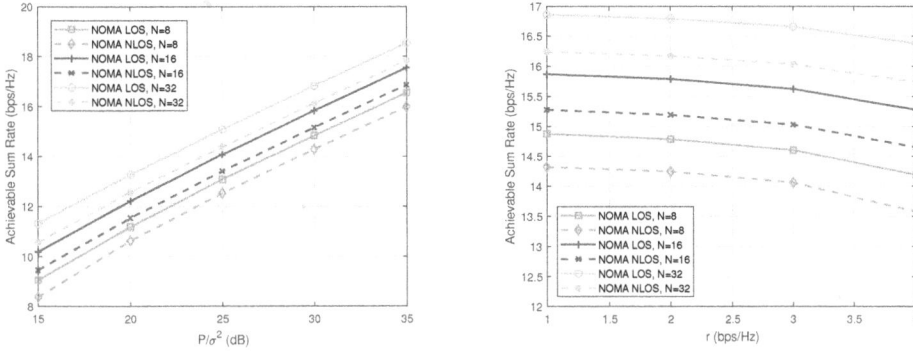

Figure 6.14 **Left**: Comparison of the ASRs between the LoS and NLoS channel models with varying total power to noise ratio, where $M = 4$, $N = 8, 16, 32$, $K = 3$ and $r_k = 1.5$ bps/Hz. **Right**: Comparison of the ASRs between the LoS and NLoS channel models with varying minimum rate constraint, where $M = 4$, $N = 8, 16, 32$, $K = 3$, $r_k = r$ and $P/\sigma^2 = 30$ dB.

BC-PSO are stable after 20 iterations, while the curves for PSO are stable within 5 iterations. The results indicate that with the acceptable expense of the computational complexity, the proposed BC-PSO algorithm can achieve a better search capability compared with the conventional PSO algorithm. The reason is that the search space for PSO is highly non-convex, the particles may converge to a sub-optimal solution fast. In contrast, the proposed BC-PSO algorithm has a convex search space. The particles move around the relaxed space and obtain more information of the solutions. Consequently, a better solution can be found in the BC-PSO algorithm. In addition, the curves of the maximal ASR, minimal ASR and mean ASR for BC-PSO are close. However, there are obvious performance differences among the PSO curves, which indicates that the proposed BC-PSO algorithm has a better convergence stability compared with the conventional PSO algorithm.

6.4 NOMA WITH HYBRID BEAMFORMING

Millimeter-wave-NOMA with hybrid beamforming was investigated in several literatures. In[14], a new transmission scheme of beamspace MIMO-NOMA was proposed, where the number of users can be larger than the number of RF chains. Based on the equivalent-channel hybrid precoding scheme, an iterative algorithm was developed to obtain the optimal power allocation for the users. In[35], a user grouping algorithm and a hybrid beamforming algorithm were proposed for millimeter-wave-MIMO-NOMA system with simultaneous wireless information and power transfer. Then, the optimization for power allocation and power splitting factors was operated to maximize the ASR. The optimal power allocation and user scheduling were obtained with the branch and bound approach in[13], where hybrid beamforming is random and fixed. In[15], the authors considered the problems of user pairing, hybrid beamforming and power allocation separately in a millimeter-wave-NOMA system.

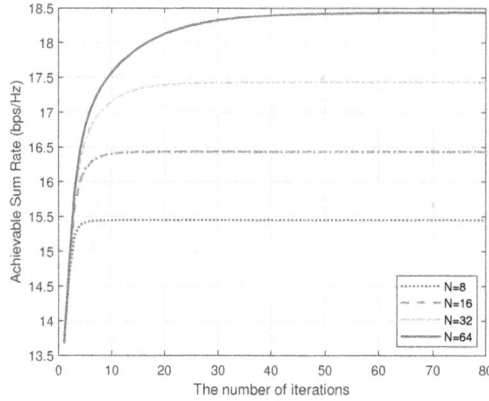

Figure 6.15 Iterations required for convergence in the BC-PSO algorithm, where $M = 4$, $N = 8, 16, 32, 64$, $K = 4$, $r_k = 1$ bps/Hz and $P/\sigma^2 = 30$ dB.

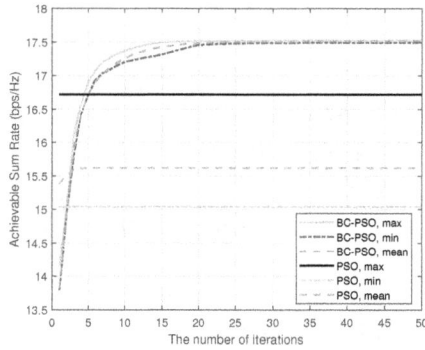

Figure 6.16 Comparison of the performance of PSO and BC-PSO, where $M = 4$, $N = 32$, $K = 4$, $r_k = 1$ bps/Hz and $P/\sigma^2 = 30$ dB.

In[36], a capacity analysis for the integrated NOMA-millimeter-wave-massive-MIMO systems was provided based on a simplified millimeter-wave channel model. In[37], a multi-beam NOMA framework for hybrid millimeter-wave systems was proposed, where a beam splitting technique was introduced to generate multiple analog beams to facilitate the NOMA transmission.

Different from the works above, we consider user grouping and jointly optimize hybrid beamforming and power allocation. The rest of the section is organized as follows. In Section 6.4.1, we present the system model. In Section 6.4.2, we first propose the user grouping algorithm and formulate the problem. Then, we provide a solution of power allocation with an arbitrary fixed hybrid beamforming in Section 6.4.3. In Section 6.4.4, we design digital beamforming and analog beamforming. In Section 6.4.5, we summarize the complete solution and provide the computational complexity.

Simulation results are given to demonstrate the performance of the proposed solution in Section 6.4.6.

6.4.1 System Model

6.4.1.1 *System Model*

In this section, we consider a single-cell downlink millimeter-wave-NOMA system. The BS is equipped with hybrid beamforming structure, where N antennas share N_{RF} RF chains. K single-antenna users are served simultaneously, where $K > N_{RF}$. The architecture of the BS is shown in Fig. 6.17, which is a fully connected hybrid beamforming structure [7]. N_S data streams in the baseband are precoded by the digital beamforming matrix $\mathbf{D}$ of size $M \times N_S$. After passing through the corresponding RF chain, the digital-domain signal from each RF chain is delivered to N PSs to perform analog beamforming. Thus, the analog beamforming matrix is $\mathbf{A}$ of size $N \times M$.

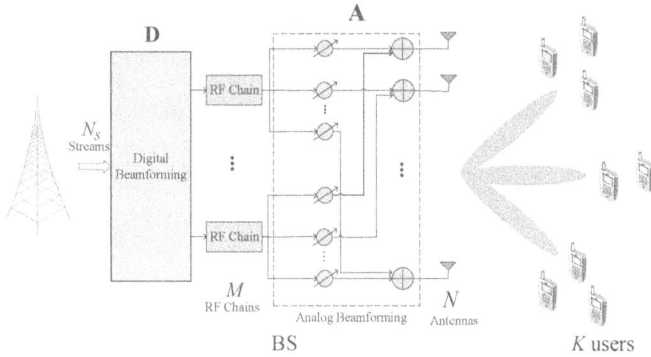

Figure 6.17 Illustration of the architecture of the BS, which is equipped with M RF chains and N antennas.

In order to achieve a high multiplexing gain, the number of data streams is assumed to be equal to the number of RF chains, i.e., $N_S = N_{RF}$. Thus, the K users should be first scheduled into $M = N_{RF}$ groups, and each group corresponds to an independent data stream. The users in the same group can perform NOMA and implement successive interference cancellation (SIC), while the signals from different groups of users are treated as interference. The details of user grouping will be shown later. Denote the user set of the m-th group as $\mathcal{G}_m$. As a result, we have $\mathcal{G}_i \cap \mathcal{G}_j = \Phi$ for $i \neq j$ and $\sum_{m=1}^{M} |\mathcal{G}_m| = K$, where $|\mathcal{G}_m|$ denotes the number of users in $\mathcal{G}_m$. Since N_{RF} RF chains can support N_{RF} data streams at most, there should be at least one user in each group to avoid the idleness of the RF resource, and thus we have $|\mathcal{G}_m| \geq 1$. Then, the received signal for the n-th user in the m-th group is

$$y_{m,n} = \mathbf{h}_{m,n}^{\mathrm{H}} \mathbf{ADPs} + u_{m,n}, \tag{6.94}$$

---

[7]It is worthy of noting that the proposed approach in this section can also be directly used for the partially connected hybrid beamforming structure[35].

where $\mathbf{h}_{m,n}$ of size $N \times 1$ is the channel response vector between the BS and the n-th user in the m-th group. $u_{m,n}$ is the Gaussian white noise at the user with average power σ^2. $\mathbf{s}$ of size $K \times 1$ is the vector of the transmission signals, where $\mathbf{s} = [s_{1,1}, \cdots, s_{1,|\mathcal{G}_1|}, \cdots, s_{M,1}, \cdots, s_{M,|\mathcal{G}_M|}]^{\mathrm{T}}$ and $E(\mathbf{ss}^{\mathrm{T}}) = \mathbf{I}_K$, and $\mathbf{P}$ is the $M \times K$ power allocation matrix: $\mathbf{P} = \mathrm{diag}\{\mathbf{p}_1, \mathbf{p}_2, \cdots, \mathbf{p}_M\}$ and $\mathbf{p}_m = [\sqrt{p_{m,1}}, \sqrt{p_{m,2}}, \cdots, \sqrt{p_{m,|\mathcal{G}_m|}}]$. $\mathbf{D}$ is the digital beamforming matrix. $\mathbf{A}$ is the analog beamforming matrix with the CM constraint of[29,30,38]

$$|[\mathbf{A}]_{i,j}| = \frac{1}{\sqrt{N}}, \ 1 \leq i \leq N, \ 1 \leq j \leq M. \tag{6.95}$$

We define the hybrid beamforming matrix as

$$\mathbf{W} = \mathbf{AD} = [\mathbf{w}_1, \mathbf{w}_2, \cdots, \mathbf{w}_M]. \tag{6.96}$$

Since we separate the transmit power from hybrid beamforming, it is without loss of generality to assume that each column of the hybrid beamforming matrix has a unit norm, i.e.,

$$\|\mathbf{w}_m\|_2 = 1, \ 1 \leq m \leq M. \tag{6.97}$$

Subject to limited scattering in the millimeter-wave band, multipath is mainly caused by reflection. As the number of the MPCs is small in general, the millimeter-wave channel has directionality and appears spatially sparse in the angle domain[17,21,22,23,24,25]. Different MPCs have different AoDs and AoAs. Without loss of generality, similar to before, we adopt the directional channel model assuming a ULA with a half-wavelength antenna spacing. For the $N \times 1$ channel response vector $\mathbf{h}_{m,n}$, we adopt the widely used Saleh-Valenzuela channel for millimeter-wave communications[14,31,35], which is [8]

$$\mathbf{h}_{m,n} = \sum_{\ell=1}^{L_{m,n}} \lambda_{m,n}^{(\ell)} \mathbf{a}_t(N, \theta_{m,n}^{(\ell)}). \tag{6.98}$$

Note that for convenience, we denote the channel coefficients in terms of both indices m and n in (6.98), where index m, $1 \leq m \leq M$, represents the m-th group, and the index n, $1 \leq n \leq |\mathcal{G}_m|$, represents the n-th user in each group. $\lambda_{m,n}^{(\ell)}$ is the complex coefficient of the ℓ-th MPC of the channel response vector for the n-th user in the m-th group. $\theta_{m,n}^{(\ell)}$, within the range $(-1, 1]$, is the cosine of the AoD[42]. $L_{m,n}$ is the total number of the MPCs. $\mathbf{a}_t(\cdot)$ is the steering vector functions defined as

$$\mathbf{a}_t(N, \theta) = [e^{j2\pi 0(d/\lambda)\theta}, e^{j2\pi(d/\lambda)\theta}, \cdots, e^{j2\pi(N-1)(d/\lambda)\theta}]^{\mathrm{T}}, \tag{6.99}$$

which depends on the array geometry. d is the antenna spacing, and λ is the signal wavelength. For a half-wavelength antenna spacing array, we have $d = \lambda/2$.

---

[8]Since we concentrate on the user grouping and resource allocation for NOMA, the channel estimation problem is not considered here. We assume that the CSI between the BS and the users is known by the BS. A number of approaches on millimeter-wave channel estimation have been proposed and could be referred, such as,[17,39,40,41].

6.4.1.2 Achievable Rate

In general, as we have seen previously, the optimal decoding order for NOMA is the increasing order of the users' channel gains[3,26]. However, for the NOMA with hybrid beamforming structure, the effective channel gains of the users are determined by both the channel gains and the beamforming gains, and the interference between different groups of users may also impact the optimal decoding order. Thus, we need to sort the effective channel gains first, and then determine the decoding order. For notational simplicity and without loss of generality, we assume that the order of the effective channel gains in the m-th group is [9]

$$|\mathbf{h}_{m,1}^{\mathrm{H}}\mathbf{w}_m|^2 \geq |\mathbf{h}_{m,2}^{\mathrm{H}}\mathbf{w}_m|^2 \geq \cdots \geq |\mathbf{h}_{m,|\mathcal{G}_m|}^{\mathrm{H}}\mathbf{w}_m|^2, \tag{6.100}$$

and thus the optimal decoding order is the increasing order of the effective channel gains[14,29,31]. Therefore, the n-th user in the m-th group can decode $s_{m,j}$, $n+1 \leq j \leq |\mathcal{G}_m|$, and then remove them from the received signal in a successive manner. The other signals are treated as interference. Thus, the SINR of the n-th user in the m-th group can be written as

$$\gamma_{m,n} = \frac{|\mathbf{h}_{m,n}^{\mathrm{H}}\mathbf{w}_m|^2 p_{m,n}}{|\mathbf{h}_{m,n}^{\mathrm{H}}\mathbf{w}_m|^2 \sum_{j=1}^{n-1} p_{m,j} + \sum_{i \neq m}\sum_{k=1}^{|\mathcal{G}_i|} |\mathbf{h}_{m,n}^{\mathrm{H}}\mathbf{w}_i|^2 p_{i,k} + \sigma^2}. \tag{6.101}$$

Note that Gaussian signalling is assumed for transmitting data here. As a result, the achievable rate of the n-th user in the m-th group is

$$R_{m,n} = \log_2(1 + \gamma_{m,n}). \tag{6.102}$$

Finally, the ASR of the proposed NOMA system is

$$R_{\mathrm{sum}} = \sum_{m=1}^{M}\sum_{n=1}^{|\mathcal{G}_m|} R_{m,n}. \tag{6.103}$$

Note that in the proposed downlink NOMA system, we assume that the CSI between the BS and the users is known by the BS, and thus user grouping, power allocation and beamforming can be accomplished at the BS. The channel-gain information and beamforming-gain information of the other users are not required at the user side. However, compared with the conventional OMA system, information about the decoding order and codebook of the prior users in the same group should be transmitted to each user to accomplish SIC, which results in extra overhead. The amount of overhead depends on the number of users within the same NOMA group. In the proposed solution of this section, a great number of users are divided into many NOMA groups, and the number of users within the same NOMA group is usually not large so as to maintain the performance. Hence, the extra overhead is, in fact, not high, especially in slow varying channel, where the decoding order and codebook are also slow varying, the overhead can be further reduced.

---

[9]We can always define the user with the n-th highest effective channel gain in the m-th group as the n-th user in this group. Thus, this simplified subscript has no influence on the solution in this section.

6.4.2 User Grouping and Problem Formulation

As the number of the users is larger than that of the RF chains, i.e., $K > M$, we need to schedule the user into M groups. To this end, we propose an intuitive algorithm for user grouping first, and then formulate a problem to jointly optimize hybrid beamforming and power allocation.

6.4.2.1 User Grouping

Due to the spatial directivity of the millimeter-wave channel, the users whose channels are highly correlated should be assigned to the same group to make full use of the multiplexing gain, while the users whose channels are uncorrelated should be assigned to different groups to decrease the interference. The normalized channel correlation between user i and user j is defined as

$$C_{i,j} = \frac{\mathbf{h}_i^{\mathrm{H}}\mathbf{h}_j}{\|\mathbf{h}_i\|_2\|\mathbf{h}_j\|_2}. \tag{6.104}$$

We use the K-means clustering algorithm to implement the user grouping, where the normalized channel correlation is defined as the measure[43]. First, we select M users randomly, denoted by $\{\Omega_1, \Omega_2, \cdots, \Omega_M\}$, as the representatives of the M clusters. Then, the other users can be assigned to the cluster according to the normalized channel correlation. For instance, user k should be assigned to the $m^\star$-th cluster, where

$$m^\star = \arg\max_{1\leq m\leq M} C_{k,\Omega_m}. \tag{6.105}$$

After that, the representative of each cluster should be updated. To further decrease the correlation of the channels between different clusters, the representative of each cluster is updated as the one with the lowest correlation with the other clusters. The correlation between a user to the other clusters is defined as the summation of the normalized channel correlation between this user to the users of the other clusters, i.e.,

$$\bar{C}_k = \sum_{1\leq j\leq K}^{j\notin \mathcal{G}^{(k)}} C_{k,j}, \tag{6.106}$$

where $\mathcal{G}^{(k)}$ denotes the cluster which includes user k, and the representative of the m-th cluster is updated as

$$\Omega_m = \arg\min_{1\leq n\leq|\mathcal{G}_m|} \bar{C}_n, \tag{6.107}$$

where $\mathcal{G}_m$ denotes the m-th cluster. After updating the representative of each cluster, the other users are reassigned to the clusters according to (6.105). The iteration is stopped if the representatives of the clusters are unchanged. The details of the proposed user grouping algorithm are summarized in Algorithm 6.3.

6.4.2.2 Problem Formulation

Generally, there are mainly two categories of optimizing the overall rate performance in a communication system. One is to maximize the ASR. However, when maximizing

---

Algorithm 6.3: User Grouping Algorithm

Input: K, M, $\{\mathbf{h}_k\}$, and $\{C_{i,j}\}$.

Output: The user grouping scheme: $\{\mathcal{G}_1, \mathcal{G}_2, \cdots, \mathcal{G}_M\}$.

1: $\mathcal{K} = \{1, 2, \cdots, K\}$.

2: Initialize $\Omega_m^{(1)} = k_m \in \mathcal{K}$ randomly for $m = 1, 2, \cdots, M$.

3: $t = 1$.

4: **while** $\{\Omega_m^{(t)}\} \neq \{\Omega_m^{(t-1)}\}$ **do**

5:    Initialize $\mathcal{G}_m = \Omega_m^{(t)}$ for $m = 1, 2, \cdots, M$.

6:    **for** $k \in \mathcal{K}/\{\Omega_m^{(t)}\}$ **do**

7:      $m^\star = \underset{1 \leq m \leq M}{\arg \max}\, C_{k, \Omega_m^{(t)}}$.

8:      $\mathcal{G}_{m^\star} = \mathcal{G}_{m^\star} \bigcup k$.

9:    **end for**

10:    $t = t + 1$.

11:    Update $\Omega_m^{(t)}$ for $m = 1, 2, \cdots, M$ according to (6.107).

12: **end while**

13: **return** $\{\mathcal{G}_1, \mathcal{G}_2, \cdots, \mathcal{G}_M\}$.

---

the sum rate, the BS tends to allocate most power and beam gains to the users with the strong channels. Then, the users with the low channel gains can not be served by the BS. The other category is to ensure the user fairness, where the max-min fairness or proportional fairness are considered to improve the performance of the users with worse channel conditions. However, the fairness among the users may result in a performance loss of the sum rate. To realize the tradeoff between the sum-rate performance and the user fairness, we maximize the ASR while ensuring the minimum achievable rate of each user, which is also adopted in the related NOMA systems[13, 14, 35]. Then, the problem is formulated as

$$\underset{\{p_{m,n}\}, \mathbf{A}, \mathbf{D}}{\text{Max}} \quad R_{\text{sum}} \tag{6.108a}$$

$$\text{s.t.} \quad C_1 \ : \ R_{m,n} \geq r_{m,n}, \quad \forall m, n, \tag{6.108b}$$

$$C_2 \ : \ p_{m,n} \geq 0, \quad \forall m, n, \tag{6.108c}$$

$$C_3 \ : \ \sum_{m=1}^{M} \sum_{n=1}^{|\mathcal{G}_m|} p_{m,n} \leq P, \tag{6.108d}$$

$$C_4 \ : \ |[\mathbf{A}]_{i,j}| = \frac{1}{\sqrt{N}}, \quad \forall i, j, \tag{6.108e}$$

$$C_5 \ : \ \|[\mathbf{AD}]_{:,m}\|_2 = 1, \quad \forall m, \tag{6.108f}$$

where the constraint C_1 is the minimum rate constraint for each user. The constraint C_2 indicates that the power allocated to each user should be non-negative. The constraint C_3 is the total transmit power constraint, where the total power at the BS is no more than P. C_4 is the CM constraint for the analog beamforming matrix, and C_5 is the unit power constraint for the hybrid beamforming matrix.

The total dimension of the variables in Problem (6.108) is $K + MN + M^2$, which is large in general. Exhaustive search for the optimal solution results in heavy computational load, which is hard to accomplish in practice. To solve Problem (6.108), there are two main challenges. One is that the optimized variables are entangled with each other, which makes the formulation non-convex. The other is that the expression of R_{sum} depends on the decoding order. In general, the optimal decoding order is the increasing order of the users' effective channel gains. However, the order of effective channel gains varies with different beamforming matrices. In other words, given different hybrid beamforming matrices, the objective function in Problem (6.108), i.e., the ASR of the users, has different expressions. The two challenges make it infeasible solve Problem (6.108) by using the existing optimization tools. Next, we will present a sub-optimal solution with promising performance but low computational complexity.

The proposed solution of Problem (6.108) can be obtained with two stages. In the first stage, we provide a low-complexity algorithm to obtain the sub-optimal power allocation with an arbitrary fixed hybrid beamforming. In the second stage, we design the hybrid beamforming, where the digital beamforming matrix and the analog beamforming matrix are obtained using the approximate zero-forcing (AZF) method and the proposed BC-PSO algorithm, respectively.

6.4.3 Solution of Power Allocation

As we have analyzed before, an essential challenge to solve Problem (6.108) is the variation of the decoding order. However, given an arbitrary fixed analog beamforming matrix $\mathbf{A}$ and an arbitrary fixed digital beamforming matrix $\mathbf{D}$, the order of the effective channel gains is fixed. For notational simplicity and without loss of generality, we assume

$$|\mathbf{h}_{m,1}^{\text{H}}\mathbf{w}_m|^2 \geq |\mathbf{h}_{m,2}^{\text{H}}\mathbf{w}_m|^2 \geq \cdots \geq |\mathbf{h}_{m,|\mathcal{G}_m|}^{\text{H}}\mathbf{w}_m|^2, \ \forall 1 \leq m \leq M, \quad (6.109)$$

where $\mathbf{w}_m = [\mathbf{AD}]_{:,m}$. The original problem can be simplified as

$$\underset{\{p_{m,n}\}}{\text{Max}} \quad R_{\text{sum}} \quad (6.110\text{a})$$

$$\text{s.t.} \quad C_1 \ : \ R_{m,n} \geq r_{m,n}, \quad \forall m, n, \quad (6.110\text{b})$$

$$C_2 \ : \ p_{m,n} \geq 0, \quad \forall m, n, \quad (6.110\text{c})$$

$$C_3 \ : \ \sum_{m=1}^{M} \sum_{n=1}^{|\mathcal{G}_m|} p_{m,n} \leq P, \quad (6.110\text{d})$$

where $\mathbf{A}$ and $\mathbf{D}$ are arbitrary but fixed.

According to the expression of the achievable rate in (6.102), a user may suffer the interference from both the intra-group users and the inter-group users. Although the hybrid beamforming matrix is fixed, the objective function and the constraint C_1 of Problem (6.110) are still non-convex. To address this problem, we divide it into two sub-problems, i.e., intra-group power allocation (intra-GPA) and inter-group power allocation (inter-GPA). Define

$$\sum_{n=1}^{|\mathcal{G}_m|} p_{m,n} = P_m, \ 1 \leq m \leq M, \tag{6.111}$$

which means the allocated power for the m-th group, and then Problem (6.110) is equivalent to

$$\underset{\{P_m\} \ \{p_{m,n}\}}{\text{Max Max}} \quad R_{\text{sum}} \tag{6.112a}$$

$$\text{s.t.} \quad C_1 : \ R_{m,n} \geq r_{m,n}, \quad \forall m, n, \tag{6.112b}$$

$$C_2 : \ p_{m,n} \geq 0, \quad \forall m, n, \tag{6.112c}$$

$$C_3 : \ \sum_{n=1}^{|\mathcal{G}_m|} p_{m,n} = P_m, \quad \forall m, \tag{6.112d}$$

$$C_4 : \ \sum_{m=1}^{M} P_m \leq P. \tag{6.112e}$$

Note that the introduced inter-GPA variables, i.e., $\{P_m\}$, have no influence on the optimality of the power allocation problem, because there is no loss of the DoF in Problem (6.112) compared with Problem (6.110), and Problem (6.112) is more tractable. First, given arbitrary and fixed inter-GPA, a closed-form sub-optimal intra-GPA can be obtained. Then, substituting the intra-GPA into Problem (6.112), we can obtain a sub-optimal inter-GPA solution. Although the proposed solution of power allocation is not globally optimal, we will prove that it is near-to-optimal when the inter-group interference is small through the theoretical analysis and simulation verification.

6.4.3.1 The Intra-GPA Problem

As shown in (6.101) and (6.102), one user may suffer the interference from the users in the same group and the users in other groups, which are called intra-group interference and inter-group interference, respectively. Considering that hybrid beamforming can be well designed in general, such that the inter-group interference is small and can be neglected. Thus, we have the following proposition to solve the intra-GPA problem.

Proposition 6.4.1. *Given an arbitrary fixed inter-GPA of $\{P_1, P_2, \cdots, P_M\}$, if the inter-group interference can be neglected, the optimal intra-GPA in Problem (6.112) should always satisfy*

$$R_{m,n} = r_{m,n}, \ 1 \leq m \leq M, \ 2 \leq n \leq |\mathcal{G}_m|. \tag{6.113}$$

Proof. If the inter-group interference is small and can be neglected, Problem (6.112) can be divided into M independent intra-GPA problems. For the m-th group, the intra-GPA problem is simplified as

$$\underset{\{p_{m,n}\}}{\text{Max}} \quad \sum_{n=1}^{|\mathcal{G}_m|} R_{m,n} \tag{6.114a}$$

$$\text{s.t.} \quad C_1 : R_{m,n} \geq r_{m,n}, \quad \forall n, \tag{6.114b}$$

$$C_2 : p_{m,n} \geq 0, \quad \forall n, \tag{6.114c}$$

$$C_3 : \sum_{n=1}^{|\mathcal{G}_m|} p_{m,n} = P_m, \tag{6.114d}$$

which is a power allocation problem without inter-group interference. This problem has been solved in[38], where the optimal power allocation always satisfies

$$R_{m,n} = r_{m,n}, \quad 2 \leq n \leq |\mathcal{G}_m|. \tag{6.115}$$

□

By solving the equation sets of $R_{m,n} = r_{m,n}$, $1 \leq m \leq M$, $2 \leq n \leq |\mathcal{G}_m|$, and $\sum_{n=1}^{|\mathcal{G}_m|} p_{m,n} = P_m$, $1 \leq m \leq M$, we can obtain a sub-optimal intra-GPA for each group of users, which is shown in (6.117) on the top of the next page, where

$$\eta_{m,n} = 2^{r_{m,n}} - 1. \tag{6.116}$$

Note that although the inter-group interference is neglected in Proposition 6.4.1, it is included when solving the equation sets. Thus, the minimal rate constraints for the users (from the 2nd one to the last one in each group) are always satisfied. The impact of the approximation on the inter-group interference will be evaluated in the simulation.

$$\begin{cases} p_{m,|\mathcal{G}_m|}^{\circ} = \dfrac{\eta_{m,|\mathcal{G}_m|}}{\eta_{m,|\mathcal{G}_m|} + 1}\left(P_m + \dfrac{\sum\limits_{i \neq m} |\mathbf{h}_{m,|\mathcal{G}_m|}^{\mathrm{H}} \mathbf{w}_i|^2 P_i + \sigma^2}{|\mathbf{h}_{m,|\mathcal{G}_m|}^{\mathrm{H}} \mathbf{w}_m|^2}\right), \\[3mm] p_{m,|\mathcal{G}_m|-1}^{\circ} = \dfrac{\eta_{m,|\mathcal{G}_m|-1}}{\eta_{m,|\mathcal{G}_m|-1} + 1}\left(P_m - p_{m,|\mathcal{G}_m|}^{\circ} + \dfrac{\sum\limits_{i \neq m} |\mathbf{h}_{m,|\mathcal{G}_m|-1}^{\mathrm{H}} \mathbf{w}_i|^2 P_i + \sigma^2}{|\mathbf{h}_{m,|\mathcal{G}_m|-1}^{\mathrm{H}} \mathbf{w}_m|^2}\right), \\[3mm] \quad\vdots \\[2mm] p_{m,2}^{\circ} = \dfrac{\eta_{m,2}}{\eta_{m,2} + 1}\left(P_m - \sum\limits_{k=3}^{|\mathcal{G}_m|} p_{m,k}^{\circ} + \dfrac{\sum\limits_{i \neq m} |\mathbf{h}_{m,2}^{\mathrm{H}} \mathbf{w}_i|^2 P_i + \sigma^2}{|\mathbf{h}_{m,2}^{\mathrm{H}} \mathbf{w}_m|^2}\right), \\[3mm] p_{m,1}^{\circ} = P_m - \sum\limits_{k=2}^{|\mathcal{G}_m|} p_{m,k}^{\circ}, \end{cases} \tag{6.117}$$

Under Proposition 6.4.1, the ASR in Problem (6.112) can be simplified as

$$R_{\text{sum}} = \sum_{m=1}^{M} R_{m,1} + \sum_{m=1}^{M} \sum_{n=2}^{|\mathcal{G}_m|} r_{m,n}. \tag{6.118}$$

Substituting (6.117) into Problem (6.112), Problem (6.112) can be transformed to

$$\underset{\{P_m\}}{\text{Max}} \quad \sum_{m=1}^{M} R_{m,1} \tag{6.119a}$$

$$\text{s.t.} \quad C_1 : R_{m,1} \geq r_{m,1}, \quad \forall m, \tag{6.119b}$$

$$C_2 : \sum_{m=1}^{M} P_m \leq P, \tag{6.119c}$$

which is an inter-GPA problem.

6.4.3.2 The Inter-GPA Problem

Due to the inter-group interference in the expression of the objective function, it is still challenging to solve Problem (6.119). We propose an iterative algorithm here. First, we initialize the group power P_m equally. Then, we start iteration. In each iteration, the inter-group interference is assumed to be invariable, and we update the inter-GPA by maximizing the ASR in Problem (6.119), where the inter-group interference is defined as

$$I_{m,n}^{(\text{inter})} \triangleq \sum_{i \neq m} \sum_{k=1}^{|\mathcal{G}_i|} |\mathbf{h}_{m,n}^{\text{H}} \mathbf{w}_i|^2 p_{i,k} = \sum_{i \neq m} |\mathbf{h}_{m,n}^{\text{H}} \mathbf{w}_i|^2 P_i. \tag{6.120}$$

Thus, the SINR for the first user in each group is linear to its signal power, i.e.,

$$\gamma_{m,1} = \frac{|\mathbf{h}_{m,1}^{\text{H}} \mathbf{w}_m|^2 p_{m,1}^{\circ}}{I_{m,1}^{(\text{inter})} + \sigma^2}, \tag{6.121}$$

where $p_{m,1}^{\circ}$ is defined in (6.117). Furthermore, according to the expression in (6.117), if the inter-group interference is invariable, $p_{m,1}^{\circ}$ is also linear to P_m. Thus, we can obtain the relationship between $\gamma_{m,1}$ and P_m as

$$\gamma_{m,1} = k_m P_m + b_m, \tag{6.122}$$

where k_m and b_m are given by

$$k_m = \frac{|\mathbf{h}_{m,1}^{\text{H}} \mathbf{w}_m|^2}{I_{m,1}^{(\text{inter})} + \sigma^2} \left(1 - \sum_{n=2}^{|\mathcal{G}_m|} \left[\eta_{m,n} \prod_{j=2}^{n} \frac{1}{(\eta_{m,j} + 1)} \right] \right), \tag{6.123}$$

$$b_m = -\frac{|\mathbf{h}_{m,1}^{\text{H}} \mathbf{w}_m|^2}{I_{m,1}^{(\text{inter})} + \sigma^2} \times \sum_{n=2}^{|\mathcal{G}_m|} \left[\eta_{m,n} \frac{I_{m,n}^{(\text{inter})} + \sigma^2}{|\mathbf{h}_{m,n}^{\text{H}} \mathbf{w}_n|^2} \prod_{j=2}^{n} \frac{1}{(\eta_{m,j} + 1)} \right].$$

It is easy to verify that $k_m > 0$ and $b_m < 0$. Then, the objective function in Problem (6.119) is equal to

$$\sum_{m=1}^{M} R_{m,1} = \sum_{m=1}^{M} \log_2(1 + \gamma_{m,1}) = \sum_{m=1}^{M} \log_2(k_m P_m + b_m + 1) \triangleq f(\{P_m\}). \quad (6.124)$$

Constraint C_1 in Problem (6.119) is equivalent to

$$R_{m,1} \geq r_{m,1} \Leftrightarrow \gamma_{m,1} \geq \eta_{m,1} \Leftrightarrow P_m \geq \frac{\eta_{m,1} - b_m}{k_m}. \quad (6.125)$$

As the objective function becomes concave now and the constraints are linear, Problem (6.119) can be directly solved by using the convex optimization tools[27]. In order to explore the essential principle of the inter-GPA for NOMA, we propose a method with low computational complexity here. We begin from the case without constraint C_1 in Problem (6.119) and give the following Lemma.

Lemma 6.4.1. *If the inter-group interference is assumed to be invariant in Problem (6.119), without the constraint C_1, the globally optimal P_m of Problem (6.119) is*

$$P_m^{\star} = \frac{P + \sum_{i=1}^{M} \frac{b_i+1}{k_i}}{M} - \frac{b_m + 1}{k_m}, \quad 1 \leq m \leq M. \quad (6.126)$$

Proof. It is obvious that $f(\{P_m\})$ defined in (6.124) is increasing with P_m, $1 \leq m \leq M$. Thus, the optimal solution for maximizing $f(\{P_m\})$ always satisfies $\sum_{m=1}^{M} P_m = P$. Then, Problem (6.119) without the constraint C_1 can be solved by Lagrange Multiplier Method, where the KKT equation set is

$$\begin{cases} \dfrac{\partial f}{\partial P_m} = \dfrac{\partial \left[\sum_{m=1}^{M} \log_2(k_m P_m + b_m + 1) \right]}{\partial P_m} = \lambda, \ 1 \leq m \leq M, \\ \sum_{m=1}^{M} P_m = P. \end{cases} \quad (6.127)$$

Solve the equation sets above and we can obtain the optimal solution of Problem (6.119) as shown in (6.126). $\square$

According to Lemma 6.4.1, if $P_m^{\star}$ in (6.126) is located in the feasible domain of the constraint C_1 in Problem (6.119), i.e.,

$$P_m^{\star} \geq \frac{\eta_{m,1} - b_m}{k_m}, \quad (6.128)$$

for all $1 \leq m \leq M$, $P_m^{\star}$ is the optimal solution of Problem (6.119). However, if $P_m^{\star}$ in (6.126) is not located in the feasible domain of the constraint C_1 in Problem (6.119), i.e.

$$P_m^{\star} < \frac{\eta_{m,1} - b_m}{k_m}, \quad (6.129)$$

for any one of $1 \leq m \leq M$, $P_m^{\star}$ is not the optimal solution of Problem (6.119). We may find the optimal solution by using the following Lemma.

Lemma 6.4.2. *If the inter-group interference is assumed to be invariant in Problem (6.119), with the constraint C_1, the globally optimal solution should always satisfy*

$$P_m^\circ = \frac{\eta_{m,1} - b_m}{k_m}, \quad \forall m \in \mathcal{U}, \tag{6.130}$$

where $\mathcal{U} = \{i | 1 \leq i \leq M, P_i^\star < \frac{\eta_{i,1} - b_i}{k_i}\}$ and $P_i^\star$ is defined in (6.126).

Proof. We prove Lemma 6.4.2 by using contradiction. Denote the optimal solution of Problem (6.119) as $\{P_m^\circ\}$. Assume there exists an index m_1, $m_1 \in \mathcal{U}$, which satisfies

$$P_{m_1}^\circ > \frac{\eta_{m_1,1} - b_{m_1}}{k_{m_1}}. \tag{6.131}$$

According to $m_1 \in \mathcal{U}$, we have

$$P_{m_1}^\circ > \frac{\eta_{m_1,1} - b_{m_1}}{k_{m_1}} > P_{m_1}^\star. \tag{6.132}$$

Then, there always exists another index m_2, $m_2 \neq m_1$, which satisfies $P_{m_2}^\circ < P_m^\star$, because of

$$\sum_{m=1}^{M} P_m^\circ \leq P = \sum_{m=1}^{M} P_m^\star. \tag{6.133}$$

Consider the power allocation of

$$\begin{cases} P_{m_1}' = P_{m_1}^\circ - \epsilon, \\ P_{m_2}' = P_{m_2}^\circ + \epsilon, \\ P_m' = P_m^\circ, \quad m \neq m_1, m_2, \end{cases} \tag{6.134}$$

where ϵ is a nonnegative and small number.

The partial derivative of the objective function is

$$\frac{\partial f}{\partial P_i} = \frac{\partial \left[\sum_{i=1}^{M} \log_2(k_i P_i + b_i + 1) \right]}{\partial P_i} = \frac{1}{\ln 2} \frac{k_i}{(k_i P_i + b_i + 1)}, \quad 1 \leq i \leq M, \tag{6.135}$$

which is a monotone decreasing function of P_i. Since $P_{m_1}^\circ > P_{m_1}^\star$ and $P_{m_2}^\circ < P_{m_2}^\star$ hold, we have

$$\begin{aligned} \left. \frac{\partial f}{\partial P_{m_1}} \right|_{\{P_m = P_m^\circ\}} &< \left. \frac{\partial f}{\partial P_{m_1}} \right|_{\{P_m = P_m^\star\}}, \\ \left. \frac{\partial f}{\partial P_{m_2}} \right|_{\{P_m = P_m^\circ\}} &> \left. \frac{\partial f}{\partial P_{m_2}} \right|_{\{P_m = P_m^\star\}}, \end{aligned} \tag{6.136}$$

where

$$\left. \frac{\partial f}{\partial P_{m_i}} \right|_{\{P_m = P_m^\circ\}}, \quad i = 1, 2, \tag{6.137}$$

represents the value of partial derivative $\frac{\partial f}{\partial P_{m_i}}$, $i = 1, 2$, at point $\{P_m = P_m^\circ\}$. Define

$$g(\epsilon) = f(\{P_m'\}) - f(\{P_m^\circ\}). \tag{6.138}$$

It is easy to verify $g(0) = 0$. The derivative of the function $g(\epsilon)$ is

$$
\begin{aligned}
\frac{d\,g}{d\,\epsilon} &= \frac{d\,(f(\{P'_m\}) - f(\{P^\circ_m\}))}{d\,\epsilon} \\[2mm]
&= \frac{d\,(\sum_{m=1}^{M} \log_2(k_m P'_m + b_m + 1) - \sum_{m=1}^{M} \log_2(k_m P^\circ_m + b_m + 1))}{d\,\epsilon} \\[2mm]
&= \frac{d\,(\log_2(k_m(P^\circ_{m_2} + \epsilon) + b_m + 1) - \log_2(k_m P^\circ_{m_2} + b_m + 1))}{d\,\epsilon} \\[2mm]
&\quad + \frac{d\,(\log_2(k_m(P^\circ_{m_1} - \epsilon) + b_m + 1) - \log_2(k_m P^\circ_{m_1} + b_m + 1))}{d\,\epsilon} \qquad (6.139) \\[2mm]
&= \left.\frac{\partial f}{\partial P_{m_2}}\right|_{\{P_m = P^\circ_m\}} - \left.\frac{\partial f}{\partial P_{m_1}}\right|_{\{P_m = P^\circ_m\}} \\[2mm]
&> \left.\frac{\partial f}{\partial P_{m_2}}\right|_{\{P_m = P^\star_m\}} - \left.\frac{\partial f}{\partial P_{m_1}}\right|_{\{P_m = P^\star_m\}} \\[2mm]
&= \lambda - \lambda = 0,
\end{aligned}
$$

which means that $g(\epsilon)$ is a monotone increasing function of ϵ. We can select a positive and sufficiently small ϵ which satisfies

$$
\begin{cases}
P'_{m_1} > P^\star_{m_1}, \\
P'_{m_2} < P^\star_{m_2},
\end{cases} \qquad (6.140)
$$

and

$$
g(\epsilon) = f(\{P'_m\}) - f(\{P^\circ_m\}) > 0. \qquad (6.141)
$$

In other words, $\{P'_m\}$ is a better solution than $\{P^\circ_m\}$. It contradicts to the assumption that $\{P^\circ_m\}$ is the optimal solution of Problem (6.119). Thus, we can conclude that for $\forall m \in \mathcal{U}$, the optimal solution of Problem (6.119) should always satisfy

$$
P^\circ_m = \frac{\eta_{m,1} - b_m}{k_m}. \qquad (6.142)
$$

$\square$

Lemma 6.4.2 provides the globally optimal power allocation for $m \in \mathcal{U}$. For $m \notin \mathcal{U}$, the optimal power allocation can be obtained by solving the following problem.

$$
\underset{\{P_m\}}{\text{Max}} \quad \sum_{m \notin \mathcal{U}} R_{m,1} \qquad (6.143a)
$$

$$
\text{s.t.} \quad C_1 : R_{m,1} \geq r_{m,1}, \quad m \notin \mathcal{U}, \qquad (6.143b)
$$

$$
C_2 : \sum_{m \notin \mathcal{U}} P_m \leq P - \sum_{j \in \mathcal{U}} P^\circ_j, \qquad (6.143c)
$$

which has a similar formulation with Problem (6.119). Thus, Lemma 6.4.1 and Lemma 6.4.2 can also be used to solve Problem (6.143), which forms a closed loop. In summary, we give Algorithm 6.4 to accomplish the inter-GPA.

Algorithm 6.4: Inter-GPA

Input: K, M, $\{\mathcal{G}_m\}$, P, $\{\mathbf{h}_k\}$, $\{r_k\}$, $\mathbf{W}$, and $F_{\max}$.
Output: Inter-GPA: $\{P_m^\circ\}$.
1: $P_m^{\circ(0)} = \frac{P}{M}$ $(1 \leq m \leq M)$.
2: **for** $t = 1 : F_{\max}$ **do**
3:     $\mathcal{M} = \{1, 2, \cdots, M\}$.
4:     $\mathcal{U} = \mathcal{M}$.
5:     **while** $\mathcal{U} \neq \Phi$ **do**
6:        Obtain k_m, b_m $(\forall m \in \mathcal{M})$ in (6.122).
7:        Obtain $P_m^\star$ $(\forall m \in \mathcal{M})$ according to (6.126).
8:        $\mathcal{U} = \{i | i \in \mathcal{M}, \ P_i^\star < \frac{\eta_{i,1} - b_i}{k_i}\}$.
9:        $P_m^{\circ(t)} = \frac{\eta_{m,1} - b_m}{k_m}$ $(\forall m \in \mathcal{U})$.
10:      $\mathcal{M} = \mathcal{M} / \mathcal{U}$.
11:     **end while**
12:     $P_m^{\circ(t)} = P_m^\star$ $(\forall m \in \mathcal{M})$.
13: **end for**
14: $P_m^\circ = P_m^{\circ(T_{\max})}$ $(1 \leq m \leq M)$.
15: **return** $\{P_m^\circ\}$.

Hereto, the power allocation is solved. Given an arbitrary fixed hybrid beamforming, we can obtain the inter-GPA using Algorithm 6.4 and obtain the intra-GPA according to (6.117). Since the proposed intra-GPA and inter-GPA solutions are both sub-optimal, we provide the following theorem to evaluate the optimality of the proposed power allocation solution.

Theorem 6.4.1. *If the inter-group interference in Problem* (6.112) *is zero (or negligibly small), the proposed solution of power allocation in Algorithm 6.4 and* (6.117) *is globally optimal.*

Proof. If the inter-group interference is zero, the intra-GPA problems are independent for different groups. According to the conclusion in Theorem 6.2.1, (6.117) is the optimal intra-GPA solution with the given fixed inter-GPA. Substituting (6.117) into Problem (6.112), the inter-GPA problem is concave and can be solved by using Algorithm 6.4 with only one iteration. Due to the concavity, the inter-GPA solution is also optimal. Thus, the globally optimal power allocation can be obtained by using the proposed scheme if the inter-group interference is zero. ∎

Based on Theorem 6.4.1, we can find that the optimality of the power allocation solution depends on the inter-group interference, which can be restrained through an elaborated beamforming design. Thus, the design of hybrid beamforming should take both decreasing the interference and increasing the ASR into account. The details will be shown in the next section.

6.4.4   Solution of Hybrid Beamforming

In this subsection, we provide the solution of hybrid beamforming in Problem (6.108). As we have analyzed previously, the design of hybrid beamforming should guarantee the suppression of the inter-group interference, as well as the improvement of the ASR. For NOMA, there may exist more than one users in each group. The traditional unidirectional beamforming cannot support all the users, because a narrow beam generated by one RF chain can only cover a small range of direction but the users may be distributed at significantly different directions with respect to the BS. Thus, a multi-directional beamforming scheme is required in the analog domain. However, the non-convex modulus constraint for analog beamforming makes the beamforming problem challenging. Besides, as shown in (6.101), due to the superposition of the inter-group interference and the intra-group interference, it is difficult to obtain the optimal hybrid beamforming solution. To this end, we propose a sub-optimal approach. First, the digital beamforming is designed using the AZF method to reduce the inter-group interference, where the analog beamomring matrix is arbitrary and fixed. Then, we use the BC-PSO algorithm in[38] to solve the analog beamforming problem, where the power allocation and digital beamforming matrix are substituted as the function of the analog beamforming matrix.

6.4.4.1   Digital Beamforming with Arbitrary Fixed Analog Beamforming

As each group of users have a unique digital beamforming vector, we may design the digital beamforming with the AZF method to reduce the inter-group interference, where the analog beamforming is arbitrary and fixed. Since the rank of the digital beamforming matrix is no more than the number of the users, i.e., $M \leq K$, the inter-group interference cannot be completely suppressed through digital beamforming. Recalling that when optimizing the power allocation, the rate gains are acquired at the first user in each group. Thus, we select the channel response vector of the user with the highest channel gain in each group as the equivalent channel vector. Note that the channel gain utilized here corresponds to the power of the channel response vector before beamforming, which differs from the effective channel gain after beamforming. Then, the $N \times M$ equivalent channel matrix is

$$\tilde{\mathbf{H}} = [\mathbf{h}_{1,1}, \mathbf{h}_{2,1}, \cdots, \mathbf{h}_{M,1}]. \tag{6.144}$$

Consequently, the digital beamforming matrix can be generated by the AZF method as [10]

$$\tilde{\mathbf{D}} = (\tilde{\mathbf{H}}^H \mathbf{A})^\dagger. \tag{6.145}$$

Due to the unit power constraint for the hybrid beamforming matrix, each column of the digital beamforming matrix should be normalized as

$$[\mathbf{D}^\circ]_{:,m} = \frac{[\tilde{\mathbf{D}}]_{:,m}}{\|\mathbf{A}[\tilde{\mathbf{D}}]_{:,m}\|_2}. \tag{6.146}$$

---

[10]Since the digital beamforming design implements an AZF method, i.e., only to the first user in each group, the inclusion of inter-group interference in the previous subsection is relevant.

Although the inter-group interference cannot be completely eliminated with digital beamforming, it can be further suppressed with analog beamforming, which has a higher degree of freedom.

6.4.4.2 Analog Beamforming Using BC-PSO Alogrithm

Given an arbitrary fixed analog beamforming matrix, we can obtain the digital beamforming matrix according to (6.145) and (6.146). Then, the inter-GPA can be obtained by Algorithm 6.4, and meanwhile the intra-GPA is given by (6.117). It is hard to optimize the analog beamforming with the conventional approaches, since the closed-form expression of R_{sum} over $\mathbf{A}$ is complicated. In addition, the analog beamforming matrix $\mathbf{A}$ with CM constraint is high-dimensional, i.e., $N \times M$, which makes the analog beamforming design difficult.

To solve this difficult problem, PSO is a good approach[34]. In the $N \times M$-dimensional search space $\mathcal{S}$, the I particles in the swarm are randomly initialized with position $\mathbf{X}$ and velocity $\mathbf{V}$. Each particle has a memory for its best found position $\mathbf{P}_{\text{best}}$ and the globally best position $\mathbf{G}_{\text{best}}$, where the goodness of a position is evaluated by the fitness function. For each iteration, the velocity and position of each particle are updated based on

$$[\mathbf{V}]_{i,j} = \omega[\mathbf{V}]_{i,j} + c_1 \times \text{rand} \times ([\mathbf{P}_{\text{best}}]_{i,j} - [\mathbf{X}]_{i,j}) + c_2 \times \text{rand} \times ([\mathbf{G}_{\text{best}}]_{i,j} - [\mathbf{X}]_{i,j}),$$
$$[\mathbf{X}]_{i,j} = [\mathbf{X}]_{i,j} + [\mathbf{V}]_{i,j},$$
(6.147)

for $i = 1, 2, \cdots, N$; $j = 1, 2, \cdots, M$. The parameter ω is the inertia weight of velocity. In general, ω is decreasing linearly from the maxima to the minima for each time of iteration to improve the convergence speed. The parameters c_1 and c_2 are the cognitive ratio and social ratio, respectively. The random number function rand returns to a number between 0.0 and 1.0 with uniform distribution.

Due to the CM constraint, the search space for $\mathbf{A}$, i.e.,

$$\{\mathbf{A} | |[\mathbf{A}]_{i,j}| = \frac{1}{\sqrt{N}}\},$$
(6.148)

is highly non-convex. It has been shown that the BC-PSO algorithm outperforms the classic PSO algorithm in the analog beamforming problem[38]. The key idea of the BC-PSO algorithm is to relax the search space as a convex set, i.e.,

$$\mathcal{S} = \{\mathbf{A} | |[\mathbf{A}]_{i,j}| \leq \frac{1}{\sqrt{N}}\},$$
(6.149)

and adjust the particles onto the boundaries for each iteration to satisfy the CM constraint. The outer boundary is defined as

$$\{\mathbf{A} | |[\mathbf{A}]_{i,j}| = d_{\text{out}}\},$$
(6.150)

where $d_{\text{out}} = \frac{1}{\sqrt{N}}$ is fixed. The inter boundary is defined as

$$\{\mathbf{A} | |[\mathbf{A}]_{i,j}| = d_{\text{in}}\},$$
(6.151)

where

$$d_{\text{in}} = \frac{t}{T_{\text{max}}} \frac{1}{\sqrt{N}}, \tag{6.152}$$

is dynamic. T_{max} is the maximum number of iterations and $t = 1, 2, \cdots, T_{\text{max}}$. For each iteration, the particles out of the boundaries are adjusted onto the boundaries. Then, after calculating the fitness function for each particle, the locally and globally best positions, i.e., $\mathbf{P}_{\text{best}}$ and $\mathbf{G}_{\text{best}}$, are updated. With this implementation, the particles can move throughout the relaxed search space and converge to satisfy the CM constraint eventually. Compared with the classic PSO algorithm, the BC-PSO algorithm has enhanced search capabilities.

6.4.5 Summary of the Complete Solution and Computational Complexity

6.4.5.1 Summary of the Complete Solution

In the above subsections, we have presented the algorithms and formulas, respectively, for user grouping, power allocation, digital beamforming and analog beamforming. Based on these algorithms and formulas, we give the complete solution to realize an arbitrary NOMA system. As shown in Algorithm 6.5, we firstly use Algorithm 6.3 to divide the users into M groups, and obtain $\{\mathcal{G}_m\}$. Then, we use the BC-PSO algorithm to iteratively optimize the position of the particle, i.e., the analog beamforming matrix, where the fitness function is defined as the ASR in (6.103). Note that in the part of power allocation and digital beamforming, we assume that the analog beamforming matrix is arbitrary and fixed. Thus, the power allocation and digital beamforming can be substituted as the function of the analog beamforming matrix in Algorithm 6.5. Given different analog beamforming matrices, we should calculate the power allocation and digital beamforming matrices first, and then obtain the ASR. In each iteration, the computations of the digital beamforming matrix $\mathbf{D}^\circ$ using (6.145) and (6.146), the inter-GPA $\{P_m^\circ\}$ using Algorithm 6.4, and the intra-GPA $\{p_{m,n}^\circ\}$ using (6.117) are performed sequentially after determining the analog beamforming matrix. Hence, after T_{max} iterations, the sub-optimal overall solution $\mathbf{A}^\circ$, $\mathbf{D}^\circ$ and $\{p_{m,n}^\circ\}$ are jointly obtained.

6.4.5.2 Computational Complexity

When operating the user grouping in Algorithm 6.3, the complexities of calculating the channel correlation and the norm channel vector are $\mathcal{O}(K^2 N)$ and $\mathcal{O}(KN)$, respectively. In each iteration, the complexities of updating the cluster representative and the user grouping are $\mathcal{O}(K^2)$ and $\mathcal{O}(KM)$, respectively. Since the number of antennas is much larger than that of the RF chains, i.e., $N \gg M$, the maximal complexity of Algorithm 6.3 is $\mathcal{O}(K^2 N)$. In Algorithm 6.4, the complexity of calculating the effective channel gains of the users is $\mathcal{O}(MKN)$. For each time of updating the inter-GPA, the maximal number of iterations to update the inter-GPA from Step 5 to 11 is M, and the complexity of computing the inter-GPA in each sub-cycle is no more than $\mathcal{O}(K^2)$. Thus, the complexity of Algorithm 6.4 is $\mathcal{O}(MKN + F_{\text{max}} MK^2)$. In Algorithm 6.5, the numbers of invoking Algorithm 6.3 and Algorithm 6.4 are

---

Algorithm 6.5: Proposed solution for millimeter-wave-NOMA

Input: K, M, N, P, $\{\mathbf{h}_k\}$, $\{r_k\}$, and parameters
　　　　for BC-PSO $\{I, T_{\max}, c_1, c_2, \omega_{\max}, \omega_{\min}\}$.
Output: $\{\mathcal{G}_m\}$, $\mathbf{A}^\circ$, $\mathbf{D}^\circ$ and $\{p_{m,n}^\circ\}$.
 1: Obtain the user grouping $\{\mathcal{G}_m\}$ using Algorithm 6.3.
 2: Initialize the position $\mathbf{A}_i$ and velocity $\mathbf{V}_i$.
 3: Find the globally best position $\mathbf{G}_{\text{best}}$.
 4: **for** $t = 1 : T_{\max}$ **do**
 5: 　$\omega = \omega_{\max} - \frac{t}{T}(\omega_{\max} - \omega_{\min})$.
 6: 　$d_{\text{out}} = \frac{1}{\sqrt{N}}$, $d_{\text{in}} = \frac{t}{T_{\max}}\frac{1}{\sqrt{N}}$.
 7: 　**for** $l = 1 : I$ **do**
 8: 　　**for** $i = 1 : N$ **do**
 9: 　　　**for** $j = 1 : M$ **do**
10: 　　　　Update $[\mathbf{V}_l]_{i,j}$ and $[\mathbf{A}_l]_{i,j}$ based on (6.147).
11: 　　　　**if** $|[\mathbf{A}_l]_{i,j}| > d_{\text{out}}$ **then**
12: 　　　　　$[\mathbf{A}_l]_{i,j} = d_{\text{out}}\frac{[\mathbf{A}_l]_{i,j}}{|[\mathbf{A}_l]_{i,j}|}$.
13: 　　　　**end if**
14: 　　　　**if** $|[\mathbf{A}_l]_{i,j}| < d_{\text{in}}$ **then**
15: 　　　　　$[\mathbf{A}_l]_{i,j} = d_{\text{in}}\frac{[\mathbf{A}_l]_{i,j}}{|[\mathbf{A}_l]_{i,j}|}$.
16: 　　　　**end if**
17: 　　　　**if** $|[\mathbf{P}_{\text{best},l}]_{i,j}| < d_{\text{in}}$ **then**
18: 　　　　　$[\mathbf{P}_{\text{best},l}]_{i,j} = d_{\text{in}}\frac{[\mathbf{P}_{\text{best},l}]_{i,j}}{|[\mathbf{P}_{\text{best},l}]_{i,j}|}$.
19: 　　　　**end if**
20: 　　　　Obtain the DBF matrix $\mathbf{D}^\circ$ according to (6.145) and (6.146).
21: 　　　　Reorder the effective channel gains of the users in each group.
22: 　　　　Obtain the inter-GPA $\{P_m^\circ\}$ using Algorithm 6.4.
23: 　　　　Obtain the intra-GPA $\{p_{m,n}^\circ\}$ according to (6.117).
24: 　　　　Obtain the fitness function R_{sum} according to (6.103).
25: 　　　**end for**
26: 　　**end for**
27: 　　Update $\mathbf{P}_{\text{best},l}$.
28: 　**end for**
29: 　Update $\mathbf{G}_{\text{best}}$.
30: **end for**
31: $\mathbf{A}^\circ = \mathbf{G}_{\text{best}}$.
32: **return** $\{\mathcal{G}_m\}$, $\mathbf{A}^\circ$, $\mathbf{D}^\circ$ and $\{p_{m,n}^\circ\}$.

---

1 and $T_{\max}IMN$, respectively. Consequently, the total computational complexity of the proposed user pairing algorithm, hybrid beamforming and power allocation algorithm is $\mathcal{O}(T_{\max}IM^2KN^2 + T_{\max}F_{\max}IM^2K^2N)$, which is a polynomial complexity. In comparison, The total computational complexity of the algorithm in[35] is $\mathcal{O}(MK^2 + MN + TK^{4.5}\log_2(1/\varepsilon))$, where T is the maximum iteration times and ε is the solution accuracy. Since the number of the antennas is much larger than those

of the users and the RF chains, i.e., $N \gg K, N \gg M$, the computational complexity in[35] is lower compared with our algorithm, because the hybrid beamforming is not jointly optimized with the power allocation.

6.4.6 Performance Simulations

In this subsection, we provide some simulation results to verify the performance of the proposed NOMA scheme. We adopt the channel model shown in (6.98), where the users are uniformly distributed from 10 m to 100 m away from the BS, and the channel gain of the node 30 m away from the BS has an average power of 0 dB to noise power. The number of MPCs for all the users is $L = 4$. Both LoS and NLoS channel models are considered. For the LoS channel, the average power of the NLoS paths is 15 dB weaker than that of the LoS path. For the NLoS channel, the coefficient of each path has an average power of $1/\sqrt{L}$. The cosine of the AoD for each path of the users is generated by a uniformly distributed random variable ranging from -1 to 1. Each point of the figures are the average performance of 100 channel realizations. The corresponding parameter settings are $I = 800, F_{\max} = 6, T_{\max} = 200, c_1 = c_2 = 1.4, \omega_{\max} = 0.9, \omega_{\min} = 0.4$.

In the simulations, we consider the following six typical millimeter-wave communication schemes: "NOMA Proposed" corresponds to the proposed joint approach, including user grouping, power allocation, and hybrid beamforming. "NOMA Ideal" is based on the proposed joint approach and with assumption of none inter-group interference, i.e., $I_{m,n}^{(\text{inter})} = 0$. Besides, "NOMA [13]" and "fully digital MIMO" correspond to the approach for NOMA with fully connected hybrid beamforming structure in[35] and the millimeter-wave-fully-digital-MIMO structure with ZF precoding, respectively. For fair comparison, the power splitting part in[35] is neglected in the simulations, which means that all the power is used for wireless information transmission. "TDMA-ZF" corresponds to the performance of millimeter-wave TDMA system, where M out of K users are served in each time slot. Each user is served by an independent analog beamformer with steering vector, and ZF and water-filling method is adopted for digital beamforming. While for "FDMA", the users are assigned into M groups, and the users in the same group perform FDMA[35]. Then, the achievable rate of the FDMA scheme for the k-th user is

$$R_k^{\text{FDMA}} = \frac{1}{|\mathcal{G}^k|} \log_2 \left(1 + \frac{|\mathbf{h}_k^{\text{H}}\mathbf{w}_k|^2 p_k}{\sum_{j \notin \mathcal{G}^k} |\mathbf{h}_k^{\text{H}}\mathbf{w}_j|^2 p_j + \frac{\sigma^2}{|\mathcal{G}^k|}} \right), \quad (6.153)$$

where $\mathcal{G}^k$ represents the group which the k-th user belongs to. The beamforming vector $\mathbf{w}_k$ and the power allocation $\{p_k\}$ are generated by using the approach in[35].

In addition, we also evaluate the performance of the energy-efficiency (EE), which is defined as the ratio between the ASR and total power consumption, i.e.,

$$EE = \frac{R_{\text{sum}}}{P + N_{\text{RF}} P_{\text{RF}} + N_{\text{PS}} P_{\text{PS}}}, \quad (6.154)$$

where R_{sum} is the ASR. P is the transmit power. P_{RF} is the power consumption of each RF chain, and N_{RF} is the number of the RF chains, where $N_{\text{RF}} = N$ for the fully digital structure and $N_{\text{RF}} = M$ for the hybrid structure. P_{PS} is the power consumption of each PS, and N_{RF} is the number of the PSs, where $N_{\text{PS}} = 0$ for the fully digital structure and $N_{\text{PS}} = MN$ for the hybrid structure. In the simulations, we select the typical parameter settings of $P = 1$ W, $P_{\text{RF}} = 250$ mW, and $P_{\text{PS}} = 1$ mW[44].

Figs. 6.18 (left) and 6.18 (right) show the ASR and EE, that is defined in (6.154), comparisons between the proposed NOMA approach, the NOMA scheme in[35], OMA and fully digital MIMO with varying minimum rate constraint under the LoS channel and the NLoS channel models, respectively. The minimum rate constraints for all the users are equal to r. Clearly, the performance of the proposed NOMA system is distinctly better than that of the OMA system, TDMA, and the solution of NOMA in[35]. Particularly, when the minimum rate constraint r ranges from 1 to 2 bps/Hz, the ASR of the proposed approach is nearly 10 bps/Hz larger than that of the scheme in[35]. The reason is as follows. When r is small, according to the NOMA principle, more beam gains and power can be allocated to the user with the highest channel gain in each group[38]. Thus, the beamforming scheme in[35] is effective, where the beam in analog domain is steering to the first user in each group. When r becomes larger, the users with worse channel conditions can only be served by the sidelobe of the beam in[35]. In contrast, the presented solution in this section can allocate more beam gains in analog domain to the users with worse channel conditions in each group. Thus,

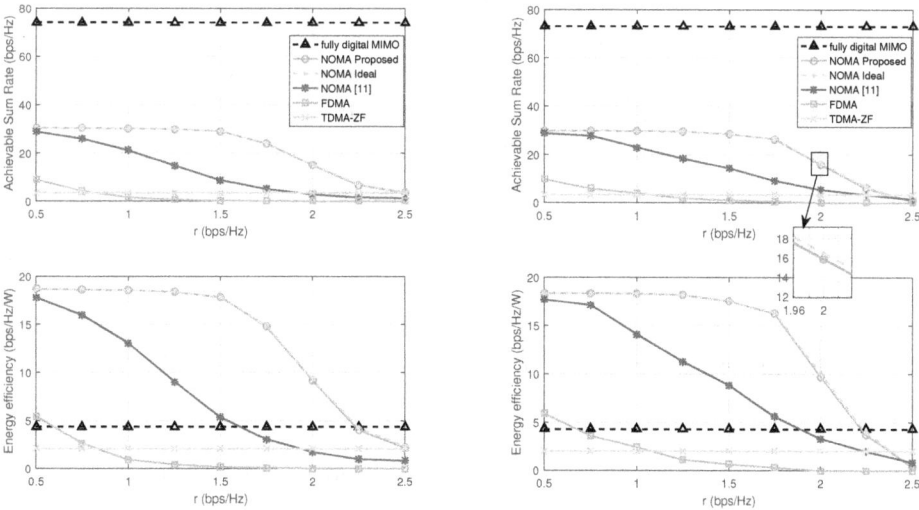

Figure 6.18 **Left**: ASR/EE comparison between the NOMA and OMA systems with varying minimum rate constraint under the LoS channel model, where $N = 64$, $M = 2$, $K = 6$, and $P/\sigma^2 = 30$ dB. **Right**: ASR/EE comparison between the NOMA and OMA systems with varying minimum rate constraint under the NLoS channel model, where $N = 64$, $M = 2$, $K = 6$, and $P/\sigma^2 = 30$ dB.

the proposed approach outperforms the scheme in[35]. However, when r is large, there may exist some channel realizations in which the minimum rate constraint cannot be satisfied. In such a case, the ASR is set to be zero. This operation is also adopted in the scheme of[35], which ensures the fairness of the comparison between the two methods. Therefore, the ASR tends to be zero for both of the two schemes, when r is sufficiently large. Since the average ASR of the proposed scheme is larger than that of the scheme in[35], it can be concluded that the presented method in this section can find a better solution and achieve a higher feasibility. Besides, we can also find that the ASR of the proposed approach is close to the ideal case, which indicates that the inter-group interference is small by using the proposed user grouping and hybrid beamforming schemes and has little influence on the ASR. This result also verifies that the approximation of neglecting the inter-group interference when optimizing the intra-GPA is reasonable. We have also provided an enlarged view of the ASR curve in Fig. 6.18 (right), it can be seen that there is a small gap between the ideal curve and the designed curve, which is caused by the inter-group interference. The performance gap is no more than 0.5 bps/Hz, which is very small compared with the total ASR. In the two figures, we can also find that, although the ASR of the fully digital MIMO structure is higher than that of both the NOMA and OMA, the EE of the fully digital MIMO structure is low compared with the hybrid beamforming structure. Particularly, the EE of the proposed NOMA scheme can achieve nearly fourfold EE compared with the fully digital MIMO structure when the minimal rate constraint is no more than 1.5 bps/Hz.

Figs. 6.19 (left) and 6.19 (right) compare the ASRs/EEs between the proposed NOMA approach, the NOMA scheme in[35], OMA and fully digital MIMO with varying total power to noise ratio under the LoS channel and the NLoS channel models, respectively. From the two figures, we can find again that the proposed NOMA approach can achieve a higher ASR than that of NOMA in[35], as well as the OMA system. Particularly, when P/σ^2 is low, i.e., the NOMA system is power limited, the superiority of the proposed algorithm is more conspicuous compared with the approach in[35]. When P/σ^2 is larger than 35 dB, the performance gap between the proposed solution and the solution in[35] stabilises around 5 bps/Hz in Fig. 6.19 (left), while the performance gap stabilises around 7.5 bps/Hz in Fig. 6.19 (right). From the two figures, we can find again that the EE of the proposed NOMA scheme with a hybrid beamforming structure is larger than that of the fully digital MIMO structure, as well as larger than the EE of OMA. When P/σ^2 becomes large, the curves of the EE for different schemes all tend to be linear, and the increasing velocity, i.e., the slope of the EE curve, for NOMA is larger than that for both fully digital MIMO and OMA.

Figs. 6.20 (left) and 6.20 (right) compare the ASRs/EEs between NOMA and OMA systems with varying number of RF chains under the LoS channel and the NLoS channel models, respectively. It can be observed that the proposed NOMA approach outperforms the OMA. In Fig. 6.20 (left), when the number of RF chains is no larger than 4, the ASR of the proposed approach is larger than that of NOMA in[35]. When the number of RF chains is 5, the scheme in[35] behaves slightly better than the proposed scheme, and both of them are close to the performance of the

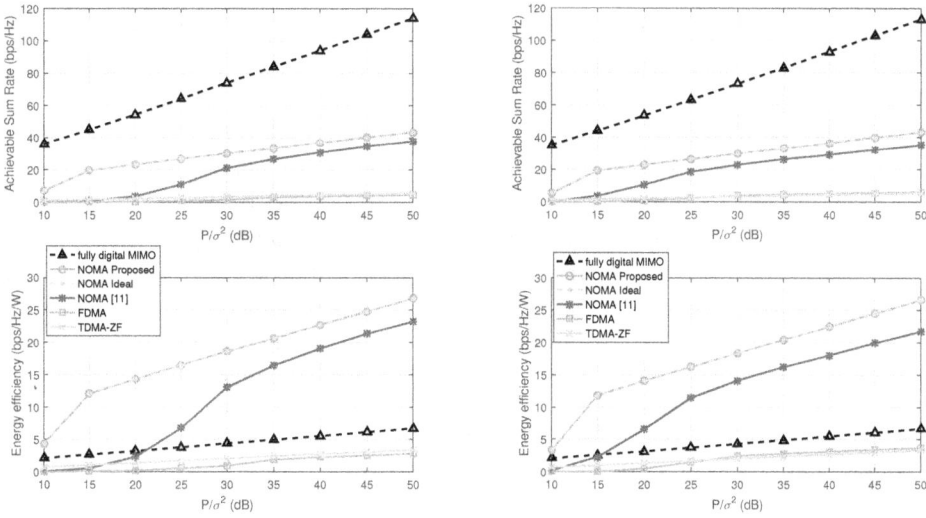

Figure 6.19 **Left**: ASR/EE comparison between the NOMA and OMA systems with varying total power to noise ratio under the LoS channel model, where $N = 64$, $M = 2$, $K = 6$, and $r_k = 1$ bps/Hz. **Right**: ASR/EE comparison between the NOMA and OMA systems with varying total power to noise ratio under the NLoS channel model, where $N = 64$, $M = 2$, $K = 6$, and $r_k = 1$ bps/Hz.

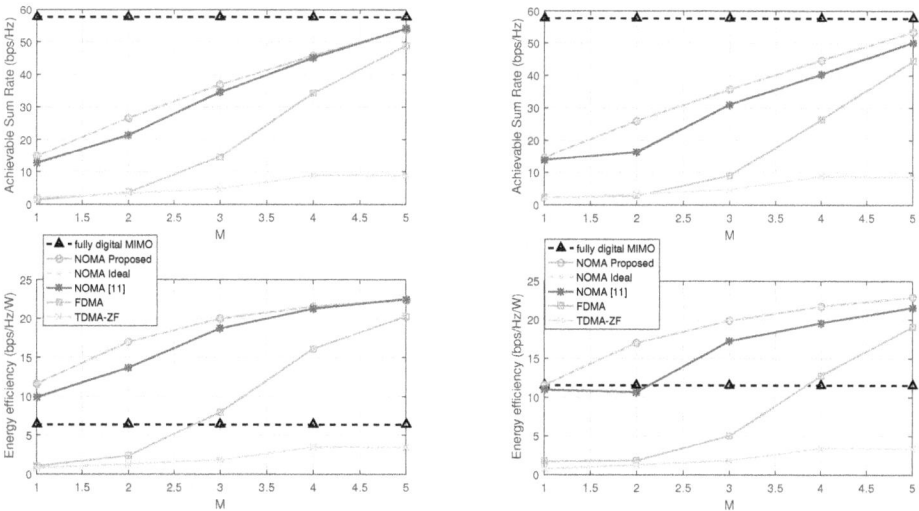

Figure 6.20 **Left**: ASR/EE comparison between the NOMA and OMA systems with varying number of RF chains under the LoS channel model, where $N = 16$, $K = 6$, $r_k = 1$ bps/Hz and $P/\sigma^2 = 30$ dB. **Right**: ASR/EE comparison between the NOMA and OMA systems with varying number of RF chains under the NLoS channel model, where $N = 16$, $K = 6$, $r_k = 1$ bps/Hz and $P/\sigma^2 = 30$ dB.

fully digital structure. The reason is that the total number of users is 6 in Fig. 6.20 (left). When the number of RF chains becomes larger, i.e., approximately equal to the number of users, the number of users in each group is usually one. Thus, the beamforming scheme in[35] is more effective, where the analog beams steer to the first user in each group. Moreover, the proposed solution always outperforms the NOMA scheme in[35] in Fig. 6.20 (right). Comparing the two figures, we can find that the ASRs of the proposed approach are almost not influenced by the channel models. In contrast, the ASRs of the NOMA scheme in[35] under the NLoS channel model is lower than that under the LoS channel model. The results indicate that the proposed approach is more robust against the channel model. We can also find that the EE of the proposed NOMA scheme increases for the number of the RF chains, and it is significantly larger than the EE of the fully digital MIMO structure.

6.5 SUMMARY

In this chapter, we investigated array beamforming enabled NOMA technologies, where three problems with different system settings and objectives were considered.

In the problem of user fairness, we formulated a joint analog beamforming and power allocation problem to maximize the minimal achievable rate among the users. In the solution, the closed-form optimal power allocation was obtained first, which converts the joint optimization problem into an equivalent beamforming problem. Then, an appropriate beamforming vector was designed.

In the problem of joint Tx-Rx beamforming and power allocation in NOMA networks, a joint Tx-Rx beamforming and power allocation problem was formulated to maximize the ASR subject to a minimum rate constraint for each user. As the problem is non-convex, we presented a sub-optimal solution with three stages. In the first stage, the optimal power allocation was obtained in closed form for an arbitrary fixed Tx-Rx beamforming. In the second stage, the optimal Rx beamforming with a closed form was designed for an arbitrary fixed Tx beamforming. In the third stage, the original joint Tx-Rx beamforming and power allocation problem was reduced to a Tx beamforming problem by using the previous results, and a BC-PSO algorithm was presented to obtain the sub-optimal solution.

In the problem of NOMA with hybrid beamforming, we considered downlink transmission with a hybrid beamforming structure. A user grouping algorithm was first presented according to the channel correlations of the users. Whereafter, a joint hybrid beamforming and power allocation problem was formulated to maximize the ASR, subject to a minimum rate constraint for each user. To solve this non-convex problem with high-dimensional variables, we first obtained the solution of power allocation under arbitrary fixed hybrid beamforming, which was divided into intra-group power allocation and inter-group power allocation. Then, given arbitrary fixed analog beamforming, we utilized the AZF method to design the digital beamforming to minimize the inter-group interference. Finally, the analog beamforming problem with the CM constraint was solved with the proposed BC-PSO algorithm.

Bibliography

[1] Jeffrey G. Andrews, Stefano Buzzi, Wan Choi, Stephen V. Hanly, Angel Lozano, Anthony C. K. Soong, and Jianzhong Charlie Zhang. What will 5G be? *IEEE J. Select. Areas Commun.*, 32(6):1065–1082, Apr. 2014.

[2] Zhiguo Ding, Xianfu Lei, George K. Karagiannidis, Robert Schober, Jinhong Yuan, and Vijay K. Bhargava. A survey on non-orthogonal multiple access for 5G networks: Research challenges and future trends. *IEEE J. Select. Areas Commun.*, 35(10):2181–2195, Oct. 2017.

[3] Linglong Dai, Bichai Wang, Zhiguo Ding, Zhaocheng Wang, Sheng Chen, and Lajos Hanzo. A survey of non-orthogonal multiple access for 5G. *IEEE Commun. Surveys Tutorials*, 20(3):2294–2323, thirdquarter 2018.

[4] Ming Xiao, Shahid Mumtaz, Yongming Huang, Linglong Dai, Yonghui Li, Michail Matthaiou, George K. Karagiannidis, Emil BjÃŭrnson, Kai Yang, Chih-Lin I, and Amitabha Ghosh. Millimeter wave communications for future mobile networks. *IEEE J. Select. Areas Commun.*, 35(9):1909–1935, Sep. 2017.

[5] Anass Benjebbour, Yuya Saito, Yoshihisa Kishiyama, Anxin Li, Atsushi Harada, and Takehiro Nakamura. Concept and practical considerations of non-orthogonal multiple access (NOMA) for future radio access. In *International Symposium on Intelligent Signal Processing and Communication Systems*, pages 770–774, Nov. 2013.

[6] Linglong Dai, Bichai Wang, Yifei Yuan, Shuangfeng Han, I. Chih-lin, and Zhaocheng Wang. Non-orthogonal multiple access for 5G: solutions, challenges, opportunities, and future research trends. *IEEE Commun. Mag.*, 53(9):74–81, Sep. 2015.

[7] Jinho Choi. Non-orthogonal multiple access in downlink coordinated two-point systems. *IEEE Commun. Lett.*, 18(2):313–316, Feb. 2014.

[8] Stelios Timotheou and Ioannis Krikidis. Fairness for non-orthogonal multiple access in 5G systems. *IEEE Signal Processing Lett.*, 22(10):1647–1651, Oct. 2015.

[9] Jinho Choi. Power allocation for max-sum rate and max-min rate proportional fairness in NOMA. *IEEE Commun. Lett.*, 20(10):2055–2058, Oct. 2016.

[10] Junpei Umehara, Yoshihisa Kishiyama, and Kenichi Higuchi. Enhancing user fairness in non-orthogonal access with successive interference cancellation for cellular downlink. In *Proc. IEEE Int. Conf. Commun. Systems*, pages 324–328, Nov. 2012.

[11] Yuanwei Liu, Maged Elkashlan, Zhiguo Ding, and George K. Karagiannidis. Fairness of user clustering in MIMO non-orthogonal multiple access systems. *IEEE Commun. Lett.*, 20(7):1465–1468, Jul. 2016.

[12] Hong Xing, Yuanawei Liu, Arumugam Nallanathan, Zhiguo Ding, and H. Vincent Poor. Optimal throughput fairness tradeoffs for downlink non-orthogonal multiple access over fading channels. *IEEE Trans. Wireless Commun.*, 17(6):3556–3571, Jun. 2018.

[13] Jingjing Cui, Yuanwei Liu, Zhiguo Ding, Pingzhi Fan, and Arumugam Nallanathan. Optimal user scheduling and power allocation for millimeter wave NOMA systems. *IEEE Trans. Wireless Commun.*, 17(3):1502–1517, Mar. 2018.

[14] Bichai Wang, Linglong Dai, Zhaocheng Wang, Ning Ge, and Shidong Zhou. Spectrum and energy efficient beamspace MIMO-NOMA for millimeter-wave communications using lens antenna array. *IEEE J. Select. Areas Commun.*, 35(10):2370–2382, Oct. 2017.

[15] Wei Wu and Danpu Liu. Non-orthogonal multiple access based hybrid beamforming in 5G mmWave systems. In *2017 IEEE 28th Annual International Symposium on Personal, Indoor, and Mobile Radio Communications (PIMRC)*, pages 1–7, Oct. 2017.

[16] Zhiqiang Wei, Lou Zhao, Jiajia Guo, Derrick Wing Kwan Ng, and Jinhong Yuan. Multi-beam NOMA for hybrid mmwave systems. *IEEE Trans. Commun.*, 67(2):1705–1719, Feb. 2019.

[17] Zhenyu Xiao, Tong He, Pengfei Xia, and Xiang-Gen Xia. Hierarchical codebook design for beamforming training in millimeter-wave communication. *IEEE Trans. Wireless Commun.*, 15(5):3380–3392, May 2016.

[18] Zhenyu Xiao, Pengfei Xia, and Xiang-Gen Xia. Codebook design for millimeter-wave channel estimation with hybrid precoding structure. *IEEE Trans. Wireless Commun.*, 16(1):141–153, Jan. 2017.

[19] Tadilo Endeshaw Bogale, Long Bao Le, Afshin Haghighat, and Luc Vandendorpe. On the number of RF chains and phase shifters, and scheduling design with hybrid analog-digital beamforming. *IEEE Trans. Wireless Commun.*, 15(5):3311–3326, May 2016.

[20] Yuan-Pei Lin. On the quantization of phase shifters for hybrid precoding systems. *IEEE Trans. Signal Processing*, 65(9):2237–2246, May 2017.

[21] Yuexing Peng, Yonghui Li, and Peng Wang. An enhanced channel estimation method for millimeter wave systems with massive antenna arrays. *IEEE Commun. Lett.*, 19(9):1592–1595, Sep. 2015.

[22] Peng Wang, Yonghui Li, Lingyang Song, and Branka Vucetic. Multi-gigabit millimeter wave wireless communications for 5G: from fixed access to cellular networks. *IEEE Commun. Mag.*, 53(1):168–178, Jan. 2015.

[23] Junho Lee, Gye-Tae Gil, and Yong H. Lee. Exploiting spatial sparsity for estimating channels of hybrid MIMO systems in millimeter wave communications. In *Proc. IEEE Global Telecommun. Conf.*, pages 3326–3331. IEEE, 2014.

[24] Zhen Gao, Chen Hu, Linglong Dai, and Zhaocheng Wang. Channel estimation for millimeter-wave massive MIMO with hybrid precoding over frequency-selective fading channels. *IEEE Commun. Lett.*, 20(6):1259–1262, Jun. 2016.

[25] Ahmed Alkhateeb, Omar El Ayach, Geert Leus, and Robert W. Heath. Channel estimation and hybrid precoding for millimeter wave cellular systems. *IEEE J. Sel. Top. Sign. Proces.*, 8(5):831–846, Oct. 2014.

[26] Yuya Saito, Yoshihisa Kishiyama, Anass Benjebbour, Takehiro Nakamura, Anxin Li, and Kenichi Higuchi. Non-orthogonal multiple access (NOMA) for cellular future radio access. In *IEEE Vehicular Technology Conference (IEEE VTC Spring)*, pages 1–5, Dresden, Germany, 2013. IEEE.

[27] Stephen Boyd and Lieven Vandenberghe. *Convex optimization*. Cambridge university press, 2004.

[28] Zhenyu Xiao, Lipeng Zhu, Zhen Gao, Dapeng Oliver Wu, and Xiang-Gen Xia. User fairness non-orthogonal multiple access (NOMA) for 5G millimeter-wave communications with analog beamforming. *IEEE Trans. Wireless Commun.*, 18(7):3411–3423, Jul. 2019.

[29] Zhenyu Xiao, Lipeng Zhu, Jinho Choi, Pengfei Xia, and Xiang-Gen Xia. Joint power allocation and beamforming for non-orthogonal multiple access (NOMA) in 5G millimeter wave communications. *IEEE Trans. Wireless Commun.*, 17(5):2961–2974, May 2018.

[30] Lipeng Zhu, Jun Zhang, Zhenyu Xiao, Xianbin Cao, Dapeng Oliver Wu, and Xiang-Gen Xia. Joint power control and beamforming for uplink non-orthogonal multiple access in 5G millimeter-wave communications. *IEEE Trans. Wireless Commun.*, 17(9):6177–6189, Sep. 2018.

[31] Zhiguo Ding, Pingzhi Fan, and H. Vincent Poor. Random beamforming in millimeter-wave NOMA networks. *IEEE Access*, 5:7667–7681, Feb. 2017.

[32] Yahya Rahmat-Samii, Dennis Gies, and Jacob Robinson. Particle swarm optimization (PSO): A novel paradigm for antenna designs. *URSI Radio Science Bulletin*, 2003(306):14–22, Sep. 2003.

[33] Jacob T. Robinson and Yahya Rahmat-Samii. Particle swarm optimization in electromagnetics. *IEEE Trans. Antennas Propagat.*, 52(2):397–407, Feb. 2004.

[34] Yoshikazu Fukuyama. Fundamentals of particle swarm optimization techniques. *Modern Heuristic Optimization Techniques: Theory and Applications to Power Systems*, pages 71–87, 2008.

[35] Linglong Dai, Bichai Wang, Mugen Peng, and Shanzhi Chen. Hybrid precoding-based millimeter-wave massive MIMO-NOMA with simultaneous wireless information and power transfer. *IEEE J. Select. Areas Commun.*, 37(1):131–141, Jan. 2019.

[36] Di Zhang, Zhenyu Zhou, Chen Xu, Yan Zhang, Jonathan Rodriguez, and Takuro Sato. Capacity analysis of NOMA with mmWave massive MIMO systems. *IEEE J. Select. Areas Commun.*, 35(7):1606–1618, Jul. 2017.

[37] Zhiqiang Wei, Lou Zhao, Jiajia Guo, Derrick Wing Kwan Ng, and Jinhong Yuan. A multi-beam NOMA framework for hybrid mmWave systems. In *2018 IEEE International Conference on Communications (ICC)*, pages 1–7, May 2018.

[38] Lipeng Zhu, Jun Zhang, Zhenyu Xiao, Xianbin Cao, Dapeng Oliver Wu, and Xiang-Gen Xia. Joint Tx-Rx beamforming and power allocation for 5G millimeter-wave non-orthogonal multiple access (mmwave-NOMA) networks. *IEEE Trans. Commun.*, 67(7):5114–5125, Jul. 2019.

[39] Matthew Kokshoorn, He Chen, Peng Wang, Yonghui Li, and Branka Vucetic. Millimeter wave MIMO channel estimation using overlapped beam patterns and rate adaptation. *IEEE Trans. Signal Processing*, 65(3):601–616, Feb. 2017.

[40] Zhenyu Xiao, Hang Dong, Lin Bai, Pengfei Xia, and Xiang-Gen Xia. Enhanced channel estimation and codebook design for millimeter-wave communication. *IEEE Trans. Veh. Technol.*, 67(10):9393–9405, Oct. 2018.

[41] Chen Hu, Linglong Dai, Talha Mir, Zhen Gao, and Jun Fang. Super-resolution channel estimation for MmWave massive MIMO with hybrid precoding. *IEEE Trans. Veh. Technol.*, 67(9):8954–8958, Sep. 2018.

[42] Constantine A. Balanis. *Antenna theory: analysis and design.* John wiley & sons, 2016.

[43] Tapas Kanungo, David M. Mount, Nathan S. Netanyahu, Christine D. Piatko, Ruth Silverman, and Angela Y. Wu. An efficient k-means clustering algorithm: Analysis and implementation. *IEEE Trans. on Pattern Analysis & Machine Intelligence*, (7):881–892, 2002.

[44] Xinyu Gao, Linglong Dai, Shuangfeng Han, Chih-Lin I, and Robert W. Heath. Energy-efficient hybrid analog and digital precoding for MmWave MIMO systems with large antenna arrays. *IEEE J. Select. Areas Commun.*, 34(4):998–1009, Apr. 2016.

CHAPTER 7

Array Beamforming Enabled UAV Communications

7.1 INTRODUCTION

During the past a few years, unmanned aerial vehicle (UAV) technologies, including the platform, communication, flying control, and surveillance techniques, have developed rapidly. Due to the high mobility, fast deployment and low cost, UAV is widely applied in military and civilian fields, e.g., reconnaissance, transportation, infrastructure inspection, agricultural irrigation, disaster rescue and so on[1, 2, 3, 4, 5]. These applications enabled by UAV can greatly reduce the labor-cost and improve the public security.

When performing different tasks, UAVs usually need to transmit mission-related information, such as sensor data and high resolution image, to the ground terminals. Thus, higher-data-rate communications are required for UAV. For high data rate communications, it is critical to improve the signal strength at Rx. Hence, antenna array is usually used to achieve array gain and overcome the path loss. It is worth noting that although the point-to-point beamforming and training methods introduced in Chapter 2 and 3 are derived from conventional terrestrial communications, it can be expanded to UAV platforms with extra characteristics, e.g., the space and energy constraints and the mobility of UAVs, etc.

For the space and energy constrains, benefiting from the short wavelength of the millimeter-wave signals, a large antenna array can be equipped in even a small area[6, 7, 8, 9, 10], which is appropriate for UAVs[11]. For a UAV with high altitude, the line of sight (LoS) path is longstanding[1, 3, 12, 13, 14, 15]. Moreover, due to the high mobility of UAV, the LoS path can be actively created on demand via the movement of UAV. The strength of the LoS component may be critically over 20 dB stronger than those of the non-LoS (NLoS) components[16]. Thus, antenna array enabled directional beamforming is quite appropriate for UAV to track the LoS path.

Besides, the directional beamforming and high propagation loss provide new opportunities to handle the dominant interference in UAV communications. The ground base station (BS) can cover UAVs with narrow beams, and a UAV can also reach the ground users with directional beams and other UAVs. The directional beams can

DOI: 10.1201/9781003366362-7

achieve higher channel gains than that of the full coverage. Thus, the spectrum efficiency can be greatly increased. Moreover, through the full-duplex (FD) technology in Chapter 4 and non-orthogonal multiple access (NOMA) technology in Chapters 5 and 6, the spectrum efficiency can be improved as well, and thus the effective capacity can be improved.

In this chapter, we will focus on antenna array beamforming enabled UAV communications, discuss its uniqueness and provide relative solutions, where antenna array plays an important role. In the second section, we will introduce the channel model of array enabled UAV communications. In the third section, we will introduce 3D beam coverage with flexible beamforming. Then in the forth and fifth sections, we will introduce single UAV deployment and multiple UAVs deployment in communication scenarios, respectively.

7.2 CHANNEL MODEL

The characteristics and channel modeling for UAV communications are more complex and challenging than that for ground communications. Accurate channel models facilitate the performance analysis of UAV-enabled wireless communications in terms of capacity and coverage[17, 18]. For antenna array enabled UAV communications, the radio propagation characteristics of millimeter-wave channel are significantly different from those for classical ground channels. Hence, it is crucial to properly model the UAV millimeter-wave channels for the ease of describing radio propagation characteristics and analyzing the performance of the communication system. In this section, we present propagation characteristics and channel modeling for UAV communications.

7.2.1 Propagation Characteristics

Compared to the microwave frequency bands, the main characters of the millimeter-wave frequency bands include short wavelength, large bandwidth, large penetration loss, and strong atmospheric attenuation. Moreover, due to the mobility of a UAV BS, the main differences between UAV and terrestrial communications include temporal variations of the non-stationary channels, dynamic change between LoS and NLoS environments, and inherent airframe shadowing and fluctuations. When a signal propagates through building walls, trees or human bodies, there are penetration losses.

If there exists an obstacle between transmitter (Tx) and receiver (Rx), the LoS path will be blocked. A widely used air-to-ground (A2G) probabilistic LoS model is given in[18] and[19], and was derived by using the statistical parameters provided by International Telecommunication Union (ITU). The LoS probability is modeled as a logistic function of the elevation angle θ as follows[18, 19]

$$P_{\text{LoS}}(\theta) = \frac{1}{1 + a \exp(-b(\theta - a))}, \tag{7.1}$$

where a and b are modeling parameters that depend on the environment. In fact, the measurement results in[20, 21, 22] have shown that A2G communication channels are mainly dominated by LoS links even if a UAV is located at a moderate altitude. For example, for a UAV operating at an altitude of 120 m, the LoS probability of A2G links in a rural environment exceeds 95%[22]. Moreover, if there are obstacles between the transceivers, the LoS path can be rapidly restored by flexibly adjusting the 3D position of the UAV. Hence, it is relatively easy for UAV to establish LoS link with ground users.

Due to the short wavelength of millimeter-wave signals and the high altitude of UAVs, A2G millimeter-wave signal propagation mechanisms differ from the conventional terrestrial propagation. The multiple components (MPCs) of millimeter-wave channels are mainly caused by reflections from ground scatters, including the earth surface, buildings and human bodies[23]. Besides, the motions of both the aerial nodes and the ground nodes introduce Doppler shifts, which result in carrier frequency offset, inter-carrier interference, and limited channel coherence time. The Doppler shift is proportional to the carrier frequency and the mobile velocity, and is also influenced by the angular dispersion[24,25]. In antenna array enabled UAV communication systems, severe Doppler shifts are caused by the high carrier frequency and high mobility. It is worth the research effort to model this property and compensate the effect for A2G communication scenarios.

7.2.2 Airframe Shadowing and Fluctuation

Airframe shadowing and hovering fluctuation are unique to the UAV communications. In A2G communications, the LoS paths may be blocked due to the UAV structure design, on-board antenna placement, and UAV flight status. Moreover, the short-wavelength signals may be more easily blocked and reflected by the metallic aircraft body, especially UAV fuselage, which needs to be considered in the modeling of UAV millimeter-wave channels. The effect of airframe shadowing cannot be eliminated by exploiting the spatial diversity at the ground node. In addition, there is no significant correlation between the airframe shadowing loss and the shadowing duration in A2G environments[26]. Specifically, the airframe shadowing loss can be modeled as a function of the aircraft roll angle, while the shadowing duration is mainly affected by the flight speed[26].

Fluctuations of the positions of the on-board antennas may be caused by the engine vibrations and wind turbulence. Strong wing and other severe weather may also cause UAV jittering. In antenna array enabled millimeter-wave-UAV communication systems, although high directional antenna gains can compensate the high path loss, the vibrations of the transceivers deteriorate the channel quality because of the narrow beamwidth. The position of the UAV also influences the degree of angle of departure (AoD) fluctuation at the UAV side[27]. One potential approach is to utilize angle of arrival (AoA) and AoD estimates to guide the beam alignment. In addition to the performance loss caused by antenna mismatch, the channel coherence time in millimeter-wave frequency bands is in the order of microsecond due to UAV jittering[28], which increases the difficulty of channel tracking and phase estimation.

7.2.3 A2G Channel Modeling

In the existing works, channel models are mainly classified into deterministic and stochastic channel models. Deterministic channel models, such as ray-tracing and map-based channel models, try to model the actual propagation characteristics of electromagnetic waves. These models rely on propagation measurements and information collected in databases regarding the environment. Stochastic channel models, such as geometry-based stochastic channel models (GSCMs) and tapped delay line (TDL) models, utilize statistical distribution models and empirical parameters to mathematically analyze the channel characteristics with a relatively low computational complexity.

The time-varying complex impulse response of the Saleh-Valenzuela model is given as follows[29]

$$h(t) = \sum_{l=1}^{L} a_l e^{-j\psi_l(t)} \delta(t - \tau_l), \tag{7.2}$$

where a_l, ψ_l and τ_l denote the time-varying amplitude, phase, and delay of the l-th MPC, respectively. For UAV communications, the fading can be modeled by the Nakagami distribution which can capture various channel fading conditions and provides a good fit with experimentally measured data[30].

For a millimeter-wave-multiple-input multiple-output (MIMO) system with N_T transmit and N_R receive antennas, the time-varying channel response, i.e., (1.136) in Chapter 1, can be expressed as

$$\mathbf{H}(t, f) = \sqrt{\frac{N_T N_R}{L}} \sum_{l=1}^{L} \beta_l e^{j2\pi(v_l t - \tau_l f)} \times \mathbf{a}_r(\theta_{r,l}, \phi_{r,l}) \mathbf{a}_t^{\mathrm{H}}(\theta_{t,l}, \phi_{t,l}), \tag{7.3}$$

where L is the total number of MPCs. For each MPC l, β_l denotes the complex gain, which includes the large-scale fading and small-scale fading. $\theta_{r,l}$, $\phi_{r,l}$, $\theta_{t,l}$, and $\phi_{t,l}$ represent the elevation AoA, azimuth AoA, elevation AoD, and azimuth AoD, respectively. Parameters τ_l and v_l are the delay and Doppler shift of the l-th MPC, respectively. $\mathbf{a}_t$ and $\mathbf{a}_r$ denote the steering vectors at Tx and Rx, respectively, which are determined by the geometry of the arrays, i.e., uniform linear array (ULA) and uniform rectangular array (URA). The steering vector of a ULA can be expressed as (1.70)

$$\mathbf{a}_{\mathrm{ULA}}(\theta) = \frac{1}{\sqrt{N}} \left[1, e^{-j\frac{2\pi}{\lambda}d\cos\theta}, \cdots, e^{-j\frac{2\pi}{\lambda}(N-1)d\cos\theta}\right]^{\mathrm{T}}, \tag{7.4}$$

and the steering vector of a URA can be expressed as (1.94)

$$\mathbf{a}_{\mathrm{URA}}(\theta, \phi) = \frac{1}{\sqrt{MN}} [1, \cdots, e^{j\frac{2\pi}{\lambda}d\sin\theta[(m-1)\cos\phi+(n-1)\sin\phi]},$$
$$\cdots, e^{j\frac{2\pi}{\lambda}d\sin\theta[(M-1)\cos\phi+(N-1)\sin\phi]}]^{\mathrm{T}}, \tag{7.5}$$

Supposing that the channel is sufficiently slow-varying over the signal duration of interest, i.e., the Doppler shifts are small, (7.3) can be simplified as follows

$$\mathbf{H}\left(f\right) = \sqrt{\frac{N_\mathrm{T}N_\mathrm{R}}{L}} \sum_{l=1}^{L} \beta_l e^{-j2\pi\tau_l f} \mathbf{a}_r\left(\theta_{r,l}, \phi_{r,l}\right) \mathbf{a}_t^\mathrm{H}\left(\theta_{t,l}, \phi_{t,l}\right). \qquad (7.6)$$

If the bandwidth is sufficiently small, the narrow-band discrete channel model is obtained as follows

$$\mathbf{H} = \sqrt{\frac{N_\mathrm{T}N_\mathrm{R}}{L}} \sum_{l=1}^{L} \beta_l \mathbf{a}_r\left(\theta_{r,l}, \phi_{r,l}\right) \mathbf{a}_t^\mathrm{H}\left(\theta_{t,l}, \phi_{t,l}\right), \qquad (7.7)$$

which is also known as the extended Saleh-Valenzuela model and has been used in the earlier chapters.

In conclusion, the channel modeling for array enabled UAV communication should consider several special characters. First, the reflection of the signals at the UAV side is rare, and thus the channel presents higher sparsity. Second, the navigation and jittering characters of a UAV make the millimeter-wave channel change frequently over time. Third, the mobility of UAV and high frequency band of millimeter-wave signals aggravate the Doppler effect in array enabled UAV communications. Moving forward, more research works and experiments are needed on the channel modeling and measurement to facilitate the investigation of array enabled UAV communications.

7.3   3D BEAM COVERAGE

Different from ground BS in cellular network, for a UAV-BS, a fixed 2D beam cannot satisfy the requirement of coverage for multiple ground user equipments (UEs). Since a UAV usually operates at a high altitude, 3D beams are required to realize the coverage from air to ground, where both the azimuth and elevation angles should be taken into account[31, 32]. Besides, due to the mobility of UAVs, the flexibility of the beam should be controlled to adapt to the varying of the environment.

7.3.1   Commonly Used 3D Beamforming Methods

In this section, we introduce the commonly used 3D beam coverage strategies for UAV-BSs.

1) Single Beam Coverage: Equipping a directional antenna at a UAV, the direction of the beam can be adjusted by using a mechanical adjustment module, such as tripod head and servo system. In addition to the directional antenna, phased array can also shape a direction beam. Compared to the fully digital beamforming and hybrid beamforming structures, phased array employs pure analog beamforming, which has a low hardware cost and power consumption. Different from the directional antenna, a phased array can change the direction

of the beam by using the electrical adjustment, i.e., changing the phase of the signal for each antenna branch. Compared to the mechanical adjustment, the electrical adjustment has a lower latency and higher efficiency.

As we have analyzed before, if we employ a steering vector as the beamforming vector, the beam pattern of a phased array is similar to that of the directional antenna. A common characteristic of the directional antenna and a small antenna array employing the steering-vector based beamforming is that they can only generate a single-directional beam in space at one time. If the number of UEs is small and the distribution of the UEs is concentrated, a single beam can cover the UEs. In this case, only the azimuth and elevation angles of the beam need to be optimized, which has a low computational complexity in general[32,33,34]. When the number of the UEs becomes large or the UEs locate dispersedly, a single narrow beam may not be able to cover all the UEs simultaneously. In such a case, the UEs which are located outside the mainlobe of the beam may face bad channel conditions. To solve this problem, an intuitive way is to dynamically adjust the direction of the beam and serve different UEs in different time slots. The advantage of the single beam coverage is that it requires a low hardware cost, low power consumption, and low computational complexity. However, the coverage efficiency is limited because of the single narrow beam.

2) Multi-Beam Coverage: An efficient way to improve the coverage efficiency is to increase the number of beams. For example, multiple directional antennas can be equipped at the UAV and each antenna steers to a specific direction. By optimizing the steering beams of the antennas, the coverage performance of the UAV can be effectively improved[35].

For large antenna arrays, the hybrid analog-digital beamforming structure is promising to be used for UAVs. In fact, the hybrid beamforming achieves a tradeoff between the high hardware-cost digital beamforming and the low-efficiency analog beamforming. In general, each radio frequency (RF) chain can generate a single analog beam. The interference between the multiple beams can be mitigated by digital beamforming, based on zero-forcing (ZF), minimum mean square error (MMSE), and other convex optimization techniques.

In addition to the phased-array-based structure, lens antenna array is another typical structure to achieve multi-beam coverage. As shown in Fig. 7.1, a lens is equipped on the front end to change the propagation characteristic of the millimeter-wave channels, which are known as beamspace MIMO[36,37]. Then, the received signals at different antennas have different gains, which depend on the AoDs/AoAs of the signals. The analog beamforming portion is realized by employing a switch-inverter network, where a small number of RF chains can select antennas that have high antenna gains. Recently, some research works have studied the application of millimeter-wave lens array for UAV-BSs to improve the communication service for UEs[31,38], where the beam selection is one of the most important issues for UE coverage in the analog domain. Compared to phase-shifter-based hybrid beamforming structure, the lens array has a low hardware cost and power consumption. It is easier to generate multi-beam to improve the coverage of the UEs. However, due to the discrete deployment of the antennas, there exists power leakage for lens arrays[37]. The selected beams may not perfectly match the AoDs/AoAs because of the low flexibility of analog beamforming, and the interference between multiple beams cannot be mitigated in the analog domain.

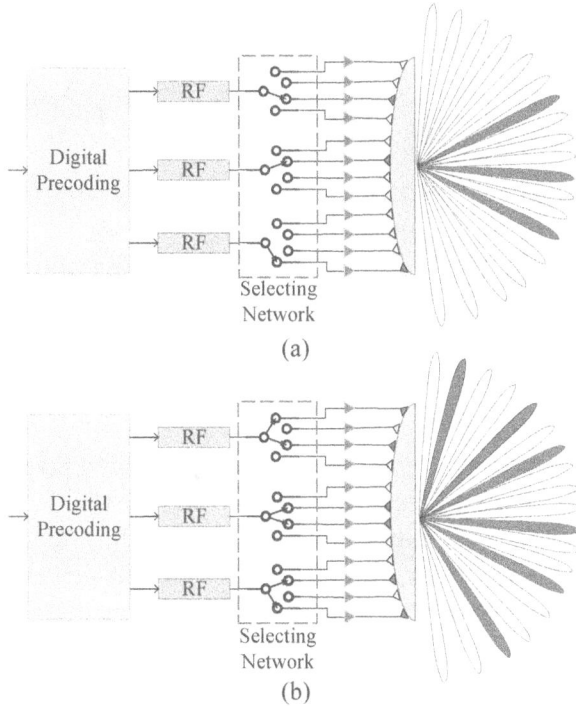

Figure 7.1 Architecture of the lens antenna array: (a) Traditional switch-based selecting network. (b) Switch and splitter/mixer based multi-beam forming network.

3) Flexible Beam Coverage: Although the 3D position of a UAV can be optimized to cover the users with a fixed beam in some cases[39], it cannot guarantee that all the ground users are always covered, especially in the cases that the UAV cannot stop, e.g., when a fixed-wing UAV is adopted or the UAV has other tasks like surveillance. One possible way to improve the coverage is to use a flexible beam, where the shape of the 3D beam can be adaptively adjusted according to the distribution of the UEs for realizing a full coverage of the target region. Hence, flexible beamforming is more suitable for a UAV-BS to accommodate its mobility. We will discuss the flexible beam coverage for array enabled UAV-BS in the next section in details.

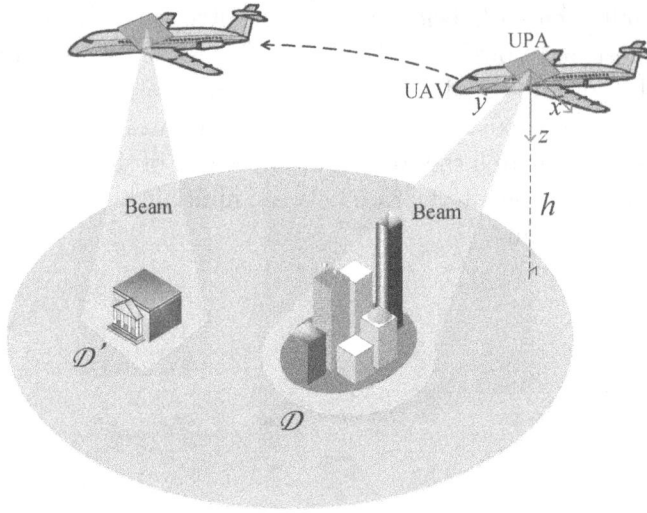

Figure 7.2 Illustration of the millimeter-wave UAV communication system.

7.3.2 Flexible Coverage System Model

As shown in Fig. 7.2, the UAV, with flight altitude h, is mounted with an $M \times N$ UPA. The target coverage area is defined as $\mathcal{D}$. The channel model between the UAV and the ground is a sparse millimeter-wave channel. For the A2G links, scattering is limited in the millimeter-wave band. Thus, the LoS path is dominant over the NLoS paths in general[40, 41]. Moreover, the presented beamforming design for coverage is only determined by the LoS path. Thus, the channel response vector with a half-wavelength-spacing UPA can be expressed as[40]

$$\mathbf{h} = \beta \mathbf{a}(M, N, \theta, \phi), \tag{7.8}$$

where β is the complex coefficient of the LoS path, θ and ϕ are the elevation and azimuth angles of the LoS path, respectively, as shown in Fig. 7.3. $\mathbf{a}(\cdot)$ is a steering vector function defined as

$$\begin{aligned}
\mathbf{a}(M, N, \theta, \phi) = [1, \cdots, e^{j\pi \sin\theta[(m-1)\cos\phi + (n-1)\sin\phi]}, \\
\cdots, e^{j\pi \sin\theta[(M-1)\cos\phi + (N-1)\sin\phi]}]^{\mathrm{T}},
\end{aligned} \tag{7.9}$$

where m and n are the coordinates of the antenna for axis x and axis y, respectively.

For a phased UPA, the elements of the antenna weight vector (AWV), i.e., the beamforming vector, have constant modulus

$$|[\mathbf{w}]_i| = \frac{1}{\sqrt{N}}, \quad i = 1, 2, ..., MN. \tag{7.10}$$

Thus, the effective channel gain between UAV and ground is

$$\left|\mathbf{h}^{\mathrm{H}}\mathbf{w}\right|^2 = |\beta|^2 \left|\mathbf{a}^{\mathrm{H}}(M, N, \theta, \phi)\mathbf{w}\right|^2 \tag{7.11}$$

As shown in Fig. 7.2, the origin of the coordinate is the position of the UAV. Axis x is parallel to an edge of the UPA, axis z is perpendicular to the ground, and axis y is perpendicular to the plane spanned by axes x and z. In this book, we assume the points in the target area have the same z-coordinate h. For a point in the target area $\mathcal{D}$, we denote its coordinate as (x, y, h). According to the geometric structure, we have

$$
\begin{cases}
\theta = \arctan \dfrac{\sqrt{x^2 + y^2}}{h}, \\
\phi = \arctan \dfrac{y}{x}.
\end{cases}
\tag{7.12}
$$

We need to design an appropriate beamforming vector, where the beam gain is concentrated on the area $\mathcal{D}$.

7.3.3 Coordinate Transformation of the Target Area

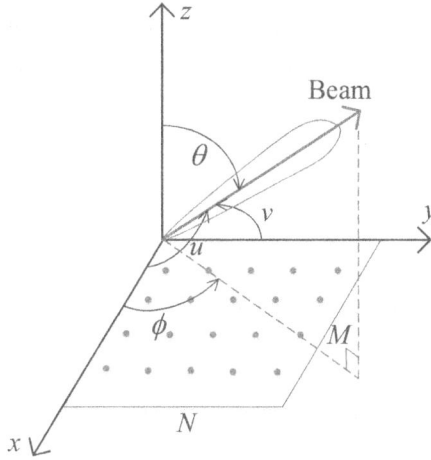

Figure 7.3 Geometric structure of the UPA.

As the shape and size of $\mathcal{D}$ are random, we need to find a minimum regular envelope to cover $\mathcal{D}$. Then the corresponding beam pattern is designed to cover the regular envelope. To simplify the design, we first take the coordinate transformation of the target area and define the regular envelope. Then, the beam pattern is designed with the new coordinate system.

We introduce a pair of *spatial angles* to decouple the conventional elevation and azimuth angles. As shown in Fig. 7.3, we define u as the inclined angle between the beam direction and the positive direction of axis x, and define v as the inclined angle between the beam direction and the positive direction of axis y. According to the geometric structure between (θ, ϕ) and (u, v), we have

$$
\begin{cases}
\cos u = \sin \theta \cos \phi, \\
\cos v = \sin \theta \sin \phi.
\end{cases}
\tag{7.13}
$$

Substituting (7.13) into (7.9), we have

$$\bar{\mathbf{a}}(M, N, u, v) = [1, \cdots, e^{j\pi[(m-1)\cos u + (n-1)\cos v]}, \cdots, e^{j\pi[(M-1)\cos u + (N-1)\cos v]}]^{\mathrm{T}}$$
$$= [1, \cdots, e^{j\pi[(M-1)\cos u]}]^{\mathrm{T}} \otimes [1, \cdots, e^{j\pi[(N-1)\cos v]}]^{\mathrm{T}},$$

$$(7.14)$$

where $\otimes$ denotes the Kronecker product. Different from the structure in (7.9), the spatial angles in (7.14) are decoupled. Thus, the beam pattern can be designed at the directions of u and v independently.

Based on (7.12) and (7.13), we have

$$\begin{cases} u = \arccos \dfrac{x}{\sqrt{x^2 + y^2 + h^2}}, \\ v = \arccos \dfrac{y}{\sqrt{x^2 + y^2 + h^2}}, \end{cases} \qquad (7.15)$$

where $x, y \in \mathbb{R}$ and $u, v \in [0, \pi]$. It can be seen as a coordinate transformation between (x, y) and (u, v), because the transformation is one-to-one correspondence given a fixed h. In other words, all the points in $\mathcal{D}$ have one and only one coordinate (u, v). As shown in Fig. 7.4, for an arbitrary target area $\mathcal{D}$, we can always search over its boundary and find the range of the two spatial-angle coordinates, which are

$$\begin{cases} u_{\min} \leq u \leq u_{\max}, \\ v_{\min} \leq v \leq v_{\max}. \end{cases} \qquad (7.16)$$

Therefore, we obtain $\bar{\mathcal{D}}$, the minimum rectangular which can cover $\mathcal{D}$, and is defined by

$$\bar{\mathcal{D}} = [u_{\min}, u_{\max}] \times [v_{\min}, v_{\max}]. \qquad (7.17)$$

Next, we will design an appropriate beam pattern to cover the area $\bar{\mathcal{D}}$.

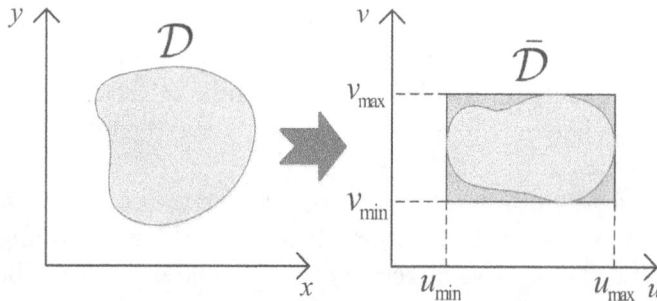

Figure 7.4 Illustration of the coordinate transformation of the target area.

7.3.4 Wide Beam Pattern Design

As we have decoupled the spatial angles of the UPA, some preliminary conclusions for ULA can be referenced. In general, given an N-element ULA, the beam width of a steering vector is $2/N$ in the cosine angle domain[42]. To obtain a broadened beam, a large array can be divided into multiple sub-arrays steering to sufficiently-spaced directions. For the direction of axis x, we need to design a beam which covers the area of $[\cos u_{\max}, \cos u_{\min}]$, while for the direction of axis y, we need to design a beam which covers the area of $[\cos v_{\max}, \cos v_{\min}]$.

We start from the direction of axis x, where the beam width of the target coverage is $\Delta_x = \cos u_{\min} - \cos u_{\max}$. Denote the number of antennas for each sub-array along axis x as M_S, so the beam width for each sub-array is $\frac{2}{M_S}$. Accordingly, the number of sub-arrays is $s_x = \lfloor \frac{M}{M_S} \rfloor$, where $\lfloor \cdot \rfloor$ denotes round down. We should ensure that the width of the designed beam pattern along axis x is no less than Δ_x, which is expressed as

$$\frac{2}{M_S} \times \left\lfloor \frac{M}{M_S} \right\rfloor \geq \Delta_x. \tag{7.18}$$

On one hand, a necessary condition for (7.18) is

$$\frac{2}{M_S} \times \frac{M}{M_S} \geq \Delta_x \Rightarrow M_S \leq \sqrt{\frac{2M}{\Delta_x}}. \tag{7.19}$$

On the other hand, a sufficient condition for (7.18) is

$$\frac{2}{M_S} \times \left(\frac{M}{M_S} - 1 \right) \geq \Delta_x \Rightarrow M_S \leq \frac{\sqrt{1 + 2\Delta_x M} - 1}{\Delta_x}. \tag{7.20}$$

As we know, a larger number of antennas can obtain a higher array gain. Thus, we should choose M_S as the maximal integer which satisfies (7.18) in the following set

$$\left[\left\lfloor \frac{\sqrt{1 + 2\Delta_x M} - 1}{\Delta_x} \right\rfloor, \cdots, \left\lfloor \sqrt{\frac{2M}{\Delta_x}} \right\rfloor \right]. \tag{7.21}$$

Then, we should choose the steering angle for each sub-array. The center of the coverage area $\bar{\mathcal{D}}$ along axis x is $c_x = (\cos u_{\min} + \cos u_{\max})/2$. We need to arrange the sub-arrays, whose beams are located at both sides of the center and can spread the coverage area. Thus, the steering angle for the p-th sub-array is

$$a_{x,p} = c_x - \frac{s_x}{M_S} + \frac{2p - 1}{M_S}, \quad 1 \leq p \leq s_x, \tag{7.22}$$

where $s_x = \lfloor \frac{M}{M_S} \rfloor$ is the number of sub-arrays along axis x. Accordingly, the beamforming vector along axis x is

$$\begin{cases} \mathbf{v}_x[(p-1)M_S + m] = \dfrac{e^{j\varphi_p} e^{j\pi[(p-1)M_S + m - 1]a_{x,p}}}{\sqrt{M}}, \\ \qquad\qquad 1 \leq p \leq s_x, \ \ 1 \leq m \leq M_S, \\ \mathbf{v}_x[(p-1)M_S + m] = 0, \ \ p > s_x \ \ \text{or} \ \ m > M_S, \end{cases} \tag{7.23}$$

where $e^{j\varphi_p}$ can be seen as the phase rotation for the p-th sub-array, which can be designed carefully to reduce the beam fluctuation between different sub-arrays.

As the beam gain for each sun-array is centered on the angle of $a_{x,p}$, the beam fluctuation at the middle of the two adjacent steering angles is most conspicuous. For this reason, we should maximize the beam gain at the angle of $\bar{a} = (a_{x,p} + a_{x,p+1})/2$.

$$
\left| \sum_{k=1}^{s_x} \sum_{m=1}^{M_{\mathrm{S}}} \frac{1}{\sqrt{M}} e^{-j\pi[(k-1)M_{\mathrm{S}}+m-1]\bar{a}} e^{j\varphi_k} e^{j\pi[(k-1)M_{\mathrm{S}}+m-1]a_{x,k}} \right|^2
$$

$$
= \frac{1}{M} \left| \sum_{k=1}^{s_x} e^{-j\pi(k-1)(2k-2p-1)} e^{j\varphi_k} \sum_{m=1}^{M_{\mathrm{S}}} e^{j\pi(m-1)\frac{2(k-p)-1}{M_{\mathrm{S}}}} \right|^2
$$

$$
\overset{(a)}{\approx} \frac{1}{M} \left| e^{j\pi(p-1)} e^{j\varphi_p} \sum_{m=1}^{M_{\mathrm{S}}} e^{\frac{-j\pi(m-1)}{M_{\mathrm{S}}}} + e^{-j\pi p} e^{j\varphi_{p+1}} \sum_{m=1}^{M_{\mathrm{S}}} e^{\frac{j\pi(m-1)}{M_{\mathrm{S}}}} \right|^2 \tag{7.24}
$$

$$
= \frac{1}{M} \left| \underbrace{\sum_{m=1}^{M_{\mathrm{S}}} e^{\frac{-j\pi(m-1)}{M_{\mathrm{S}}}}}_{\Sigma_1} - e^{j(\varphi_{p+1}-\varphi_p)} \underbrace{\sum_{m=1}^{M_{\mathrm{S}}} e^{\frac{j\pi(m-1)}{M_{\mathrm{S}}}}}_{\Sigma_2} \right|^2
$$

$$
\overset{\triangle}{=} \frac{1}{M} \left| \Sigma_1 - e^{j(\varphi_{p+1}-\varphi_p)} \Sigma_2 \right|^2,
$$

where (a) is according to the fact that when $|2(k-p)-1| \geq 2$,

$$
\left| \sum_{m=1}^{M_{\mathrm{S}}} e^{j\pi(m-1)\frac{2(k-p)-1}{M_{\mathrm{S}}}} \right|^2, \tag{7.25}
$$

is small and can be neglected if M_{S} is not small, and thus only the components of $k = p, p+1$ are considered. To maximize (7.24), we should let Σ_1 and $-e^{j(\varphi_{p+1}-\varphi_p)}\Sigma_2$ have the same phase, and we have

$$
\varphi_{p+1} - \varphi_p = \angle\Sigma_1 - \angle\Sigma_2 + \pi \overset{\triangle}{=} \Delta\varphi, \tag{7.26}
$$

where $\angle(\cdot)$ denotes the phase of a complex number. Thus, we can choose the phase rotation for the p-th sub-array along axis x of

$$
\varphi_p = p\Delta\varphi. \tag{7.27}
$$

Similarly, we can obtain the beamforming vector along axis y, which is

$$
\begin{cases} \mathbf{v}_y[(q-1)N_{\mathrm{S}}+n] = \dfrac{e^{j\psi_q} e^{j\pi[(q-1)N_{\mathrm{S}}+n-1]a_{y,q}}}{\sqrt{N}}, \\ \qquad\qquad 1 \leq q \leq s_y, \quad 1 \leq n \leq N_{\mathrm{S}}, \\ \mathbf{v}_y[(q-1)N_{\mathrm{S}}+n] = 0, \quad q > s_y \quad \text{or} \quad n > N_{\mathrm{S}}. \end{cases} \tag{7.28}
$$

As the beamforming vectors along axis x and axis y are independent, we can multiply them directly and obtain the beamforming vector of the UPA, i.e.,

$$
\mathbf{w} = \mathbf{v}_x \otimes \mathbf{v}_y. \tag{7.29}
$$

7.3.5 Performance Evolution

We assume that the modulus of the complex coefficient β in the channel model (7.8) is proportional to $1/d$, where d is the distance between the UAV and the target point, and β has a unit power at the node 100 m away from the UAV. The flight altitude of the UAV is 100 m, and the origin of the coordinates is the location of the UAV. The number of the antennas for UPA is $M \times N = 64 \times 64$.

Figs. 7.5 (left) and 7.5 (right) show the performance of the beam coverage. In Fig. 7.5 (left), the target coverage area is $\mathcal{D}_1 = [-30, 30]$ m $\times [-2, 2]$ m, which is directly under the projection of the UAV. It can be seen that the designed beam pattern can achieve high effective channel gain in the coverage area, and the size and shape of the real coverage area are close to the target coverage area. In Fig. 7.5 (right), the target coverage area is $\mathcal{D}_2 = [40, 70]$ m $\times [40, 70]$ m, which is below the diagonal of the UAV. The effective channel gain in the coverage area is high, where the target coverage area is included in the real coverage area. Interestingly, the shape of the real coverage area is distortional compared with the target area. The reason is that we design the beam pattern based on the spatial angles of u and v, which are the inclined angle between the beam direction and the positive direction of axis x and axis y, respectively. The coordinate transformation from (x, y) to (u, v) results in the distortion of the coverage area. In addition to flexible coverage, from both the two figures we can observe that the beamforming gain is mainly concentrated in the target coverage area, which shows that the presented 3D beamforming is also effective.

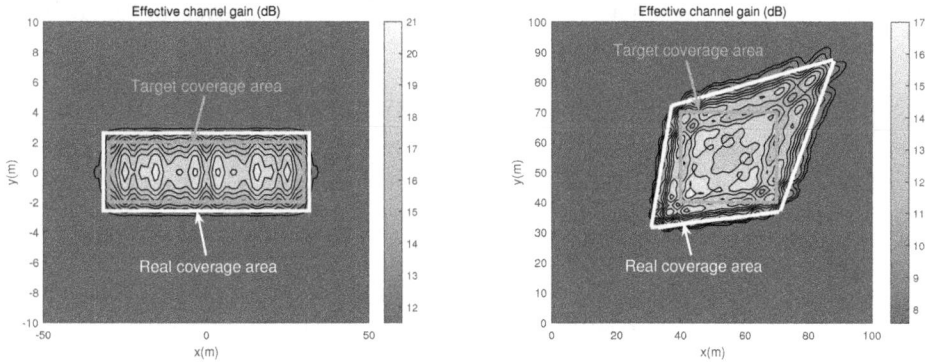

Figure 7.5 **Left**: The beam coverage of the presented approach, where $\mathcal{D}_1 = [-30, 30]$m $\times [-2, 2]$m is right under the projection of the UAV. **Right**: The beam coverage of the presented approach, where $\mathcal{D}_2 = [40, 70]$m $\times [40, 70]$m is below the diagonal of the UAV.

In this section, we presented a 3D beamforming approach to achieve effective and flexible coverage in UAV communications. The simulation results show that the target area can be well covered and the beamforming gain is mainly concentrated in the target coverage area.

7.4   SINGLE UAV DEPLOYMENT

Benefiting from the mobility and controllability, a UAV may be fast deployed in area without terrestrial infrastructure, which could serve as an air BS to compensate for defects of ground BS and improve the communication quality of service (QoS). Particularly, a UAV-BS can be used to collect high-speed wireless data form the ground users and sensors, which is with high significance for future Internet of Things (IoT). Different from traditional single antenna in a microwave system, large antenna arrays can be deployed in millimeter-wave band to meet the requirement of high data rate for UAV-BS. In this section, we jointly optimize UAV-BS deployment and beamforming to maximize the ASR in a multi-user millimeter-wave-UAV system, subject to a minimum rate constraint for each user, a position constraint for the UAV-BS and a constant modulus (CM) constraint for the beamforming vector.

7.4.1   Single UAV System Model

Without loss of generality, Fig. 7.6 shows a downlink multi-user scenario considered, where a single UAV-BS equipped with an N-element half-wavelength ULA serves K single-antenna users on the ground. To obtain the position of the UAV/users and the spacing geometry relationship, a 3D rectangular coordinate system is established first, where multiple users are distributed on the horizontal plane with coordinates $(x_i, y_i, 0)$, $i = 1, 2, ..., K$, and the UAV-BS is located at (x, y, h_U). h_U is the flying altitude. The phased antenna array with analog beamforming structure is utilized at the UAV-BS, where all antennas share a single RF chain and each antenna branch

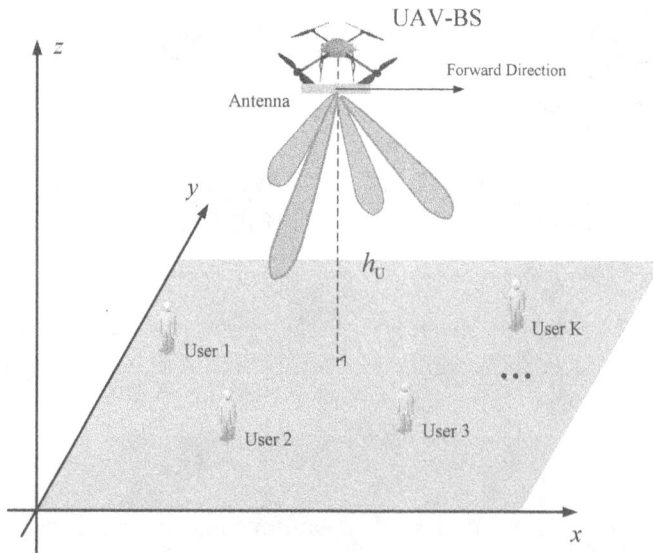

Figure 7.6 Illustration of a array enabled UAV cellular system, where one UAV-BS with N antennas serves multiple users.

has a phase shifter with a power amplifier (PA) to drive the antenna. Generally, all the PAs have the same scaling factor. Therefore, the beamforming vector, i.e., the AWV, has CM elements. For ULA structure, we denote a CM beamforming vector with $|[\mathbf{w}]_n| = \frac{1}{\sqrt{N}}$ for $n = 1, 2, ..., N$.

Then the received symbol of user i adopting an analog beamforming structure on the UAV-BS is expressed as

$$y_i = \mathbf{h}_i^{\mathrm{H}} \mathbf{w} \sqrt{P} s_i + z_i, \tag{7.30}$$

where s_i is a transmitted symbol for user i, P is the total transmit power at the UAV-BS, and z_i is the Gaussian white noise at user i. $\mathbf{h}_i$ is the millimeter-wave channel response vector between user i and the UAV-BS, discussed in next subsection.

7.4.2 Single UAV-BS Downlink Channel Model

The channel between user i and the UAV-BS is a millimeter-wave channel. It is known that millimeter-wave channels have spatial sparsity in the angle domain, and the limited MPCs are mainly generated by reflections rather than scattering[42,43]. Different MPCs have different AoDs. Then, similar to the previous chapters, the general channel response vector between user i and the UAV-BS can be expressed as[42,44]

$$\bar{\mathbf{h}}_i = \sum_{\ell=1}^{L_i} \beta_{i,\ell} \mathbf{a}(N, \theta_{i,\ell}), \tag{7.31}$$

where $\beta_{i,\ell}$ is the channel gain coefficient of the ℓ-th MPC for user i, $\theta_{i,\ell}$ is the steering angle of the ℓ-th MPC for user i, L_i is the total number of MPCs for user i. $\mathbf{a}(\cdot)$ is the steering vector function defined as

$$\mathbf{a}(N, \theta) = [1, \ e^{j\pi\theta}, \ e^{j\pi 2\theta}, ..., e^{j\pi(N-1)\theta}]^{\mathrm{T}}, \tag{7.32}$$

which depends on the array geometry. Let $\bar{\theta}_{i,\ell}$ denote the real AoD of the ℓ-th MPC for user i, then we have $\theta_{i,\ell} = \cos(\bar{\theta}_{i,\ell})$ called the steering angle. Therefore, $\theta_{i,\ell}$ is within the range $[-1, 1]$. For convenience, in the rest of this section, $\theta_{i,\ell}$ is also called AoD.

Affected by the UAV-BS altitude, the blockage effect will impact the probability of existing an LoS component. The probability model can be described as a function of the elevation angle θ as (7.1)[45]. The elevation angle of user i is given by

$$\theta_i = \frac{180}{\pi} \tan^{-1}(\frac{h_{\mathrm{U}}}{D_i}), \tag{7.33}$$

with the horizontal distance from the UAV-BS to user i denoted as

$$D_i = \sqrt{(x - x_i)^2 + (y - y_i)^2}. \tag{7.34}$$

We can find that the probability of existing an LoS component increases as the elevation angle increases, and it approaches 1 when h_{U} is large enough.

Under the scenario of no blockage, the general channel response vector between user i and the UAV-BS in (7.31) can be re-described as

$$\mathbf{h}_i = \underbrace{\beta_{i,1}\mathbf{a}(N, \theta_{i,1})}_{\text{LOS component}} + \underbrace{\sum_{\ell=2}^{L_i} \beta_{i,\ell}\mathbf{a}(N, \theta_{i,\ell})}_{\text{NLOS components}}, \qquad (7.35)$$

which is defined as an LoS channel model, consisting of one LoS component and $L_i - 1$ NLoS components. The AoD for the LoS component $\theta_{i,1}$ is defined as the cosine value of the direction vector between the UAV-BS and user i and the forward direction vector $[1, 0, 0]$ in Fig. 7.6. Then $\theta_{i,1}$ in (7.35) is expressed as

$$\theta_{i,1} = \frac{[x_i - x, \ y_i - y, \ -h_{\mathrm{U}}] \cdot [1, \ 0, \ 0]}{|[x_i - x, \ y_i - y, \ -h_{\mathrm{U}}]| \times |[1, \ 0, \ 0]|} = \frac{x_i - x}{\sqrt{(x_i - x)^2 + (y_i - y)^2 + h_{\mathrm{U}}^2}}, \qquad (7.36)$$

AoDs of NLoS components $\theta_{i,\ell}$ are random values within the range $[-1, 1]$. For the LoS component of user i, the channel gain coefficient $\beta_{i,1}$ is a constant, depending on propagation loss affected by propagation distance and carrier frequency. It can be expressed as[32]

$$\beta_{i,1} = \frac{1}{(\frac{4\pi f}{c}) \cdot d_i^{\alpha_{\mathrm{LoS}}}}, \qquad (7.37)$$

where d_i is the propagation distance from the UAV-BS to user i, denoted as

$$d_i = \sqrt{(x - x_i)^2 + (y - y_i)^2 + h_{\mathrm{U}}^2}. \qquad (7.38)$$

α_{LoS} is the LoS path loss exponent, c is the constant of light speed, f is the carrier frequency of the transmitted signal, and c/f is the wavelength of the carrier.

For a certain NLoS component of user i, the channel gain coefficient is a random variable expressed as

$$\beta_{i,\ell} = \frac{\zeta_f}{(\frac{4\pi f}{c}) \cdot d_i^{\alpha_{\mathrm{NLoS}}}}, \qquad (7.39)$$

where α_{NLoS} is the NLoS path loss exponent, ζ_f is the small scale Rayleigh fading factor[46].

Regarding the parameter settings, the carrier frequency can be set as a typical value $f = 28$ GHz. According to the millimeter-wave channel measurement results in[43], the path loss exponents in (7.37) and (7.39) can be set as $\alpha_{\mathrm{LoS}} = 0.95$ and $\alpha_{\mathrm{NLoS}} = 2.25$, respectively.

Correspondingly, the probability of not existing an LoS component, i.e., an NLoS channel, is given by

$$P_{\mathrm{NLOS}}(\theta) = 1 - P_{\mathrm{LOS}}(\theta). \qquad (7.40)$$

Compared with the LoS channel model in (7.35), all MPCs are NLoS components. Then an NLoS channel model can be expressed as

$$\mathbf{h}_i = \underbrace{\sum_{\ell=2}^{L_i} \beta_{i,\ell}\mathbf{a}(N, \theta_{i,\ell})}_{\text{NLOS components}}, \qquad (7.41)$$

where the definitions of $\beta_{i,\ell}$ and $\theta_{i,\ell}$ are the same as that in (7.35).

7.4.3 UAV-BS Deployment with Array Beamforming

An immediate and basic problem is finding the best UAV-BS position to maximize the achievable sum rate (ASR) of multiple users, where the channel is assumed to be known *a priori*. In practice, some state-of-the-art channel estimation solutions[47,48,49], especially the ones using compressive sensing schemes, can be used for robust synchronization and channel estimation. For each user, the achievable rate R_i is denoted by[1]

$$R_i = \log_2\left(1 + \frac{P \cdot |\mathbf{h}_i^H \mathbf{w}|^2}{\sigma^2}\right), \tag{7.42}$$

where P is the total transmit power at the UAV-BS, σ^2 is the power of Gaussian white noise at user i. $|\mathbf{h}_i^H \mathbf{w}|^2$ denotes the effective channel gain between UAV-BS and user i.

In this problem, there are also minimal rate constraints for the users. The UAV-BS deployment intertwines with the beamforming design, which makes the problem complicated to solve. Taking the rate and 2D-beamforming constraints into consideration, the problem is formulated by

$$\underset{x,y,\mathbf{w}}{\text{Maximize}} \quad \sum_{i=1}^{K} R_i \tag{7.43a}$$

$$\text{Subject to} \quad R_i \geq r_i, \quad i = 1, 2, ..., K, \tag{7.43b}$$

$$|[\mathbf{w}]_n| = \frac{1}{\sqrt{N}}, n = 1, 2, ...N, \tag{7.43c}$$

where K is the total number of users, r_i denotes the minimal rate constraint for user i, $|[\mathbf{w}]_n| = \frac{1}{\sqrt{N}}$ is the CM constraint due to using the phase shifters in each antenna branch at the UAV-BS. The optimization variables are the projected coordinates of UAV-BS (x, y) and the beamforming vector $\mathbf{w}$. The above Problem (7.43) is challenging, not only due to the non-convex constraints, but also due to that the parameters to be optimized are entangled with each other. We can solve the original problem with two tractable steps. First, by introducing the approximate beam pattern, the original problem can be simplified as a deployment and beam gain allocation problem. After finding the optimal position of the UAV-BS, the beamforming problem can be solved independently by using the artificial bee conlony (ABC) algorithm.

7.4.3.1 Solution of the UAV-BS Deployment Problem

For an arbitrary user i, the UAV-BS should form a narrow beam steering along its strongest multipath component to maximize the effective channel gain. Let $|\beta_i| = \max_\ell |\beta_{i,\ell}|$. The effective channel gain can be approximated as

$$|\mathbf{h}_i^H \mathbf{w}|^2 \approx |\beta_i|^2 |\mathbf{a}_i^H \mathbf{w}|^2, \tag{7.44}$$

where $|\mathbf{a}_i^H \mathbf{w}|^2$ denotes the antenna beam gain for user i. Let $c_i = |\mathbf{a}_i^H \mathbf{w}|^2$, we introduce the following lemma to simplify the original problem.

---

[1]Note that OMA, e.g, FDMA, is utilized here, where each user occupies an independent time/frequency/code slot receiving data from the UAV-BS.

Lemma 7.4.1. *With approximate beamforming, i.e., beam gains are zeros along non-user directions and significant toward user directions, the sum of beam gains satisfies*[2]

$$\sum_{i=1}^{K} c_i = N, \tag{7.45}$$

where N is the number of antennas for a ULA.

Proof. In the case of approximate beamforming, it is assumed no sidelobe, i.e., beam gains concentrate on the AoDs of strongest MPC for each user defined as θ_i and the beam pattern is flat along each direction. In addition, for ULA with N antennas, the beam width of each user's beam pattern is $2/N$[42]. It is worth mentioning that the beam width is not strictly equal to $2/N$, i.e., practical beam width may greater than or less than this value. Under such precondition, the average power of $\mathbf{w}$ in the angle domain is

$$
\begin{aligned}
&\frac{1}{2} \int_{-1}^{1} |\mathbf{a}(N, \theta)^{\mathrm{H}} \mathbf{w}|^2 \, d\theta \\
&= \frac{1}{2} \times \frac{2}{N} (|\mathbf{a}(N, \theta_1)^{\mathrm{H}} \mathbf{w}|^2 + |\mathbf{a}(N, \theta_2)^{\mathrm{H}} \mathbf{w}|^2 + \ldots + |\mathbf{a}(N, \theta_K)^{\mathrm{H}} \mathbf{w}|^2) \\
&= \frac{1}{N} (c_1 + c_2 + \ldots + c_K).
\end{aligned} \tag{7.46}
$$

On the other hand, in terms of (7.32), we can expand the integral term as follows

$$
\begin{aligned}
\frac{1}{2} \int_{-1}^{1} |\mathbf{a}(N, \theta)^{\mathrm{H}} \mathbf{w}|^2 \, d\theta &= \frac{1}{2} \int_{-1}^{1} \sum_{m=1}^{N} [\mathbf{w}]_m e^{-j\pi(m-1)\theta} \sum_{n=1}^{N} [\mathbf{w}]_n^* e^{-j\pi(n-1)\theta} \, d\theta \\
&= \frac{1}{2} \int_{-1}^{1} \sum_{m=1}^{N} \sum_{n=1}^{N} [\mathbf{w}]_m e^{-j\pi(m-1)\theta} [\mathbf{w}]_n^* e^{-j\pi(n-1)\theta} \, d\theta
\end{aligned} \tag{7.47}
$$

$$
\begin{aligned}
&= \sum_{m=1}^{N} [\mathbf{w}]_m [\mathbf{w}]_m^* + \frac{1}{2} \sum_{m=1}^{N} \sum_{n=1, n\neq m}^{N} [\mathbf{w}]_m [\mathbf{w}]_n^* \int_{-1}^{1} e^{j\pi(n-m)\theta} \, d\theta \\
&= \|\mathbf{w}\|_2^2.
\end{aligned}
$$

We separate transmit power from beamforming design. $\mathbf{w}$ is the normalized beamforming vector. We denote that power is equal to 1, so we have $\|\mathbf{w}\|_2^2 = 1$. Under the CM constraint, the modulus of each entry for $\mathbf{w}$ is $\frac{1}{\sqrt{N}}$. Besides, based on (7.46) and (7.47), we have

$$\frac{1}{N}(c_1 + c_2 + \ldots + c_K) = \|\mathbf{w}\|_2^2 = 1. \tag{7.48}$$

□

---

[2]For a UPA with $M \times N$ antenna elements, Lemma 7.4.1 becomes $\sum_{i=1}^{K} c_i = M \times N$.

By using Lemma 7.4.1, Problem (7.43) can be re-described as

$$\underset{x,y,c_i}{\text{Maximize}} \quad \sum_{i=1}^{K} \log_2(1 + \frac{P \cdot |\beta_i|^2 c_i}{\sigma^2}) \tag{7.49a}$$

$$\text{Subject to} \quad \log_2(1 + \frac{P \cdot |\beta_i|^2 c_i}{\sigma^2}) \geq r_i, \quad i = 1, 2, ..., K, \tag{7.49b}$$

$$c_1 + c_2 + ... + c_K = N, \tag{7.49c}$$

where $|\mathbf{a}_i^H \mathbf{w}|^2$ are replaced by the approximate gain c_i, $i = 1, 2, ..., K$.

Problem (7.49) is designed to solve the UAV-BS deployment problem under the setup of approximate beam gain. The CM constraint is not involved in the above problem, but will be considered in the following beamforming problem. It is noteworthy that sidelobes exist in beam pattern practically, and the precondition of Lemma 7.4.1 is approximate beamforming. Therefore, practical beam gains are not equal but close to the values of c_i solved by the above model, where c_i can be considered as reference values of the original problem.

As we can see, the dimension of the optimization variables in Problem (7.49) is $K + 2$, while the dimension of the position is only 2. Thus, it is possible to use the exhaustive searching method to find the UAV-BS position with the maximum sum rate. With the fixed UAV-BS altitude, the specific implementation is dividing the user-distributed area into a grid with a certain precision, where each point represents the projection coordinate of the UAV-BS. We calculate the maximum ASR on each grid point, i.e., solving the optimization problem once on each grid point[3]. On each point, the channel gain coefficient for each user is constant. Then the objective function on a certain grid point can be written as

$$\begin{aligned} f(c_i) &= \log_2 \left(1 + \frac{P \cdot |\beta_1|^2 c_1}{\sigma^2}\right) \cdots \left(1 + \frac{P \cdot |\beta_K|^2 c_K}{\sigma^2}\right) \\ &\triangleq \log_2 \left(1 + m_1 c_1\right) \cdots \left(1 + m_K c_K\right), \end{aligned} \tag{7.50}$$

where the coefficients in front of c_i are substituted by constants m_i.

It is easy to verify that the objective function of Problem (7.49) with a fixed UAV-BS position is concave, and the constraints are linear. Thus, it can be solved by using the standard convex optimization tools.

Hereto, we have solved the first step, i.e., UAV-BS deployment problem and obtained an optimal position of the UAV-BS under the assumption of approximate beamforming. Certainly, the higher precision of the grid has, the better the solution is. The presented solution of Problem (7.49) is summarized in Algorithm 7.1.

---

[3]Note that the study in this part can be generalized to the 3D optimization, e.g., it may be a possible way to divide both the user-distributed area and the height of UAV-BS into a 3D grid. In this case, the height of UAV-BS could be further optimized by 3D grid searching with Algorithm 7.1. The optimization of the height of the UAV-BS can be further considered.

Algorithm 7.1: Solution of the UAV-BS Deployment and Beam Gain Allocation Problem

---

Input: G; /*Precision of the grid. */

K; /*The number of ground users. */

ValueMatrix $(G \times G, 1 + K)$; /*Generate matrix to save solutions.*/

GridMatrix (G, G); /*Grid points.*/

Output: GridMatrix$_{i,j}$ as the UAV-BS near-optimal projected position (x^*, y^*).

1: **for** $i = 1 : G$ **do**
2:    /*Obtain solution on each grid.*/
3:    **for** $j = 1 : G$ **do**
4:       Solve Problem (7.49) and obtain objective function value V, optimal beam gain **c**;

5:       /*Save the values in the matrix. */
6:       **ValueMatrix**$_{(i-1) \times G + j, \ 1} = V$;
7:       $[$**ValueMatrix**$]_{(i-1) \times G + j, \ 2:end} = $ **c**;
8:    **end for**
9: **end for**

10: Find the maximal value V_{max} in the first column of **ValueMatrix**;
11: Record the beam gain **c** at this time;
12: Record i and j at this time;

13: **return** **GridMatrix**$_{i,j}$ as the UAV-BS near-optimal projected position (x^*, y^*).

7.4.3.2 Solution of the Beamforming Problem

Substituting the obtained optimal position of the UAV-BS to the original problem, we obtain the beamforming problem, i.e.,

$$\underset{\mathbf{w}}{\text{Maximize}} \quad \sum_{i=1}^{K} \log_2(1 + m_i \cdot |\mathbf{a}_i^H \mathbf{w}|^2) \tag{7.51a}$$

$$\text{Subject to} \quad \log_2(1 + m_i \cdot |\mathbf{a}_i^H \mathbf{w}|^2) \geq r_i, \ i = 1, 2, ..., K, \tag{7.51b}$$

$$|[\mathbf{w}]_n| = \frac{1}{\sqrt{N}}, n = 1, 2, ...N, \tag{7.51c}$$

where $m_i = \frac{P \cdot |\beta_i|^2}{\sigma^2}$, $|\beta_i|$ is the channel gain coefficient along the strongest MPC. The beamforming vector **w** needs to be carefully designed so that the beam gains are significantly along the users' directions, and meanwhile, the minimal rate constraint for each user is satisfied.

The dimension of the beamforming vector **w** is high and each element of **w** has a CM constraint, which leads to non-convexity of this problem. Thus, solving this high-dimensional problem is challenging, and the conventional methods are incapable of obtaining satisfying results. Some swarm-based algorithms can be considered here, e.g., particle swarm optimization (PSO) algorithm[50], ant colony optimization (ACO) algorithm[51] and ABC algorithm[52]. Several existing studies have showed that the PSO and ACO algorithms are likely falling into local optimal solution[53], which is a fatal weakness to solve the problem in (7.51), because the CM constraint leads to many sub-optimal points of the objective function. In contrast, ABC algorithm has great advantages in finding global optimal solution rather than a local optimal solution, and it has a fast convergence rate. Hence, the ABC algorithm is more suitable to solve the optimization problem with high-dimensional variables in (7.51).

In this section, we use ABC algorithm to solve (7.51). The background derives from the nectar-gathering behavior of bee colonies, where two self-organizing cluster models are positive feedback for fine nectar sources and negative feedback for inferior ones, respectively. Therefore, from an optimization perspective, the optimal solution of problems can be found via two models.

Due to the non-convexity of the equality CM constraint $|[\mathbf{w}]_n| = \frac{1}{\sqrt{N}}$, the problem is still hard to solve. To guarantee the value of each element is $\frac{1}{\sqrt{N}}$, ABC algorithm is adopted to search the phase φ of each element, where $\mathbf{w}$ can be expressed as $\mathbf{w} = \frac{1}{\sqrt{N}} \cdot e^{j\varphi}$. Then Problem (7.51) can be re-described as

$$\underset{\varphi}{\text{Maximize}} \quad \sum_{i=1}^{K} \log_2\left(1 + m_i \cdot \frac{1}{N}|\mathbf{a}_i^{\mathrm{H}} e^{j\varphi}|^2\right) \tag{7.52a}$$

$$\text{Subject to} \quad \log_2\left(1 + m_i \cdot \frac{1}{N}|\mathbf{a}_i^{\mathrm{H}} e^{j\varphi}|^2\right) \geq r_i, \ i = 1, 2, ..., K. \tag{7.52b}$$

Since ABC algorithm can solve the unconstrained optimization problem more easily, we transform the above constraint optimization problem into an unconstrained one by means of the penalty function. In the light of the penalty function theory, we re-describe the constraints in (7.52) as

$$p_i(\varphi) = \log_2\left(1 + m_i \cdot \frac{1}{N}|\mathbf{a}_i^{\mathrm{H}} e^{j\varphi}|^2\right) - r_i \geq 0, \tag{7.53}$$

where the number of constraints is K. Then the beamforming problem is tantamount to (7.52) with two parts as

$$\underset{\varphi}{\text{Minimize}} \quad -\sum_{i=1}^{K} \log_2\left(1 + m_i \cdot \frac{1}{N}|\mathbf{a}_i^{\mathrm{H}} e^{j\varphi}|^2\right) + \varrho \sum_{i=1}^{K} [\max\{0, -p_i(\varphi)\}]^2, \tag{7.54}$$

where ϱ is a positive number whose order of magnitude is much larger than the first part in (7.54). If φ is a feasible solution, the value of $\max\{0, -p_i(\varphi)\}$ is 0. If not, the value of $\max\{0, -p_i(\varphi)\}$ is $-p_i(\varphi)$. Under the influence of the penalty factor ϱ, the value of the objective function in (7.54) will be large if the constraints are not satisfied. Hereto, Problem (7.52) is transformed into an unconstraint problem.

Based on the transformed objective function, we next introduce ABC algorithm to acquire the phase $\boldsymbol{\varphi}$ of beamforming vector, which is shown in Algorithm 7.2. The method is mainly divided into three stages: Foraging bees discovering new nectar sources, on-looker bees finding new sources based on foragers, and scout bees seeking new sources to replace the ones which have not been updated for multiple times. The algorithm is demonstrated as follows.

- *Initialization.* We firstly initialize a matrix as the initial search location, where each row represents a feasible solution vector $\boldsymbol{\varphi}$ (nectar source) and the number of columns is the dimension of the optimization variable. For the i-th row, the solution is initialized as

$$\mathbf{s}_{i,j} = \mathbf{s}_j^{\min} + \phi_{i,j} \times (\mathbf{s}_j^{\max} - \mathbf{s}_j^{\min}), \tag{7.55}$$

where $i = 1, 2, ..., N_S$, $j = 1, 2, ..., D_S$, $\mathbf{s}_j^{\min} = 0$ and $\mathbf{s}_j^{\max} = 2\pi$. $\phi \in [0, 1]$ is a random number. N_S is the number of sources and D_S is the dimension of the solution. We calculate the fitness function value F_i for each source as

$$F_i = \begin{cases} \dfrac{1}{1 + g(x_i)}, & g(x_i) \geq 0, \\ 1 + \mathrm{abs}(g(x_i)), & \text{otherwise,} \end{cases} \tag{7.56}$$

where $g(x_i)$ is the i-th objective function value. It can be seen from the formula that the smaller the value of $g(x_i)$, the larger the fitness function value which represents the quality of solution. Based on the fitness function value F_i, we can find the best solution vector and the corresponding objective function value. Moreover, we define a threshold to limit the iteration times. If a source has not been updated for many iteration times even exceeding the threshold, it will be discarded.

- *Search for new sources.* At the beginning of process, according to the existing sources, a new source is randomly generated around each source by a random number Γ within a range precision $(-\gamma, \gamma)$ in Algorithm 7.2. If the fitness value of new source $[\mathbf{v}]_{i,:}$ is larger than that of $[\mathbf{s}]_{i,:}$, $[\mathbf{v}]_{i,:}$ will replace $[\mathbf{s}]_{i,:}$ via the greedy selection method. Otherwise, $[\mathbf{s}]_{i,:}$ is kept. After this treatment, the better source will be preserved. Meanwhile, the number of times that each source has not been updated is recorded.

- *Generate new sources using roulette method.* For each source, they are updated with probability as

$$p_i = \frac{F_i}{\sum\limits_{i}^{N_S} F_i}, \tag{7.57}$$

where N_S is the number of sources. It can be seen that the probability is related to the fitness value of each source. The greater the value is, the more likely the source is updated. Meanwhile, the best solution vector and the corresponding objective function value are also updated.

Algorithm 7.2: ABC Algorithm for Solving the Beamforming Vector.

Input: $N_S = 200$; /*The number of food source. */

$D_S = N$; /*The dimension of optimization variable φ.*/

SourceMatrix (N_S, D_S); /*According to (7.55).*/

FitnessVector $(N_S, 1)$; /*Calculate the fitness value.*/

BestSolution; /*The maximum fitness function value and φ at this time.*/

Limit $= 300$; /*Threshold.*/

Loop $= 1000$; /*Number of iterations.*/

$\mathbf{t} = \text{zeros}\ (N_S, 1)$;

Output: *BestSolution*

1: **for** $\ell = 1 : Loop$ **do**

2:    /*Search for new sources.*/

3:    **for** $i = 1 : N_S$ **do**

4:       $[\mathbf{NewSource}]_{i,:} = [\mathbf{SourceMatrix}]_{i,:} +$
      $\Gamma \times ([\mathbf{SourceMatrix}]_{i,:} - [\mathbf{SourceMatrix}]_{k,:})$; /*$i \neq k, \Gamma \in (-\gamma, \gamma)$*/

5:       /*Greedy Selection. */

6:       **if** NewFitness> **FitnessVector**$_i$ **then**

7:          **NewSource**$[i]$; /*Replace.*/

8:       **else**

9:          **SourceMatrix**$[i]$; /*Remain.*/

10:          $[\mathbf{t}]_i = [\mathbf{t}]_i + 1$;

11:       **end if**

12:    **end for**

13:    Calculate **ProbMatrix**; /* According to (7.57).*/

14:    $j = 1;\ \ k = 1$;

15:    **while** $j \leq$ NumSource **do**

16:       **if** $rand <$ **ProbMatrix**$_k$ **then**

17:          $j = j + 1$;

18:          Search for new source;

19:          $k = k + 1$;

20:       **end if**

21:    **end while**

22:    /*Record optimal value and φ.*/

23:    *BestSolution*;

24:    /*Discard some sources.*/

25:    **if** $[\mathbf{t}]_i > Limit$ **then**

26:       $[\mathbf{t}]_i = 0$;

27:       Reinitialize this source;

28:    **end if**

29: **end for**

30: **return** *BestSolution*

- *Determine if there is a discarded source.* Count the number of times that each source is updated. If the maximum number of not-updated times exceeds the predetermined threshold, the source will be initialized randomly to replace the original one. The algorithm finishes until reaching the maximum number of iterations. Then the optimal source and the corresponding objective function value will be output. If not, Step 2 will be executed.

According to Algorithm 7.2, the phase φ of $\mathbf{w}$ for each element can be found. The ultimate solution $\mathbf{w}$ is calculated by $\mathbf{w} = \frac{1}{\sqrt{N}} \cdot e^{j\varphi}$, which satisfies the CM constraint. Hereto, we have solved the beamforming problem, i.e., have found the feasible solution and obtained the beamforming vector $\mathbf{w}$.

To sum up, firstly, we solve the optimization problems under the approximate beamforming lemma by Algorithm 7.1, jointly optimizing the UAV-BS deployment and approximate beam gain allocation for each user. We can obtain the position of UAV-BS. Then, given the coordinates of the UAV-BS solved by Algorithm 7.1, we design the beamforming vector to realize practical beamforming by Algorithm 7.2, approaching the approximate beam pattern and the approximate beam gain to maximize the system ASR.

7.4.4 Performance Evaluation

In this section, we evaluate the performance of the presented UAV-BS deployment with array beamforming approach. As aforementioned, the problem is divided into two steps, and different methods are used to find the feasible solutions. In the simulation, we consider a scenario that one UAV-BS serves multiple ground users with the millimeter-wave carrier frequency 28 GHz, a typical frequency band in an urban area[43]. The corresponding parameters in (7.1) are $a = 11.95$ and $b = 0.14$[19]. The positions of ground users are randomly generated, and the ASR in each figure is averaged over 100 user distributions. In addition, for each user distribution and UAV-BS positioning, the performance is averaged over 100 channel realizations. For each user, the number of MPCs is set as $L_i = 4$. We start from the performance evaluation of the designed 2D and 3D beam patterns.

First, we evaluate the performance of the designed beam pattern. Fig. 7.7 compares the approximate beam pattern solved in (7.49) with designed beam pattern using the beamforming vector $\mathbf{w}$ by solving Problem (7.54), where we assume the minimum rate constraints for three ground users are $8, 4, 4$ bps/Hz, respectively. With regard to the approximate beam pattern, it is assumed no beam overlap or sidelobe. Beam gains are concentrated on the user directions and the beam pattern is flat along each direction. The approximate beam pattern can be treated as a design reference to evaluate the performance of our presented solution. The beamforming vector $\mathbf{w}$ is designed to approach the approximate beam gain of each user. Fig. 7.7 shows the comparison results with $N = 16, 32, 64$, and from this figure, we can observe that the beam gains are significant along the user directions. The result in Fig. 7.7 demonstrates that the presented beamforming approach is effective, although there are small gaps between the approximate beam gains and the designed beam gains along the user directions.

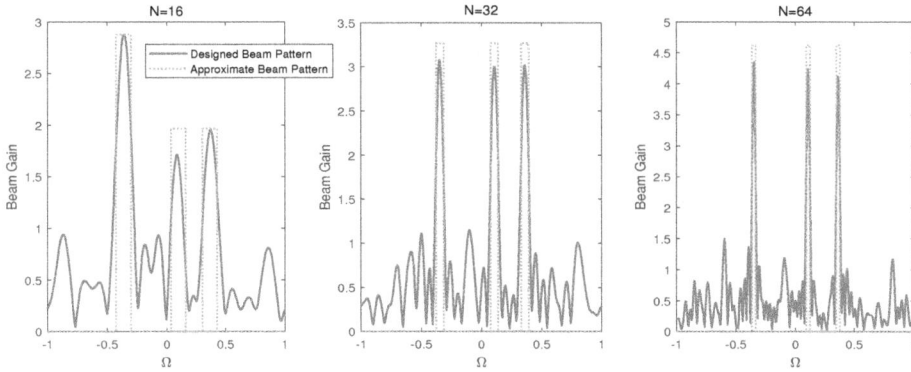

Figure 7.7 Comparison between the approximate beam patterns and the designed beam patterns with ULA configuration for $K = 3$.

Fig. 7.8 shows the designed UPA 3D beam pattern in the scenario of six users, where $M = 16$, $N = 16$. The minimum constraints for the users are all 2bps/Hz. The near-optimal position of UAV-BS and the beamforming vector have been solved. We can observe that the beam gains are significant along the user directions, i.e., elevation and azimuth angles.

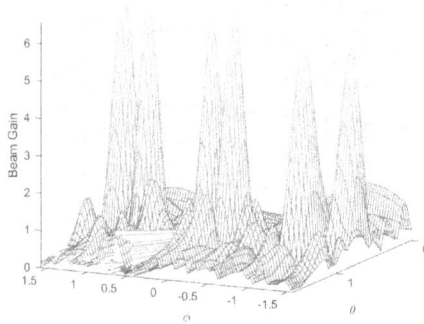

Figure 7.8 Designed UPA 3D beam pattern for $M = 16$, $N = 16$ and $K = 6$.

Then, we evaluate the performance of the presented UAV-BS deployment approach. Fig. 7.9 shows the ASR comparison between the solved position and random position for UAV-BS in the scenario of three and four users against P with 32 antennas. The power of Gaussian white noise σ^2 is set as -100 dBm (uniformly hereinafter). The latter results are based on the average performance of 100 random positions in the surrounding area of the ground users. The minimum achievable rate constraints for the users are all 1 bps/Hz. From Fig. 7.9 we can observe that the ASR of the optimal UAV-BS position is larger than the average sum rate of a random position.

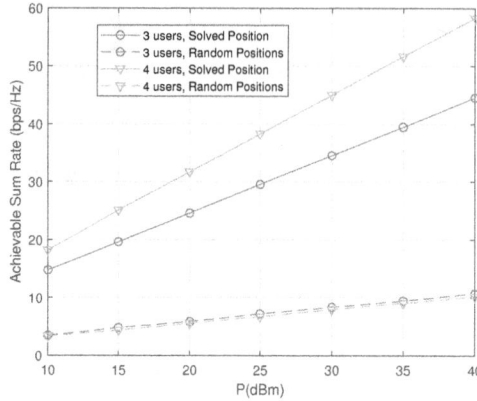

Figure 7.9 Achievable sum rate comparison between the solved position and random positions for UAV-BS versus P for $N = 32$.

Then, we explore the solved position of UAV-BS against the minimum rate constraint for the scenario of five and six users as Fig. 7.10 shows, where the number of antennas is 64. In order to reflect the minimum rate constraint impact on the position of the UAV-BS, user positions are randomly generated once. The plane-coordinate system is applied to denote the coordinates of the users and the projected coordinate of UAV-BS as well. In both cases, user 1 with variant minimum rate constraint is randomly generated as $(196.29, 182.53)$ at first. We increase the rate constraint of user 1 from 2 to 9 bps/Hz as shown in the vertical coordinates, while the values of the constraints for other users are all fixed as 2 bps/Hz. For the case of five users, the other four coordinates are randomly generated as $(108.68, 0.94)$, $(55.67, 24.31)$,

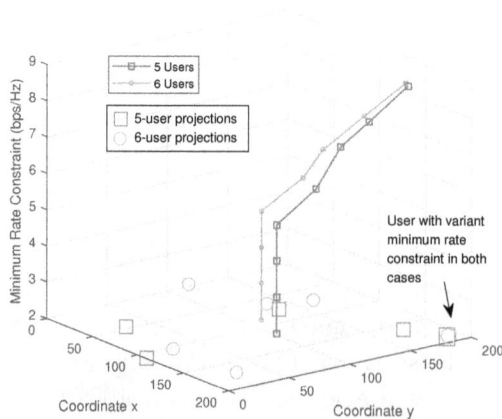

Figure 7.10 The near-optimal position of the UAV-BS versus the minimum rate constraint for the scenarios of five and six users, $N = 64$.

(84.90, 134.15) and (168.96, 165.17), respectively. The user projections are shown in Fig. 7.10. We can find that as the minimum rate constraint for user 1 increases, the optimal position of UAV-BS moves from the center of five users to user 1. Similarly for the case of six users, the other five coordinates are randomly generated as (110.76, 18.27), (62.56, 140.33), (78.45, 158.64), (177.59, 27.34) and (0.94, 115.02), respectively. We still control the other five users' minimum rate constraints invariantly as 2 bps/Hz and increase the constraint for user 1 as above. Likewise, the optimal position of UAV-BS moves from the center of six users to user 1. Fig. 7.10 illustrates that the optimal position of UAV-BS is not constant if the minimum achievable rate constraint for each user alters. If a certain user has an extremely higher minimum rate constraint than the other users, the optimal position of UAV-BS will be much closer to this user as expected.

Next, we evaluate the quality of the solution solved by ABC algorithm in the beamforming problem (7.54). For the feasible solution for vector $\mathbf{w}$, ABC algorithm is used for solving the phase φ of each element. After the phase vector φ is solved, the beamforming vector can be computed by $\mathbf{w} = \frac{1}{\sqrt{N}}e^{j\varphi}$.

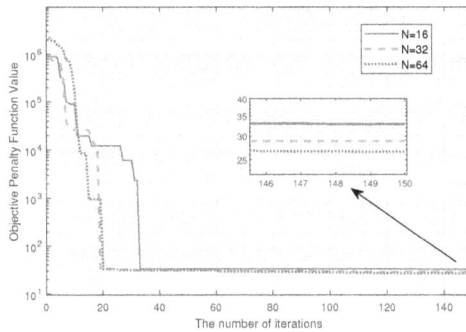

Figure 7.11 Comparison of the convergence properties of ABC algorithm for $N = 16, 32, 64$ and $K = 3$.

Fig. 7.11 shows the convergence properties of objective penalty function (7.54) computed by ABC algorithm with the cases of $N = 16, 32, 64$, where the vertical axis is represented in logarithmic scale. The range precision Γ illustrated in Algorithm 7.2 is set as a random number within the range of $(-5, 5)$. Due to the existence of negative values, we add 50 on the results of each iteration. The minimum achievable rates for three users are set as $8, 4, 4$ bps/Hz respectively. In such three cases, we set the penalty factor in (7.54) as $\varrho = 10^6$. Fig. 7.11 shows the comparison results among different numbers of antennas, where all of the function values converge, and the number of iterations is not large. From this figure, we can find that in the initial stage of the iteration, under the effect of the penalty factor in three cases, the object function values are on the 10^6 order of magnitudes, which means the constraints of the optimization problem are not satisfied. After convergence, the function values are all between 10 and 10^2 order of magnitudes, which demonstrates that the optimization variable has been solved by ABC algorithm, meanwhile satisfying the constraints.

The vertical coordinate in this figure denotes the value of the objective function in (7.54), where we observe that the final convergent value under $N = 64$ is less than that of the other two cases, which demonstrates that the ASR under $N = 64$ is the largest.

Fig. 7.12 shows the effect of the UAV-BS height on the ASR calculated by three kinds of beam patterns, i.e., the approximate beam pattern, the designed beam pattern and the random beam pattern. Relevant parameter settings are $N = 32, 64$, $P = 40$ dBm, $r_i = 0.5$ bps/Hz, $i = 1, 2, ..., 6$. The approximate value is calculated by the approximate beam gains c_i of each user. The random beam pattern is formed by an undesigned beamforming vector, where each element of $\mathbf{w}$ has a phase randomly distributed within the range of $[0, 2\pi]$ to fairly satisfy each user's rate constraint. It is considered to be the classic beamforming strategy without a specific design. According to Fig. 7.12, we can find that the ASR presents a trend of rising first and then falling for the three methods. This is because when h_U is lower than a certain altitude, the probability of having LoS component P_{LoS} in (7.1) is small. In this scope, as h_U increases, P_{LoS} also increases. And P_{LoS} has a greater impact on the system achievable rate than the propagation loss. Nevertheless, when h_U is sufficiently large, the ASR decreases as h_U increases. This is because the LoS path loss becomes larger as h_U rises. In this scope, LoS component always exists, and the impact of LoS path loss is more significant than the probability of existing an LoS component. The performance of random beam pattern is worse than those of the other two approaches. This is because beam gains may not concentrate on user directions of the strongest MPC. The UAV-BS altitude is set as 200m in the following simulations.

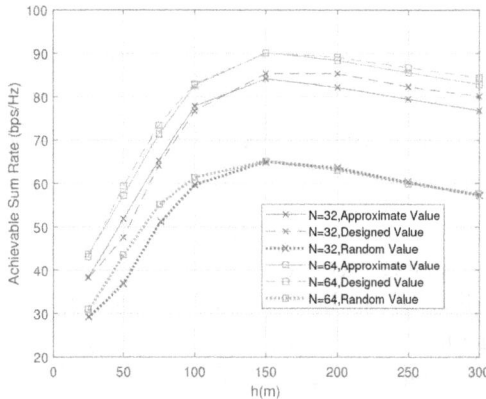

Figure 7.12   The ASR performance versus the UAV-BS height, where $N = 32$, 64 and $K = 6$.

Fig. 7.13 shows the comparison of the ASR between approximate beam pattern and designed beam pattern with varying rate constraints in the case of three users. Relevant parameter settings are $P = 20$ dBm, $h_U = 200$ m, $r_2 = r_3 = 2$ bps/Hz, $N = 16, 32, 64$. The minimum rate constraint of user 1, r_1, varies from 4 to 8 bps/Hz. We can find that the designed values are close to the approximate values with different

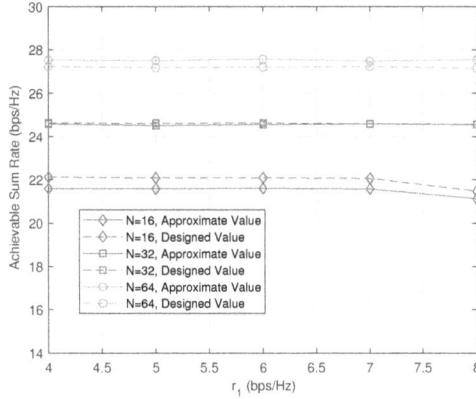

Figure 7.13 Comparison of the ASR between the approximate value and designed value versus rate constraint for $N = 16,\ 32,\ 64$ and $K = 3$.

numbers of antennas, which demonstrates that the presented solution of the original problem is effective. As r_1 increases, there exists a slight decline. This is because the UAV-BS is getting closer to user 1 to satisfy its rate constraint, which results in the decline of the system ASR. On the other hand, we can conclude that no matter how to set r_i, the original problem can always be solved by the presented algorithms, and meanwhile the designed value is close to the approximate one.

Fig. 7.14 shows the comparison of the ASR via approximate beam pattern, designed beam pattern and random beam pattern with different numbers of antennas against P. We compare two kinds of antenna structures with millimeter-wave transmission, e.g., a single antenna $(N = 1)$ and the analog beamforming structure adopted in this book, respectively. With regard to such two structures, data transmission for each user proceeds in independent slots complying with OMA strategy with full power P. The single antenna has the characteristic of omnidirectional radiation and there is no beam gain in this structure, i.e. the value of beam gain for each user is 1. The ASR with single antenna structure is calculated in the sum of all slots by

$$R_{single} = \sum_{i=1}^{K} \log_2(1 + \frac{P \cdot |\beta_i|^2 \times 1}{\sigma^2}) = \sum_{i=1}^{K} \log_2(1 + \frac{P \cdot |\beta_i|^2}{\sigma^2}). \qquad (7.58)$$

The minimum achievable rates for three users are all set as 0.5 bps/Hz against all transmit powers. As shown, the designed values are close to the approximate values. The random beam pattern performs worse than the other two patterns. From Figs. 7.7, 7.11 and 7.14, we can find that the performance of the presented solution in beamforming problem is fine, where the beam gain are concentrated on the strongest MPC of each user, and meanwhile the minimum achievable rate constraints of users are satisfied. Moreover, the ASR with the analog designed beamforming structure is larger than that with a single antenna.

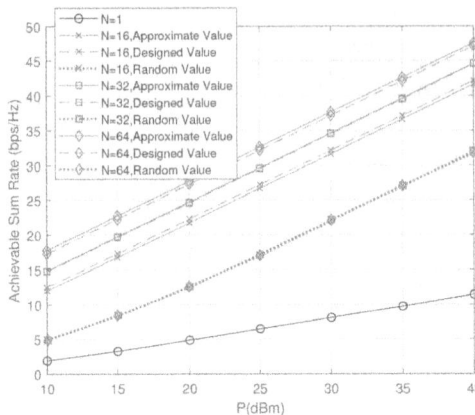

Figure 7.14 Comparison of the ASR between the approximate value, the value calculated by designed beam pattern and the value calculated by random beam pattern versus P, $N = 1$, 16, 32, 64, $K = 3$.

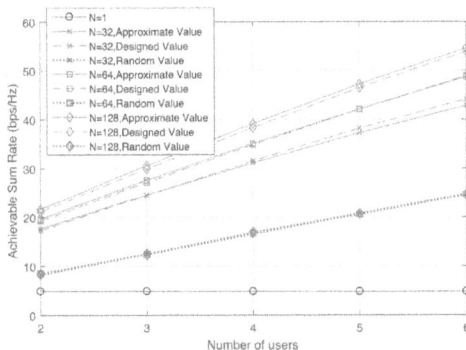

Figure 7.15 Comparison of the ASR between the approximate value, the value calculated by designed beam pattern and the value calculated by random beam pattern versus the number of users, $N = 1$, 32, 64, 128.

Next, Fig. 7.15 depicts the results of multi-user scenario, where we also compare the ASR performance via three kinds of beam patterns with different numbers of antennas. P is fixed at 30 dBm. The number of antennas is set as $1, 32, 64, 128$. In despite of different numbers of users, the minimum achievable rate for each user is all set as 2 bps/Hz. From this figure we also observe that the ASR of the designed beam pattern is close to the approximate rate. It can be concluded that our presented solution is suitable for multi-user scenario. Moreover, the ASR of random beam pattern similarly exhibits weakness compared with the other two beam patterns. For a single antenna, as the number of users increases, the ASR changes little. From Figs. 7.14 and 7.15, the analog designed beamforming structure shows priority on the ASR than a single antenna, no matter against different transmit powers or against different numbers of users.

We have demonstrated above evaluations to show that the solution of the UAV-BS deployment and beamforming problem is reasonable as desired. In addition to the ASR performance for different antenna structures, a more practical metric to evaluate the performance is the energy-efficiency (EE) of antennas[54], which is defined as the ratio between the ASR and total power consumption

$$\text{EE} = \frac{R_{\text{sum}}}{P + N_{\text{RF}} P_{\text{RF}} + N_{\text{PS}} P_{\text{PS}} + P_{\text{BB}}} \ (\text{bps/Hz/W}), \tag{7.59}$$

where P is the total transmit power, P_{PS} denotes the power consumption of each phase shifter, P_{RF} denotes the power consumed by each RF chain, and P_{BB} denotes the baseband's power consumption. Generally, we set the typical values $P_{\text{PS}} = 40$ mW where 4-bit phase shifters are utilized, $P_{\text{RF}} = 300$ mW, and $P_{\text{BB}} = 200$ mW given in[54]. N_{PS} is the number of phase shifters, where for the analog beamforming structure, it is equal to the number of antennas, N. While for a single antenna, there is no phase shifter. N_{RF} is the number of RF chains, and for both structures, $N_{\text{RF}} = 1$.

Fig. 7.16 shows EE against transmit power P, where $K = 3$ and the minimum achievable rates for three users are all set as 0.5 bps/Hz. We can observe that for both analog beamforming structure and single antenna, the value of EE increases when $P<25$ dBm and then decreases. This is because as P rises, the impact of total power consumption on EE will become more significant than the ASR in (7.59). Therefore, it is worth setting P reasonably to ensure that the antenna energy efficiency remains at a high level.

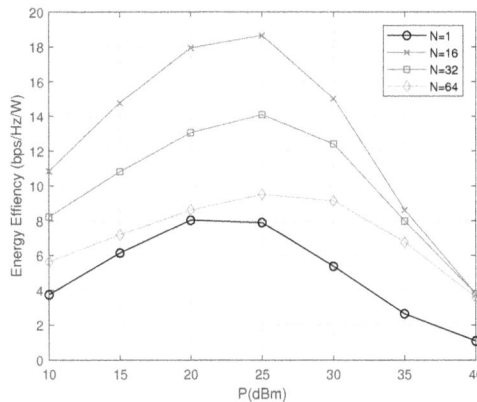

Figure 7.16 Energy efficiency versus P for $K = 3$.

In addition, the performance comparison in terms of EE against the number of users is shown in Fig. 7.17, where we set P as 30 dBm and the minimum achievable rates for users are all 2 bps/Hz. We can find that as the number of users increases, EE of a single antenna keeps stable, while analog beamforming structure shows an upward trend. From Figs. 7.16 and 7.17, with regard to the overall EE performance, analog beamforming structure shows superiority compared to a single antenna. Moreover, EE decreases along with the increase of antenna numbers. This is because one RF chain is connected to all phase shifters and $N_{\mathrm{PS}} = N$, which leads to higher power consumption as N increases. In summary, we can leverage the half-wavelength antenna space structure to mount massive antenna arrays, considering a tradeoff between the ASR and the energy efficiency. Multi-directional beam gains concentrating on the user directions can be reaped by designing the beamforming vector. Therefore, the analog beamforming structure is more appropriate to be mounted on the UAV-BS in millimeter-wave communication system.

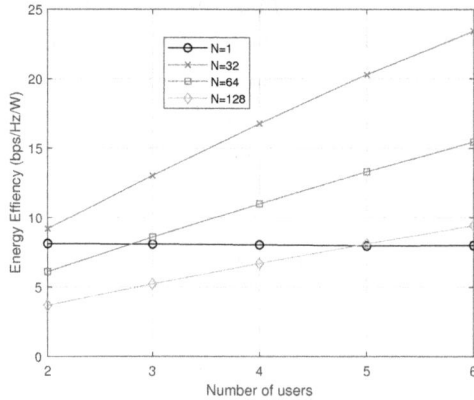

Figure 7.17 Energy efficiency versus the number of users for $P = 30$ dBm.

In this section, we have investigated the problem of maximizing the ASR of a multi-user single UAV communication system, where we should find the beamforming vector of the analog beamforming structure using ULA to steer to the strongest MPC and meanwhile satisfy the minimal achievable rate constraint for each user. The simulation results demonstrate that the presented solutions could achieve good performance in terms of the ASR and energy efficiency. And this joint optimizing UAV-BS deployment and beamforming strategy could be generalized for different types of antenna arrays.

7.5 MULTIPLE UAVS DEPLOYMENT

The works above focus on UAV communication system that employs a single UAV in a single cell and utilizes directional antenna array with fixed beam patterns. Different from these works, in this section, we consider a downlink MIMO network employing multiple UAV-BSs to serve multiple ground users, which can be potentially used in the hot-spot and/or remote areas to provide large-bandwidth communication service. To compensate for the high path loss of millimeter-wave channels and mitigate intra-cell and/or inter-cell interference, hybrid beamforming structure is utilized at the UAV-BSs and analog beamforming structure is utilized at the users.

7.5.1 Multiple UAVs System Model

As shown in Fig. 7.18, we consider a downlink communication network where M UAV-BSs are deployed to serve K users on the ground. The UAV-BSs and users are equipped with UPAs employing $N_\mathrm{B} = N_\mathrm{B}^x \times N_\mathrm{B}^y$ and $N_\mathrm{U} = N_\mathrm{U}^x \times N_\mathrm{U}^y$ antennas, respectively. At each UAV, N_RF RF chains are employed and fully connected with the N_B antennas via $N_\mathrm{RF} N_\mathrm{B}$ phase shifters and $N_\mathrm{RF} \leq N_\mathrm{B}$[55], while only one RF chain is equipped at the user because of the hardware constraint. The sets of UAV-BSs and users are denoted as $\mathcal{M}$ and $\mathcal{K}$, respectively. The user set served by UAV-BS m is denoted as $\mathcal{K}_m$, where we assume that the number of users served by each UAV-BS is no larger than that of the RF chains, i.e., $|\mathcal{K}_m| \leq N_\mathrm{RF}$. In addition, each user is connected with one UAV-BS only. Thus, we have $\bigcup_{m \in \mathcal{M}} \mathcal{K}_m = \mathcal{K}$ and $\mathcal{K}_m \cap \mathcal{K}_j = \Phi$ for $m \neq j$. The horizontal coordinates of user k are given by $\mathbf{u}_k = [x_k, y_k] \in \mathbb{R}^{1 \times 2}$ on the ground plane. We assume that the altitudes of the UAV-BSs are fixed to H, and the horizontal coordinates of UAV-BS m are denoted as $\mathbf{v}_m = [X_m, Y_m] \in \mathbb{R}^{1 \times 2}$.

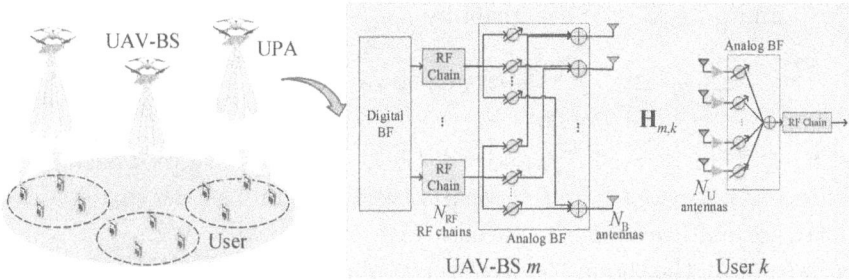

Figure 7.18 Illustration of the considered multi-UAV aided downlink communication network, where hybrid beamforming and analog beamforming structures are employed at the UAV-BSs and ground users, respectively.

At each user $k \in \mathcal{K}_m$, the received signal is given by

$$\bar{y}_k = \mathbf{w}_k^\mathrm{H} \mathbf{H}_{m,k} \mathbf{A}_m \mathbf{D}_m \mathbf{s}_m + \sum_{j \neq m} \mathbf{w}_k^\mathrm{H} \mathbf{H}_{j,k} \mathbf{A}_j \mathbf{D}_j \mathbf{s}_j + \mathbf{w}_k^\mathrm{H} \mathbf{n}_k, \qquad (7.60)$$

where $\mathbf{H}_{m,k} \in \mathbb{C}^{N_\mathrm{B} \times N_\mathrm{U}}$ is the channel matrix between UAV-BS m and user k. $\mathbf{A}_m \in \mathbb{C}^{N_\mathrm{B} \times |\mathcal{K}_m|}$ and $\mathbf{D}_m \in \mathbb{C}^{|\mathcal{K}_m| \times |\mathcal{K}_m|}$ are the analog beamforming and digital beamforming matrices of UAV-BS m, respectively. $\mathbf{w}_k \in \mathbb{C}^{N_\mathrm{U} \times 1}$ denotes the analog beamforming vector of user k. $\mathbf{s}_m \sim \mathcal{CN}(\mathbf{0}, \mathbf{I}_{|\mathcal{K}_m|})$ is the signal transmitted by UAV-BS m, and each element of $\mathbf{s}_m$ represents an independent data stream intended for one user. $\mathbf{n}_k \sim \mathcal{CN}(\mathbf{0}, \sigma^2 \mathbf{I}_{|\mathcal{K}_m|})$ is the white Gaussian noise at user k, and σ^2 is the average power of the noise at each antenna branch.

According to (7.7), the channel matrix between UAV-BS m and user k is modeled as follows

$$\mathbf{H}_{m,k} = \sum_{\ell=0}^{L_{m,k}} \beta_{m,k}^{(\ell)} \sqrt{N_\mathrm{B} N_\mathrm{U}} \mathbf{a}_\mathrm{U}(\hat{\theta}_{m,k}^{(\ell)}, \hat{\phi}_{m,k}^{(\ell)}) \mathbf{a}_\mathrm{B}^\mathrm{H}(\theta_{m,k}^{(\ell)}, \phi_{m,k}^{(\ell)}), \qquad (7.61)$$

where indices $\ell = 0$ and $\ell \geq 1$ represent the LoS and NLoS components, respectively. $L_{m,k}$ is the total number of NLoS path components of the channel between UAV-BS m and user k. $\theta_{m,k}^{(\ell)}$, $\phi_{m,k}^{(\ell)}$, $\hat{\theta}_{m,k}^{(\ell)}$, and $\hat{\phi}_{m,k}^{(\ell)}$ are, respectively, the elevation AoD, the azimuth AoD, the elevation AoA, and the azimuth AoA of the ℓ-th path between UAV-BS m and user k. $\beta_{m,k}^{(\ell)}$ is the complex coefficient of the ℓ-th path. Especially for the LoS component, the amplitude of the complex coefficient is given by

$$|\beta_{m,k}^{(0)}| = \frac{\chi_{m,k}\beta}{(H^2 + \|\mathbf{v}_m - \mathbf{u}_k\|_2^2)^{\alpha/4}}, \tag{7.62}$$

where $\beta = \frac{c_0}{4\pi f_c}$ denotes the channel gain at reference distance $d_0 = 1$ m. c_0 is the speed of light, and f_c is the carrier frequency. $\alpha \geq 2$ is the large-scale path loss exponent for millimeter-wave signals. Random variable $\chi_{m,k}$ is equal to 1 if the LoS path exists between UAV-BS m and user k, and equal to 0 otherwise. In general, the probability that an LoS path exists is determined by the environment and the altitude of the UAV (or the elevation angle between the UAV and user)[1]. The steering vectors $\mathbf{a}_\mathrm{B} \in \mathbb{C}^{N_\mathrm{B} \times 1}$ and $\mathbf{a}_\mathrm{U} \in \mathbb{C}^{N_\mathrm{U} \times 1}$ are given by

$$\mathbf{a}_\tau(\theta, \phi) = \frac{1}{\sqrt{N_\tau}} [e^{j2\pi \frac{d}{\lambda} \cos\theta[(n_x-1)\cos\phi+(n_y-1)\sin\phi]}]^\mathrm{T}, \tag{7.63}$$

with $1 \leq n_x \leq N_\tau^x, 1 \leq n_y \leq N_\tau^y$, and $\tau \in \{\mathrm{B}, \mathrm{U}\}$. d is the distance between adjacent antennas in both horizontal and vertical directions and λ is the carrier wavelength. Particularly, for half-wavelength spaced arrays, we have $d = \lambda/2$.

Then, we obtain the signal-to-interference-plus-noise ratio (SINR) of the n-th user served by UAV-BS m as follows

$$\gamma_{m,n} = \frac{\left|\mathbf{w}_{k_{m,n}}^\mathrm{H} \mathbf{H}_{m,k_{m,n}} \mathbf{A}_m \mathbf{d}_{m,n}\right|^2}{I_{m,n}^\mathrm{intra} + I_{m,n}^\mathrm{inter} + \sigma^2}, \tag{7.64}$$

where $k_{m,n}$ is the index of the n-th user served by UAV-BS m and

$$\mathbf{D}_m = [\mathbf{d}_{m,1}, \mathbf{d}_{m,2}, \cdots, \mathbf{d}_{m,N_\mathrm{RF}}]. \tag{7.65}$$

$$\begin{cases} I_{m,n}^\mathrm{intra} = \sum_{i \neq n} \left|\mathbf{w}_{k_{m,n}}^\mathrm{H} \mathbf{H}_{m,k_{m,n}} \mathbf{A}_m \mathbf{d}_{m,i}\right|^2, \\ I_{m,n}^\mathrm{inter} = \sum_{j \neq m} \left\|\mathbf{w}_{k_{m,n}}^\mathrm{H} \mathbf{H}_{j,k_{m,n}} \mathbf{A}_j \mathbf{D}_j\right\|_2^2, \end{cases} \tag{7.66}$$

represent the intra-cluster and inter-cluster interference, respectively. As a result, the achievable rate of user $k_{m,n}$ is given by

$$R_{m,n} = \log_2\left(1 + \gamma_{m,n}\right). \tag{7.67}$$

To maximize the ASR of the ground users[4], we formulate the following problem for joint optimization of the UAV positioning, user clustering, Tx-BF at the UAV-BSs, and Rx-BF at the users:

$$\max_{\{\mathbf{v}_m, \mathcal{K}_m, \mathbf{A}_m, \mathbf{D}_m, \mathbf{w}_k\}} \sum_{m \in \mathcal{M}} \sum_{1 \leq n \leq |\mathcal{K}_m|} R_{m,n} \tag{7.68a}$$

$$\text{s.t.} \quad |\mathcal{K}_m| \leq N_{\text{RF}}, \ \forall m, \tag{7.68b}$$

$$\bigcup_{m \in \mathcal{M}} \mathcal{K}_m = \mathcal{K}, \tag{7.68c}$$

$$\mathcal{K}_m \bigcap \mathcal{K}_j = \Phi, \ \forall m \neq j, \tag{7.68d}$$

$$\left| [\mathbf{A}_m]_{i,j} \right| = \frac{1}{\sqrt{N_{\text{B}}}}, \ \forall m, \ i, \ j, \tag{7.68e}$$

$$\|\mathbf{A}_m \mathbf{D}_m\|_F^2 \leq P, \ \forall m, \tag{7.68f}$$

$$|[\mathbf{w}_k]_l| = \frac{1}{\sqrt{N_{\text{U}}}}, \ \forall k, l, \tag{7.68g}$$

$$R_{m,n} \geq r_{m,n}, \ \forall m, \ 1 \leq n \leq |\mathcal{K}_m|, \tag{7.68h}$$

where constraint (7.68b) ensures that the number of users connected to each UAV-BS does not exceed N_{RF}. Constraint (7.68c) indicates that all the users should be served by UAV-BSs. Constraint (7.68d) guarantees that each user is served by one UAV-BS at most. Constraint (7.68e) is the CM constraint on the analog beamforming matrices of the UAV-BSs. Constraint (7.68f) ensures that the transmit power of each UAV-BS cannot exceed a maximum value P. Constraint (7.68g) is the CM constraint on the analog beamforming vectors of the users. Constraint (7.68h) guarantees that the achievable rate of each user is no less than the required (minimum) rate $r_{m,n}$. As we can see, Problem (7.68) involves high-dimensional variable matrices/vectors and combinatorial programming variables. This highly non-convex problem cannot be directly solved via standard convex optimization tools. Thus, we develop a sub-optimal solution for (7.68) in next subsection.

7.5.2 Positioning, Clustering, and Beamforming for Multiple UAV-BSs

Due to the directivities of the millimeter-wave channels influenced by the positions of UAV-BSs and the directional feature in spatial domain for millimeter-wave beam generated by large antenna arrays, the UAV-BS positioning and beamforming designs are highly coupled. If we adopt an alternating optimization between the positioning and beamforming variables, the solution may be trapped in a local optimum. To address this problem, we first optimize the UAV-BS positioning and user clustering under the assumption of ideal beam patterns, which effectively decouples the positioning and beamforming variables. Then, given the positions of the UAV-BSs and the user clustering, we present an alternating optimization algorithm for designing Tx-beamforming and Rx-beamforming.

---

[4]The results in this section can be readily extended to the general case of users' weighted sum-rate maximization with given equal or different weights of the users for their fairness and/or priority control.

7.5.2.1 Joint UAV-BS Positioning and User Clustering

As we discussed in Section 7.2, the existence of the LoS path is highly likely for A2G links. Especially for millimeter-wave channels, the NLoS components have much smaller powers compared to the LoS components, and have small effects on the LoS components because of the spatial sparsity. Besides, the AoDs and AoAs of the NLoS components are unpredictable for different positions of the UAV-BSs. Hence, we omit the NLoS components of the channels in this subsection, and define the LoS channel as[5]

$$\mathbf{H}_{m,k}^{\mathrm{LoS}} = \beta_{m,k}^{(0)} \sqrt{N_{\mathrm{B}} N_{\mathrm{U}}} \mathbf{a}_{\mathrm{U}}(\hat{\theta}_{m,k}^{(0)}, \hat{\phi}_{m,k}^{(\ell)}) \mathbf{a}_{\mathrm{B}}^{\mathrm{H}}(\theta_{m,k}^{(0)}, \phi_{m,k}^{(0)}). \tag{7.69}$$

Since the high-dimensional and highly coupled beamforming matrices/vectors make Problem (7.68) intractable, we assume ideal beam patterns to simplify the original problem. According to antenna theory, the array gain of a user is maximized if and only if the steering vectors for the channel between the user and its serving UAV-BS are used for beamforming[56]. However, this steering-vector based analog beamforming results in non-negligible interference at the ground users. To make the problem tractable, we define ideal beam patterns as follows.

Definition 7.5.1. (Ideal Beam Patterns) *For ideal beam patterns (which may not be realizable), the full array gains are obtained for the target signals, while all interference is completely eliminated, i.e.,*

$$\begin{cases} \left| \bar{\mathbf{w}}_{k_{m,n}}^{\mathrm{H}} \mathbf{H}_{m,k_{m,n}}^{\mathrm{LoS}} \bar{\mathbf{A}}_m \bar{\mathbf{d}}_{m,n} \right|^2 = \dfrac{\beta^2 N_{\mathrm{B}} N_{\mathrm{U}} p_{m,n}}{\left(H^2 + \left\| \mathbf{v}_m - \mathbf{u}_{k_{m,n}} \right\|_2^2 \right)^{\alpha/2}}, \\[4mm] \displaystyle\sum_{i \neq n} \left| \bar{\mathbf{w}}_{k_{m,n}}^{\mathrm{H}} \mathbf{H}_{m,k_{m,n}}^{\mathrm{LoS}} \bar{\mathbf{A}}_m \bar{\mathbf{d}}_{m,i} \right|^2 = 0, \\[4mm] \displaystyle\sum_{j \neq m} \left\| \bar{\mathbf{w}}_{k_{m,n}}^{\mathrm{H}} \mathbf{H}_{j,k_{m,n}}^{\mathrm{LoS}} \bar{\mathbf{A}}_j \bar{\mathbf{D}}_j \right\|_2^2 = 0, \end{cases} \tag{7.70}$$

where

$$p_{m,n} = \left\| \bar{\mathbf{A}}_m \bar{\mathbf{d}}_{m,n} \right\|_2^2, \tag{7.71}$$

is the signal power for user $k_{m,n}$. $\bar{\mathbf{A}}_m$ and

$$\bar{\mathbf{D}}_m = [\bar{\mathbf{d}}_{m,1}, \bar{\mathbf{d}}_{m,2}, \cdots, \bar{\mathbf{d}}_{m,|\mathcal{K}_m|}], \tag{7.72}$$

are the ideal analog beamforming and digital beamforming matrices of the UAV-BS, respectively. $\bar{\mathbf{w}}_{k_{m,n}}$ is the ideal analog beamforming vector of user $k_{m,n}$.

---

[5]We assume the LoS path always exists between a UAV and user, i.e., $\chi_{m,k} = 1$. In the beamforming stage, the original channel model in (7.61) will be adopted. The influence of this assumption will be evaluated in the simulation section.

Substituting (7.70) into (7.64) and (7.67), we obtain an approximate upper bound for the achievable rate of user $k_{m,n}$ as follows

$$\bar{R}_{m,n} = \log_2 \left(1 + \frac{\beta^2 N_{\mathrm{B}} N_{\mathrm{U}} p_{m,n}}{\left(H^2 + \| \mathbf{v}_m - \mathbf{u}_{k_{m,n}} \|_2^2 \right)^{\alpha/2} \sigma^2} \right). \tag{7.73}$$

Next, we optimize the UAV-BS positioning and user assignment for the maximization of the upper bound in (7.73). Let $\mathbf{p}_m = [p_{m,1}, p_{m,2}, \cdots, p_{m,|\mathcal{K}_m|}]^{\mathrm{T}}$ denote the power allocation vector of UAV-BS m. Problem (7.68) is simplified as follows

$$\max_{\{\mathbf{v}_m, \mathcal{K}_m, \mathbf{p}_m\}} \sum_{m \in \mathcal{M}} \sum_{1 \leq n \leq |\mathcal{K}_m|} \bar{R}_{m,n} \tag{7.74a}$$

$$\text{s.t.} \quad (7.68b), \ (7.68c), \ (7.68d), \tag{7.74b}$$

$$p_{m,n} \geq 0, \ \forall m, \ n, \tag{7.74c}$$

$$\sum_{1 \leq n \leq |\mathcal{K}_m|} p_{m,n} \leq P, \ \forall m, \tag{7.74d}$$

$$\bar{R}_{m,n} \geq r_{m,n}, \ \forall m, \ n. \tag{7.74e}$$

Problem (7.74) is still a combinatorial programming problem and involves highly coupled variables. We present an efficient iterative algorithm for solving this non-convex problem. First, the users are clustered into M groups based on their horizontal coordinates by employing the K-means algorithm, where the summation of the Euclidean distances from the cluster centers to their associated users is minimized. The initial horizonal positions of the UAV-BSs are set as the coordinates of the cluster centers, while the initial user clustering corresponds to the users in the same group. Then, we start an iterative process as follows.

1) Power Allocation: For given $\{\mathbf{v}_m^{(t-1)}, \mathcal{K}_m^{(t-1)}\}$ obtained in the $(t-1)$-th iteration, the power allocation vectors of different UAV-BSs are mutually independent. For UAV-BS m, we solve the following problem:

$$\max_{\mathbf{p}_m} \sum_{1 \leq n \leq |\mathcal{K}_m|} \bar{R}_{m,n} \tag{7.75a}$$

$$\text{s.t.} \quad (7.74c), (7.74d), (7.74e). \tag{7.75b}$$

Problem (7.75) is a standard concave problem with respect to $\mathbf{p}_m$. It can be solved by utilizing the water-filling algorithm. The optimal solution for power allocation in closed form is given by

$$p_{m,n}^{(t)} = \max \left\{ \lambda_m - \frac{1}{g_{m,n}}, \frac{2^{r_{m,n}} - 1}{g_{m,n}} \right\}, \tag{7.76}$$

where

$$g_{m,n} = \frac{\beta^2 N_{\mathrm{B}} N_{\mathrm{U}}}{(H^2 + \left\| \mathbf{v}_m^{(t-1)} - \mathbf{u}_{k_{m,n}} \right\|_2^2)^{\alpha/2} \sigma^2}, \tag{7.77}$$

and λ_m is chosen to satisfy

$$\sum_{1 \le n \le |\mathcal{K}_m|} p_{m,n}^{(t)} = P. \tag{7.78}$$

2) UAV-BS Positioning: For given $\{\mathbf{v}_m^{(t-1)}, \mathcal{K}_m^{(t-1)}, \mathbf{p}_m^{(t)}\}$, the positions of the UAV-BSs are optimized by solving the following problem:

$$\max_{\{\mathbf{v}_m\}} \sum_{m \in \mathcal{M}} \sum_{1 \le n \le |\mathcal{K}_m|} \bar{R}_{m,n} \tag{7.79a}$$

$$\text{s.t.} \quad \bar{R}_{m,n} \ge r_{m,n}, \ \forall m, \ n. \tag{7.79b}$$

Problem (7.79) is not convex due to the non-concave form of $\bar{R}_{m,n}$. To tackle this problem, we exploit successive convex optimization. The achievable rate in (7.79) is rewritten as

$$\bar{R}_{m,n} = \log_2 f_{m,n} + \hat{R}_{m,n}, \tag{7.80}$$

where

$$f_{m,n} = \frac{(H^2 + \|\mathbf{v}_m - \mathbf{u}_{k_{m,n}}\|_2^2)^{\alpha/2} + g_0 p_{m,n}}{(H^2 + \|\mathbf{v}_m^{(t-1)} - \mathbf{u}_{k_{m,n}}\|_2^2)^{\alpha/2}},$$

$$\hat{R}_{m,n} = -\frac{\alpha}{2} \log_2 \frac{H^2 + \|\mathbf{v}_m - \mathbf{u}_{k_{m,n}}\|_2^2}{H^2 + \|\mathbf{v}_m^{(t-1)} - \mathbf{u}_{k_{m,n}}\|_2^2}, \tag{7.81}$$

$$g_0 = \frac{\beta^2 N_{\mathrm{B}} N_{\mathrm{U}}}{\sigma^2}.$$

As can be observed, both $f_{m,n}$ and $\hat{R}_{m,n}$ are convex with respect to $\|\mathbf{v}_m - \mathbf{u}_{k_{m,n}}\|_2^2$. Furthermore, $\|\mathbf{v}_m - \mathbf{u}_{k_{m,n}}\|_2^2$ is convex with respect to $\mathbf{v}_m$. Thus, $f_{m,n}$ and $\hat{R}_{m,n}$ can be lower-bounded by their first-order Taylor expansions at point $\mathbf{v}_m^{(t-1)}$, i.e.,

$$f_{m,n} \ge A_{m,n} (\mathbf{v}_m^{(t-1)} - \mathbf{u}_{k_{m,n}})^{\mathrm{T}} (\mathbf{v}_m - \mathbf{v}_m^{(t-1)}) + B_{m,n} \triangleq \Gamma_{m,n}, \tag{7.82}$$

$$\hat{R}_{m,n} \ge C_{m,n} (\mathbf{v}_m^{(t-1)} - \mathbf{u}_{k_{m,n}})^{\mathrm{T}} (\mathbf{v}_m - \mathbf{v}_m^{(t-1)}) \triangleq \Upsilon_{m,n}, \tag{7.83}$$

where

$$A_{m,n} = \alpha (H^2 + \|\mathbf{v}_m^{(t-1)} - \mathbf{u}_{k_{m,n}}\|_2^2)^{-1},$$
$$B_{m,n} = 1 + g_0 p_{m,n}^{(t)} (H^2 + \|\mathbf{v}_m^{(t-1)} - \mathbf{u}_{k_{m,n}}\|_2^2)^{\alpha/2}, \tag{7.84}$$
$$C_{m,n} = -\alpha (H^2 + \|\mathbf{v}_m^{(t-1)} - \mathbf{u}_{k_{m,n}}\|_2^2)^{-1} \ln 2.$$

Then, Problem (7.79) is relaxed as follows

$$\max_{\{\mathbf{v}_m\}} \sum_{m \in \mathcal{M}} \sum_{1 \leq n \leq |\mathcal{K}_m|} \log_2 \left(\Gamma_{m,n} \right) + \Upsilon_{m,n} \tag{7.85a}$$

$$\text{s.t.} \quad \log_2 \left(\Gamma_{m,n} \right) + \Upsilon_{m,n} \geq r_{m,n}, \ \forall m, \ n, \tag{7.85b}$$

$$\left\| \mathbf{v}_m - \mathbf{v}_m^{(t-1)} \right\|_2 \leq d^{(t)}, \ \forall m. \tag{7.85c}$$

Note that inequalities (7.82) and (7.83) are only valid in a small neighborhood of point $\mathbf{v}_m^{(t-1)}$. Hence, constraint (7.85c) is introduced, where parameter $d^{(t)}$ represents a radius of a spherical neighborhood and gradually decreases during the iterations to guarantee convergence. One possible choice is $d^{(t)} = d^{(t-1)}/\kappa_1$, where $\kappa_1 > 1$ is the step size for the reduction of the radius. Problem (7.85) is convex and the optimal solution $\mathbf{v}_m^{(t)}$ can be obtained by using convex optimization tools, such as CVX[57].

3) User Clustering: For given $\{\mathbf{v}_m^{(t)}, \mathcal{K}_m^{(t-1)}, \mathbf{p}_m^{(t)}\}$, the user assignment variables are optimized by solving the following problem:

$$\max_{\{\mathcal{K}_m\}} \sum_{m \in \mathcal{M}} \sum_{1 \leq n \leq |\mathcal{K}_m|} \bar{R}_{m,n} \tag{7.86a}$$

$$\text{s.t.} \quad (7.68\text{b}), \ (7.68\text{c}), \ (7.68\text{d}), \tag{7.86b}$$

$$\bar{R}_{m,n} \geq r_{m,n}, \ \forall m, \ n. \tag{7.86c}$$

Problem (7.86) is a combinatorial programming problem, which is difficult to obtain a globally optimal solution under polynomial complexity. To address this problem, we define two matching operations for the user clustering, namely handover and swap. For handover operation $\psi_{m,n}^j$, $m \neq j$, the n-th user served by UAV-BS m is reassigned to UAV-BS j. There is an implicit condition that before the handover, the number of served users for the latter UAV-BS is not equal to the maximum, i.e., $|\mathcal{K}_j| < N_{\mathrm{RF}}$. For swap operation $\varphi_{m,n}^{j,q}$, $m \neq j$, user $k_{m,n}$ and user $k_{j,q}$ switch their serving UAV-BSs while the other users' clusters remain unchanged.

If a handover operation $\psi_{m,n}^j$ increases the objective function (7.86a) for given $\{\mathbf{v}_m^{(t)}, \mathbf{p}_m^{(t)}\}$ and satisfies constraint (7.86c), we call $\psi_{m,n}^j$ a valid handover, and the corresponding user clustering is changed, i.e.,

$$\mathcal{K}_j \leftarrow \mathcal{K}_m(n), \ \text{if} \ \varphi_{m,n}^{j,q} \ \text{is a valid handover}, \ \forall m \neq j. \tag{7.87}$$

Similarly, if a swap operation $\varphi_{m,n}^{j,q}$ increases the objective function (7.86a) for given $\{\mathbf{v}_m^{(t)}, \mathbf{p}_m^{(t)}\}$ and satisfies constraint (7.86c), we call $\varphi_{m,n}^{j,q}$ a valid swap, and the corresponding user clustering is changed, i.e.,

$$\mathcal{K}_m(n) \rightleftharpoons \mathcal{K}_j(q), \ \text{if} \ \varphi_{m,n}^{j,q} \ \text{is a valid swap}, \ \forall m \neq j. \tag{7.88}$$

Algorithm 7.3: UAV-BS Positioning and User Clustering

1) Initialize:

Initialize $\{\mathbf{v}_m^{(0)}, \mathcal{K}_m^{(0)}\}$ using K-means algorithm.

Let $t = 1$.

2) Iteration:

Update $\{\mathbf{p}_m^{(t)}\}$ according to (7.76) for given $\{\mathbf{v}_m^{(t-1)}, \mathcal{K}_m^{(t-1)}\}$.

Update $\{\mathbf{v}_m^{(t)}\}$ by solving (7.85) for given $\{\mathbf{v}_m^{(t-1)}, \mathcal{K}_m^{(t-1)}, \mathbf{p}_m^{(t)}\}$.

Update $\{\mathcal{K}_m^{(t)}\}$ according to (7.87) and (7.88) for given $\{\mathbf{v}_m^{(t)}, \mathcal{K}_m^{(t-1)}, \mathbf{p}_m^{(t)}\}$.

Update $t \leftarrow t + 1$.

3) Result:

The increase of the ASR is below a threshold ϵ_1.

Note that for both handover and swap operations, the power allocation vectors of the two corresponding UAV-BSs should be updated according to (7.76). If all handover and swap matching operations are considered but no valid operation is found, a sub-optimal solution for Problem (7.86) is obtained and given by $\mathcal{K}_m^{(t)}$.

The overall algorithm for solving Problem (7.74) is summarized in Algorithm 7.3.

7.5.2.2 Beamforming Design

After obtaining the solution for the UAV-BS positioning and user clustering, we optimize the hybrid beamforming matrices of the UAV-BSs and the analog beamforming vectors of the users to approach ideal beam patterns. Note that for fixed positions of the UAV-BSs, the channel state information (CSI) can be acquired via channel estimation. Thus, in this subsection, we adopt the original channel model in (7.61), and assume perfect CSI knowledge is available at the UAV-BSs and users. For given $\{\mathbf{v}_m, \mathcal{K}_m\}$, Problem (7.68) simplifies as follows

$$\max_{\{\mathbf{A}_m, \mathbf{D}_m, \mathbf{w}_k\}} \sum_{m \in \mathcal{M}} \sum_{1 \leq n \leq |\mathcal{K}_m|} R_{m,n} \tag{7.89a}$$

$$\text{s.t.} \quad (7.68e), (7.68f), (7.68g), (7.68h). \tag{7.89b}$$

In Problem (7.89), the analog/digital beamforming matrices of the UAV-BSs and the analog beamforming vectors of the users are highly coupled. Besides, the CM constraints on the analog beamforming matrices and vectors in (7.68e) and (7.68g) are highly non-convex. These two aspects pose the main challenges for solving Problem (7.89). To address this issue, we present an efficient algorithm which alternately optimizes the analog beamforming matrices of the UAV-BSs, the digital beamforming matrices of the UAV-BSs, and the analog beamforming vectors of the users.

First, we initialize the analog beamforming matrices and vectors. For the n-th user served by UAV-BS m, we obtain the left and right singular vectors corresponding to the maximum singular value of channel matrix $\mathbf{H}_{m,k_{m,n}}$ as $\mathbf{u}_{m,n}$ and $\mathbf{r}_{m,n}$, respectively. The initial values of the n-th column of analog beamforming matrix $\mathbf{A}_m$ and the analog beamforming vector of user $k_{m,n}$ are given by

$$[\mathbf{A}_m^{(0)}]_{:,n} = \frac{e^{j\angle\mathbf{r}_{m,n}}}{\sqrt{N_{\mathrm{B}}}}, \quad \forall m, \ 1 \le n \le |\mathcal{K}_m|,$$

$$\mathbf{w}_{k_{m,n}}^{(0)} = \frac{e^{j\angle\mathbf{u}_{m,n}}}{\sqrt{N_{\mathrm{U}}}}, \quad \forall m, \ 1 \le n \le |\mathcal{K}_m|, \tag{7.90}$$

which can increase the effective channel gain between user $k_{m,n}$ and its serving UAV-BS in the analog domain. The initial digital beamforming (DBF) matrix of UAV-BS is given by

$$\mathbf{D}_m^{(0)} = \frac{\sqrt{P}\mathbf{I}_{|\mathcal{K}_m|}}{\|\mathbf{A}_m^{(0)}\|_F}, \quad \forall m. \tag{7.91}$$

Then, we start an iterative process as follows.

1) Analog beamforming for UAV-BSs: For given

$$\{\mathbf{A}_m^{(t-1)}, \mathbf{D}_m^{(t-1)}, \mathbf{w}_k^{(t-1)}\}, \tag{7.92}$$

obtained in the $(t-1)$-th iteration, we optimize each column of the analog beamforming matrices of the UAV-BSs in a successive manner, where each column is designed to maximize the effective channel gain of the target user, as well as decrease the interference for other users. Specifically, for the n-th column of $\mathbf{A}_m$, i.e., $\mathbf{a}_{m,n}$, we formulate the following problem:

$$\max_{\mathbf{a}_{m,n}} \left| \mathbf{w}_{k_{m,n}}^{(t-1)\mathrm{H}} \mathbf{H}_{m,k_{m,n}} \hat{\mathbf{A}}_m \mathbf{d}_{m,n}^{(t-1)} \right| \tag{7.93a}$$

$$\text{s.t.} \quad \left| \mathbf{w}_{k_{m,i}}^{(t-1)\mathrm{H}} \mathbf{H}_{m,k_{m,i}} \hat{\mathbf{A}}_m \mathbf{d}_{m,n}^{(t-1)} \right| \le \eta_{m,n,m,i}^{(t)}, \ \forall i \ne n, \tag{7.93b}$$

$$\left| \mathbf{w}_{k_{j,q}}^{(t-1)\mathrm{H}} \mathbf{H}_{m,k_{j,q}} \hat{\mathbf{A}}_m \mathbf{d}_{m,n}^{(t-1)} \right| \le \eta_{m,n,j,q}^{(t)}, \ \forall j \ne m, \ \forall q, \tag{7.93c}$$

$$\left| [\mathbf{a}_{m,n}]_i \right| \le \frac{1}{\sqrt{N_{\mathrm{B}}}}, \ \forall i, \tag{7.93d}$$

where

$$\hat{\mathbf{A}}_m = [\mathbf{a}_{m,1}^{(t)}, \cdots, \mathbf{a}_{m,n-1}^{(t)}, \mathbf{a}_{m,n}, \mathbf{a}_{m,n+1}^{(t-1)}, \cdots, \mathbf{a}_{m,|\mathcal{K}_m|}^{(t-1)}]. \tag{7.94}$$

The objective function in (7.93a) is designed to maximize the effective channel gain of the n-th user served by UAV-BS m. Constraints (7.93b) and (7.93c) limit the intra-cell and inter-cell interference, respectively. Parameter $\eta_{m,n,j,q}^{(t)}$ is an upper bound on the interference at user $k_{j,q}$ caused by the signal intended for user $k_{m,n}$, which gradually decreases in the course of the iterations. One possible choice is

$$\eta_{m,n,j,q}^{(t)} = \frac{\sigma}{\sqrt{100K}} + \left| \mathbf{w}_{k_{j,q}}^{(t-1)\mathrm{H}} \mathbf{H}_{m,k_{j,q}} \mathbf{A}_m^{(t-1)} \mathbf{d}_{m,n}^{(t-1)} \right| / \kappa_2, \tag{7.95}$$

where $\kappa_2 > 1$ is the step size for the reduction of the interference, and $\frac{\sigma}{\sqrt{100K}}$ is a lower bound for $\eta_{m,n,j,q}^{(t)}$ such that the interference has a much smaller power compared to the noise. The CM constraint for the analog beamforming matrices is relaxed to the convex constraint shown in (7.93d). In fact, this relaxation has little impact on the performance of the achievable rate as shown in the following theorem.

Theorem 7.5.1. *If the feasible region of Problem* (7.93) *is not empty, there always exists an optimal solution, where at most* $r = \text{rank}(\tilde{\mathbf{H}}_{m,n})$ *elements do not satisfy the CM constraint with*

$$
\begin{aligned}
\tilde{\mathbf{H}}_{m,n} &= [\mathbf{H}_{j,q}^{\text{H}} \mathbf{w}_{k_{j,q}}]_{(j,q) \in \mathcal{J}_{m,n}}, \\
\mathcal{J}_{m,n} &= \{(j,q) \mid k_{j,q} \in \mathcal{K}, k_{j,q} \neq k_{m,n}\}.
\end{aligned}
\tag{7.96}
$$

Proof. For notational simplicity, we omit the superscripts and subscripts of the variables in Problem (7.93) and give a general form as follows

$$
\max_{\mathbf{a}} \quad \left| \mathbf{w}_1^{\text{H}} \mathbf{H}_1 \mathbf{A} \mathbf{d} \right| \tag{7.97a}
$$

$$
\text{s.t.} \quad \left| \mathbf{w}_k^{\text{H}} \mathbf{H}_k \mathbf{A} \mathbf{d} \right| \leq \eta_k, \; 2 \leq k \leq K, \tag{7.97b}
$$

$$
\left| [\mathbf{a}]_i \right| \leq \frac{1}{\sqrt{N_{\text{B}}}}, \; 1 \leq i \leq N_{\text{B}}, \tag{7.97c}
$$

with $\mathbf{w}_k \in \mathbb{C}^{N_{\text{U}} \times 1}$ and $\mathbf{H}_k \in \mathbb{C}^{N_{\text{U}} \times N_{\text{B}}}$ for $1 \leq k \leq K$, and

$$
\begin{aligned}
\mathbf{A} &= [\mathbf{a}_1, \cdots, \mathbf{a}_{n-1}, \mathbf{a}, \mathbf{a}_{n+1}, \cdots, \mathbf{a}_G] \in \mathbb{C}^{N_{\text{B}} \times G}, \\
\mathbf{d} &= [d_1, d_2, \cdots, d_G]^{\text{T}} \in \mathbb{C}^{G \times 1}, \\
G &= |\mathcal{K}_m|.
\end{aligned}
\tag{7.98}
$$

Constraints (7.93b) and (7.93c) are unified as (7.97b). Then, we rewrite Problem (7.97) by separating the n-th column of matrix $\mathbf{A}$, i.e.,

$$
\max_{\mathbf{a}} \quad \left| \mathbf{h}_1^{\text{H}} \mathbf{a} d_n + \mathbf{h}_1^{\text{H}} \hat{\mathbf{A}}_n \hat{\mathbf{d}}_n \right| \tag{7.99a}
$$

$$
\text{s.t.} \quad \left| \mathbf{h}_k^{\text{H}} \mathbf{a} d_n + \mathbf{h}_k^{\text{H}} \hat{\mathbf{A}}_n \hat{\mathbf{d}}_n \right| \leq \eta_k, \; 2 \leq k \leq K, \tag{7.99b}
$$

$$
\left| [\mathbf{a}]_i \right| \leq \frac{1}{\sqrt{N_{\text{B}}}}, \; 1 \leq i \leq N_{\text{B}}, \tag{7.99c}
$$

where $\hat{\mathbf{A}}_n$ is the sub-matrix of $\mathbf{A}$ with n-th column $\mathbf{a}$ removed, and $\hat{\mathbf{d}}_n$ is the sub-vector of $\mathbf{d}$ with n-th element d_n removed. $\mathbf{h}_k = \mathbf{H}_k^{\text{H}} \mathbf{w}_k$ is the equivalent channel vector after receiver-beamforming (Rx-BF) for $1 \leq k \leq K$. Note that if $d_n = 0$, the objective function in (7.99a) and the left-hand side of (7.99b) are both constant. In such a case, any candidate vector in the feasible region can be the optimal solution of $\mathbf{a}$. Thus, we only need to consider the case for

$d_n \neq 0$. By scaling the arguments of (7.99a) and (7.99b) with factor $1/|d_n|$, Problem (7.99) is equivalent to

$$\max_{\mathbf{a}} \quad \left| \mathbf{h}_1^{\mathrm{H}} \mathbf{a} + b_1 \right| \tag{7.100a}$$

$$\text{s.t.} \quad \left| \mathbf{h}_k^{\mathrm{H}} \mathbf{a} + b_k \right| \leq \bar{\eta}_k, \ 2 \leq k \leq K, \tag{7.100b}$$

$$|[\mathbf{a}]_i| \leq \frac{1}{\sqrt{N_{\mathrm{B}}}}, \ 1 \leq i \leq N_{\mathrm{B}}, \tag{7.100c}$$

where $b_k = \mathbf{h}_k^{\mathrm{H}} \hat{\mathbf{A}}_n \hat{\mathbf{d}}_n / d_n$ and $\bar{\eta}_k = \eta_k / |d_n|$ for $1 \leq k \leq K$.

Let $\mathbf{a}^\circ$ denote the optimal solution for Problem (7.100), which makes

$$\begin{cases} \mathbf{h}_1^{\mathrm{H}} \mathbf{a}^\circ + b_1 = c_1, \\ \mathbf{H}^{\mathrm{H}} \mathbf{a}^\circ + \mathbf{b} = \mathbf{c}, \end{cases} \tag{7.101}$$

with $\mathbf{H} = [\mathbf{h}_2, \cdots, \mathbf{h}_K]$, $\mathbf{b} = [b_2, \cdots, b_K]^{\mathrm{T}}$, and $\mathbf{c} = [c_2, \cdots, c_K]^{\mathrm{T}}$. According to the formulation of Problem (7.100), we know that $|c_1|$ is the maximum of the objective function.

Let r denote the rank of matrix $\mathbf{H}$. Assume that $\mathbf{a}^\circ$ has $r+1$ elements which do not satisfy the CM constraint[6], i.e.,

$$|[\mathbf{a}^\circ]_i| < \frac{1}{\sqrt{N_{\mathrm{B}}}}, \ for \ i = 1, 2, \cdots, r+1. \tag{7.102}$$

In the following, we aim to find a new solution in a subset of the feasible region of Problem (7.100), which can be obtained by solving the following problem:

$$\max_{[\mathbf{a}]_i(1 \leq i \leq r+1)} \quad \left| \mathbf{h}_1^{\mathrm{H}} \mathbf{a} + b_1 \right| \tag{7.103a}$$

$$\text{s.t.} \quad \mathbf{H}^{\mathrm{H}} \mathbf{a} + \mathbf{b} = \mathbf{c}, \tag{7.103b}$$

$$|[\mathbf{a}]_i| \leq \frac{1}{\sqrt{N_{\mathrm{B}}}}, \ 1 \leq i \leq r+1, \tag{7.103c}$$

$$[\mathbf{a}]_q = [\mathbf{a}^\circ]_q, \ r+2 \leq q \leq N_{\mathrm{B}}. \tag{7.103d}$$

To separate different elements of $\mathbf{a}$, constraint (7.103b) is rewritten as

$$[\mathbf{H}_1^{\mathrm{H}}, \mathbf{H}_2^{\mathrm{H}}][\mathbf{a}_1^{\mathrm{T}}, \mathbf{a}_2^{\mathrm{T}}]^{\mathrm{T}} = \mathbf{c} - \mathbf{b}, \tag{7.104}$$

where $\mathbf{H}_1 = [\mathbf{H}]_{1:r,:}$ and $\mathbf{H}_2 = [\mathbf{H}]_{r+1:N_{\mathrm{B}},:}$ are the sub-matrices of $\mathbf{H}$. $\mathbf{a}_1 = [\mathbf{a}]_{1:r}$ and $\mathbf{a}_2 = [\mathbf{a}]_{r+1:N_{\mathrm{B}}}$ are the sub-vectors of $\mathbf{a}$. As a result, constraint (7.103b) is equivalent to

$$\mathbf{H}_1^{\mathrm{H}} \mathbf{a}_1 = -\mathbf{H}_2^{\mathrm{H}} \mathbf{a}_2 + \mathbf{c} - \mathbf{b}. \tag{7.105}$$

---

[6] Any elements of $\mathbf{a}^\circ$ can be exchanged with the first $r+1$ elements, which can be realized by adjusting the order of $\mathbf{a}^\circ$'s elements, and simultaneously exchanging the corresponding columns of $\mathbf{H}$ and the corresponding elements of $\mathbf{h}_1$. Thus, the adopted assumption is without loss of generality.

Since $\mathbf{H}_1$ is a sub-matrix of $\mathbf{H}$, the rank of $\mathbf{H}_1$, denoted as r_s, is no larger than r. By conducting elementary column transformation on matrix $\mathbf{H}_1$, we can always find a matrix $\mathbf{P}$ which makes $\tilde{\mathbf{H}}_1^H = \mathbf{P}^H \mathbf{H}_1^H$ an echelon matrix. The equation set in (7.105) is equivalent to

$$\tilde{\mathbf{H}}_1^H \mathbf{a}_1 = -\mathbf{P}^H \mathbf{H}_2^H \mathbf{a}_2 + \mathbf{P}^H (\mathbf{c} - \mathbf{b}). \tag{7.106}$$

Define $\{\pi_i,\ 1 \le i \le r\}$ as the sequence of the indices from 1 to r, and let $\pi_i,\ 1 \le i \le r_s$, denote the index of the first non-zero element in the i-th row of $\tilde{\mathbf{H}}_1^H$. According to Gaussian elimination, to solve equation set (7.106), elements $[\mathbf{a}]_{\pi_i},\ 1 \le i \le r_s$, can always be expressed as the affine functions of the other elements of $\mathbf{a}$, i.e.,

$$[\mathbf{a}]_{\pi_i} = g_{r+1}^{\pi_i}[\mathbf{a}]_{r+1} + \sum_{r_s+1 \le j \le r} g_{\pi_j}^{\pi_i}[\mathbf{a}]_{\pi_j} \sum_{r+2 \le q \le N_B} g_q^{\pi_i}[\mathbf{a}]_q + \varsigma_{\pi_i}$$

$$\triangleq g_{r+1}^{\pi_i}[\mathbf{a}]_{r+1} + v_{\pi_i},\ 1 \le i \le r_s, \tag{7.107}$$

where $g_\tau^{\pi_i}$ and ς_{π_i} denote the coefficient of $[\mathbf{a}]_\tau$ and the constant term in the expression of $[\mathbf{a}]_{\pi_i}$, respectively, which can be obtained according to Gaussian elimination. Next, we keep $[\mathbf{a}]_\tau = [\mathbf{a}^\circ]_\tau$ fixed for $\tau = \pi_{r_s+1}, \pi_{r_s+2}, \cdots, \pi_r$ and $\tau = r+2, r+3, \cdots, N_B$, and try to construct a new solution by adjusting $[\mathbf{a}]_{r+1}$. In this way, the objective function in (7.103a) is written as an affine function of $[\mathbf{a}]_{r+1}$, i.e.,

$$\left| \mathbf{h}_1^H \mathbf{a} + b_1 \right| = \left| \hat{k}[\mathbf{a}^\circ]_{r+1} + \hat{b} \right|, \tag{7.108}$$

where

$$\begin{cases} \hat{k} = \sum_{1 \le i \le r_s} [\mathbf{h}_1]_{\pi_i}^* g_{r+1}^{\pi_i} + [\mathbf{h}_1]_{r+1}^*, \\ \hat{b} = \sum_{1 \le i \le r_s} [\mathbf{h}_1]_{\pi_i}^* v_{\pi_i} + \sum_{r_s+1 \le j \le r} [\mathbf{h}_1]_{\pi_j}^*[\mathbf{a}]_{\pi_j} + \sum_{r+2 \le q \le N_B} [\mathbf{h}_1]_q^*[\mathbf{a}]_q + b_1, \end{cases} \tag{7.109}$$

are both constants.

As we have assumed $|[\mathbf{a}^\circ]_i| < \frac{1}{\sqrt{N_B}}$ for $i = 1, 2, \cdots, r+1$, and from (7.107), we can always find a real number δ, which is positive and small enough to satisfy

$$\begin{cases} |[\mathbf{a}^\circ]_{r+1} \pm \delta| < \dfrac{1}{\sqrt{N_B}}, \\ |g_{r+1}^{\pi_i}([\mathbf{a}^\circ]_{r+1} \pm \delta) + v_{\pi_i}| < \dfrac{1}{\sqrt{N_B}},\ 1 \le i \le r_s. \end{cases} \tag{7.110}$$

This means that $[\mathbf{a}^\circ]_{r+1} + \delta$ and $[\mathbf{a}^\circ]_{r+1} - \delta$ are both located in the feasible region of Problem (7.103). Since $\mathbf{a}^\circ$ is the optimal solution for Problem (7.103), the objective function at $[\mathbf{a}^\circ]_{r+1} + \delta$ and $[\mathbf{a}^\circ]_{r+1} - \delta$ is no larger than at $[\mathbf{a}^\circ]_{r+1}$, i.e.,

$$\left| \hat{k}([\mathbf{a}^\circ]_{r+1} \pm \delta) + \hat{b} \right| \le \left| \hat{k}[\mathbf{a}^\circ]_{r+1} + \hat{b} \right|$$

$$\Rightarrow \pm 2\mathrm{Re}\left((\hat{k}[\mathbf{a}^\circ]_{r+1} + \hat{b})^* \hat{k}\delta \right) + |\hat{k}\delta|^2 \le 0 \tag{7.111}$$

$$\Rightarrow |\hat{k}\delta|^2 \le 0 \Rightarrow \hat{k} = 0.$$

In other words, for fixed $[\mathbf{a}]_\tau = [\mathbf{a}^\circ]_\tau$, $\tau = \pi_{r_s+1}, \pi_{r_s+2}, \cdots, \pi_r, r+2, r+3, \cdots, N_{\mathrm{B}}$, and under constraint (7.103b), the objective function in (7.103a) is a constant, i.e., $|\mathbf{h}_1^{\mathrm{H}}\mathbf{a} + b_1| \equiv |\hat{b}| = |c_1|$. Then, we can keep the phase of $[\mathbf{a}]_{r+1}$ fixed and gradually increase its amplitude. Due to the continuity of complex values, the values of $[\mathbf{a}]_{\pi_i}$, $1 \le i \le r_s$, also continuously change according to (7.107). We can always find an amplitude of $[\mathbf{a}]_{r+1}$ to ensure that at least one element in

$$\{[\mathbf{a}]_{\pi_i}, \ 1 \le i \le r_s\} \bigcup \{[\mathbf{a}]_{r+1}\} \tag{7.112}$$

has amplitude $\frac{1}{\sqrt{N_{\mathrm{B}}}}$ and the other elements have amplitudes smaller than $\frac{1}{\sqrt{N_{\mathrm{B}}}}$. In other words, for any $r+1$ elements of $\mathbf{a}^\circ$ which do not satisfy the CM constraint, we can always adjust the values of these $r+1$ elements and find a new solution where at most r elements do not satisfy the CM constraint. This thus completes the proof. $\qquad\qquad\qquad\qquad\qquad\qquad\qquad\qquad\qquad\qquad\Box$

Problem (7.93) is not a convex optimization problem because a convex objective function is maximized but not minimized. To address this issue, the objective function in (7.93a) is rewritten as

$$\left| \mathbf{w}_{k_{m,n}}^{(t-1)\mathrm{H}} \mathbf{H}_{m,k_{m,n}} \mathbf{a}_{m,n} [\mathbf{d}_{m,n}^{(t-1)}]_n + \rho_{m,n} \right|, \tag{7.113}$$

where the constant component is

$$\rho_{m,n} = \sum_{i \le n-1} \mathbf{w}_{k_{m,n}}^{(t-1)\mathrm{H}} \mathbf{H}_{m,k_{m,n}} \mathbf{a}_{m,i}^{(t)} [\mathbf{d}_{m,n}^{(t-1)}]_i + \sum_{i \ge n+1} \mathbf{w}_{k_{m,n}}^{(t-1)\mathrm{H}} \mathbf{H}_{m,k_{m,n}} \mathbf{a}_{m,i}^{(t-1)} [\mathbf{d}_{m,n}^{(t-1)}]_i. \tag{7.114}$$

Based on the triangle inequality, we have

$$\left| \mathbf{w}_{k_{m,n}}^{(t-1)\mathrm{H}} \mathbf{H}_{m,k_{m,n}} \bar{\mathbf{A}}_m \mathbf{d}_{m,n}^{(t-1)} \right| \le \left| \mathbf{w}_{k_{m,n}}^{(t-1)\mathrm{H}} \mathbf{H}_{m,k_{m,n}} \mathbf{a}_{m,n} [\mathbf{d}_{m,n}^{(t-1)}]_n \right| + |\rho_{m,n}|, \tag{7.115}$$

where the equality holds if and only if

$$\mathbf{w}_{k_{m,n}}^{(t-1)\mathrm{H}} \mathbf{H}_{m,k_{m,n}} \mathbf{a}_{m,n} [\mathbf{d}_{m,n}^{(t-1)}]_n, \tag{7.116}$$

and $\rho_{m,n}$ have the same phase. Thus, we can first maximize

$$\left| \mathbf{w}_{k_{m,n}}^{(t-1)\mathrm{H}} \mathbf{H}_{m,k_{m,n}} \mathbf{a}_{m,n} [\mathbf{d}_{m,n}^{(t-1)}]_n \right|, \tag{7.117}$$

and then make a phase rotation on $\mathbf{a}_{m,n}$ (multiplied by $e^{j\nu}$) so that

$$\mathbf{w}_{k_{m,n}}^{(t-1)\mathrm{H}} \mathbf{H}_{m,k_{m,n}} \mathbf{a}_{m,n} [\mathbf{d}_{m,n}^{(t-1)}]_n, \tag{7.118}$$

and $\rho_{m,n}$ have the same phase. Note that the phase notation on $\mathbf{a}_{m,n}$ does not impact the value of

$$\left| \mathbf{w}_{k_{m,n}}^{(t-1)\mathrm{H}} \mathbf{H}_{m,k_{m,n}} \mathbf{a}_{m,n} [\mathbf{d}_{m,n}^{(t-1)}]_n \right|. \tag{7.119}$$

Hence, the objective function in (7.93a) is equivalent to

$$\mathrm{Re}(\mathbf{w}_{k_{m,n}}^{(t-1)\mathrm{H}} \mathbf{H}_{m,k_{m,n}} \mathbf{a}_{m,n} [\mathbf{d}_{m,n}^{(t-1)}]_n + \rho_{m,n} e^{-j\nu_{m,n}}). \tag{7.120}$$

where $\nu_{m,n}$ represents the phase of $\rho_{m,n}$. With this modification, Problem (7.93) is equivalently transformed into

$$\max_{\mathbf{a}_{m,n}} \ \mathrm{Re}(\mathbf{w}_{k_{m,n}}^{(t-1)\mathrm{H}} \mathbf{H}_{m,k_{m,n}} \mathbf{a}_{m,n} [\mathbf{d}_{m,n}^{(t-1)}]_n + \rho_{m,n} e^{-j\nu_{m,n}}) \tag{7.121a}$$

$$\text{s.t.} \quad (7.93b), (7.93c), (7.93d). \tag{7.121b}$$

As can be observed, Problem (7.121) is a convex problem because the objective function becomes an affine function, and the optimal solution $\mathbf{a}_{m,n}^\circ$ can be obtained by using CVX[57]. After solving Problem (7.121) for all $\mathbf{a}_{m,n}$, the modulus normalization is performed as follows

$$[\mathbf{A}_m^{(t)}]_{:,n} = \frac{e^{j\angle(\mathbf{a}_{m,n}^\circ)}}{\sqrt{N_\mathrm{B}}}, \quad \forall m, \ n. \tag{7.122}$$

According to Theorem 7.5.1, the normalization operation on the analog beamforming matrices has a little impact on the optimality of the solution because only a small number of elements in each column are adjusted.

2) Digital beamforming for UAV-BSs: For given $\{\mathbf{A}_m^{(t)}, \mathbf{D}_m^{(t-1)}, \mathbf{w}_k^{(t-1)}\}$, we optimize the digital beamforming matrices of the UAV-BSs by solving the following problem

$$\max_{\{\mathbf{D}_m\}} \ \sum_{m\in\mathcal{M}} \sum_{1\le n\le |\mathcal{K}_m|} R_{m,n} \tag{7.123a}$$

$$\text{s.t.} \quad (7.68f), (7.68h). \tag{7.123b}$$

Problem (7.123) is a non-convex problem because $R_{m,n}$ is not convex with respect to $\{\mathbf{D}_m\}$. We present to use the following approach to address this problem, which is based on the important relation between the SINR and the MMSE[58].

If a single-tap equalizer is employed at the users, the MSE of user $k_{m,n}$ can be expressed as

$$\varepsilon_{m,n} = \mathbb{E}\left[\|c_{m,n}\bar{y}_{m,n} - s_{m,n}\|_2^2\right]$$
$$= \left\|c_{m,n}\hat{\mathbf{h}}_{m,k_{m,n}}^{\mathrm{H}} \mathbf{D}_m - \mathbf{e}_{m,n}^{\mathrm{T}}\right\|_2^2 + \sum_{j\ne m} \left\|c_{m,n}\hat{\mathbf{h}}_{j,k_{m,n}}^{\mathrm{H}} \mathbf{D}_j\right\|_2^2 + |c_{m,n}\sigma|^2, \tag{7.124}$$

where $c_{m,n}$ is the equalization coefficient of the single-tap equalizer at user $k_{m,n}$, and

$$\hat{\mathbf{h}}_{j,k_{m,n}} = \mathbf{A}_j^{(t)\mathrm{H}} \mathbf{H}_{j,k_{m,n}}^{\mathrm{H}} \mathbf{w}_{k_{m,n}}^{(t-1)}, \tag{7.125}$$

represents the equivalent channel after analog beamforming. $\mathbf{e}_{m,n} \in \mathbb{R}^{|\mathcal{K}_m|\times 1}$ is a vector with 1 as the n-th element and 0 elsewhere. The MMSE can be achieved as follows

$$\left. \frac{\partial \varepsilon_{m,n}}{\partial c_{m,n}} \right|_{c_{m,n}=c_{m,n}^\circ} = 0 \Rightarrow$$

$$c_{m,n}^\circ = \left(\hat{\mathbf{h}}_{m,k_{m,n}}^{\mathrm{H}} \mathbf{d}_{m,n}^{(t-1)}\right)^* \left(\left|\hat{\mathbf{h}}_{m,k_{m,n}}^{\mathrm{H}} \mathbf{d}_{m,n}^{(t-1)}\right|^2 + \xi_{m,n}\right)^{-1}, \tag{7.126}$$

where

$$\xi_{m,n} = \sum_{i \neq n} |\hat{\mathbf{h}}_{m,k_{m,n}}^{\mathrm{H}} \mathbf{d}_{m,i}^{(t-1)}|^2 + \sum_{j \neq m} \|\hat{\mathbf{h}}_{j,k_{m,n}}^{\mathrm{H}} \mathbf{D}_j^{(t-1)}\|_2^2 + \sigma^2. \tag{7.127}$$

Substituting (7.126) into (7.124), we can find that the following equation always holds, i.e.,

$$\varepsilon_{m,n}|_{c_{m,n}=c_{m,n}^{\circ}} = (1 + \gamma_{m,n})^{-1}. \tag{7.128}$$

Let $\mathbf{C} = [c_{m,n}]_{1 \leq m \leq M, 1 \leq n \leq N_{\mathrm{RF}}}$. Then Problem (7.123) is equivalent to

$$\min_{\{\mathbf{D}_m\}, \mathbf{C}} \sum_{m \in \mathcal{M}} \sum_{1 \leq n \leq |\mathcal{K}_m|} \log_2 \varepsilon_{m,n} \tag{7.129a}$$

$$\text{s.t.} \quad (7.68\mathrm{f}), \tag{7.129b}$$

$$\varepsilon_{m,n} \leq 2^{-r_{m,n}}. \tag{7.129c}$$

Problem (7.129) is still non-convex. We introduce an auxiliary function

$$\psi(u_{m,n}) = 2^{u_{m,n}-1} \varepsilon_{m,n} - u_{m,n}, \tag{7.130}$$

which is minimized at the following point

$$u_{m,n}^{\circ} = -\log_2 \varepsilon_{m,n} + 1, \tag{7.131}$$

with the minimum value $\psi(u_{m,n}^{\circ}) = \log_2 \varepsilon_{m,n}$.

Let $\mathbf{U} = [u_{m,n}]_{1 \leq m \leq M, 1 \leq n \leq N_{\mathrm{RF}}}$. Then, Problem (7.129) is equivalent to

$$\min_{\{\mathbf{D}_m\}, \mathbf{C}, \mathbf{U}} \sum_{m \in \mathcal{M}} \sum_{1 \leq n \leq |\mathcal{K}_m|} \left(2^{u_{m,n}-1} \varepsilon_{m,n} - u_{m,n}\right) \tag{7.132a}$$

$$\text{s.t.} \quad (7.68\mathrm{f}), \ (7.129\mathrm{c}). \tag{7.132b}$$

To find a sub-optimal solution for Problem (7.132), we alternately optimize $\mathbf{C}$, $\mathbf{U}$, and $\{\mathbf{D}_m\}$ in the course of the iterations. For given $\{\mathbf{A}_m^{(t)}, \mathbf{D}_m^{(t-1)}, \mathbf{w}_k^{(t-1)}\}$, we obtain the optimal $\mathbf{C}^{(t)}$ according to (7.126). For given $\mathbf{C}^{(t)}$, we obtain the optimal $\mathbf{U}^{(t)}$ according to (7.131). For given $\{\mathbf{A}_m^{(t)}, \mathbf{w}_k^{(t-1)}\}$, $\mathbf{C}^{(t)}$, and $\mathbf{U}^{(t)}$, Problem (7.132) is a convex problem with respect to $\{\mathbf{D}_m\}$ and the optimal solution $\{\mathbf{D}_m^{(t)}\}$ can be obtained by using CVX[57].

3) Analog beamforming for Users: For given $\{\mathbf{A}_m^{(t)}, \mathbf{D}_m^{(t)}, \mathbf{w}_k^{(t-1)}\}$, the receive beamforming vectors for different users can be optimized separately. Thus, we can optimize these analog beamforming vectors independently to maximize the achievable rate of each user. Specifically, for the n-th user served by UAV-BS m, the problem is given by

$$\max_{\mathbf{w}_{k_{m,n}}} R_{m,n} \tag{7.133a}$$

$$\text{s.t.} \quad (7.68\mathrm{g}), \ (7.68\mathrm{h}). \tag{7.133b}$$

Due to the non-convex CM constraint for the analog beamforming vector, it is difficult to obtain the optimal solution for Problem (7.133). To address this problem, we optimize each element of the analog beamforming vector in a successive manner for the maximization of the corresponding user's achievable rate. Next, we show that the optimal solution for an element of the analog beamforming vector can be derived if the other elements are fixed. Denoting the l-th element of $\mathbf{w}_{k_{m,n}}$ as $\frac{e^{j\omega}}{\sqrt{N_U}}$, the SINR of user $k_{m,n}$ is written as

$$
\gamma_{m,n} = \frac{\mathbf{w}_{k_{m,n}}^{\mathrm{H}} \mathbf{B}_{m,n}^{(t)} \mathbf{w}_{k_{m,n}}}{\mathbf{w}_{k_{m,n}}^{\mathrm{H}} \mathbf{Q}_{m,n}^{(t)} \mathbf{w}_{k_{m,n}} + \sigma^2}
$$

$$
\triangleq \frac{\bar{a}_{m,n,l}^{(t)} + \bar{b}_{m,n,l}^{(t)*} e^{j\omega} + \bar{b}_{m,n,l}^{(t)} e^{-j\omega}}{\bar{c}_{m,n,l}^{(t)} + \bar{d}_{m,n,l}^{(t)*} e^{j\omega} + \bar{d}_{m,n,l}^{(t)} e^{-j\omega}},
$$

(7.134)

where the corresponding constant matrices and scalars are given by

$$
\mathbf{Q}_{m,n}^{(t)} = \sum_{i \neq n} \mathbf{H}_{m,k_{m,n}} \mathbf{A}_m^{(t)} \mathbf{d}_{m,i}^{(t)} \left(\mathbf{H}_{m,k_{m,n}} \mathbf{A}_m^{(t)} \mathbf{d}_{m,i}^{(t)} \right)^{\mathrm{H}}
$$

$$
+ \sum_{j \neq m} \mathbf{H}_{j,k_{m,n}} \mathbf{A}_j^{(t)} \mathbf{D}_j^{(t)} \left(\mathbf{H}_{j,k_{m,n}} \mathbf{A}_j^{(t)} \mathbf{D}_j^{(t)} \right)^{\mathrm{H}},
$$

$$
\mathbf{B}_{m,n}^{(t)} = \mathbf{H}_{m,k_{m,n}} \mathbf{A}_m^{(t)} \mathbf{d}_{m,n}^{(t)} \left(\mathbf{H}_{m,k_{m,n}} \mathbf{A}_m^{(t)} \mathbf{d}_{m,n}^{(t)} \right)^{\mathrm{H}},
$$

$$
\bar{a}_{m,n,l}^{(t)} = \frac{\left[\mathbf{B}_{m,n}^{(t)} \right]_{l,l}}{N_U} + \sum_{r \neq l, s \neq l} \left[\mathbf{w}_{k_{m,n}} \right]_r^* \left[\mathbf{B}_{m,n}^{(t)} \right]_{r,s} \left[\mathbf{w}_{k_{m,n}} \right]_s,
$$

$$
\bar{b}_{m,n,l}^{(t)} = \frac{1}{\sqrt{N_U}} \sum_{q \neq l} \left[\mathbf{w}_{k_{m,n}} \right]_q^* \left[\mathbf{B}_{m,n}^{(t)} \right]_{q,l},
$$

$$
\bar{c}_{m,n,l}^{(t)} = \frac{\left[\mathbf{Q}_{m,n}^{(t)} \right]_{l,l}}{N_U} + \sum_{r \neq l, s \neq l} \left[\mathbf{w}_{k_{m,n}} \right]_r^* \left[\mathbf{Q}_{m,n}^{(t)} \right]_{r,s} \left[\mathbf{w}_{k_{m,n}} \right]_s + \sigma^2,
$$

$$
\bar{d}_{m,n,l}^{(t)} = \frac{1}{\sqrt{N_U}} \sum_{q \neq l} \left[\mathbf{w}_{k_{m,n}} \right]_q^* \left[\mathbf{Q}_{m,n}^{(t)} \right]_{q,l}.
$$

Theorem 7.5.2. *An optimal solution of ω for the maximization of $\gamma_{m,n}$ in (7.134) is given by*

$$
\omega_{m,n,l}^{(t)} = \arcsin \left(z_{m,n,l}^{(t)} \right) - \varpi_{m,n,l}^{(t)},
$$

(7.135)

where

$$
z_{m,n,l}^{(t)} = \frac{2\mathrm{Im} \left(\bar{b}_{m,n,l}^{(t)*} \bar{d}_{m,n,l}^{(t)} \right)}{\left| h_{m,n,l}^{(t)} \right|},
$$

$$
h_{m,n,l}^{(t)} = \bar{b}_{m,n,l}^{(t)} \bar{c}_{m,n,l}^{(t)} - \bar{a}_{m,n,l}^{(t)} \bar{d}_{m,n,l}^{(t)},
$$

$$
\varpi_{m,n,l}^{(t)} = \angle \left(h_{m,n,l}^{(t)} \right).
$$

(7.136)

Proof. For notational simplicity, we omit the superscripts and subscripts of the variables in (7.134), which is written as

$$\gamma(\omega) = \frac{a + b^*e^{j\omega} + be^{-j\omega}}{c + d^*e^{j\omega} + de^{-j\omega}}. \tag{7.137}$$

In this proof, we will show that the maximum point of $\gamma(\omega)$ can be uniquely obtained in closed form.

Since $\gamma(\omega)$ represents the SINR of a user, the denominator of $\gamma(\omega)$, i.e., the power of interference plus noise, is positive. Besides, $\gamma(\omega)$ is periodic with period 2π and smooth, and thus its maximum and minimum are achieved at its extreme points over one period, which can be found by solving the inequation:

$$\frac{d\gamma(\omega)}{d\omega} \le 0,$$
$$\stackrel{(a)}{\Leftrightarrow} \frac{(jbe^{j\omega} - jb^*e^{-j\omega})(c + de^{j\omega} + d^*e^{-j\omega})}{(c + d^*e^{j\omega} + de^{-j\omega})^2}$$
$$- \frac{(jde^{j\omega} - jd^*e^{-j\omega})(a + be^{j\omega} + b^*e^{-j\omega})}{(c + d^*e^{j\omega} + de^{-j\omega})^2} \le 0$$
$$\stackrel{(b)}{\Leftrightarrow} \text{Im}\left((bc - ad)e^{j\omega}\right) - 2\text{Im}(b^*d) \ge 0 \tag{7.138}$$
$$\stackrel{(c)}{\Leftrightarrow} \sin(\omega + \varpi) - z \ge 0$$
$$\stackrel{(d)}{\Leftrightarrow} 2n\pi + \arcsin(z) - \varpi \le \omega$$
$$\le (2n + 1)\pi + \arcsin(z) - \varpi, \ n = \cdots, -1, 0, 1, \cdots,$$

where step (a) is obtained by calculating the differential. Step (b) is obtained by expending the equation and combining similar terms. In step (c), the parameters are given by $\varpi = \angle(bc - ad)$ and

$$z = \frac{2\text{Im}(b^*d)}{|bc - ad|}. \tag{7.139}$$

Step (d) is obtained according to basic properties of sine function. As can be observed, $\gamma(\omega)$ has two extreme points over one period. It is easy to find that $\gamma(\omega)$ increases with ω in the left neighborhood of $\arcsin(z) - \varpi$ and decreases with ω in the right neighborhood of $\arcsin(z) - \varpi$. Thus, we conclude that $\arcsin(z) - \varpi$ is an optimal solution for the maximization of $\gamma(\omega)$. The proof is thus completed. □

As a result, the optimal solution to the l-th element of $\mathbf{w}_{k_{m,n}}$ in the t-th iteration is

$$\left[\mathbf{w}_{k_{m,n}}^{(t)}\right]_l = \frac{e^{j\omega_{m,n,l}^{(t)}}}{\sqrt{N_U}}. \tag{7.140}$$

For each step of the optimization of Rx analog beamforming, the achievable rate of the corresponding user is non-decreasing. Thus, constraint (7.68h) always holds during the iteration.

The overall algorithm for solving Problem (7.89) is summarized in Algorithm 7.4.

Algorithm 7.4: BF Design

---

1) **Initialize:**

Initialize $\{\mathbf{A}_m^{(0)}, \mathbf{D}_m^{(0)}, \mathbf{w}_k^{(0)}\}$ according to (7.90) and (7.91).
Let $t = 1$.

2) **Iteration:**

Update $\{\mathbf{a}_{m,n}^{(t)}\}$ by solving (7.121) for given $\{\mathbf{A}_m^{(t-1)}, \mathbf{D}_m^{(t-1)}, \mathbf{w}_k^{(t-1)}\}$.
Normalization of analog beamforming matrices according to (7.122).
Update $\mathbf{C}^{(t)}$ according to (7.126) for given $\{\mathbf{A}_m^{(t)}, \mathbf{D}_m^{(t-1)}, \mathbf{w}_k^{(t-1)}\}$.
Update $\{\varepsilon_{m,n}\}$ according to (7.124) for given $\mathbf{C}^{(t)}$ and
$\{\mathbf{A}_m^{(t)}, \mathbf{D}_m^{(t-1)}, \mathbf{w}_k^{(t-1)}\}$.
Update $\mathbf{U}^{(t)}$ according to (7.131) for given $\{\varepsilon_{m,n}\}$.
Update $\{\mathbf{D}_m^{(t)}\}$ by solving (7.132) for given $\{\mathbf{A}_m^{(t)}, \mathbf{w}_k^{(t-1)}\}$, $\mathbf{C}^{(t)}$, and $\mathbf{U}^{(t)}$.
Update $\mathbf{w}_k^{(t)}$ according to (7.140) for given $\{\mathbf{A}_m^{(t)}, \mathbf{D}_m^{(t)}, \mathbf{w}_k^{(t-1)}\}$.
Update $t \leftarrow t + 1$.

3) **Result:**

The increase of the ASR is below a threshold ϵ_2.

7.5.3   Performance Evaluation

In this section, we provide simulation results to evaluate the performance of the presented UAV-BS positioning, user clustering, and beamforming scheme for multi-UAV aided millimeter-wave communication networks.

7.5.3.1   *Simulation Setup and Benchmark Schemes*

The users are randomly distributed in a 2×2 km^2 region following uniform distribution. Half-wavelength spacing UPAs are used at the UAV-BSs and users. We adopt the channel model in (7.61), where the LoS component is assumed present and the numbers of NLoS components of the channels are assumed the same, i.e., $L_{m,k} = L$. The average strength of the NLoS components is set to be 20 dB smaller than that of the LoS path[59, 60]. The adopted simulation parameter settings are provided as follows, similar to[43, 61], unless specified otherwise. Except for Figs. 7.19 and 7.20, each simulation point in the simulation figures is averaged over 10^3 user distributions and channel realizations.

$H = 100$m	Altitude of the UAV-BSs
$P = 35$dBm	Maximum transmit power of the UAV-BSs
$\sigma^2 = -110$dBm	Power of the noise at users
$f_c = 38$GHz	Carrier frequency

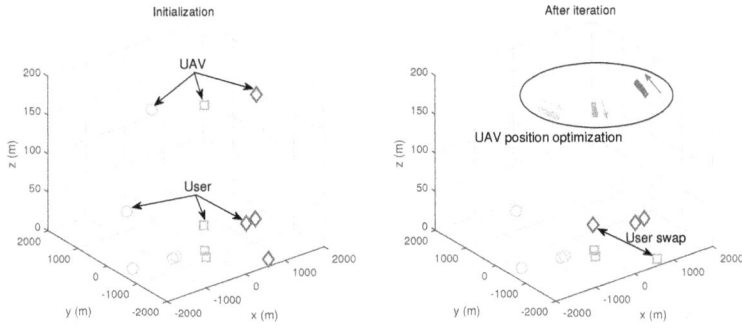

Figure 7.19 Demonstration of the presented solution for the UAV-BS positioning and user clustering in Algorithm 7.3.

$\alpha_{\text{LoS}} = 2.3$	Path loss exponent for LoS paths
$L = 4$	Number of NLoS components
$N_B^x \times N_B^y = 16 \times 16$	Antenna array size at UAV-BSs
$N_U^x \times N_U^y = 4 \times 4$	Antenna array size at users
$N_{\text{RF}} = 4$	Number of RF chain at UAV-BSs
$M = 3$	Number of the UAV-BSs
$K = 10$	Number of the users
$r_{m,n} = 1\text{bps/Hz}$	Required achievable rate of user $k_{m,n}$
$d^{(0)} = 200\text{m}$	Initial radius for positioning in (7.85c)
$\kappa_1 = 1.4$	Step size for positioning in (7.85c)
$\kappa_2 = 2$	Step size for beamforming in (7.93)
$\epsilon_1 = 0.05\text{bps/Hz}$	Threshold for convergence of Algorithm 7.3
$\epsilon_2 = 0.05\text{bps/Hz}$	Threshold for convergence of Algorithm 7.4

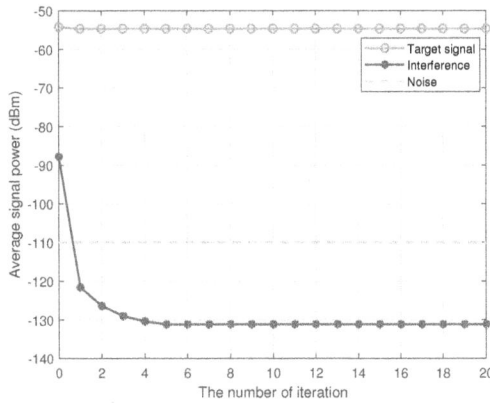

Figure 7.20 Demonstration of the presented solution for transmit and receive beamforming in Algorithm 7.4.

Four benchmark schemes are used for comparison in the considered multi-UAV aided mmWave communication networks, namely "ProPos-ProClu-FDMIMO", "ProPos-ProClu-SLNR", "ProPos-ProClu-KDBF", and "BenchPos-BenchClu-ProBF". For the first three schemes, the presented UAV-BS positioning and user clustering are adopted, while different beamforming strategies are employed. For the "ProPos-ProClu-FDMIMO" scheme, FDMIMO structures are utilized at the UAV-BSs and users. For the "ProPos-ProClu-SLNR" scheme, signal-to-leakage-plus-noise ratio (SLNR) based hybrid beamforming presented in[55] is employed at the UAV-BSs. For the "ProPos-ProClu-KDBF" scheme, Kronecker decomposition (KD) based hybrid beamforming presented in[62] is employed at the UAV-BSs. While for the last "BenchPos-BenchClu-ProBF" scheme, the UAV-BSs are deployed right over M users randomly selected from K users. The users are then successively connected to the nearest UAV-BS under the constraint of maximum number of serving users for each UAV-BS. Then, the presented beamforming algorithm is employed under the above heuristic UAV-BS positioning and user clustering.

7.5.3.2 *Demonstration of Presented Solution*

First, we provide a demonstration of the presented solution for the UAV-BS positioning and user clustering in Fig. 7.19. The users are marked by '□' and randomly distributed on the ground, while the UAV-BSs are marked by '○'. The users in the same cluster and the corresponding serving UAV-BS are marked with the same color. As shown in Fig. 7.19 (left), the initial position of a UAV-BS is given by the average coordinates of the users served by this UAV. By employing the presented joint UAV-BS positioning and user clustering algorithm, the upper bound on the ASR of the users given by (7.74a) increases from 152.06 bps/Hz to 168.71 bps/Hz, which demonstrates the effectiveness of Algorithm 7.1. In Fig. 7.19 (right), we find that the UAV-BS is inclined to be deployed over the area with a higher density of users. This positioning scheme may improve the channel qualities of more numbers of users and thus increases the ASR. As the second UAV-BS (marked with red color) and the third UAV-BS (marked with blue color) move toward the better positions, the corresponding user clustering is also updated such that the ASR is maximized.

In Fig. 7.20, we provide a demonstration of the presented solution for beamforming, where the powers of the target signal, interference, and noise are averaged over all users. The positions of the users, the UAV-BS positioning, and user clustering are the same with that in Fig. 7.19 (right). As can be observed, the power of the target signal is almost unchanged during the iteration, while the average power of interference quickly decreases to a value much smaller than the noise power. The results in Fig. 7.20 demonstrate the rational behind the presented BF algorithm, which can effectively mitigate the intra-cluster and inter-cluster interference, as well as achieve closely the optimal rates of the users under the ideal beam patterns.

7.5.3.3 Convergence Evaluation and Performance Comparison

Fig. 7.21 illustrates the convergence of the two presented algorithms. As can be observed, both presented algorithms converge after 8 iterations. The performance gap between the upper bound on the ASR obtained with Algorithm 7.3 (i.e., the summation of $\bar{R}_{m,n}$ in (7.73)) and the practical ASR obtained with Algorithm 7.4 (i.e., the summation of $R_{m,n}$ in (7.67)) is very small. These results demonstrate that the optimization of the UAV-BS positioning and user clustering under the assumption ideal beam patterns is reasonable for LoS channels, and the presented beamforming strategy can effectively approach the performance of the ideal beam patterns. Furthermore, the modulus normalization of the analog beamforming matrices in (7.122) has a negligible influence on the ASR. This is because Theorem 7.5.1 guarantees that the relaxation and normalization of the analog beamforming matrices impact only a small number of their elements.

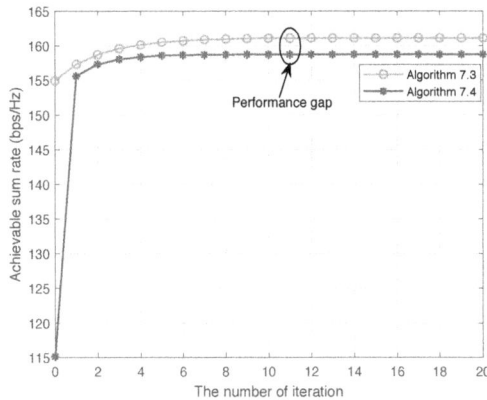

Figure 7.21 Convergence of the two presented algorithms.

Fig. 7.22 compares the ASR performance of different methods versus the transmit power at the UAV-BSs. As can be observed, the presented solution achieves an ASR performance very close to the system with fully-digital MIMO (FDMIMO), and outperforms the other three benchmark schemes. The results in Fig. 7.22 demonstrate that the presented beamforming method can achieve a near optimal rate performance as compared to the more costly FDMIMO. In particular, for larger transmit powers, the performance gain of the presented hybrid beamforming scheme compared to SLNR-based and KD-based hybrid beamforming becomes significant.

Figs. 7.23 (left) and 7.23 (right) compare the ASR performance of different methods versus the number of users and the antenna array size at the UAV-BSs, respectively. We observe again that the presented beamforming solution closely approaches the upper-bound provided by the FDMIMO system and outperforms the SLNR-based and KD-based hybrid beamforming schemes. In Fig. 7.23 (left), when the number of users is small, the performance gap between the presented solution and the heuristic UAV-BS positioning and user clustering in the "BenchPos-BenchClu-ProBF" scheme

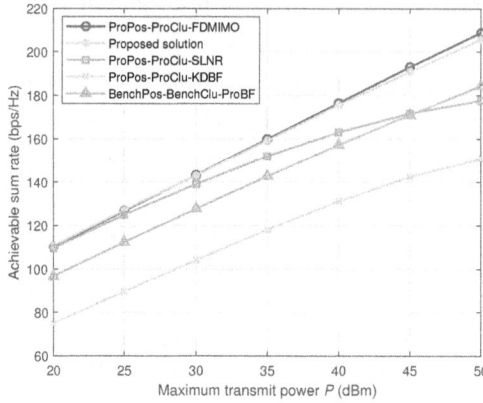

Figure 7.22 ASRs of different methods versus transmit powers at the UAV-BSs.

is small. This is because deploying a UAV-BS right over a user can effectively increase the achievable rate of this user. However, as the number of users increases, the achievable rates of the other users that are located far away from the serving UAV-BS cannot be guaranteed. This result demonstrates that the optimization of the UAV-BS positioning and user clustering is important when the number of users is large. Besides, the KD-based hybrid beamforming scheme is highly depended on the prime factorization of N_{B}. If the number of antennas is not in an exponential form of two and the number of users is large, the DoF for KD-based analog beamforming is not sufficient to suppress the interference from all path components, and thus the effective channel gains of the target users cannot be guaranteed. In contrast, the presented analog beamforming strategy can be used for any antenna size and efficiently mitigate the interference without an obvious loss of the received powers of the target signals.

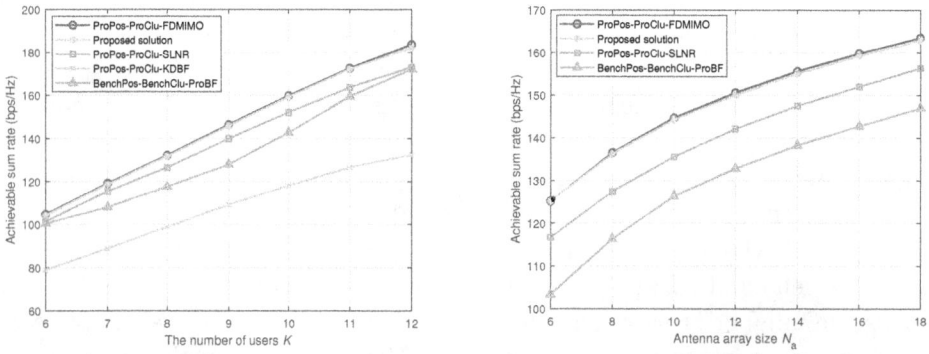

Figure 7.23 **Left**: ASRs of different methods versus numbers of users. **Right**: ASRs of different methods versus antenna array sizes at the UAV-BSs with $N_{\mathrm{B}}^x = N_{\mathrm{B}}^x = N_{\mathrm{a}}$.

7.5.3.4 Performance Evaluation under Practical Factors

In the UAV-BS positioning stage, we assume that an LoS path always exists for the channels between UAV-BSs and users. However, if an obstacle is located between a UAV and a user, the LoS path may be blocked. In such a case, the communication links can only be maintained via NLoS paths and a performance declines is caused. Thus, we evaluate the ASR performance of the presented solution under the practical channel model given by (7.61), where the probability that an LoS path does not exist is denoted as P_{NLoS}. In other words, $\chi_{m,k}$ has a probability P_{NLoS} equal to zero. Fig. 7.24 shows the ASR performance versus P_{NLoS}. As the probability of NLoS channels increases, the performance of ASR decreases, and about 12% performance loss is caused for $P_{\mathrm{NLoS}} = 30\%$. Note that the presented beamforming algorithm can be used for any conditions of the channels. Thus, to resolve the issue of LoS-path blockage, further research may focus on improving the UAV-BS positioning and user clustering. For example, the UAVs may first move over the target region and acquire the environment information via sensing technologies. Then, based on the prior knowledge of the environment, the candidate positions of UAVs can be restrained in the coordinates which are more likely to establish LoS links with the served users. Besides, the user clustering can be adjusted after UAV-BS positioning such that more users have an LoS connection with their serving UAV-BSs.

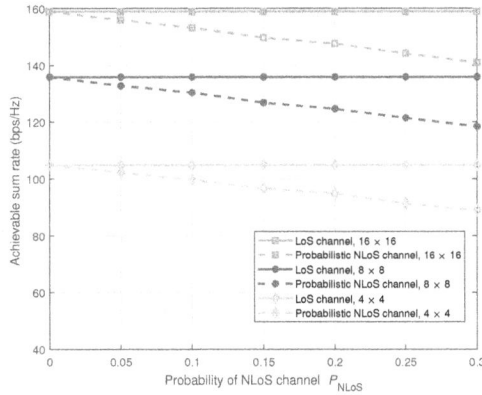

Figure 7.24 ASR of the presented solution versus probabilities of the NLoS channels due to potential obstacles between the UAV-BSs and users, with array sizes 16 × 16, 8 × 8, or 4 × 4 at the UAV-BSs.

As a UAV is hovering in the air, the airflow disturbances and the vibration of engine cause unavoidable body jittering, which may result in negative impacts on the beam steering. The CSI acquired at the current time slot may be inaccurate for the next time slot. Thus, we evaluate the impact of UAV jittering on the ASR performance for the considered multi-UAV aided millimeter-wave communication system in Fig. 7.25. The estimation error of the elevation and azimuth AoDs is assumed randomly distributed within $[-\vartheta, \vartheta]$ following uniform distribution, where ϑ is the maximum of the angle error[63, 64]. As can be observed, the ASR linearly decreases as the AoD error increases. In particular, for a larger number of antennas, the performance loss becomes more severe. The results in Fig. 7.25 show the necessity to investigate robust beamforming strategies for compensation of the impact of UAV jittering in future work. For example, an efficient beam tracking strategy may be designed based on the modeling of practical UAV jittering to deal with the AoD errors.

In this section, we presented the deployment of multiple UAV-BSs to provide communication service for ground users in the millimeter-wave frequency bands, where large antenna arrays are employed for compensation of the high path loss and for mitigation of intra-cell and inter-cell interference. We formulated an optimization problem which jointly optimizes the UAV-BS positioning, user clustering, Tx-BF, and Rx-BF for the maximization of the ASR, subject to a minimum rate constraint for each user. Simulation results demonstrated that the presented solution outperforms three other benchmark schemes, and the presented hybrid beamforming algorithm closely approaches the performance upper bound provided by FDMIMO. Furthermore, we also evaluated the impact of two practical factors for multi-UAV aided millimeter-wave networks and suggested that further investigation may focus on handling the issues of the LoS-path blockage and robust beamforming under imperfect CSI.

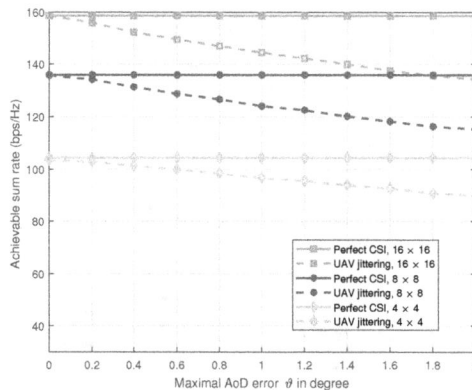

Figure 7.25 ASR of the presented solution versus the AoD errors caused by UAV jittering for different array sizes at the UAV-BSs.

7.6 SUMMARY

In this chapter, we discussed the array beamforming enabled UAV communications, where a large antenna array can be equipped in a small area of a UAV.

First, we provided an overview of the channel characteristics in UAV communications. Compared to terrestrial communications, the propagation characteristics in UAV communications are very unique because of obstacles, large Doppler shifts, aircraft shadowing, and UAV fluctuation. In particular, in real-world applications, the moving UAV may cause transceivers to become blocked by the fuselage itself.

Moreover, even a slight jitter can have a significant impact on the performance of the highly directional beam. These unique properties introduce great challenges for channel modeling for array enabled UAV communications. We provided a typical analytical channel model that can be used in UAV communications, i.e., Saleh-Valenzuela model. It has been extended and widely used for massive MIMO systems to show the delay spread, amplitudes, and phase angles of the MPCs.

Then we discussed antenna array enabled 3D beam coverage for UAV-BS. If the number of users is small or the users are concentrated, a single directional beam can cover the users by equipping the UAV-BS with directional array and adjusting the azimuth and elevation angles of the beam. If the number of the users is large or the users are widely dispersed, a single narrow beam may not be able to cover all the users at the same time, which limits the the coverage efficiency. Therefore, it is crucial to increase the number of beams. For large antenna array, hybrid beamforming technology can be used to generate multiple beams and eliminate the interference among the beams. Besides, the RF chain can also be connected to a lens antenna array to generate multiple beams. The RF chain can select different antennas to connect in order to generate beams steering to different directions after lens refraction, which greatly reduces beamforming complexity and hardware cost. For the area with dense users, the number of the users is much larger than that of the RF chains. In such a case, it may not be possible to shape so many beams to steer to individual users. Instead, a flexible beam coverage approach can be used to fully cover this area. The key challenge is that the target area may have different sizes and shapes. Moreover, due to the movement of the UAV, the target area changes frequently. Hence, a low-complexity 3D beamforming method is required to achieve real-time full coverage. Equipped with a UPA, a flexible 3D beam at UAV was provided. Besides, the influence of the obstacles on the ground and dynamic beam adjustment can be further discussed in the future.

Then, we introduced array beamforming enabled single UAV communications and multi-UAV communications, respectively, where large antenna arrays are employed for compensation of the high path loss and for mitigation of intra-cell and inter-cell interference. Both deployment and beamforming design have essential impact on the throughput of the system. Hence, jointly optimizing the deployment and beamforming is important for UAV communication systems. In single UAV-BS communication scenario, we formulated a problem to maximize the ASR of all the users, subject to a minimum rate constraint for each user, a position constraint of the UAV-BS, and a CM constraint for the beamforming vector. We solved the non-convex problem with two steps. First, by introducing the approximate beam pattern, we solved the deployment and beam gain allocation sub-problem. Then, we utilized ABC algorithm to solve the beamforming sub-problem. For the global optimization problem, we find the near-optimal position of the UAV-BS and the beamforming vector to steer toward each user, subject to an analog beamforming structure. In multiple UAV-BS communication scenario, we formulated a problem to jointly optimize the UAV positioning, user clustering, and hybrid beamforming for the maximization of user ASR, subject to a minimum rate constraint for each user. Since the problem is highly non-convex and involves high-dimensional variable matrices and combinatorial programming variables, we developed a sub-optimal solution via alternating optimization, successive convex optimization, and combinatorial optimization. First, we designed the UAV positioning and user clustering under the assumption of ideal beam patterns, which significantly decouples the UAV positioning and directional beamforming. Then, the transmit and receive beamforming variables are successively optimized to approach the ideal beam patterns.

In conclusion, the antenna array is quite suitable for UAV communications and can play an important role in increasing communication capacity, reducing interference and improving the QoS with beamforming technologies.

Bibliography

[1] Yongs Zeng, Qingqing Wu, and Rui Zhang. Accessing from the sky: A tutorial on UAV communications for 5G and beyond. *Proc. IEEE*, 107(12):2327–2375, Dec. 2019.

[2] Lav Gupta, Raj Jain, and Gabor Vaszkun. Survey of important issues in UAV communication networks. *IEEE Commun. Surveys Tuts.*, 18(2):1123–1152, 2016.

[3] Yong Zeng, Rui Zhang, and Teng Joon Lim. Wireless communications with unmanned aerial vehicles: opportunities and challenges. *IEEE Commun. Mag.*, 54(5):36–42, May 2016.

[4] Zhenyu Xiao, Pengfei Xia, and Xiang-Gen Xia. Enabling UAV cellular with millimeter-wave communication: potentials and approaches. *IEEE Commun. Mag.*, 54(5):66–73, 2016.

[5] Jianwei Zhao, Feifei Gao, Guoru Ding, Tao Zhang, Weimin Jia, and Arumugam Nallanathan. Integrating communications and control for UAV systems: Opportunities and challenges. *IEEE Access*, 6:67519–67527, 2018.

[6] Jiayi Zhang, Emil Björnson, Michail Matthaiou, Derrick Wing Kwan Ng, Hong Yang, and David J. Love. Prospective multiple antenna technologies for beyond 5G. *IEEE J. Select. Areas Commun.*, 38(8):1637–1660, Aug. 2020.

[7] Wenyan Ma, Chenhao Qi, and Geoffrey Ye Li. High-resolution channel estimation for frequency-selective mmWave massive MIMO systems. *IEEE Trans. Wireless Commun.*, 19(5):3517–3529, May 2020.

[8] Wenyan Ma, Chenhao Qi, Zaichen Zhang, and Julian Cheng. Sparse channel estimation and hybrid precoding using deep learning for millimeter wave massive MIMO. *IEEE Trans. Commun.*, 68(5):2838–2849, May 2020.

[9] Xuyao Sun, Chenhao Qi, and Geoffrey Ye Li. Beam training and allocation for multiuser millimeter wave massive MIMO systems. *IEEE Trans. Wireless Commun.*, 18(2):1041–1053, Feb. 2019.

[10] Zhenyu Xiao, Lipeng Zhu, Yanming Liu, Pengfei Yi, Rui Zhang, Xiang-Gen Xia, and Robert Schober. A survey on millimeter-wave beamforming enabled UAV

communications and networking. *IEEE Commun. Surveys Tuts.*, 24(1):557–610, 1st Quart. 2022.

[11] Lu Yang and Wei Zhang. Beam tracking and optimization for UAV communications. *IEEE Trans. Wireless Commun.*, 18(11):5367–5379, Nov. 2019.

[12] Chiya Zhang, Weizheng Zhang, Wei Wang, Lu Yang, and Wei Zhang. Research challenges and opportunities of UAV millimeter-wave communications. *IEEE Wireless Commun.*, 26(1):58–62, Feb. 2019.

[13] Long Zhang, Hui Zhao, Shuai Hou, Zhen Zhao, Haitao Xu, Xiaobo Wu, Qiwu Wu, and Ronghui Zhang. A survey on 5G millimeter wave communications for UAV-assisted wireless networks. *IEEE Access*, 7:117460–117504, 2019.

[14] Zhiyong Feng, Lei Ji, Qixun Zhang, and Wei Li. Spectrum management for mmWave enabled UAV swarm networks: Challenges and opportunities. *IEEE Commun. Mag.*, 57(1):146–153, Jan. 2019.

[15] Aziz Altaf Khuwaja, Yunfei Chen, Nan Zhao, Mohamed-Slim Alouini, and Paul Dobbins. A survey of channel modeling for UAV communications. *IEEE Commun. Surveys Tuts.*, 20(4):2804–2821, Fourthquarter 2018.

[16] Ming Xiao, Shahid Mumtaz, Yongming Huang, Linglong Dai, Yonghui Li, Michail Matthaiou, George K. Karagiannidis, Emil Bjornson, Kai Yang, Chih-Lin I, and Amitabha Ghosh. Millimeter wave communications for future mobile networks. *IEEE J. Select. Areas Commun.*, 35(9):1909–1935, Sep. 2017.

[17] Jaroslav Holis and Pavel Pechac. Elevation dependent shadowing model for mobile communications via high altitude platforms in built-up areas. *IEEE Trans. Antennas Propagat.*, 56(4):1078–1084, Apr. 2008.

[18] Akram Al-Hourani, Sithamparanathan Kandeepan, and Abbas Jamalipour. Modeling air-to-ground path loss for low altitude platforms in urban environments. In *Proc. IEEE Global Commun. Conf.*, pages 2898–2904, Dec. 2014.

[19] Akram Al-Hourani, Sithamparanathan Kandeepan, and Simon Lardner. Optimal LAP altitude for maximum coverage. *IEEE Wireless Commun. Lett.*, 3(6):569–572, Dec. 2014.

[20] Bertold Van Der Bergh, Alessandro Chiumento, and Sofie Pollin. LTE in the sky: trading off propagation benefits with interference costs for aerial nodes. *IEEE Commun. Mag.*, 54(5):44–50, May 2016.

[21] David W. Matolak and Ruoyu Sun. Air–ground channel characterization for unmanned aircraft systems—part iii: The suburban and near-urban environments. *IEEE Trans. Veh. Technol.*, 66(8):6607–6618, Aug. 2017.

[22] Xingqin Lin, Vijaya Yajnanarayana, Siva D. Muruganathan, Shiwei Gao, Henrik Asplund, Helka-Liina Maattanen, Mattias Bergstrom, Sebastian Euler, and Y.-P. Eric Wang. The sky is not the limit: LTE for unmanned aerial vehicles. *IEEE Commun. Mag.*, 56(4):204–210, Apr. 2018.

[23] Wahab Khawaja, Ismail Guvenc, David W. Matolak, Uwe-Carsten Fiebig, and Nicolas Schneckenburger. A survey of air-to-ground propagation channel modeling for unmanned aerial vehicles. *IEEE Commun. Surveys Tuts.*, 21(3):2361–2391, 3rd Quart. 2019.

[24] Ibrahim A. Hemadeh, Katla Satyanarayana, Mohammed El-Hajjar, and Lajos Hanzo. Millimeter-wave communications: Physical channel models, design considerations, antenna constructions, and link-budget. *IEEE Commun. Surveys Tuts.*, 20(2):870–913, 2nd Quart. 2018.

[25] Sundeep Rangan, Theodore S. Rappaport, and Elza Erkip. Millimeter-wave cellular wireless networks: Potentials and challenges. *Proc. IEEE*, 102(3):366–385, Mar. 2014.

[26] Ruoyu Sun, David W. Matolak, and William Rayess. Air-ground channel characterization for unmanned aircraft systems—part iv: Airframe shadowing. *IEEE Trans. Veh. Technol.*, 66(9):7643–7652, Sep. 2017.

[27] Wei Wang and Wei Zhang. Jittering effects analysis and beam training design for UAV millimeter wave communications. *IEEE Trans. Wireless Commun.*, pages 1–1, 2021.

[28] Morteza Banagar, Harpreet Dhillon, and Andreas Molisch. Impact of UAV jittering on the air-to-ground wireless channel. *arXiv Preprint arXiv:2004.02771v2.*, 2020.

[29] Adel A. M. Saleh and Reinaldo A. Valenzuela. A statistical model for indoor multipath propagation. *IEEE J. Select. Areas Commun.*, 5(2):128–137, Feb. 1987.

[30] Niklas Goddemeier and Christian Wietfeld. Investigation of air-to-air channel characteristics and a UAV specific extension to the rice model. In *Proc. IEEE Globecom Workshops*, pages 1–5, Dec. 2015.

[31] Qianqian Cheng, Lixin Li, Kaiyuan Xue, Huan Ren, Xu Li, Wei Chen, and Zhu Han. Beam-steering optimization in multi-UAVs mmWave networks: A mean field game approach. In *Proc. Int. Conf. Wireless Commun. Sign. Proces.*, pages 1–5, Oct. 2019.

[32] Nikita Tafintsev, Mikhail Gerasimenko, Dmitri Moltchanov, Mustafa Akdeniz, Shu-Ping Yeh, Nageen Himayat, Sergey Andreev, Yevgeni Koucheryavy, and Mikko Valkama. Improved network coverage with adaptive navigation of mmWave-based drone-cells. In *Proc. IEEE Global Commun. Conf. Workshops*, pages 1–7, Dec. 2018.

[33] Hossein Vaezy, Mehdi Salehi Heydar Abad, Ozgur Ercetin, Halim Yanikomeroglu, Mohammad Javad Omidi, and Mohammad Mahdi Naghsh. Beamforming for maximal coverage in mmWave drones: A reinforcement learning approach. *IEEE Commun. Lett.*, 24(5):1033–1037, May 2020.

[34] Weizheng Zhang, Wei Zhang, and Jun Wu. UAV beam alignment for highly mobile millimeter wave communications. *IEEE Trans. Veh. Technol.*, 69(8):8577–8585, Aug. 2020.

[35] Ke Li, Xu Zhu, Yufei Jiang, and Fu-Chun Zheng. Closed-form beamforming aided joint optimization for spectrum- and energy-efficient UAV-BS networks. In *Proc. IEEE Global Commun. Conf.*, pages 1–6, Dec. 2019.

[36] Yong Zeng and Rui Zhang. Millimeter wave MIMO with lens antenna array: A new path division multiplexing paradigm. *IEEE Trans. Commun.*, 64(4):1557–1571, Apr. 2016.

[37] Tian Xie, Linglong Dai, Derrick Wing Kwan Ng, and Chan-Byoung Chae. On the power leakage problem in millimeter-wave massive MIMO with lens antenna arrays. *IEEE Trans. Signal Process.*, 67(18):4730–4744, Sep. 2019.

[38] Huan Ren, Lixin Li, Wenjun Xu, Wei Chen, and Zhu Han. Machine learning-based hybrid precoding with robust error for UAV mmWave massive MIMO. In *Proc. IEEE Int. Conf. Commun.*, pages 1–6, May 2019.

[39] Mohammad Mozaffari, Walid Saad, Mehdi Bennis, and Mérouane Debbah. Efficient deployment of multiple unmanned aerial vehicles for optimal wireless coverage. *IEEE Commun. Lett.*, 20(8):1647–1650, 2016.

[40] Jianwei Zhao, Feifei Gao, Qihui Wu, Shi Jin, Yi Wu, and Weimin Jia. Beam tracking for UAV mounted SatCom on-the-move with massive antenna array. *IEEE J. Select. Areas Commun.*, 36(2):363–375, Feb. 2018.

[41] Zhenyu Xiao, Lipeng Zhu, Jinho Choi, Pengfei Xia, and Xiang-Gen Xia. Joint power allocation and beamforming for non-orthogonal multiple access (NOMA) in 5G millimeter wave communications. *IEEE Trans. Wireless Commun.*, 17(5):2961–2974, May 2018.

[42] Zhenyu Xiao, Tong He, Pengfei Xia, and Xiang-Gen Xia. Hierarchical codebook design for beamforming training in millimeter-wave communication. *IEEE Trans. Wireless Commun.*, 15(5):3380–3392, May 2016.

[43] Theodore S. Rappaport, George R. MacCartney, Mathew K. Samimi, and Shu Sun. Wideband millimeter-wave propagation measurements and channel models for future wireless communication system design. *IEEE Trans. Commun.*, 63(9):3029–3056, Sep. 2015.

[44] Ahmed Alkhateeb, Omar El Ayach, Geert Leus, and Robert W. Heath. Channel estimation and hybrid precoding for millimeter wave cellular systems. *IEEE J. Sel. Top. Sign. Proces.*, 8(5):831–846, Oct. 2014.

[45] Mohammad Mozaffari, Walid Saad, Mehdi Bennis, and Mérouane Debbah. Mobile unmanned aerial vehicles (UAVs) for energy-efficient internet of things communications. *IEEE Trans. Wireless Commun.*, 16(11):7574–7589, Nov. 2017.

[46] Mathew K. Samimi, George R. MacCartney, Shu Sun, and Theodore S. Rappaport. 28 GHz millimeter-wave ultrawideband small-scale fading models in wireless channels. In *2016 IEEE 83rd Vehicular Technology Conference (VTC Spring)*, pages 1–6, May 2016.

[47] Zhenyu Xiao, Hang Dong, Lin Bai, Pengfei Xia, and Xiang-Gen Xia. Enhanced channel estimation and codebook design for millimeter-wave communication. *IEEE Trans. Veh. Technol.*, 67(10):9393–9405, Oct. 2018.

[48] Zhen Gao, Linglong Dai, Shuangfeng Han, Chih-Lin I, Zhaocheng Wang, and Lajos Hanzo. Compressive sensing techniques for next-generation wireless communications. *IEEE Wireless Commun.*, 25(3):144–153, Jun. 2018.

[49] Zhen Gao, Chao Zhang, and Zhaocheng Wang. Robust preamble design for synchronization, signaling transmission, and channel estimation. *IEEE Trans. Broadcast.*, 61(1):98–104, Mar. 2015.

[50] Maurice Clerc and James Kennedy. The particle swarm - explosion, stability, and convergence in a multidimensional complex space. *IEEE Trans. Evol. Comput.*, 6(1):58–73, Feb. 2002.

[51] Marco Dorigo, Vittorio Maniezzo, and Alberto Colorni. Ant system: optimization by a colony of cooperating agents. *IEEE Trans. Syst., Man, Cybern. B*, 26(1):29–41, Feb. 1996.

[52] Dervis Karaboga and B. Basturk. On the performance of artificial bee colony (ABC) algorithm. *Applied Soft Computing*, 8(1):687–697, Jan. 2008.

[53] Dervis Karaboga and Bahriye Akay. A comparative study of artificial bee colony algorithm. *Applied Mathematics and Computation*, 214(1):108 – 132, Aug. 2009.

[54] Linglong Dai, Bichai Wang, Mugen Peng, and Shanzhi Chen. Hybrid precoding-based millimeter-wave massive MIMO-NOMA with simultaneous wireless information and power transfer. *IEEE J. Select. Areas Commun.*, 37(1):131–141, Jan. 2019.

[55] Shu Sun, Theodore S. Rappaport, and Mansoor Shaft. Hybrid beamforming for 5G millimeter-wave multi-cell networks. In *Proc. IEEE Conf. Comput. Commun. Workshops*, pages 589–596, Apr. 2018.

[56] Constantine A. Balanis. *Antenna Theory: Analysis and Design*. Hoboken, NJ, USA: Wiley, 2016.

[57] Stephen Boyd and Lieven Vandenberghe. *Convex Optimization*. Cambridge, U.K.: Cambridge Univ. Press, 2004.

[58] Wei Xu, Yuke Cui, Hua Zhang, Geoffrey Ye Li, and Xiaohu You. Robust beamforming with partial channel state information for energy efficient networks. *IEEE J. Sel. Areas Commun.*, 33(12):2920–2935, Dec. 2015.

[59] Theodore S. Rappaport, Eshar Ben-Dor, James N. Murdock, and Yijun Qiao. 38 GHz and 60 GHz angle-dependent propagation for cellular peer-to-peer wireless communications. In *Proc. IEEE Int. Conf. Commun.*, pages 4568–4573, Jun. 2012.

[60] Gilwon Lee, Youngchul Sung, and Junyeong Seo. Randomly-directional beamforming in millimeter-wave multiuser MISO downlink. *IEEE Trans. Wireless Commun.*, 15(2):1086–1100, Feb. 2016.

[61] Akram Al-Hourani, Sithamparanathan Kandeepan, and Simon Lardner. Optimal LAP altitude for maximum coverage. *IEEE Wireless Commun. Lett.*, 3(6):569–572, Dec. 2014.

[62] Guangxu Zhu, Kaibin Huang, Vincent Kin Nang Lau, Bin Xia, Xiaofan Li, and Sha Zhang. Hybrid beamforming via the kronecker decomposition for the millimeter-wave massive MIMO systems. *IEEE J. Sel. Areas Commun.*, 35(9):2097–2114, Sep. 2017.

[63] Dongfang Xu, Yan Sun, Derrick Wing Kwan Ng, and Robert Schober. Multiuser MISO UAV communications in uncertain environments with no-fly zones: Robust trajectory and resource allocation design. *IEEE Trans. Commun.*, 68(5):3153–3172, May 2020.

[64] Morteza Banagar, Harpreet S. Dhillon, and Andreas F. Molisch. Impact of UAV jittering on the air-to-ground wireless channel. *arXiv Preprint arXiv:2004.02771*, 2020.

CHAPTER 8

Array Beamforming Enabled UAV Networking

8.1 INTRODUCTION

With unique characteristics, e.g., the mobility of UAVs, the array beamforming enabled communications on the UAV platform can be a lot different from conventional terrestrial communications. In Chapter 7, we have introduced the array beamforming enabled single UAV communications and multi-UAV communications have been presented, including channel model, 3D beam coverage, single UAV deployment and multiple UAVs deployment. Nevertheless, the interaction among the UAVs is not contained in Chapter 7, while UAV-to-UAV communication is a key technology to support multiple UAVs to carry out complex mission collaboratively in practical communication systems. In fact, the multi-UAV system is organized in a mesh manner, which is referred to as a flying ad hoc network (FANET). Due to the characteristics of high autonomy, flexibility, and self-healing, FANETs have broad applications in military and civil domains[1,2].

Compared to terrestrial ad hoc networks, the network organization and link maintenance for FANETs are more challenging because of fast changing links, especially for harsh environments with strong electromagnetic interference. Unique requirements, such as high throughput, low probability of intercept, and high anti-interference ability pose further challenges for FANETs. Millimeter-wave communication technologies with abundant spectrum resources are promising to support high-rate transmission in FANETs. Moreover, the directivity of millimeter-wave channels and the narrow beamwidth provide significant potential for enhancing the security and anti-interference capabilities of FANETs. However, most of the existing works on FANETs place emphasis on employing sub-6 GHz frequency bands for networking[1,3,4,5,6,7,8,9].

In this chapter, we comprehensively discuss and analyze crucial issues and the corresponding potential solutions for antenna-array enabled FANETs. Firstly, the network architecture for antenna-array UAV FANETs is presented. In particular, software defined networking (SDN) will play an important role in network management due to the resulting flexibility and programmability. Secondly, we analyze how to establish and maintain links in antenna-array UAV FANETs. Specifically, the challenges and solutions for directional neighbor discovery, the comparative analysis of existing UAV routing strategies, and the potential of advanced technologies for resource allocation are comprehensively surveyed. Then, we reveal the significant benefits of joint millimeter-wave and sub-6 GHz band networking. Finally, security threats and potential solutions to these threats are discussed.

8.2 NETWORK ARCHITECTURE

The network architecture for antenna-array enabled FANETs is a foundation for network management and application. In an antenna-array enabled FANET, the network topology and routing stability may change frequently due to the 3D motions of UAVs and the directional transmission of signals, and thus the efficiency of the network management may be affected. Architectures of antenna-array enabled FANETs are still under research. Some common architectures and corresponding solutions that may be adopted for FANETs are introduced as follows.

8.2.1 Network Topology

Typical network topologies include star and mesh networks[3], as shown in Fig. 8.1. The star topology for antenna-array enabled FANETs is a relatively simple structure, where a UAV (or a ground BS) serves as the control and data-forwarding center, and other UAVs only connect with the control center. For a small-scale antenna-array UAV communication network, this structure is easy for control and management because of the short distance and less airframe shadowing. However, as the number and range of UAVs get larger, UAVs may suffer from severe path loss and much more airframe shadowing, and the center UAV may face nasty link congestion and interference. Once the center UAV breaks down, the entire network loses control and becomes paralyzed. Comparing to the star topology, the mesh topology has higher autonomy, flexibility, and invulnerability. A UAV node can associate with any other nodes via either single-hop or multi-hop routing to tackle the problems of high path loss and possible shadowing. Therefore, the mesh topology owns the features of highly-resilient reorganization and malfunction tolerance. However, the data packet transmission from a UAV node to the destination node may require high-complexity

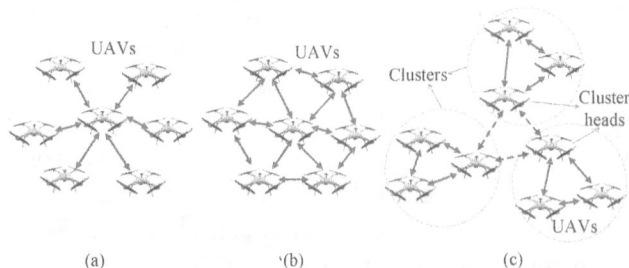

Figure 8.1 Three typical network topologies: (a) Star topology. (b) Mesh topology. (c) Hybrid hierarchical topology.

and high-overhead communication protocols, especially in large-scale mesh networks. In contrast, by employing a cluster-based scheme, the hybrid hierarchical topology is a good approach to reduce the complexity. Moreover, the inter-cluster and intra-cluster networks can select any type of topology according to application scenarios and management strategies. However, it also brings new problems. For example, the cluster size, cluster number, and cluster head should be carefully designed[10]. Meanwhile, the directional transmission raises the difficulties for topology discovery and management, which will be discussed later.

8.2.2 SDN-Based Network

Due to the heterogeneity of different aerial nodes, the reorganization of an FANET is limited by the hardware and protocol constraints. This problem can be resolved with the SDN architecture by programmatically controlling the network[11, 12]. The introduction of SDN to FANETs helps different aerial nodes efficiently acquire the network state, and this scheme caters to the requirements of the routing and resource scheduling design in dynamic environments. Specifically, UAVs can equip programmable SDN switches (e.g., openflow[13]), which contain flow tables and protocols for communicating with controllers. Some of the UAVs are installed with control facilities, which indicates that the control plane can be centralized, distributed, or hybrid. A typical centralized SDN-based FANET is shown in Fig. 8.2. The control UAVs have a global view of the network, and thus global resources and traffic requests can be efficiently scheduled. Moreover, by decoupling control and data planes, the SDN architecture can increase the visibility and availability of the UAV network topology. It also enhances the abilities of the routing selection and network configuration. However, there are still some challenges to be addressed when using antenna array in millimeter-wave frequency bands. First, in a large-scale antenna-array enabled FANET, the number of the control links is large, which makes the antenna configuration and beam management of the controller more difficult. The UAVs are usually equipped with limited number of RF chains, which means the number of the accessed nodes is limited for an SDN controller in a specific time slot. If a distributed architecture is used, the association assignment is an inevitable issue, but multi-beam may clash and lead to strong interference. Besides, due to the separation of the control and data links, the spectrum efficiency may be affected. For different rate requirements of control and data information, careful channel allocation and bandwidth selection are required.

8.2.3 Summary and Discussion

In the above, we have focused on the issues and challenges arising for antenna-array enabled FANETs for various network architectures. The network architecture partly determines the characteristics and possible applications of the network. Although ad hoc networks are not yet fully embedded in traditional cellular networks, there is no doubt that more and more new applications and intelligent technologies will facilitate the development of ad hoc networks driven by 6G. To the authors' best knowledge, research on the design of antenna-array enabled FANETs has not been conducted yet. As discussed before, compared to the centralized and distributed topologies, the

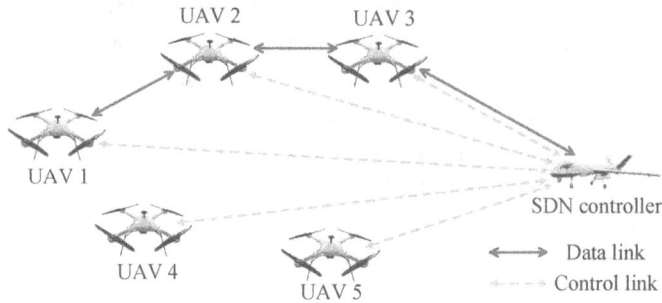

Figure 8.2   Architecture of a typical centralized SDN-based FANET.

hybrid topology is more suitable for large-scale antenna-array enabled FANETs, as it achieves a compromise between manageability and autonomy[10]. In addition, deploying SDN controllers on UAVs has great potential. Specifically, the programmability and flexibility of SDN can support the UAVs to improve the communication protocols to meet different communication requirements[11, 12, 13]. Therefore, future research on SDN-assisted antenna-array enabled FANETs will be important.

8.3   LINK ESTABLISHMENT AND MAINTENANCE

Although the introduction of millimeter-wave communications to FANETs may tremendously enhance the data rate and the anti-jamming capability of communication links, the directional characters of beams and the 3D motions of UAVs greatly challenge the link establishment and maintenance. Significant performance degradation and link outage can be observed from a large-scale on-the-moving UAV network in a millimeter-wave frequency band[14]. Due to the directional antennas and dynamic network topology, both beam steering and transmission path selection should be taken into consideration for designing the communication protocols. So far, the research works on antenna-array enabled FANETs are limited. We next analyze the specific issues in the link establishment and maintenance of antenna-array enabled FANETs and discuss possible solutions in the following, in order to inspire future research.

8.3.1   Neighbor Discovery

Before establishing a communication link between two nodes in a network, it is necessary to perceive and maintain the connection with each other, which is known as neighbor discovery (also called routing discovery). An effective neighbor discovery accelerates the implementation efficiency of upper-layer protocols and serves as a key foundation of the topology and networking. It usually requires the communicating parties to complete the Hello package transmission as agreed. Due to the 3D-space and mobility characteristics, UAVs may need frequent neighbor detections to maintain network connectivity. The easiest way is to always do neighbor discovery throughout the mission. However, the long-term neighbor discovery consumes much energy and resource, and unnecessary substantial overhead is generated. Therefore, the frequency of the neighbor-discovery operations should be carefully designed to balance the efficiency and overhead. Using directional transmit antennas and omnidirectional receive antennas, the authors in[8] developed a two-way handshaking discovery scheme in 3D UAV networks, considering the deployment and mobility of UAVs. The Markov process was adopted to analyze the efficiency of the proposed scheme, and extensive simulation results showed that the overhead of neighbor discovery can be balanced. In fact, when the future motion information of a UAV formation is already known or can be correctly predicted, the frequency of neighbor discovery can be reduced. In addition, most of the existing schemes for neighbor discovery adopt synchronous clock, such as TDMA. For FANET, the distributed nodes may not have perfect synchronization, and thus, it is necessary to support the asynchronous scenarios[7].

The neighbor discovery in antenna-array enabled FANETs is shown in Fig. 8.3. Compared to the network employing omnidirectional antennas, the utilization of directional antennas can avoid signal conflicts and interference problems, and increases the distance of neighbor discovery. However, the characteristic of the directional transmission induces a serious misalignment problem (also called spatial rendezvous problem[15]), which costs much more scanning time in a 3D space as compared to 2D. Specifically, all the control frames, e.g., request-to-send (RTS) or clear-to-send (CTS), have to transmit in a sector sweep manner. As a result, the real-time beam alignment causes serious delays and overheads. It is impractical to implement exhaustive beam scanning in a highly dynamic 3D UAV network. In addition to beam alignment in the spatial domain, perfect alignment and synchronization strategies in the time and frequency domains are also required, but further increase the overhead and delay. To reduce the adverse influence of the alignment problem, a possible solution is that the Rx nodes work in a quasi-omnidirectional manner in the transmission, similar to the directional neighbor discovery in the traditional sub-6 GHz frequency bands[16]. However, quasi-omnidirectional Rxs may receive multiple beacons from different directions, resulting in collisions. For directional transmission in frequency bands, it is beneficial for nodes to have partial prior knowledge (e.g., the directions of the potential neighbor nodes) in order to rapidly discover neighbors and achieve fast convergence. However, acquiring real-time 3D positions and velocities of the moving UAVs is also difficult. Hence, the 3D motion prediction of UAVs is vital to simplify the neighbor discovery and maintain the network connection[17].

In addition, the problems of the deafness and hidden terminals challenge the design of neighbor discovery and upper-layer protocols for directional communications[18,19]. When the nodes are working in a directional transmission mode, they can only receive the signals from the mainlobe direction. Due to the weak signal gain, the destination node may not receive RTS from the sidelobe direction and reply CTS in time. Thus, the sender nodes that locate in the sidelobe direction of the Rx become deaf nodes. On the other hand, for two sender nodes located in the same sector/beam of a destination node, one sender may not accomplish the directional RTS/CTS handshake with the destination node because the other sender is communicating with the destination node. This circumstance results in a deafness problem for the first sender node. The deafness problem leads to the

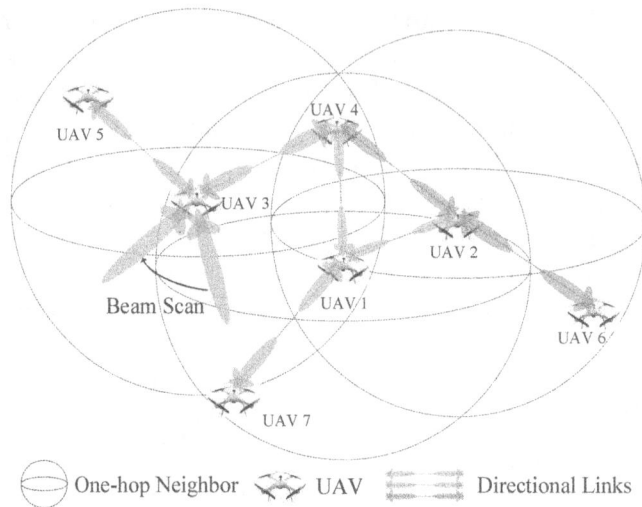

Figure 8.3 Neighbor discovery in antenna-array enabled FANET.

short-term inequity and seriously influences the utilization of the space resource[20]. The hidden terminal problem in directional ad hoc networks is caused by two conflicting nodes that cannot listen to each other and send data to the same terminal at the same time[21,22]. Due to the directional transmission, some nodes may be unaware of the existing communication links in the network, and the conflict is induced when a node tries to send RTS to another node who is communicating. The hidden terminal problem may also be caused by the asymmetry in antenna gain because the communication ranges between the directional and omnidirectional modes are different[21,22].

For a network using directional antennas in frequency bands, the existing neighbor discovery schemes can be categorized into two main classes, i.e., deterministic and probabilistic schemes. The probabilistic approach is memoryless, where the nodes randomly choose directions to steer their beams for neighbor discovery[23,24]. This approach usually performs better in terms of average discovery delay[23]. However, the main problem is that it does not ensure the successful discovery. In deterministic approach, the nodes steer their beams in accordance with predefined sequences[15,25,26]. Although the average delay is usually larger than that in probabilistic approach, the neighbor discovery is guaranteed within a bounded time.

To tackle the spatial rendezvous and deafness problems, the authors in[15] proposed a deterministic approach, namely hunting-based directional neighbor discovery algorithm. The nodes rotate their beams to scan the potential neighbors in a 2D plane and the nodes with a faster angular velocity will catch up with the slower nodes. The condition of successful beacon-ACK handshake and an upper-bound on the worst discovery time were derived. To achieve a desirable tradeoff between the average and worst-case delays, the authors in[25] and[26] developed pseudo-deterministic schemes such that the process of generating scanning sequences includes both a deterministic and a random component in 60 GHz networks. Under the case with heterogeneous

TABLE 8.1   Comparison of different schemes for neighbor discovery with directional antennas.

Reference	Scheme	Main idea	Pros	Cons
[15]	Deterministic	Steer beams in accordance with predefined sequences	Ensures successful discovery	High average delay
[23, 24]	Probabilistic	Memoryless beam scanning	Low average delay compared to deterministic schemes	Possible discovery failure
[25, 26]	Pseudo-deterministic	Generate scanning sequences including both a deterministic and a random component	Tradeoff between low average delay and high successful discovery rate	May not fit in a 3D environment
[28]	ML-based	Interact with the environment and learn from the experience direction based state space and successful discovery based reward	High successful discovery rate and low total delay for many neighbors	Possibly low scalability and high learning cost

beamwidths of the nodes and without any prior coordination or synchronisation, the authors in[25] proposed an oblivious neighbor discovery algorithm. This strategy adopts a short extended address for the nodes, which means that the latency of scanning operations is small. Subsequently, the authors in[26] used the above method for constructing extended identity sequences, and then utilized the Polya's enumeration theorem and Fredricksen, Kessler and Maiorana algorithm to find the shorter and efficient scanning sequences for the nodes. In[27], the authors proposed a stochastic multi-armed bandit (MAB) online learning solution to resolve the neighbor discovery and selection in antenna-array device-to-device (D2D) networks. To reduce the discovery time and maximize the average throughput, a group of energy-constraint MAB based algorithms were developed. Since the networks are assumed quasi-static in 2D space, these solutions[15, 23, 24, 25, 26, 27] should be improved for being applied in an antenna-array enabled FANET. In Table 8.1, the different neighbor discovery schemes with directional antennas are compared. More research efforts and advanced technologies for neighbor discovery are required to meet the requirements of high throughput and low latency under complex 3D environments.

Machine learning emerges as a powerful tool and has been used for various purposes in UAV networks. Specifically, Q-Learning theory was applied in[28] for directional neighbor discovery in ad hoc networks. The Q-Learning based algorithm takes the antenna steering direction as the *state*, the transmitting/receiving strategy as the *action*, and the successful neighbor discovery as the *reward*. By interacting with the environment and learning from the experience, the proposed algorithm shows better performance than conventional sequential scanning. However, the learning-based neighbor discovery needs a certain of successful samples, which tests the timeliness. Besides, it may be hard to update the changing topology for UAV nodes in FANETs. An alternative way is that each UAV only maintains the state of its neighbors and employs distributed intelligent decisions according to the available information. Hence, machine learning-based neighbor discovery methods in antenna-array enabled FANETs are worthy of more studies.

8.3.2 Routing

In antenna-array enabled FANET, the routing design owns unique requirements. When a UAV needs to transmit its data to other UAVs or ground BSs, the data transfer paths should be selected under the consideration of the QoS requirements, associated data traffic, and network topology. Unlike traditional ad hoc networks, the 3D-mobility character of UAVs has to be considered in the routing design for FANETs. Moreover, the size, weight and power (SWAP) limitation, the unstable link management, and the frequent removal and addition of UAV nodes are relevant for routing design. Especially for antenna-array enabled FANET, it is essential to consider the information from different layers, such as the channel conditions and interference from physical layer, fault tolerance and hop count from the network layer, throughput and delay from the data link layer, and QoS requirements and reliability from the application layer[29]. Since the high-speed UAV may cause high-dynamic topology, attention should also be paid to the link stability in routing. Under this circumstance, the authors in[30] introduced Gaussian Markov moving model to describe the movement of UAVs, which achieved better packet delivery rate and lower end-to-end delay.

The potential routing methods for FANETs can be classified into four types based on the strategies used, namely, topology-based, geographic, hybrid (topology-based and geographic), and bio-inspired routing, as shown in Table 8.2. Specifically, in topology-based routing, the routing information from the source to the destination must be obtained from the network topology information before data transmission starts. Based on the assumption that each UAV knows its own location from an on-board positioning system, geographic routing utilizes the local geographic locations of the UAVs to make the data packet forwarding decisions. Hybrid routing generally combines reactive routing and geographic routing. Bio-inspired routing is inspired by biological systems. Based on[9], we illustrate the different routing categories in more details, along with their main ideas, performance comparisons, and application scenarios in Table 8.2. For more information on FANET routing protocols, we refer to[9]. However, the applicability of these routing methods needs further study. In antenna-array enabled FANETs, beam scanning for directional neighbor discovery may increase the signaling overhead, delay, and energy consumption of proactive routing and flooding-based reactive routing. Geographic routing strategies require additional hardware at the UAVs[9]. In addition, how to match the intermittent contact time of high-speed UAVs with the beam scanning time is also an important issue. To address these challenges, routing combined with reliable mobility prediction strategies may be a potential solution, but requires further investigation. In addition, for control signaling, a low-frequency omnidirectional strategy for routing is preferable. We note that, compared to traditional routing methods, bio-inspired routing algorithms can utilize their own self-organizing means to manage the dynamic features of FANETs.

TABLE 8.2 Classification, main idea, and comparison of routing protocols in FANETs[9].

Classification	Subclassification	Main idea	Signaling overhead	Communication latency	Bandwidth and energy consumption	Memory requirement	Application scenarios
Topology-based	Proactive	Each UAV locally stores and periodically refreshes a routing table	Very high (periodical update)	Very low	Very high (frequent update)	Very high (storage of network information)	Small-scale and real-time applications
	Reactive (on-demand)	UAVs generally use the flooding technique to find routes only when there is a data transmission requirement	High (flooding)	High (route discovery)	High (flooding)	High (on-demand storage)	Data collection or remote sensing for small-scale to medium-scale UAV networks
	Hybrid	Depending on the type, characteristics and requirements of UAVs, different proactive or reactive protocols are adopted by different UAVs	Medium	Medium to high	Medium	High	Reconnaissance search and rescue for small-scale to large-scale UAV networks
	Cluster-based	Aiming at the network scalability issue, this routing strategy integrates the cluster head selection, clustering and real-time management schemes into traditional protocols	High (cluster maintenance)	Medium	Medium (different inter and intra protocols)	Medium	Military confrontation and network coverage for small-scale to large-scale UAV networks
Geographic	Non DTN	Using the local mobility prediction and the neighbor node closest to the destination as the routing metric, the UAV makes the decision on selecting the next hop	Low	Low	Low	Low (only neighbor information)	Cooperative monitoring, reconnaissance, and battlefield applications for small-scale to large-scale UAV networks
	DTN	Aiming at the problem of intermittent network connections, the UAV selects the appropriate next hop based on node movement and the store-carry-forward mechanism	High	High	High	Low (only neighbor information)	Delay tolerant applications, such as video making and data collection, for small-scale to large-scale UAV networks
Hybrid (topology-based and geographic)		Based on the local node information and movement prediction obtained by the reactive routing of local flooding, the failure links use geographic routing to select alternative routing options	Medium	Medium	Medium	High	Network coverage and multi-task cooperation for small-scale to large-scale UAV networks
Bio-inspired		Through local communication with less complex interactions and the cooperative response ability to internal and external disturbances, evolutionary or swarm-based bionic algorithms are used to make routing decisions	High	Medium	High	Medium	Intelligent searching and battlefield applications for for small-scale to large-scale UAV networks

DTN = delay tolerant networking.

So far, there have been only a few works on bio-inspired routing protocols. As an example, artificial bee colony and ant colony algorithms have been used for routing in FANETs[30, 31, 32], where the routing discovery process in FANETs is modeled as the honey collection process in a bee colony or the food finding process in an ant colony. The obtained performance are very promising. However, these methods have a relatively high computational complexity. Thus, bio-inspired strategies need to be explored more in detail to verify their true potential.

In antenna-array enabled FANET, the network topology and link quality are known to UAVs after accomplishing neighbor discover. That is to say, some of traditional routing protocols can be applied in antenna-array enabled FANET. The remaining issue is to select the data transfer paths, which can be modeled as a multi-commodity flow problem[33]. Since the FANET is a non-delay tolerant network, the commodity flows are time-dependent, which is an NP-hard problem and is different from that in static networks. A possible method is to use graph theory based strategies[34]. However, the modeling process of the routing path selection may be difficult because routing and resource allocation are coupled in general[35,36]. Hence, how to decouple the corresponding problem and find a global solution is valuable to investigate.

8.3.3 Resource Allocation

To enhance the performance of a network, the wireless resources need to be carefully allocated for mitigating the interference and improving the throughput. Compared to the sub-6 GHz frequency bands, the design of the MAC protocols for antenna-array enabled FANETs is more challenging due to the directional transmission mode, dynamical link fluctuation, beam management, time-consuming beam alignment, etc. To maintain the high-efficiency link connection, it is vital to realize reasonable resource allocation and sharing of the space, time, frequency, and other limited resources for different nodes. IEEE 802.11aj[37] and IEEE 802.11ay[38] have provided detailed MAC designs of wireless networks for supporting the frequency bands above 45 GHz, but they are more suitable in low-mobility indoor environments. The existing research works in antenna-array enabled UAV networks mainly focus on the physical layer design, while the MAC layer research is still in its infancy.

In the time domain, the millimeter-wave transmission challenges the frame design. Most present antenna-array enabled UAV communication networks adopt the half-duplex (HD) mode. Thus, in ad hoc networks, it is important to reasonably design the frame to guarantee efficient transmission and avoid collisions as multi-path and multi-hop routing generates[39]. Applying the full-duplex (FD) mode doubles the spectrum efficiency and decreases the network delay, but the scheduling algorithm should be carefully designed to handle the interference[40]. In the space domain, although beamforming technologies in frequency bands bring high spectrum efficiency and anti-jamming ability, the beam management is challenging due to the frequent change of the topology and connection. Specifically, as the relative direction and communication range change, the transmit beams must be realigned in a real time and the number of connected neighbor UAVs may change. Thus, fast beam tracking and resource reconfiguration methods should be used. Besides, the multiplexing of the beams in ad hoc networks will cause serious collision and interference, which should be properly designed. In the frequency domain, frequency-division assisted SDMA can avoid interference, but it reduces the spectrum efficiency and transmission bandwidth. Moreover, to maximize the network throughput, the global spectrum management and real-time allocation should be optimized according to the 3D network topology and time-varying interference. In addition, the transmit power control is also very crucial for antenna-array enabled FANETs in terms of the energy-efficiency. As the scale of a antenna-array enabled FANET increases, the computational complexity for resource optimization exponentially increases, which challenges the timeliness of the network management.

In order to minimize the number of time slots for multi-path multi-hop transmissions, the authors in[39] utilized two heuristic algorithms for traffic distribution and transmission scheduling to determine the frame structure. The scheme achieves a superior performance in terms of the delay and throughput comparing to the other directional MAC protocols, subject to the minimum traffic demand of all flows. Similarly, to improve the efficiency of concurrent transmission for millimeter-wave networks, the authors in[40] redesigned the time slot to enlarge the scheduling space and proposed an efficient time-slot adjustment scheduling algorithm in the multi-hop packet forwarding process. However, the above works in millimeter-wave Wireless Personal Area Networks (WPANs) may not be perfectly suited to antenna-array enabled FANET. In[41], the authors proposed a fast beam tracking algorithm in antenna-array UAV mesh networks, where a self-healing request/response frame was designed to ensure the network robustness, and an efficient algorithm for the re-selection of the UAV group leader was developed to ensure high link quality between the group leader and ground BS. The proposed self-healing mechanism improves the robustness of the antenna-array UAV mesh networks and reduces the overhead in establishing the directional communication links comparing to existing MAC protocols. In[5], using directional antennas in FANET, the authors proposed a position-prediction-based directional MAC protocol, including the position prediction, communication control, and data transmission phases. In the first phase, each UAV can be a position sender and directionally transmits its GPS-coordinate vector clockwise. The position packet brings only an extra 17 bytes of overhead. When a node acts as a sender, the other nodes are working as listeners until receiving the position packets. In the second phase, three control packet handshakings, i.e., RTS, CTS and wait-to-send (WTS), are executed. In the third phase, the Tx UAVs steer their antennas and transmit data to the Rxs. For channel scheduling, the authors in[42] modeled the interaction of adjacent links as a 3D time-varying interference graph and utilized graph coloring method to allocate the millimeter-wave channels in UAV swarm networks. This approach is a potential way to solve the channel allocation in antenna-array enabled FANET. However, each UAV needs to periodically carry out the interference measurement and channel estimation, which may require high system overhead.

Traditional approaches for allocating the wireless resource are usually based on optimization techniques, e.g., greedy heuristic search[43], iterative methods for local optimum[44], hyper-graph coloring[42], matching theory[45], polyblock-based optimization[46]. However, all these optimization methods require accurate CSI and may not perform well for a large-scale antenna-array enabled FANET. In contrast, machine learning is a promising approach for network optimization. In[47], the authors developed a deep learning approach, which bypassed the channel estimation and scheduled the links efficiently based on the geographic spatial information. The generalization ability of the neural network was demonstrated for different link density, which revealed the advantages compared to traditional optimization methods and heuristic algorithms. In a centralized wireless network with imperfect CSI, the authors in[48] proposed a joint user scheduling and resource-block allocation scheme via federated learning, in which a Gaussian process regression based method and Lyapunov optimization framework were utilized to learn and track the wireless channel and to solve the stochastic optimization problem, respectively.

8.3.4 Summary and Discussion

Neighbor discovery is the basis for achieving self-organization, and is also the premise of routing and resource scheduling decisions. First, the frequency of neighbor discovery in FANETs should be optimized to balance efficiency and overhead. Although directional neighbor discovery can reduce interference and increase the detection range, it will introduce a serious spatial rendezvous problem[15], which can cause intolerable latency because of beam scanning in 3D space. A potential solution is to use prior knowledge of location or mobility prediction to assist directional neighbor discovery based on a pseudo-deterministic approach. In particular, neighbor discovery based on machine learning is promising for predicting the mobilities of the UAVs[28]. However, this method needs a certain amount of data samples. Besides, the existing directional neighbor discovery strategies, as shown in Table 8.1, are based on 2D scenes, which cannot be well extended to 3D scenes. Moreover, the topology of the ad hoc network changes frequently, which implies that highly time efficient directional neighbor discovery schemes are needed.

The design of routing and resource allocation schemes for antenna-array enabled FANETs is a coupled decision-making process. We have compared different routing strategies for FANETs in Table 8.2. Since directional beams pose new challenges, it remains to be seen whether these routing strategies are applicable to antenna-array UAV communications. When routing discovery is finished, the transmission path selection of the data links evolves into a network flow problem. However, different from traditional ad hoc networks, beamforming and the dynamic topology should be taken into account in antenna-array enabled FANETs. In fact, resource allocation greatly affects the performance of multi-flow and multi-hop data transmission. The resource allocation for antenna-array enabled FANETs needs to consider not only the original decision domains of time, frequency, and power, but also the beam domain. Although this will improve the network performance, it will make resource scheduling more complex. In addition, since routing and resource allocation significantly influence each other, it is usually difficult to obtain a globally optimal solution. Instead, suboptimal heuristic methods with low complexity are of interest for joint routing and resource allocation in antenna-array enabled FANETs.

8.4 INTEGRATION OF SUB-6 GHZ AND MILLIMETER-WAVE BANDS

Although the use of millimeter-wave frequency bands brings new potentials for FANET, there are some challenging problems induced by the directional transmission as we discussed before. In practice, different frequency bands can be jointly utilized according to their characteristics and the requirements of different applications, such as the control link, data link, and target detection. The advantages of sub-6 GHz and millimeter-wave frequency bands can be combined for networking. In the following, the integration of sub-6 GHz and millimeter-wave frequency bands in FANETs will be discussed.

First, the omnidirectional communications in sub-6 GHz frequency bands can be utilized for network controlling. Due to the fluctuating communication links and complex beam-alignment operations in antenna-array enabled FANET, it is difficult to initialize and maintain stable connections. Meanwhile, since the directional beams of UAVs have to frequently scan, a high latency is induced and the delay-sensitive control messages cannot be delivered in time. Hence, utilizing the control channels under sub-6 GHz is a good solution. For handling the mobility, UAVs can periodically exchange their location information via the control channel and conduct location prediction for neighbors[6, 49]. In addition, the resource allocation and routing commands can also be transmitted in the low-frequency control channel for improving the management efficiency of the network. Besides, the auxiliary information from the sub-6 GHz channels can be used to assist to accomplish beam management and establish data links in the millimeter-wave frequency bands. Second, the directional communications can be utilized to transmit high-rate and delay-tolerant data. Since the periodic prediction of the location information produces large amounts of control messages, the controlling on mobility management can be partly transferred to the data plane for releasing the overload on the low-capacity control channel[50].

In[51], the authors proposed a cooperative neighbor discovery procedure, in which the 2.4 GHz link was used to assist neighbor discovery and the 60 GHz link was applied for high data transmission in ad hoc networks. Compared to the conventional directional neighbor discovery procedure, this scheme can reduce the average discovery time by 69%-78%. Furthermore, they expanded the scheme to integrate omnidirectional neighbor discovery and directional data transmission[52], namely the multi-band directional neighbor discovery. In particular, the proposed scheme provided compatible superframe structure to the IEEE 802.11a and IEEE 802.11ad specifications.

However, resource allocation in such multi-band integration ad hoc networks still faces great challenges. First, resource allocation and interference control are imperative in the control plane. Channel allocation policies help avoid the interference by assigning different time slots or frequency channels to UAVs within the interference range. The channel allocation problem can be modeled as a graph coloring problem[53, 54], where the colors (i.e., channels) are assigned to the UAVs to avoid conflict. In addition, game-theoretic approaches have attracted a lot of attention to resolve the resource allocation problem[55, 56]. In[57], the authors combined the graph coloring and game-theoretic approaches to reduce the co-channel interference and improve the channel reuse capability. Meanwhile, the authors proved the existence of a Nash equilibrium in the proposed graph coloring game and the convergence under the proposed distributed message-passing protocol. In addition to conventional optimization methods, online learning-based adaptive resource allocation approaches are also potential ways, but the dynamic modeling and low-complexity design are required for further investigation. Second, in the data plane, the beam management for achieving reliable and high-rate communications in multi-hop data transmission is challenging. In particular, due to the UAV jittering, there is a tradeoff between directional antenna gain and beamwidth[58], which poses a challenging requirement for beamforming.

Summary and Discussion: Considering the advantages of omnidirectional coverage and directional high-gain transmission, the integration of the sub-6 GHz and millimeter-wave bands holds great potential for FANETs. Omnidirectional transmissions of control information can enhance the efficiency of network management, and the high-rate, low-delay data can be transmitted through millimeter-wave links. However, it should be noted that the ground cellular network will introduce serious interference for aerial networks operating in sub-6 GHz bands. Therefore, the proposed approach needs to consider the potential unavailability of low-frequency links in real-world scenarios, and carry out resource scheduling and interference management in the control plane. In addition, the nature of antenna-array UAV links can also pose some challenges for directional data transmission, such as beam scheduling and beam design.

8.5 SECURITY

The utilization of millimeter-wave frequency bands for FANETs improves the network security. In general, there are three primary types of attackers in wireless communication networks, i.e., eavesdroppers, jammers, and untrusted nodes[59, 60]. In the first scenario, given the open nature of wireless communications, eavesdroppers have the chance to intercept some confidential data, which leads to information leakage. In the second scenario, jammers intend to degrade the channel quality by transmitting jamming signals to the legitimate Rxs, which leads to information loss. In the third scenario, an untrusted node in the network may be unauthenticated or have a lower level of security than the other nodes, which provides an opportunity for eavesdropping if manipulated by criminals. Fig. 8.4 demonstrates an example of UAV communication network which includes three types of security attackers. In this scenario, UAVs form an aerial ad hoc network to transmit information to legitimate ground terminals, while the jammer transmits jamming signals to the relay UAV to degrade its channel, an eavesdropper is trying to intercept the classified information, and an untrusted UAV may try to decode the confidential information that they are relaying.

8.5.1 Security Metrics

Secrecy capacity is the most widely used metric in physical layer security as given by[61, 62, 63]

$$S_{m,k} = \mathbb{E}[\log_2(1 + \gamma_{m,k}) - \log_2(1 + \gamma_{\text{eav}})]^+, \tag{8.1}$$

where γ_{eav} is the equivalent signal-to-interference-plus-noise ratio (SINR) of the wire-tap channel between the serving UAV and the eavesdroppers, which is defined as the difference between the capacity of legitimate pair and that of the eavesdropping link. Another important metric is the *secrecy outage probability*, which measures the probability that the rate redundancy of an information encoding scheme is smaller than

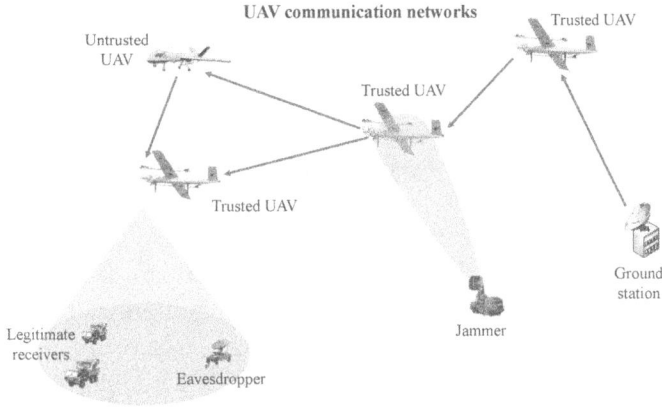

Figure 8.4 Security attackers in UAV communication networks.

the capacity of the eavesdropping link[64]. Besides, there are three important metrics for measuring the performance of network-wide security, namely *area secure link number (ASLN)* [65], *network-wide secrecy throughput (NST)* [66], and *network-wide secrecy energy efficiency (NSEE)* [67]. ASLN is defined as the average number of secure links per unit area, where a secure link refers to a communication link that neither connection outage nor secrecy outage occurs. Correspondingly, under the required connection outage and secrecy outage probabilities, the achievable rate of successful information transmission per unit area is called NST. Further, NSEE is defined as the ratio of NST to the power consumption per unit area, which is used to evaluate the energy efficiency.

8.5.2 Anti-Eavesdropping Techniques

Security issues for millimeter-wave communication networks [66,68,69,70,71], UAV communication networks [59,72,73,74,75], and antenna-array UAV communication networks[76,77] have attracted widespread attention. In addition to conventional anti-eavesdropping techniques, such as artificial noise and cooperative jamming[72,73,78], the flexibility of UAV positioning and the directionality of millimeter-wave transmission can be exploited for improving secure communications.

For the links between two UAVs, the trajectory of UAVs can be cooperatively designed with the resource allocation to enhance the security [79]. It was demonstrated that the passive eavesdroppers may be detected by legitimate transceivers from the local oscillator powers that are inadvertently leaked from eavesdroppers' RF front end[80]. With this information, a UAV could adjust its transmit power in different waypoints to increase secrecy rate and change its position and velocity to bypass or quickly pass through the eavesdroppers. Inevitably, this trajectory design leads to more energy consumption for propulsion. Considering millimeter-wave communications, the A2A or A2G channels are very sparse in the angular domain, which entails highly directional transmission. Considering the distribution of the eavesdroppers, the secrecy rate can be effectively improved by setting a protected zone around users[77].

Besides, beamforming also improves transmission security by suppressing the signal power in the directions of the eavesdroppers [81, 82]. UAV mounted beamforming can make this process more efficient because the high altitude of the UAV can provide more elevation separation, while its horizontal position can be adjusted to achieve a better azimuth separation [73]. It was demonstrated that by adjusting the altitude of a UAV relay, the secrecy rate of the legitimate links can be improved[76]. Furthermore, the deployment of intelligent reflecting surfaces (IRSs) also has a great potential for enhancing communication security[83]. Specifically, an IRS can be deployed on the buildings to reflect the signals towards locations of eavesdroppers to create destructive interference. What's more, untrusted UAV can configure IRSs to relaying data of the high security requirement, because there are no information transmissions between untrusted UAVs and trusted UAVs.

The network-layer technologies can also be used for the security design in a antenna-array enabled FANET. First, the multi-hop relaying transmission is an effective way to avoid information leakage by designing a route which bypasses eavesdroppers[84]. This strategy is also effective to antagonize jamming attacks, in which the most secure route needs to be selected. Second, UAV-assisted artificial jamming is a promising approach, where a number of UAVs in the network can be utilized to send artificial jamming signals to the eavesdroppers for deteriorating their wiretap channels[85]. However, jamming signals not only interfere eavesdropping links but also impact legitimate transmissions. Thus, the transmit power of the cooperative jamming signal needs to be carefully designed to achieve high secrecy performance[76]. Third, coordinated multipoint (CoMP) transmission can be utilized to enhance the network security, where multiple UAVs form virtual antenna arrays to improve the beam energy in the direction of legitimate Rx while degrading that in the other directions[59].

8.5.3 Summary and Discussion

In Table 8.3, we summarize the main types of attackers and potential responses for antenna-array enabled FANETs. In particular, the directionality of beams reduces the vulnerability of the network. Physical layer security techniques, such as artificial noise injection and cooperative jamming, have been proposed for UAV-assisted networks[72, 73, 78]. Antenna-array UAV communications offers additional security. A single UAV can adjust its 3D position to avoid the jamming area. Meanwhile, beamforming can suppress the signal power in the direction of the eavesdroppers. Besides, cooperative jamming with directional jamming signals can be used by multiple UAVs to improve the secrecy throughput. Furthermore, in antenna-array enabled FANETs, security aspects can be taken into account in the routing decisions in order to select a safe data transmission path. In the future, by combining UAVs with millimeter-wave technology, novel anti-eavesdropping techniques can be developed to improve security.

Table 8.3 Description of and potential responses to main types of attackers in millimeter-wave-enabled FANETs.

Type of attackers	Description	Solutions for improving security
Eavesdropper	Intercepts data around legitimate receivers	Resource allocation and trajectory optimization, artificial noise, cooperative jamming, beamforming, CoMP
Jammer	Degrades channel quality by transmitting jamming signals to legitimate UAVs	CoMP, routing design
Untrusted UAV	May be unauthenticated or have a lower level of security than other nodes	Cooperative jamming, CoMP, routing design, IRS assisted methods

8.6 SUMMARY

In this chapter, we presented the issues and challenges for antenna array enabled FANETs for various network architectures, which partly determine the characteristics and possible applications of the network. Specifically, we introduced the network topology and the SDN-based network. Then, in the third section, we considered the link establishment and maintenance, i.e., neighbor discovery, routing and resource allocation. Specifically, neighbor discovery is the basis for achieving self-organization. After neighbor discovery, the design of routing and resource allocation schemes for antenna array enabled FANETs is a coupled decision-making process which is usually difficult to obtain a globally optimal solution. Hence, sub-optimal heuristic methods with low complexity are usually considered. Next, in the forth section, we considered the integration of the sub-6 GHz and millimeter-wave bands. Omnidirectional transmission can be used for control information while high-rate data can be transmitted through millimeter-wave bands. Finally, we introduced the network security, including security metrics and anti-eavesdropping techniques.

Bibliography

[1] Wajiya Zafar and Bilal Muhammad Khan. Flying ad-hoc networks: Technological and social implications. *IEEE Technol. Soc. Mag.*, 35(2):67–74, Jun. 2016.

[2] Zhenyu Xiao, Lipeng Zhu, Yanming Liu, Pengfei Yi, Rui Zhang, Xiang-Gen Xia, and Robert Schober. A survey on millimeter-wave beamforming enabled UAV communications and networking. *IEEE Commun. Surveys Tuts.*, 24(1):557–610, 1st Quart. 2022.

[3] Lav Gupta, Raj Jain, and Gabor Vaszkun. Survey of important issues in UAV communication networks. *IEEE Commun. Surveys Tuts.*, 18(2):1123–1152, 2nd Quart. 2016.

[4] Xianbin Cao, Peng Yang, Mohamed Alzenad, Xing Xi, Dapeng Wu, and Halim Yanikomeroglu. Airborne communication networks: A survey. *IEEE J. Select. Areas Commun.*, 36(9):1907–1926, Sep. 2018.

[5] Zhigao Zheng, Arun Kumar Sangaiah, and Tao Wang. Adaptive communication protocols in flying ad hoc network. *IEEE Commun. Mag.*, 56(1):136–142, Jan. 2018.

[6] Ganbayar Gankhuyag, Anish Prasad Shrestha, and Sang-Jo Yoo. Robust and reliable predictive routing strategy for flying ad-hoc networks. *IEEE Access*, 5:643–654, 2017.

[7] Bo Yang, Min Liu, and Zhongcheng Li. Rendezvous on the fly: Efficient neighbor discovery for autonomous UAVs. *IEEE J. Select. Areas Commun.*, 36(9):2032–2044, Sep. 2018.

[8] Zhiqing Wei, Xinyi Liu, Chenyang Han, and Zhiyong Feng. Neighbor discovery for unmanned aerial vehicle networks. *IEEE Access*, 6:68288–68301, 2018.

[9] Demeke Shumeye Lakew, Umar Sa'ad, Nhu-Ngoc Dao, Woongsoo Na, and Sungrae Cho. Routing in flying ad hoc networks: A comprehensive survey. *IEEE Commun. Surveys Tuts.*, 22(2):1071–1120, 2nd Quart. 2020.

[10] Jingjing Wang, Chunxiao Jiang, Zhu Han, Yong Ren, Robert G. Maunder, and Lajos Hanzo. Taking drones to the next level: Cooperative distributed unmanned-aerial-vehicular networks for small and mini drones. *IEEE Veh. Technol. Mag.*, 12(3):73–82, Sep. 2017.

[11] Bruno Astuto A. Nunes, Marc Mendonca, Xuan-Nam Nguyen, Katia Obraczka, and Thierry Turletti. A survey of software-defined networking: Past, present, and future of programmable networks. *IEEE Commun. Surveys Tuts.*, 16(3):1617–1634, 3rd Quarter. 2014.

[12] Diego Kreutz, Fernando Manuel Valente Ramos, Paulo Esteves Veríssimo, Christian Esteve Rothenberg, Siamak Azodolmolky, and Uhlig. Software-defined networking: A comprehensive survey. *Proc. IEEE*, 103(1):14–76, Jan. 2015.

[13] Fei Hu, Qi Hao, and Ke Bao. A survey on software-defined network and openflow: From concept to implementation. *IEEE Commun. Surveys Tuts.*, 16(4):2181–2206, 4th Quart. 2014.

[14] Zhangyu Guan and Tejas Kulkarni. On the effects of mobility uncertainties on wireless communications between flying drones in the mmWave/THz bands. In *Proc. IEEE Conf. Comput. Commun. Workshops*, pages 768–773, Apr. 2019.

[15] Yu Wang, Shiwen Mao, and Theodore S. Rappaport. On directional neighbor discovery in mmWave networks. In *Proc. Int. Conf. Distrib. Comput. Syst.*, pages 1704–1713, Jun. 2017.

[16] Gabriel Astudillo and Michel Kadoch. Neighbor discovery and routing schemes for mobile ad-hoc networks with beamwidth adaptive smart antennas. *Telecommun. Syst.*, 66:17–27, Jan. 2017.

[17] Javier Rodriguez-Fernandez, Nuria Gonzalez-Prelcic, and Robert W. Heath. Position-aided compressive channel estimation and tracking for millimeter wave multi-user MIMO air-to-air communications. In *Proc. IEEE Int. Conf. Commun. Workshops*, pages 1–6, May 2018.

[18] Li Yan, Haichuan Ding, Lan Zhang, Jianqing Liu, Xuming Fang, Yuguang Fang, Ming Xiao, and Xiaoxia Huang. Machine learning-based handovers for sub-6 GHz and mmWave integrated vehicular networks. *IEEE Trans. Wireless Commun.*, 18(10):4873–4885, Oct. 2019.

[19] Hong-Ning Dai, Kam-Wing Ng, Minglu Li, and Min-You Wu. An overview of using directional antennas in wireless networks. *Int. J. Commun. Syst.*, 26(4):413–448, Apr. 2013.

[20] Masanori Takata, Masaki Bandai, and Takashi Watanabe. A MAC protocol with directional antennas for deafness avoidance in ad hoc networks. In *Proc. IEEE Global Telecommun. Conf.*, pages 620–625, Nov. 2007.

[21] Md. Nasre Alam, Md. Asdaque Hussain, and Kyung Sup Kwak. Neighbor initiated approach for avoiding deaf and hidden node problems in directional MAC protocol for ad-hoc networks. *Wirel. Netw.*, 19, Jul. 2013.

[22] Romit Roy Choudhury, Xue Yang, Ram Ramanathan, and Nitin H. Vaidya. On designing MAC protocols for wireless networks using directional antennas. *IEEE Trans. Mob. Comput.*, 5(5):477–491, May 2006.

[23] Zhensheng Zhang and Bo Li. Neighbor discovery in mobile ad hoc self-configuring networks with directional antennas: Algorithms and comparisons. *IEEE Trans. Wireless Commun.*, 7(5):1540–1549, May 2008.

[24] Xueli An, R. Venkatesha Prasad, and Ignas Niemegeers. Impact of antenna pattern and link model on directional neighbor discovery in 60 GHz networks. *IEEE Trans. Wireless Commun.*, 10(5):1435–1447, May 2011.

[25] Lin Chen, Yong Li, and Athanasios V. Vasilakos. On oblivious neighbor discovery in distributed wireless networks with directional antennas: Theoretical foundation and algorithm design. *IEEE/ACM Trans. Netw.*, 25(4):1982–1993, Aug. 2017.

[26] Amjad Riaz, Sajid Saleem, and Syed Ali Hassan. Energy efficient neighbor discovery for mmWave D2D networks using polya's necklaces. In *Proc. IEEE Glob. Commun. Conf.*, pages 1–6, Dec. 2018.

[27] Sherief Hashima, Kohei Hatano, Eiji Takimoto, and Ehab Mahmoud Mohamed. Neighbor discovery and selection in millimeter wave D2D networks using stochastic MAB. *IEEE Commun. Lett.*, 24(8):1840–1844, Aug. 2020.

[28] Shengbo Huang, Mo Li, and Liang Zhao. An intelligent neighbor discovery algorithm for ad hoc networks with directional antennas. In *Proc. Int. Conf. Mechatron. Sci., Electric Eng. Comput.*, pages 302–305, Dec. 2013.

[29] Xiangrui Fan, Wenlong Cai, and Jinyong Lin. A survey of routing protocols for highly dynamic mobile ad hoc networks. In *Proc. IEEE Int. Conf. Commun. Technol.*, pages 1412–1417, Oct. 2017.

[30] Baozhi Zhao and Qing Ding. Route discovery in flying ad-hoc network based on bee colony algorithm. In *Proc. IEEE Int. Conf. Artif. Intell. Comput. Appl.*, pages 364–368, Mar. 2019.

[31] Alexey V. Leonov. Modeling of bio-inspired algorithms AntHocNet and BeeAd-Hoc for flying ad hoc networks (FANETS). In *Proc. Int. Sci.-Tech. Conf. Actual Problems Electron. Instrum. Eng.*, volume 03, Oct. 2016.

[32] Alexey V. Leonov. Application of bee colony algorithm for FANET routing. In *Proc. Int. Conf. Young Spl. Micro/Nanotechnol. Electron Devices*, pages 124–132, Jun. 2016.

[33] Vinay Kolar and Nael B. Abu-Ghazaleh. A multi-commodity flow approach for globally aware routing in multi-hop wireless networks. In *Proc. Fourth Annu. IEEE Int. Conf. Pervasive Comput. Com.*, pages 1–10, Mar. 2006.

[34] Tao Zhang, Hongyan Li, Jiandong Li, Shun Zhang, and Haiying Shen. A dynamic combined flow algorithm for the two-commodity max-flow problem over delay-tolerant networks. *IEEE Trans. Wireless Commun.*, 17(12):7879–7893, Dec. 2018.

[35] Pradeep Chathuranga Weeraddana, Marian Codreanu, Matti Latva-aho, and Anthony Ephremides. Resource allocation for cross-layer utility maximization in wireless networks. *IEEE Trans. Veh. Technol.*, 60(6):2790–2809, Jul. 2011.

[36] Amr A. El-Sherif and Amr Mohamed. Joint routing and resource allocation for delay minimization in cognitive radio based mesh networks. *IEEE Trans. Wireless Commun.*, 13(1):186–197, Jan. 2014.

[37] IEEE-802.11aj. Wireless LAN medium access control (MAC) and physical layer (PHY) specifications amendment 3: Enhancements for very high throughput to support chinese millimeter wave frequency bands (60 GHz and 45 GHz). pages 1–306, Apr. 2018.

[38] IEEE-P802.11ay. Wireless LAN medium access control (MAC) and physical layer (PHY) specifications–amendment: Enhanced throughput for operation in license-exempt bands above 45 GHz. pages 1–791, Jul. 2019.

[39] Yong Niu, Chuhan Gao, Yong Li, Depeng Jin, Li Su, and Dapeng Wu. Boosting spatial reuse via multiple-path multihop scheduling for directional mmWave WPANs. *IEEE Trans. Veh. Technol.*, 65(8):6614–6627, Aug. 2016.

[40] Wenson Chang, Chien-Wen Wu, and Lin Yi-Xin. Efficient time-slot adjustment and packet-scheduling algorithm for full-duplex multi-hop relay-assisted mmWave networks. *IEEE Access*, 6:39273–39286, 2018.

[41] Pei Zhou, Xuming Fang, Yuguang Fang, Rong He, Yan Long, and Gaoyong Huang. Beam management and self-healing for mmWave UAV mesh networks. *IEEE Trans. Veh. Technol.*, 68(2):1718–1732, Feb. 2019.

[42] Zhiyong Feng, Lei Ji, Qixun Zhang, and Wei Li. Spectrum management for mmWave enabled UAV swarm networks: Challenges and opportunities. *IEEE Commun. Mag.*, 57(1):146–153, Jan. 2019.

[43] Xinzhou Wu, Saurabha Tavildar, Sanjay Shakkottai, Tom Richardson, Junyi Li, Rajiv Laroia, and Aleksandar Jovicic. FlashLinQ: A synchronous distributed scheduler for peer-to-peer ad hoc networks. *IEEE/ACM Trans. Netw.*, 21(4):1215–1228, Aug. 2013.

[44] Kaiming Shen and Wei Yu. FPLinQ: A cooperative spectrum sharing strategy for device-to-device communications. In *Proc. IEEE Int. Symp. Inf. Theor.*, pages 2323–2327, Jun. 2017.

[45] Yanming Liu, Kai Liu, Jinglin Han, Lipeng Zhu, Zhenyu Xiao, and Xiang-Gen Xia. Resource allocation and 3-D placement for UAV-enabled energy-efficient IoT communications. *IEEE Internet Thing J.*, 8(3):1322–1333, Feb. 2021.

[46] Li Ping Qian and Ying Jun Zhang. S-MAPEL: Monotonic optimization for non-convex joint power control and scheduling problems. *IEEE Trans. Wireless Commun.*, 9(5):1708–1719, May 2010.

[47] Wei Cui, Kaiming Shen, and Wei Yu. Spatial deep learning for wireless scheduling. *IEEE J. Select. Areas Commun.*, 37(6):1248–1261, Jun. 2019.

[48] Madhusanka Manimel Wadu, Sumudu Samarakoon, and Mehdi Bennis. Federated learning under channel uncertainty: Joint client scheduling and resource allocation. In *Proc. IEEE Wireless Commun. Networking Conf.*, pages 1–6, May 2020.

[49] Han Peng, Abolfazl Razi, Fatemeh Afghah, and Jonathan Ashdown. A unified framework for joint mobility prediction and object profiling of drones in UAV networks. *J. Commun. Netw.*, 20(5):434–442, Oct. 2018.

[50] Bowen Zeng, Tian Song, and Jianping An. A dual-antenna collaborative communication strategy for flying ad hoc networks. *IEEE Commun. Lett.*, 23(5):913–917, May 2019.

[51] Hyunhee Park, Yongsun Kim, Insun Jang, and Sangheon Pack. Cooperative neighbor discovery for consumer devices in mmWave ad-hoc networks. In *Proc. IEEE Int. Conf. Consum. Electron.*, pages 100–101, Jan. 2012.

[52] Hyunhee Park, Yongsun Kim, Taewon Song, and Sangheon Pack. Multiband directional neighbor discovery in self-organized mmWave ad hoc networks. *IEEE Trans. Veh. Technol.*, 64(3):1143–1155, Mar. 2015.

[53] Imrich Chlamtac and Shlomit S. Pinter. Distributed nodes organization algorithm for channel access in a multihop dynamic radio network. *IEEE Trans. Comput.*, C-36(6):728–737, Jun. 1987.

[54] Mishra Vishram, Lau Chiew Tong, and Chan Syin. A channel allocation based self-coexistence scheme for homogeneous ad-hoc networks. *IEEE Wireless Commun. Lett.*, 4(5):545–548, Oct. 2015.

[55] Fan Wang, Marwan Krunz, and Shuguang Cui. Price-based spectrum management in cognitive radio networks. *IEEE J. Sel. Top. Signal Process.*, 2(1):74–87, Feb. 2008.

[56] Pedro B. F. Duarte, Zubair Md. Fadlullah, Athanasios V. Vasilakos, and Nei Kato. On the partially overlapped channel assignment on wireless mesh network backbone: A game theoretic approach. *IEEE J. Select. Areas Commun.*, 30(1):119–127, Jan. 2012.

[57] Kai-Ju Wu, Yao-win Peter Hong, and Jang-Ping Sheu. Coloring-based channel allocation for multiple coexisting wireless body area networks: A game-theoretic approach. *IEEE Trans. Mobile Comput.*, 2020 (Early access).

[58] Mohammad Taghi Dabiri, Hossein Safi, Saeedeh Parsaeefard, and Walid Saad. Analytical channel models for millimeter wave UAV networks under hovering fluctuations. *IEEE Trans. Wireless Commun.*, 19(4):2868–2883, Apr. 2020.

[59] Xiaofang Sun, Derrick Wing Kwan Ng, Zhiguo Ding, Yanqing Xu, and Zhangdui Zhong. Physical layer security in UAV systems: Challenges and opportunities. *IEEE Wireless Commun.*, 26(5):40–47, Oct. 2019.

[60] Xiang He and Aylin Yener. Cooperation with an untrusted relay: A secrecy perspective. *IEEE Trans. Inf. Theory*, 56(8):3807–3827, Aug. 2010.

[61] Ying Ju, Hui-Ming Wang, Tong-Xing Zheng, Qinye Yin, and Moon Ho Lee. Safeguarding millimeter wave communications against randomly located eavesdroppers. *IEEE Trans. Wireless Commun.*, 17(4):2675–2689, Apr. 2018.

[62] Yongxu Zhu, Gan Zheng, and Michael Fitch. Secrecy rate analysis of UAV-enabled mmWave networks using matérn hardcore point processes. *IEEE J. Select. Areas Commun.*, 36(7):1397–1409, Jul. 2018.

[63] Yang Wu, Weiwei Yang, and Xiaoli Sun. Securing UAV-enabled millimeter wave communication via trajectory and power optimization. In *Proc. IEEE Int. Conf. Comput. Commun.*, pages 970–975, Dec. 2018.

[64] Tong-Xing Zheng and Hui-Ming Wang. Optimal power allocation for artificial noise under imperfect CSI against spatially random eavesdroppers. *IEEE Trans. Veh. Technol.*, 65(10):8812–8817, Oct. 2016.

[65] Chao Wang and Hui-Ming Wang. Physical layer security in millimeter wave cellular networks. *IEEE Trans. Wireless Commun.*, 15(8):5569–5585, Aug. 2016.

[66] Xi Zhang, Xiangyun Zhou, and Matthew R. McKay. Enhancing secrecy with multi-antenna transmission in wireless ad hoc networks. *IEEE Trans.Inf. Forensics Secur.*, 8(11):1802–1814, Nov. 2013.

[67] Xiaoming Chen and Lei Lei. Energy-efficient optimization for physical layer security in multi-antenna downlink networks with QoS guarantee. *IEEE Commun. Lett.*, 17(4):637–640, Apr. 2013.

[68] Wen-Qin Wang and Zhi Zheng. Hybrid MIMO and phased-array directional modulation for physical layer security in mmWave wireless communications. *IEEE J. Select. Areas Commun.*, 36(7):1383–1396, Jul. 2018.

[69] Yongxu Zhu, Lifeng Wang, Kai-Kit Wong, and Robert W. Heath. Secure communications in millimeter wave ad hoc networks. *IEEE Trans. Wireless Commun.*, 16(5):3205–3217, May 2017.

[70] Ahmed F. Darwesh and Abraham O. Fapojuwo. Secrecy rate analysis of mmWave MISO ad hoc networks with null space precoding. In *Proc. IEEE Wireless Commun. Netw. Conf.*, pages 1–6, May 2020.

[71] Yongxu Zhu, Lifeng Wang, Kai-Kit Wong, and Robert W. Heath. Physical layer security in large-scale millimeter wave ad hoc networks. In *Proc. IEEE Glob. Commun. Conf.*, pages 1–6, Dec. 2016.

[72] Wei Wang, Xinrui Li, Miao Zhang, Kanapathippillai Cumanan, Derrick Wing Kwan Ng, Guoan Zhang, Jie Tang, and Octavia A. Dobre. Energy-constrained UAV-assisted secure communications with position optimization and cooperative jamming. *IEEE Trans. Commun.*, 68(7):4476–4489, Jul. 2020.

[73] Qingqing Wu, Weidong Mei, and Rui Zhang. Safeguarding wireless network with UAVs: A physical layer security perspective. *IEEE Wireless Commun.*, 26(5):12–18, Oct. 2019.

[74] Zhichao Sheng, Hoang Duong Tuan, Ali Arshad Nasir, Trung Q. Duong, and H. Vincent Poor. Secure UAV-enabled communication using Han-Kobayashi signaling. *IEEE Trans. Wireless Commun.*, 19(5):2905–2919, May 2020.

[75] Jia Ye, Chao Zhang, Hongjiang Lei, Gaofeng Pan, and Zhiguo Ding. Secure UAV-to-UAV systems with spatially random UAVs. *IEEE Wireless Commun. Lett.*, 8(2):564–567, Apr. 2019.

[76] Xiaowei Pang, Mingqian Liu, Nan Zhao, Yunfei Chen, Yonghui Li, and F. Richard Yu. Secrecy analysis of UAV-based mmWave relaying networks. *IEEE Trans. Wireless Commun.*, 20(8):4990–5002, Aug. 2021.

[77] Yavuz Yapici, Nadisanka Rupasinghe, İsmail Güvenç, Huaiyu Dai, and Arupjyoti Bhuyan. Physical layer security for NOMA transmission in mmWave drone networks. *IEEE Trans. Veh. Technol.*, 70(4):3568–3582, Apr. 2021.

[78] Amitav Mukherjee, S. Ali A. Fakoorian, Jing Huang, and A. Lee Swindlehurst. Principles of physical layer security in multiuser wireless networks: A survey. *IEEE Commun. Surveys Tuts.*, 16(3):1550–1573, 3rd Quart. 2014.

[79] Guangchi Zhang, Qingqing Wu, Miao Cui, and Rui Zhang. Securing UAV communications via joint trajectory and power control. *IEEE Trans. Wireless Commun.*, 18(2):1376–1389, Feb. 2019.

[80] Amitav Mukherjee and A. Lee Swindlehurst. Detecting passive eavesdroppers in the MIMO wiretap channel. In *Proc. IEEE Int. Conf. Acoust., Speech and Sign. Process.*, pages 2809–2812, Mar. 2012.

[81] Huici Wu, Yang Wen, Jiazhen Zhang, Zhiqing Wei, Ning Zhang, and Xiaofeng Tao. Energy-efficient and secure air-to-ground communication with jittering UAV. *IEEE Trans. Veh. Technol.*, 69(4):3954–3967, Apr. 2020.

[82] Chao Wang, Zan Li, Tong-Xing Zheng, Hongyang Chen, and Xiang-Gen Xia. Robust hybrid precoding design for securing millimeter wave IoT networks under secrecy outage constraint. *IEEE Internet Thing J.*, 8(16):13024–13038, Aug. 2021.

[83] Marco Di Renzo, Alessio Zappone, Merouane Debbah, Mohamed-Slim Alouini, Chau Yuen, Julien de Rosny, and Sergei Tretyakov. Smart radio environments empowered by reconfigurable intelligent surfaces: How it works, state of research, and the road ahead. *IEEE J. Select. Areas Commun.*, 38(11):2450–2525, Nov. 2020.

[84] Leonardo Jimenez Rodriguez, Nghi H. Tran, Trung Q. Duong, Tho Le-Ngoc, Maged Elkashlan, and Sachin Shetty. Physical layer security in wireless cooperative relay networks: state of the art and beyond. *IEEE Commun. Mag.*, 53(12):32–39, Dec. 2015.

[85] An Li, Qingqing Wu, and Rui Zhang. UAV-enabled cooperative jamming for improving secrecy of ground wiretap channel. *IEEE Wireless Commun. Lett.*, 8(1):181–184, Feb. 2019.

CHAPTER 9

Antenna Array Enabled Space/Air/Ground Communications

9.1 INTRODUCTION

In Chapter 7, we introduced the array beamforming enabled communications on UAV platforms, where 3D beam coverage, single UAV and multiple UAVs deployment were presented. Moreover, the FANET, where multiple UAVs can collaboratively carry out complex missions, was introduced in Chapter 8. In addition to the antenna array enabled communication on the UAV platform, the space/air/ground communication network is becoming a promising paradigm for next generation communication network that needs the support of array beamforming. In addition to ground wireless communications, the satellites can provide globally seamless communication coverage, while the aircraft can achieve on demand deployment and wide-area communication coverage in emergencies. Meanwhile, the application of antenna array and the mobility of space/air/ground platforms poses substantial new characteristics to the antenna array enabled space/air/ground communication systems.

Nevertheless, there are also many challenges and research directions worth exploring. Taking a wide variety of application tasks, antenna array enabled space/air/ground communication networks are becoming increasingly complicated, decentralized, and autonomous. As a result, it may be challenging to employ mathematical model-based theories to solve problems in large-scale and dynamic cases. In contrast, artificial intelligence (AI), with model-free, data-driven, adaptive, scalable, and distributed characteristics, shows great potential to achieve significant performance enhancement for space/air/ground communication networks. Besides, AI is a potential solution to solve the complex resource scheduling problem for antenna array enabled communication systems, where not only the original decision domains of time, frequency, and power are considered, but also the beam domain is involved.

On top of that, as the demand for communication in dense urban areas increases, it is inevitably to integrate satellite communications and airborne communications with ground communications, i.e., forming the space-air-ground integrated network

DOI: 10.1201/9781003366362-9

(SAGIN). In SAGIN, ground infrastructures cooperate with aircraft and satellites to solve coverage limitations, access restrictions, and timeliness requirements, and provide users with better and more real-time services. High dynamic scenarios result in more complex and difficult routing and handover management. Future resources should pay more attention to the integrated system, such as the assignment allocation, power allocation, spectrum allocation and equipment management. The integration of large-scale antenna array provides a significant technique support for SAGIN.

In this chapter, we first introduce the Low Earth Orbit (LEO) satellite communication and its unique features, i.e., beam coverage, beamforming and beam management and handover. The next is the airborne communication network. In this section, from the perspective of communications, we mainly focus on beam tracking, Doppler effects and joint positioning. And from the perspective of networking, directional neighbor discovery, routing and resource management are addressed. Last, we mainly focus on the applications of antenna array in ground cellular communications, i.e., cellular massive multiple-input multiple-output (MIMO), cell-free MIMO and vehicle-to-everything (V2X) communication.

9.2   LEO SATELLITE COMMUNICATION

Satellite communications usually mean the communications between a satellite platform and a ground station or different satellite platforms. Satellites can operate in a Geostationary Earth Orbit (GEO) constellation, a Medium Earth Orbit (MEO) constellation, and a LEO constellation, according to the orbital height. Compared to terrestrial networks and airborne networks, satellite communication networks have a much larger coverage area. However, long-distance communication between satellites and ground leads to much larger link loss and transmission delay. Satellite communication networks enabled by antenna arrays can make up for the above shortcomings and obtain more flexible beam coverage to meet the needs of users to access the network anytime and anywhere. Meanwhile, the particular characters of satellite altitude, frequency and movement bring several future research directions in adaptive multi-beam pattern and footprint planning and multi-spot beam arrangement.

1) Adaptive Multi-Beam Pattern and Footprint Planning

When providing services to users in remote areas, the uneven distribution of user terminals and dynamic changes in traffic demand, and satellite network access requirements will vary with the user's access time and geographic location. Therefore, in order to meet the ubiquitous access needs of users anytime and anywhere, adaptive multi-beam patterns and footprint planning represent an important research direction. Beam patterns and fingerprints of satellite array antennas are susceptible to the uneven distribution of user and traffic requirements, channel conditions, user quality of service (QoS) requirements, and wireless resources. Therefore, uniform traffic load distribution, simplified radio resource management, effective load and frequency distribution need to be emphasized. The above problem is usually modeled as a compromise between unlimited resource management, load balancing, and user demand. However, this problem is generally a highly non-convex optimization problem, which is challenging to deal with.

2) Multi-Spot beam arrangement

The requirement for broadband satellite capabilities has been diversified because of the continuous expansion of the scope of people activities and the rapid growth of traffic demand. However, the spectrum for satellite communications is becoming increasingly scarce. It is necessary to effectively use the limited spectrum resources to share resources with other communication systems. Digital beamforming has a high degree of flexibility and can be used to allocate power resources. Besides, in satellite communication systems equipped with digital beamforming technology, the theoretical relationship between multi-point beam placement and throughput is an important research direction in the future. The internal mechanism of the distance between spot beams in the same frequency band and the distance between adjacent spot beams in different frequency bands and the overall system throughput is still unclear. To improve the overall system throughput through multi-point beam placement is usually constructed as a 0-1 non-convex optimization model. Therefore, solving this problem is challenging.

In addition to the research directions, there are also several unique features to the satellite communication networks in beam coverage, beamforming, beam management and handover, as described below.

9.2.1 Various Beam Patterns

In satellite communications, a variety of service scenarios may require different coverage schemes, thus calling for various beam patterns. Generally speaking, broad coverage requirements are usually accomplished by wide beams, which include global beams, hemispherical beams and regional beams. However, wider beams are usually accompanied by smaller antenna gains. Therefore, wide beams are more suitable for transmitting/receiving user control signals or broadcasting communications. On the other hand, spot beams are presented to improve antenna gains and promote multiplexing gains. The more concentrated beams can reduce transmit power, and increase communication capacity, but with smaller coverage area. Therefore, spot beams are more suitable for providing high-speed data services. Besides, to balance the stable transmission requirements of control signals and high-speed requirements of data signals, a hybrid wide-spot beam is presented in[1], which is essentially the combination of wide beams and spot beams.

For wide beam, one of the main technologies that provides such kind of beam pattern is reconfigurable antennas. According to their electrical performance, reconfigurable antennas can be divided into three main categories: reconfigurable frequency, reconfigurable pattern, and reconfigurable polarization. In[2], a type of antenna with a frequency bandwidth from 1.15 GHz to 1.6 GHz was designed for wide-bandwidth beam global navigation satellite system (GNSS). By adjusting the effective aperture of the antenna, the radiation pattern of the antenna can be reconstructed, thereby achieving wide beam coverage. In[3], a beamwidth reconfigurable microstrip patch antenna of H-plane pattern was designed to achieve wide beam coverage, where the beam width can be continuously adjusted from 50° to 112°. However, a single wide-beam antenna usually results in the loss of gain as the antenna beam width increases, thereby reducing the QoS for users. To solve this problem, a left-bias pattern and a right-bias pattern were combined through pattern reconfigurable technology[4,5,6], where the wide beam coverage area of the reconfigurable pattern antenna is the union of the coverage provided by the left-bias pattern and coverage of right-bias pattern.

For spot beam, it is necessary to flexibly adjust the center point of the beam to ensure that the communication target is within the coverage area, due to the limited coverage area of the spot beam and the mobility of both satellites and users. In different traffic scenarios, the distribution of business volume is not uniform, for example, metropolis regions and emergency communications during disasters. Therefore, traffic-based dynamic coverage schemes are needed to adjust the size of a spot beam and resource allocation[7]. To support the non-uniform distribution of users and varying traffic requirements, adaptive multi-beam pattern and footprint planning were developed[8], where spot beams with flexible sizes and positions were designed according to user spatial clustering to improve the flexibility of satellite communication systems. In[9], a coverage metric was presented to measure the average coverage level of satellite constellations of different orbital altitudes for backhaul. Among spot beams, time division multiple access (TDMA) spot-beam communication process was further formulated as a discrete-time queuing problem to calculate the quantity of accessed equipments in a unit area. In addition, the relationship between the equipment density, maximum tolerable delay, and satellite constellation coverage level was derived. A steerable spot-beam reflector antenna was explored in[10], where the steerable spot beam can be quickly repositioned to provide flexible coverage by rotating the reflector around its apex (referred to as vertex rotation). In[11], an effective optimization method of multiple-feed per beam antenna based on genetic algorithm was presented to improve the coverage performance of spot beams, where the orthogonality constraint introduced by the lossless beamforming network was taken into account.

The main idea of the hybrid wide-spot beam is to provide a wide beam and multiple spot beams at the same time. The wide beam, with fixed direction and coverage, is utilized to cover the whole service area for the transmission of control signals such as mobility management, session management, bearer establishment and mapping. On the other hand, the spot beams are always steered to the users for the high-speed transmission of user data. In order to enable efficient modulation and coding techniques for data transmission, spot beams usually require much higher power consumption than that of the wide beams. Note that spot beams are more flexible for planning the system capacity and resource configuration according to the needs of users, due to the steerable beams. In summary, with the hybrid wide-spot beam strategy, the structure of the satellite access network is actually reconstructed, that is, the separation of the control plane and the user plane is realized.

9.2.2 MBA

Under the circumstance of exponentially increasing communication demands, designing a satellite system with high throughput is becoming a hot-spot in both academia and industry[12]. However, the limited resources available for satellite make it challenging to fulfill the requirements. Multiple beam array (MBA) and the corresponding multi-beam forming techniques are promising solutions[13]. MBA is an antenna that uses the same aperture to generate multiple beams with different directions simultaneously. By achieving polarization isolation and space isolation effectively, MBA can realize spectrum multiplexing thus increase communication throughput. Moreover, a global or regional beam coverage can be split into several small cells and covered by independent spot beams. In this way, the ground terminal may use a small aperture antenna to realize high-speed data transmission. To avoid interference, different beams work in different frequency bands or adopt different polarization modes. By proper beamforming schemes, the multi-beam forming can help to achieve high gains in the target areas, while leaking low gains outside the serving areas. Therefore, the transmitting power can be reduced.

MBA can be reflector-based architectures, phased array architectures, and lens-based architectures[14]. Reflector antennas and lens-based antennas leverage optical elements such as reflectors and lens to reach higher gains. Therefore, they are applied in MEO/GEO satellites to serve for remote transmission. On the other hand, the phased array architecture is more suitable for LEO satellites with high-flexibility requirements, by means of beamforming. The multi-beam forming in phased array MBA includes analog beamforming and digital beamforming. Globalstar leveraged analog beamforming in its MBA with the beamforming network composed of power dividers. Iridium utilized the beamforming network composed of Butler matrix. Once the beamforming network is determined, the beam shape, the intersection level and beam direction of adjacent beams are fixed and difficult to change. Notably, if the number of beams increases, the beamforming network of analog beamforming will be complex to realize. In addition, the fixed scheme is difficult to be adaptive. Thus, digital beamforming is attracting more interests. The radio frequency (RF) signals received by multiple antenna array elements are respectively converted to baseband through multiple channels, and beamforming is realized through the digital signal processor. Supported by digital architecture, the adaptive beamforming can be applied in satellite MBAs. The possibility of using digital beamforming network to design satellite antenna systems with adaptive beamforming was discussed in[15]. Aiming to reduce the complexity of beamforming design for antennas with large number of emitters, a low complexity algorithm was presented in[16]. The authors in[17] presented an adaptive beamforming method based on user locations. The locations could be provided by users, whose terminals were equipped with the navigation subsystem.

It is worth noting that no matter the analog or digital beamforming, after dividing cells, the shape of beam for the cell needs to be decided. Therefore, it is necessary to find the appropriate amplitude and phase weighting values for each element of the array. This problem can be formulated as the optimization problem. With proper algorithms, the required beam pattern can be obtained. Multi-beam forming can also be combined with rate-splitting multiple access (RSMA) and on-board processing to boost performance and better manage interference between users compared to spital devision multiple access (SDMA) and non-orthogonal multiple access (NOMA).

9.2.3   Beam Management and Handover

Satellite systems provided a wide range of communication service coverage. LEO satellite has the characteristics of low orbit height and short electromagnetic wave round-trip time, which can effectively solve the delay problem for satellite communication. However, the rapid movement of LEO satellite may cause frequent handover of user calls, which challenges the beam management technology for LEO systems[18]. The beam management mainly consists of beam handover and beam scheduling. Beam handover is also called cellular handover or intra-satellite handover, which refers to the handover of links between adjacent beams within the coverage area of the same satellite.

Beam handover technologies mainly include the non-priority handover, queuing priority handover and reserved channel strategies. The non-priority handover strategy employs a fixed channel allocation method to allocate a fixed number of channels to each cell and each type of service. Although this strategy is simple, it can not adapt to the dynamic changes of the network traffic, which reduces the efficiency of resource utilization for the system. It is generally used in combination with other strategies[19]. The queuing priority handover strategy[20,21,22,23,24,25] is based on queuing technology to distinguish the priority of various types of calls or requests and determine the network resource allocation accordingly. When the satellite receives a new call or handover request, while there is no channel available for the next beam, the request will be placed in a special queue for waiting. If the channel is idle at a specific time, then the channel can be scheduled by the next beam, otherwise, the channel will be forcibly interrupted. Calls or requests in the same queue are allocated according to the first-in-first-out principle. Different priorities can also be set for different queues. The queues with higher priority get more network resources. The reserved channel strategy uses the concept of a protected channel, which is set up in each cell specifically for handover services[26]. The key issue of the reserved channel strategy is to set a reasonable threshold so that the reserved channel resources conform to the actual situation of the network, so as to avoid a waste of network resources or affect the effectiveness of the strategy. In addition to being a fixed value, the threshold of the reserved channel can be dynamically adjusted according to the network status, which may improve the network resource usage[18]. The existing methods of dynamically adjusting the threshold are presented as follows. Adjustment strategies based on forecast, probability models or state quantities were used to predict different types of requests, and dynamically adjust the threshold of the reserved channel according to the prediction results[27]. In[28], an adaptive dynamic channel allocation strategy was presented to reduce the overall handover blocking probability. An opportunistic call admission protocol was presented to avoid the cost of researching resources for users in a series of beams along the predicted user trajectory in[29]. In time-based adjustment strategy, the reserved channel of the next beam can be adjusted according to the time the user stayed in the current beam or the expected channel usage time[28]. In[27], a time-based channel reservation algorithm was presented to ensure the probability of zero handover failure. The LEO satellite communication network usually adopts multiple earth orbit satellites with limited coverage to form a specific constellation. To form a communication link, the user needs to connect to one of the serving satellites. Due to the fast moving characteristics of LEO satellites, inter-satellite handover occurs frequently. Once handover occurs, it involves the problem of beam scheduling. The user terminal always selects the maximum instantaneous elevation when handover happens in[18]. In[29], an adjustment strategy based on QoS was presented to dynamically adjust reserved channels, and overcome the low bandwidth utilization rate problem in the reserved channel mechanism. Moreover, user satisfaction was utilized to measure system QoS. An inter-satellite handover algorithm based on the position and signal strength of the active user terminal in[27] was presented to maximize the user throughput. By measuring the transmission delay and Doppler shift of user terminal, the network can estimate and measure the position of user terminal during the call process, so as to reserve channel resources for the user.

In practical applications, the non-priority handover strategy, queuing priority strategy, and channel reservation strategy can be selected according to the actual situation, or multiple strategies can be utilized simultaneously. In general, the queuing priority strategy has a better performance in terms of the blocking rate and the drop rate for LEO satellite communication networks[26]. Nevertheless, the comparative analysis of these strategies and other QoS indicators and system capacity needs to be further studied.

9.3 AIRBORNE COMMUNICATIONS AND NETWORKING

Airborne communication systems utilize various aircraft equipped with transceivers and sensors, to build communication access platforms[30]. These aircraft mainly include UAVs, airships, and balloons, making up the low-altitude platforms (LAPs) and high-altitude platform (HAPs). Compared with ground communication systems, airborne communication systems can be flexibly deployed in a cost-effective manner, irrespective of terrain. Compared with satellite communication systems, airborne communication systems have much shorter range LoS links, resulting in lower latency and less propagation loss. Therefore, airborne communication is a key part of space/air/ground communications. Enabled by antenna array, the system can obtain new benefits. For instance, antenna array provides considerable beam gains to compensate propagation loss through directional transmission, which improves the channel quality. Besides, the directional transmission is beneficial to the reuse of spectrum resource in the spatial domain. At the same time, antenna array enabled airborne communication systems have varieties of distinct characteristics and challenges in both communications and networking. Fig. 9.1 illustrates the typical scenarios for the airbone systems, where varieties of aircraft form an aerial ad hoc network and accomplish missions collaboratively.

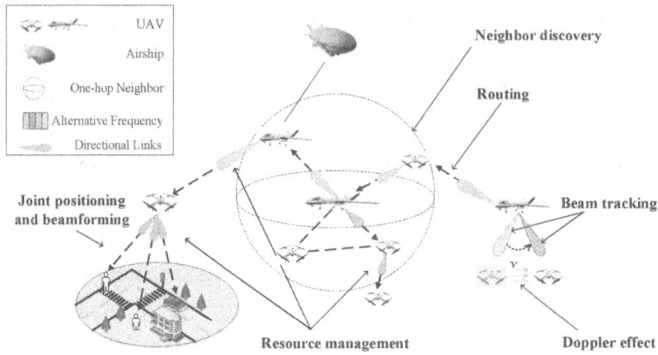

Figure 9.1 Illustration of the typical scenarios for antenna array enabled airborne communications and networking.

9.3.1 Beam Tracking

Compared with terrestrial systems, 3D mobility with very high dynamic is one of the typical characters for airborne systems, especially for large-scale UAVs. Due to the mobility, the directional transmission mechanism enabled by antenna array brings serious beam misalignment problem, which leads to the deterioration of communication performance or even the interruption of connection. To maintain the beam alignment, typical beam tracking scheme for the conventional terrestrial systems is to train the beam direction periodically through transmitting pilot signals. However, the high mobility as well as platform constraints (e.g., SWAP) of airborne system may result in unacceptable burden of pilot transmission, challenging the training-based beam tracking scheme from being applied to that. Therefore, beam tracking is practically important while quite a challenging problem for antenna array enabled airborne communication system. A distinct property for air-to-ground (A2G)/air-to-air (A2A)/air-to-satellite (A2S) communications is that their channels are dominated by line of sight (LoS) paths. By exploiting this property, mobility prediction-based beam tracking schemes are more efficient when LoS paths exist between airborne platforms and other platforms. The angular velocity estimation, and angular domain information, i.e., elevation angle and azimuth angle can be utilized to save pilot overhead, rapidly establish and reliably maintain communication links for A2G communications[31,32] and A2A communications[33]. Airborne sensors which provide movement state information, like global positioning system (GPS) and micro inertial measurement unit, can assist the coarse beam alignment[34]. Besides, machine learning-based schemes, such as Q-learning[35], long short-term memory recurrent neural network[36] were exploited to assist beam tracking, for their ability to predict beam alignment based on sequential beam tracking experience. In summary, the high dynamic of aircraft makes the beam tracking a challenging problem for antenna array enabled airborne communication systems. Both efficiency and overhead need to be balanced when designing solutions. Available information and new techniques that help predict the mobility of airborne platforms can be exploited to assist beam tracking and help reduce the overhead.

9.3.2 Doppler Effect

For airborne communication systems, an inevitable problem is the Doppler Effect, which can introduce carrier frequency offset and inter-carrier interference. It is known that the Doppler shift f_D of a received signal is a function of carrier frequency f_c, relative velocity v, angle of arrival (AoA) θ, and angle of the relative velocity θ_v, expressed as $f_D = (v/c) f_c (\theta - \theta_v)$, where c is the speed of light. If , MPCs arrive at the Rx with different AoAs, e.g., with large angular dispersion, the resulting different Doppler shifts will produce spectral broadening, called Doppler spread[37]. In high dynamic airborne systems, one might initially think that Doppler spread would be high and cause catastrophic effects on communications. In fact, as airborne platforms operate in high altitudes, the channels are mainly dominated by LoS paths. MPCs are expected to have very similar Doppler shifts with relatively small angular spread. This is especially true for high carrier frequency, such as millimeter-wave frequency[38]. Besides, directional transmission enabled by antenna array will further reduce the number of MPCs and in turn reduce the angular spread[39, 40]. Large Doppler shift with small Doppler spread can be well mitigated by frequency synchronization. Doppler power spectrum is an important statistical property to characterize the Doppler spread, which expresses the power spectral density of the received signals as a function of the Doppler shift[40]. As a result, Doppler power spectrum were derived and analyzed in many studies on wideband non-stationary A2A/A2G channel model[41, 42, 43, 44]. It was found that the UAV rotations significantly affect channel correlations[44]. To reduce the Doppler effect, it is necessary to perform Doppler frequency shift (DFS) estimation and compensation. The idea of coarse estimation plus fine estimation may be favorable to achieve a fast and accurate DFS estimation and compensation[45]. In a word, the high mobility of airborne platforms, the use of higher carrier frequency and the directional transmission make the Doppler effect of airborne communications different from that of conventional terrestrial communications. It is worth the research effort to model this property and compensate the effect for multiple airborne communication scenarios.

9.3.3 Joint Positioning and Beamforming

With 3D mobility, airborne platforms can design their positions or trajectories according to the mission to improve communication performance. Enabled by antenna array, beamforming can be designed not only to improve the received signal power but also to mitigate mutual interference. Therefore, there is a great deal of freedom for antenna array enabled airborne communication systems for jointly positioning (also trajectory) and beamforming design. However, the joint design is challenging. Different from the positioning of an aircraft with single antenna, positioning and directional beamforming are coupled for antenna array enabled airborne systems. The channel state among the Tx and Rx may change according to the aircraft's position and posture. Because of the coupling variables, the optimization problem is non-convex and difficult to solve. Moreover, the positioning design of multiple airborne platforms is more tricky, since interference between different terminals needs to be properly considered. To solve the challenging joint positioning and beamforming problem, a feasible solution is the iterative algorithm, where beamforming and trajectory are alternately optimized[46]. Specifically, in each iteration, the trajectory is optimized by fixing beamforming direction, and then beamforming is optimized with fixed trajectory. Alternatively, the ideal beam pattern was introduced and the joint optimization problems were solved in two steps[47,48]. The ideal beam pattern states that the summation of the beam gains in different directions is approximately equal to the number of antennas[49]. After substituting the ideal beam gain, a more tractable joint deployment and beam gain allocation problem is obtained, followed by approaching the ideal beam pattern through multi-beam forming techniques. In addition, by applying a modified cosine antenna pattern approximation of uniform linear array (ULA), the UAV trajectory and directional beamforming can be jointly optimized in a single convex optimization problem[50]. Besides, to ensure a robust joint trajectory and transmit beamforming design, practical considerations such as UAV jittering, user location uncertainty, wind speed uncertainty, and no-fly zones are necessary[51]. In summary, to give full play to the unique advantages of the antenna array enabled airborne communication systems, flexible positioning needs to be simultaneously designed with effective beamforming. As an appealing and challenging research direction, it is worth the effort for exploiting both optimization strategy and practical communication scenario.

On top of joint positioning and beamforming, joint deployment and beamforming should be considered, since the deployment and beamforming are highly coupled as well. A distinct superiority of airborne platforms is on-demand deployment. Moreover, a distinct superiority of antenna array is beamforming. Thus, antenna array enabled airborne communication systems have a great degree of freedom (DoF) to perform joint deployment and beamforming design to improve communication performance. However, the deployment and beamforming problem should be jointly considered. The channels are affected by different positions of airborne platforms. Moreover, when considering the dynamic scenario such as the movement of users, the design is more challenging due to the trajectory and beamforming. Besides, practical factors such as aircraft jittering may cause beam misalignment thus deteriorating the communication link quality. As a result, the robust joint deployment and beamforming design for airborne communication networks, which concentrates on both optimization strategy and practical scenario, is an appealing future research direction.

9.3.4 Antenna Array Enabled Aerial Ad Hoc Network

Aerial ad hoc networks refer to multi-aircraft systems organized in an ad hoc fashion, aiming to accomplish complex missions cooperatively. Compared to single aircraft aerial system, aerial ad hoc networks are more flexible, reliable and survivable through redundancy. Therefore, aerial ad hoc networks have broad military, civilian, and commercial applications such as remote sensing, traffic monitoring, border surveillance, and relay networks[52]. At the same time, aerial ad hoc networks have distinct characteristics such as a high level of network heterogeneity, highly dynamic, frequently changed network topologies, weakly connected communication links, and vulnerable to jamming and eavesdropping[30]. Directional communication enabled by antenna array provides significant performance gain for aerial ad hoc networks. By focusing electromagnetic energy only in the intended direction, antenna array can enlarge transmission distance for a given power level, which improves network connectivity. On the other hand, directional beams increase spatial reuse, which allows more simultaneous transmissions and enhances anti-jamming/eavesdropping abilities, thus providing higher network capacity and security[53]. Nevertheless, these benefits are accompanied by certain unique challenges. Mechanisms that were designed for terrestrial ad hoc networks or with omnidirectional communications need to be redesigned for the antenna array enabled aerial ad hoc networks. We provide an overview on the important issues and potential solutions, mainly about neighbor discovery, routing, and resource management.

Neighbor discovery, also known as routing discovery, refers to the process of finding one-hop neighbors, which is a crucial initial step for establishing connections among the nodes[54]. For omnidirectional antenna enabled networks, simple broadcast can reach all neighbors. The problem is more challenging for antenna array enabled networks, since nodes need to determine when and where to point their directional beams simultaneously to discover each other. A natural approach to contour the challenge is to use omnidirectional antenna in the neighbor discovery process[55,56]. For example, a dual-antenna collaborative communication strategy was presented in[55] for aerial ad hoc networks, where neighbor discovery is based on low-frequency heartbeat location information piggybacked on control frames enabled by omnidirectional antenna. The main drawback of this approach is that an additional omnidirectional antenna is required. Following the similar idea, an antenna array can work in a quasi-omnidirectional manner by omnidirectional beamforming to perform neighbor discovery[57,58]. However, wider beam means lower beamforming gain, thus causing shorter discovery range. Without synchronization and any available information, probabilistic approach can be performed, where each node randomly chooses a direction to steer its beam. Obviously, this approach lacks performance guarantee in terms of discovery delay[59]. With time synchronization among nodes, e.g., with satellite positioning system as common clock source, deterministic approach can be developed, where the beam of each node is steered based on a predefined sequence. For example, the antenna scans its beam clockwise to perform neighbor discovery in[53,57]. In this case, neighbors can be discovered within one cycle with a high probability. With partial prior information available, such as the location of other nodes piggybacked through routing updates[53] or the location/motion prediction[55], neighbor discovery may be performed more rapidly and achieve fast convergence, known as informed discovery[53].

After neighbor discovery, an aerial ad hoc network requires mechanisms for discovering routes and forwarding packets along these routes. Routing plays the role, and has a major impact on network throughput and packet delay. Compared to conventional ad hoc networks, the 3D high mobility of aircraft brings intermittent connections and frequent topology changes for aerial ad hoc networks, which need to be emphatically considered during routing design. The routing schemes for aerial ad hoc network can be categorized into topology-based[60, 61, 62], geographic/location-based[63, 64], and bio-inspired[65, 66, 67]. Topology-based routing requires to obtain the routing path before data transmission begins, which has high transmission efficiency but may cause high overhead for routing discovery and maintenance. Geographic routing utilizes geographic positions of the aircraft for routing decisions, which requires hardware installations of aircraft. Bio-inspired routing is inspired by collective behavior of biological systems, such as the honey collection in a bee colony, or food finding in an ant colony. Since there is no significant difference from routing in omni-directional aerial ad hoc networks. Routing schemes for antenna array enabled aerial ad hoc networks can draw lessons from that designed for omnidirectional aerial ad hoc networks[64, 68].

To encourage the quality of communication in a network, there is a need for a framework to dynamically manage various resources including time domain, frequency domain, power domain, space domain, and so on[69]. Therefore, resource management plays a key role in aerial ad hoc networks. Typically, resource management includes spectrum management, task assignment, interference management, power control, and so on. The goal for spectrum management is to improve spectrum utilization as well as to reduce mutual interference, ensuring efficient and robust wireless communication for a network[70]. Control-data separation architecture may achieve both stable and high-rate communication for aerial ad hoc networks[55]. Specifically, lower frequency was utilized for one omnidirectional antenna enabled control channel, ensuring stable control frames transmission. Higher frequencies were utilized for directional antenna enabled data channels, enabling broadband data transmission. Oppositely, control-data sharing the same bandwidth may achieve higher spectrum utilization but also a potential interference problem[71]. Due to the directional transmission characteristic and platform restriction, particular attention should be paid to resources in space domain and power domain for antenna array enabled aerial ad hoc networks. Benefiting from antenna array, narrow beams increase spatial reuse, and thus enable more simultaneous transmissions and decrease mutual inferences[53, 72]. Efficient beam management scheme is necessary to guarantee network performance. At the same time, beam misalignment problem should be addressed considering the high mobility of aircraft. Besides, the onboard energy of aircraft, especially for small UAVs, is usually limited. Thus, energy-efficient operations such as transmit power control, load balancing and node sleep are essential for aerial ad hoc networks[71].

The directional transmission of antenna array and high dynamics of airborne platforms have brought new challenges for both resource management and routing. Besides, as a matter of fact, the resource management in physical and media access control (MAC) layers and the routing in network layer are highly coupled. Thus, the joint design of resource management and routing for airborne ad hoc networks is necessary and challenging. According to resource management, the multiple dimensional resources such as time slot, spectrum, spatial beam, and power should be carefully managed according to the communication tasks. In addition, due to the high dynamics of aircraft, the airborne ad hoc network's topology is rapidly changing, resulting in not only the change of routing paths but also time-varying available communication resources. To enhance the overall system performance when facing multiple concurrent tasks, it is promising to perform real-time cross layer optimization to allocate the resources in an active manner and update the routing paths according to the network state. Hence, the joint resource management and routing problem should be considered.

In summary, the high-dynamics of aircraft and directional transmission bring unique challenges on antenna array enabled airborne ad hoc networks. Prior information and geographic positions are helpful, and can be exploited to facilitate the process of neighbor discovery and routing. Bio-inspired routing scheme is a promising routing solution worth exploring. Besides, directional transmission and platform constraints bring more considerations regarding resource management in space domain and power domain.

9.4 GROUND CELLULAR COMMUNICATIONS

Massive antenna array technology has been widely used in the fifth generation (5G) communication systems nowadays, such as beamforming technology based on antenna array[73,74,75]. Beyond 5G (B5G) and the sixth generation (6G) communications need to address more challenges on high data rate, low latency, massive connectivity, seamless coverage and high-speed mobility. Antenna array will be one of the key technologies to support ground communications by providing high beamforming gain and multiplexing of users. We will introduce the applications of antenna array in ground communications in details as follows.

Nevertheless, there are some practical considerations for ground communications as well. As the number of antenna elements arises, hardware cost will be a challenging problem for MIMO communication system. It is urgent to improve the traffic capacity and reduce the cost at the same time. Besides, channel state information (CSI) estimation is one of the major challenges in large-scale antenna array enabled communication system. How to design low-complexity pilot training in MIMO system to achieve channel estimation is an important research topic. As the number of access users increases, inter-cell and intra-cell interferences become much severer, especially in dense urban areas, and effective interference management methods are needed. Moreover, beam handover and beam tracking methods are supposed to be used in hotspot areas. For the V2X network, various problems and challenges have been presented in such highly dynamic vehicular communications. Machine learning may be a potential candidate in the handover process design.

9.4.1 Cellular Massive MIMO

In order to deal with massive connectivity and provide better service for users, cellular networks with smaller cells compared to 4G are widely used in 5G communication networks. The combination of millimeter-wave and large-scale antenna array brings new solutions for high throughput. Nevertheless, it brings new challenges at the same time. To overcome high path loss and blockage of high-frequency band signals, an effective approach is that dividing smaller cells to provide better user QoS by getting the Txs and Rxs closer. The small cells, which are defined as low-power wireless access points (APs) operated in licensed spectrum, can provide improved cellular coverage, capacity and applications for homes, enterprises and other connectivity[76], compensate millimeter-wave pass loss and contribute to seamless coverage.

Although multi-cell systems can provide better performance for users, they may suffer severe inter-cell interference caused by frequency reuse, especially for cell-edge users. Inference management and elimination is one of the most significant challenges for multi-cell transmissions, which needs the cooperation among base stations (BSs) in different cells. The coordinated beamforming designed for massive MIMO multi-cell networks, where BSs are equipped with a large antenna array, has attracted great concern to achieve interference suppression. There are two important downlink multi-cell interference mitigation techniques, i.e., large-scale MIMO (LS-MIMO) and network MIMO[77]. In a LS-MIMO system, BSs equipped with multiple antennas not only serve their scheduled users, but also null out interference caused to other users within cooperating cluster using zero-forcing (ZF) beamforming. In a network MIMO system, BSs eliminate interference through data and CSI exchange over the backhaul links and joint transmission using ZF beamforming. It was proved that LS-MIMO can be the preferred approach for multi-cell interference mitigation in wireless networks. To improve the throughput of cell-edge users, two interference alignments, termed interfering channel alignment based coordinated beamforming and interference alignment based coordinated beamforming, can be used[78]. Two BSs jointly optimize their beamforming to improve the data rates of cell-edge users without data sharing between two cells.

As the number of antennas increases, one of the immediate problems is that the spatial limitations at the top of BS tower limit the use of massive linear antenna array. For example, the length of 64 half-wave antennas in linear array paradigm will reach 4 m at the carrier frequency of 2.4 GHz. Hence, it is crucial to limit massive antenna array in a smaller form factor. To overcome this problem, full dimension MIMO that utilizes a large number of antennas placed in a 2D antenna array at BSs has attracted substantial research attention from both wireless industry and academia in the past few years[79]. It is defined in 3rd Generation Partnership Project (3GPP) and is considered as a critical technology for 5G cellular systems to improve network capacity as it allows cellular systems to support a large number of users by using multi-user MIMO technology. It allows the extension of spatial separation to elevation domain as well as traditional azimuth domain as shown in Fig. 9.2, which can reduce the form factor of antenna array at the same time[80]. Both azimuth and elevation angles of the downlink beams can be steered dynamically[79,81], which exploit full DoFs. Benefiting from the additional DoF of full dimension MIMO, flexible 3D beamforming can be employed in BSs to achieve effective interference coordination in cellular networks[82,83]. Nevertheless, both the works in[82] and[83] utilize statistical CSI in order to reduce the feedback overhead of channel estimation. How to obtain instantaneous CSI is a great challenge for largescale antenna array. Fortunately, the use of RSMA in largescale antenna array and massive MIMO systems has been shown to boost the performance over conventional massive MIMO in the presence of imperfect CSI due to frequency-division duplex (FDD) quantization[84,85], time-division duplex

(TDD) pilot contamination[86], phase noise and hardware impairments[87], or due to mobility and latency[88].

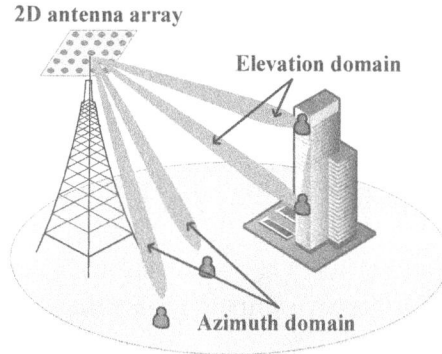

Figure 9.2   Full dimension MIMO.

Although massive MIMO further improves spectral efficient and link reliability, it comes at the cost of significantly increased computational complexity compared to small-scale MIMO systems. In particular, uplink signal detection becomes inefficient and has high complexity because of the large increase of dimensions caused by massive antennas. Conventional optimal method, such as maximum-likelihood detection, is not suitable anymore for high complexity. Massive MIMO systems at BSs requires novel detection algorithms that fit for high-dimensional problems with low complexity [89,90]. There has been several reduced-complexity linear minimum mean square error (LMMSE)-based detectors presented, but still require high hardware complexity and power consumption as the number of transmit antenna increases[91] or the number of users increases[92]. An iterative data detection algorithm based on the coordinate descent method can be used to further reduce complexity[90], which is able to achieve the same or even higher bit error rate performance compared with the classical LMMSE algorithm. To reduce the signal processing pressure at BSs, distributed algorithm, where BS antennas are divided into different clusters and each cluster has independent computing hardware is an effective method[93].

Except for data detection, interference management is quite important for uplink transmission in multi-cell MIMO systems, where a large number of small cells result in severe uplink interference for pilot reuse in channel estimation. In[94], the simulation results showed that higher level of pilot reuse results in lower achievable sum spectral efficiency and an uplink detector was developed to suppress both intra-cell and inter-cell interference based on minimum mean square error (MMSE). In[95], the authors improved uplink performance in massive MIMO macrocells through uplink power control and cell range extension in a two-tier massive heterogeneous cellular network, which ingrates both cellular network and massive MIMO. In multi-cell systems, the cooperation of BSs that regards BSs as distributed antennas is common and effective method to achieve interference elimination, but requires a large amount of CSI between BSs and users among cooperating cells. Interference suppression approach that does not require cell cooperation is convenient and the novel semi-blind uplink interference suppression method for multi-cell multi-user massive

MIMO systems in[96] is confirmed to be the most effective solution evolving spectral use for future wireless networks.

However, most presented data detection and interference management methods depend on perfect CSI at BSs, which is impractical. Hence, how to design low-complexity data detector at BSs to reduce power consumption and design effective pilot to suppress interference under imperfect CSI is quite important in uplink networks and needs further study.

9.4.2 Cell-Free MIMO

5G cellular communication networks can provide much higher peak data rates and traffic throughput and lower latency compared to previous cellular technologies. However, this outstanding performance can only be achieved by the users nearby BSs. For severe inter-cell interference, the experience performance of edge-users can be much worse. In a conventional cellular network, each user is connected to the BS in one of the cells and the BSs have multiple active users to serve at a certain time, which causes inter-cell interference inevitably[97]. All the service antennas are located in a compact area and have low backhaul requirements.

In contrast, in a cell-free network, there are a large number of distributed antennas, called APs, that serve a much smaller number of users over the same time/frequency resources[98]. "Cell-free" signifies that there are no cell boundaries during data downlink transmissions from the user perspective. An AP will cooperate with different sets of APs when serving different users. The conventional cellular network is shown in Fig. 9.3(left) and cell-free network is shown in Fig. 9.3(right). It is users that select the set of APs that can provide the best service for itself, instead of the network. Namely, cell-free network is a user-centric paradigm[99].

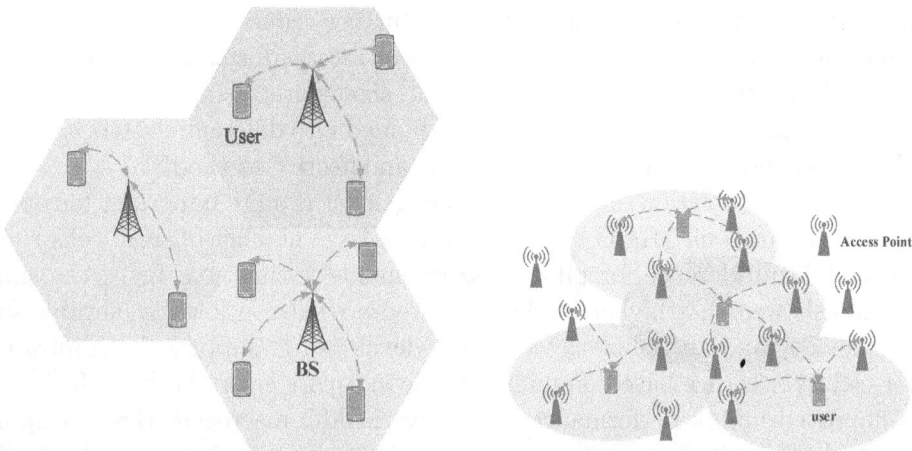

Figure 9.3 **Left**: Conventional Cellular Network. **Right**: Cell-free Network

The APs are connected via fronthaul to central processing units that are responsible for the coordination and are seen as the enabler of cell-free massive MIMO[97]. Each AP uses local channel estimation based on received uplink user pilot and applies conjugate beamforming to transmit data to users[100]. Although conjugate beamforming only requires local CSI, its design is based on a large-scale nonconvex problem with very high computational complexity. Hence, conjugate beamforming that admits a low-scale optimization formulation for computational tractability is required[101]. The presented method can improve both the Shannon function rate and ultra-reliable and low-latency communication (URLLC).

Most of the traffic congestion happens at the cell edges in cellular networks so that user-experienced performance is poor. The purpose of the cell-free paradigm is not to achieve high peak performance, but to provide more uniform performance. It is proved that the cell-free massive MIMO significantly outperforms small-cell in both median and 95%-likely performance. What is worth noting is that the cell-free massive MIMO system can provide almost 20-fold increase in 95%-likely per-user throughput compared with small-cell system[98]. Moreover, the simulation results in[102] shown that the 95%-likely per-user throughput of cell-free system can be further improved through increasing antenna number.

One of the main challenges to design cell-free massive MIMO is how to achieve a network that is scalable in the sense of being implementable in a large network. Specifically, how to achieve the benefits of cell-free massive MIMO in a practicable way under high computational complexity and fronthaul capacity requirements should be considered. Motivated by this purpose, the framework for scalable cell-free systems should be developed and the method to make the network scalable is needed[103]. Although the presented method in[103] is nearly optimal, power allocation for centralized and distributed operation was not considered. Power control is very important to protect users from strong interference. There have been many heuristic power allocation schemes[104, 105], how to perform effective and scalable power allocation in cell-free systems still needs further study.

9.4.3 V2X Communication

Autonomous driving has been an innovative technology for future intelligent transport systems, where V2X communications can enhance the safety and efficiency[106], including vehicle-to-vehicle (V2V), vehicle-to-infrastructure (V2I), vehicle-to-pedestrain (V2P) and vehicle-to-network (V2N). Future 5G cellular systems will support vehicular networks and high data transmission rates among fully connected vehicles, where vehicles will be equipped with more sensors and generate gigabit data per second. Besides, 5G is supposed to support high-speed terminals such as highspeed trains. The V2X network is shown in Fig. 9.4. Massive connectivity, explosive data, low latency, high-speed terminals, frequent handover and user infotainment bring great challenges for 5G vehicular communications.

Millimeter-wave technology with large-scale antenna array will play a significant role in vehicular networks by providing high data transmission rates. It is proved that millimeter-wave massive MIMO can deliver Gbps data rates for next-generation vehicular networks[107]. However, it has high requirements for LoS. If transmission link is blocked, link quality will be greatly reduced. It is practical to model a theoretical highway communication model, where vehicles are served by BSs alongside the road

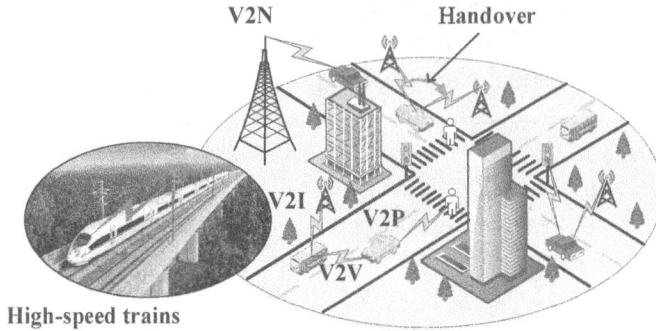

Figure 9.4 V2X network.

and blockage was particularly considered[108]. In the presented practical application scenario, heavy vehicles in low lanes may obstruct the LoS paths between vehicles in fast lanes and BSs, through which analyzing how blockage densities impact the user achievable data rate can be studied.

One of the challenges for vehicular networks is that high-speed terminals will make widely-adopted technology used in static scenarios or low-speed scenarios inefficient due to Doppler spread. With frequent handover and rapidly changing CSI, fast beam alignment techniques should be developed to meet vehicle's mobility requirements. However, frequent beam sweeping will suffer large overhead and is ineffective in high mobility environment. It is necessary to design a beam alignment scheme for millimeter-wave V2V communication between neighbor vehicles on highway with high speed without any searching steps in beam training[109]. The presented beam alignment method can provide significant throughput improvements compared to general car-following scenarios on the high-way.

What is more remarkable is that for high speed trains with speed of over 400 km/h in the future, providing services satisfying traffic demand for numerous passengers is a great challenge. Compared to highway scenarios, high-speed trains run at higher speeds, which can lead to more frequent handovers and severe inter-carrier interference due to Doppler frequency spread. Therefore many existing beam tracking methods fail to apply to high-speed trains. In term of this issue, dynamic beam tracking strategy for millimeter-wave high-speed railway communications that can adjust the beam direction and beam width jointly should be applied[110]. Doppler frequency offset can be compensated through beam alignment and data transmission can be realized through hybrid beamforming for high-speed train communications[111]. Undoubtedly, massive antenna array will play an important role in such a scenario.

Much existing research on beam tracking is based on known vehicle positions and statistics CSI. Effective channel estimation is necessary for such a dynamic scenario. Wide beam design may be a feasible method against high-speed mobility. Besides, millimeter-wave signal is sensitive to LoS path. To provide strong connection between BSs and vehicles, IRSs can be intelligently applied.

9.5 SUMMARY

In this chapter, we gave an overview on antenna array enabled space/air/ground communication network. In the first section, i.e., LEO satellite communication, we introduced several unique features to the LEO satellite communications, i.e., beam coverage, beamforming, beam management and handover. For beam coverage, we introduced the wide beam, spot beam and hybrid wide-spot beam and analyzed their features and application scenarios. Wide beams which have a wider coverage and smaller antenna gains, are more suitable for control signals and broadcasting signals. Spot beams are more suitable for providing high-speed data services. As for hybrid wide-spot beam, it is the combination of wide beams and spot beams. In beamforming, we introduced multiple beam array, i.e., MBA, which is able to generate multiple beams with different directions simultaneously. MBA can be reflector-based architectures, phased array architectures and lens-based architectures. And as for beam management and handover, the non-priority handover, queuing priority handover and reserved channel strategies are introduced. In practice, the handover technologies can be selected according to the actual situation, or multiple strategies can be utilized simultaneously.

The second section is airborne communications and networking. In this section, from the perspective of communications, we focused on beam tracking, Doppler effects, and joint positioning and beamforming. From the perspective of networking, aerial ad hoc network was addressed. The beam tracking is to maintain the beam alignment, which may result in unacceptable burden of pilot transmission in conventional beam tracking method. Doppler effect is an inevitable problem for airborne communication, which requires more research efforts to model this property and compensate the effect for multiple airborne communication scenarios. And to give full play to the unique advantages of the antenna array enabled airborne communication systems, flexible positioning needs to be simultaneously designed with effective beamforming, i.e., joint positioning and beamforming. As for aerial ad hoc network, we gave an overview on neighbor discovery, routing, resource management and joint resource management and routing. Prior information and geographic positions can be exploited to facilitate the process of neighbor discovery and routing. Besides, directional transmission and platform constraints bring more considerations regarding resource management in space domain and power domain.

In the last section, we mainly introduced three applications of antenna array in ground cellular communication, i.e., cellular massive MIMO, cell-free MIMO and V2X communication. In cellular massive MIMO, users may suffer severe inter-cell interference. The coordinated beamforming is a method to achieve interference suppression. In downlink scenarios, there are two interference mitigation techniques, i.e., LS-MIMO and network MIMO. As for the spatial limitations at the top of BS tower, full dimension MIMO has attracted substantial research attention. To reduce the significantly increased computational complexity of massive MIMO, LMMSE-based detectors, iterative data detection algorithm based on the coordinate descent method and distributed algorithms is used. In cell-free MIMO, it is users that select the set of APs that can provide the best service for itself, which is a user-centric paradigm. And the purpose of the cell-free paradigm is not to achieve high peak performance, but to provide more uniform performance. In V2X communication, we analyzed two scenarios, i.e., highway communication and high speed trains, and addressed their issues. In the highway scenario, it is practical to model a theoretical highway communication model, which can be used to analyze the blockage. In addition, the Doppler spread caused by high-speed terminals and beam alignment method were introduced. As for the high-speed train scenario, dynamic beam tracking strategy are applied to compensate Doppler frequency offset. Moreover, data transmission can be realized through hybrid beamforming.

Bibliography

[1] Yongtao Su, Yaoqi Liu, Yiqing Zhou, Jinhong Yuan, Huan Cao, and Jinglin Shi. Broadband LEO satellite communications: Architectures and key technologies. *IEEE Wireless Commun.*, 26(2):55–61, Apr. 2019.

[2] Dau-Chyrh Chang, Po-Wei Cheng, Chih-Hung Lee, and Chia-Tsung Wu. Broadband wide-beam circular polarization antenna for global navigation satellite systems application. In *Proc. Asia-Pacific Symposium on Electromagnetic Compatibility*, Taipei, Taiwan, China, May 2015.

[3] Ahmed Khidre, Fan Yang, and Atef Z. Elsherbeni. Reconfigurable microstrip antenna with tunable radiation beamwidth. In *Proc. IEEE Antennas and Propagation Society Int. Symposium*, Orlando, FL, Jul. 2013.

[4] Muhammad Saeed Khan, Antonio-Daniele Capobianco, Sajid Mehmood Asif, Adnan Iftikhar, Benjamin D. Braaten, and Raed M. Shubair. A pattern reconfigurable printed patch antenna. In *Proc. IEEE Int. Symposium on Antennas and Propagation*, Fajardo, PR, Jun. 2016.

[5] Xiu-Yin Zhang, Di Xue, Liang-Hua Ye, Yong-Mei Pan, and Yao Zhang. Compact dual-band dual-polarized interleaved two-beam array with stable radiation pattern based on filtering elements. *IEEE Trans. Antennas Propagat.*, 65(9):4566–4575, Sep. 2017.

[6] Wanchen Yang, Lizheng Gu, Wenquan Che, Qian Meng, Quan Xue, and Cao Wan. A novel steerable dual-beam metasurface antenna based on controllable feeding mechanism. *IEEE Trans. Antennas Propagat.*, 67(2):784–793, Feb. 2019.

[7] Cen Qian, Sihai Zhang, and Wuyang Zhou. Traffic-based dynamic beam coverage adjustment in satellite mobile communication. In *Proc. Int. Conf. Wireless Commun. and Signal Process.*, Hefei, China, Oct. 2014.

[8] Puneeth Jubba Honnaiah, Nicola Maturo, Symeon Chatzinotas, Steven Kisseleff, and Jens Krause. Demand-based adaptive multi-beam pattern and footprint planning for high throughput GEO satellite systems. *IEEE Open J. Commun. Society*, 2:1526–1540, Jul. 2021.

[9] Haotian Zhou, Liang Liu, and Huadong Ma. Coverage and capacity analysis of LEO satellite network supporting internet of things. In *Proc. IEEE Int. Conf. Commun.*, Shanghai, China, May 2019.

[10] JiXiang Wan, ShaoPeng Lu, XuDong Wang, and YongQiang Ai. A steerable spot beam reflector antenna for geostationary satellites. *IEEE Antennas Wireless Propagat. Lett.*, 15:89–92, May 2015.

[11] Long Zhang, Wei-Bing Zhang, Jin-Wen Shi, Qi Gong, and Xiu-Ji Chen. Research of beam optimization for multiple feeds per beam (MFB) antennas based on genetic algorithm. In *Proc. IEEE Asia-Pacific Microwave Conf.*, Singapore, Dec. 2019.

[12] Seong-Mo Moon, Sohyeun Yun, In-Bok Yom, and Han Lim Lee. Phased array shaped-beam satellite antenna with boosted-beam control. *IEEE Trans. Antennas Propagat.*, 67(12):7633–7636, Dec. 2019.

[13] José M. Montero, Ana M. Ocampo, and Nelson Jorge G. Fonseca. C-band multiple beam antennas for communication satellites. *IEEE Trans. Antennas Propagat.*, 63(4):1263–1275, Apr. 2015.

[14] Qinghua Lai, Chu Gao, Tianjie Peng, Zhenhua Liu, Xiaotao Wang, Handong Wu, and Junmei Ma. A digital beam-forming multiple-beam reflector antenna subsystem for GEO communication satellites. In *Proc. European Microwave Conf.*, London, UK, 2016.

[15] Garmy Sow, Olivier Besson, Marie-Laure Boucheret, and Cécile Guiraud. Beamforming for satellite communications in emergency situations. *Eur. Trans. Telecomm.*, 19(2):161–171, Jan. 2008.

[16] Julien Montesinos, Olivier Besson, and C. Larue De Tournemine. Adaptive beamforming for large arrays in satellite communications systems with dispersed coverage. *IET Communications*, 5(3):350–361, Feb. 2011.

[17] Dunmin Zheng and Santanu Dutta. Adaptive beamforming for mobile satellite systems based on user location/waveform. In *Proc. IEEE Veh. Technol. Conf.*, Honolulu, HI, Sep. 2019.

[18] Jihyung Kim, Mi Young Yun, Dukhyun You, and Moon-Sik Lee. Beam management for 5G satellite systems based on NR. In *Proc. Int. Conf. on Inf. Networking*, 2020.

[19] Enrico Del Re, Romano Fantacci, and Giovanni Giambene. Different queuing policies for handover requests in low earth orbit mobile satellite systems. *IEEE Trans. Veh. Technol.*, 48(2):448–458, Mar. 1999.

[20] Mona M. Riad and Mohamed M. Elsokkary. Fixed channel allocation handover strategies in LEO satellite systems. In *Proc. National Radio Science Conf.*, volume 2, Mansoura, Egypt, Mar. 2001.

[21] Evangelos Papapetrou and Foteini-Niovi Pavlidou. Analytic study of Doppler-based handover management in LEO satellite systems. *IEEE Trans. Aerosp. Electron. Syst.*, 41(3):830–839, Jul. 2005.

[22] M. Zhao, Y. Jiang, and G. X. Li. New channel assignment strategy based on DPRQ in LEO system. *Journal of System Simulation*, 21(13):4038–4027, Mar. 2009.

[23] Dong Yan. Spotbeam handover scheme for LEO satellite mobile communication systems by using utility function. *Journal of Huazhong University of Science and Technology (Nature Science Edition)*, 36(5):5–8, May 2008.

[24] Stylianos Karapantazis and Foteini-Niovi Pavlidou. Design issues and QoS handover management for broadband LEO satellite systems. *IEE Proceedings - Communications*, 152(6):1006–1014, Dec. 2005.

[25] Lila Boukhatem, André-Luc Beylot, Dominique Gaiti, and Guy Pujolle. TCRA : a resource reservation scheme for handover issue in LEO satellite systems. In *Proc. IEEE Wireless Commun. Networking Conf.*, New Orleans, Louisiana, Mar. 2003.

[26] Neh Hedjazi, Malika Ouacifi, Rachida Bouchouareb, Meriem Ourghi, Messaoud Gareh, and Djamel Benatia. The handover in the constellations of satellites in low orbit. *Int. Journal of Advanced Science and Technology*, 41(11):19–24, May 1972.

[27] Stephan Olariu, Rajendra Shirhatti, and Albert Y. Zomaya. OSCAR - an opportunistic call admission protocol for LEO satellite networks. In *Proc. Int. Conf. on Parallel Processing*, Montreal, Canada, 2004.

[28] A. Bottcher and R. Werner. Strategies for handover control in low earth orbit satellite systems. In *Proc. IEEE Veh. Technol. Conf.*, Stockholm, Sweden, Jun. 1994.

[29] W. Zhao, R. Tafazolli, and B. G. Evans. Combined handover algorithm for dynamic satellite constellations. *Electronics Letters*, 32(7):622–624, Mar. 1996.

[30] Xianbin Cao, Peng Yang, Mohamed Alzenad, Xing Xi, Dapeng Wu, and Halim Yanikomeroglu. Airborne communication networks: A survey. *IEEE J. Select. Areas Commun.*, 36(9):1907–1926, Sep. 2018.

[31] Lu Yang and Wei Zhang. Beam tracking and optimization for UAV communications. *IEEE Trans. Wireless Commun.*, 18(11):5367–5379, Nov. 2019.

[32] Yi Huang, Qingqing Wu, Ting Wang, Guohua Zhou, and Rui Zhang. 3D beam tracking for cellular-connected UAV. *IEEE Wireless Commun. Lett.*, 9(5):736–740, May 2020.

[33] Wenjun Xu, Yongning Ke, Chia-Han Lee, Hui Gao, Zhiyong Feng, and Ping Zhang. Data-driven beam management with angular domain information for mmWave UAV networks. *IEEE Trans. Wireless Commun.*, 2021 (Early access).

[34] Jianwei Zhao, Feifei Gao, Linling Kuang, Qihui Wu, and Weimin Jia. Channel tracking with flight control system for UAV mmWave MIMO communications. *IEEE Commun. Lett.*, 22(6):1224–1227, Jun. 2018.

[35] Hsiao-Lan Chiang, Kwang-Cheng Chen, Wolfgang Rave, Mostafa Khalili Marandi, and Gerhard Fettweis. Machine-learning beam tracking and weight optimization for mmWave multi-UAV links. *IEEE Trans. Wireless Commun.*, 2021 (Early access).

[36] Weijie Yuan, Chang Liu, Fan Liu, Shuangyang Li, and Derrick Wing Kwan Ng. Learning-based predictive beamforming for UAV communications with jittering. *IEEE Wireless Commun. Lett.*, 9(11):1970–1974, Nov. 2020.

[37] Wahab Khawaja, Ismail Guvenc, David W. Matolak, Uwe-Carsten Fiebig, and Nicolas Schneckenburger. A survey of air-to-ground propagation channel modeling for unmanned aerial vehicles. *IEEE Commun. Surveys Tuts.*, 21(3):2361–2391, Third quarter 2019.

[38] Sundeep Rangan, Theodore S. Rappaport, and Elza Erkip. Millimeter-wave cellular wireless networks: Potentials and challenges. *Proc. IEEE*, 102(3), Mar. 2014.

[39] Zhenyu Xiao, Pengfei Xia, and Xiang-Gen Xia. Enabling UAV cellular with millimeter-wave communication: potentials and approaches. *IEEE Commun. Mag.*, 54(5):66–73, May 2016.

[40] Javier Lorca, Mythri Hunukumbure, and Yue Wang. On overcoming the impact of Doppler spectrum in millimeter-wave V2I communications. In *Proc. IEEE Globecom Workshops*, Singapore, Dec. 2017.

[41] Zhangfeng Ma, Bo Ai, Ruisi He, Gongpu Wang, Yong Niu, and Zhangdui Zhong. A wideband non-stationary air-to-air channel model for UAV communications. *IEEE Trans. Veh. Technol.*, 69(2):1214–1226, Feb. 2020.

[42] Michael Walter, Dmitriy Shutin, David W. Matolak, Nicolas Schneckenburger, Thomas Wiedemann, and Armin Dammann. Analysis of non-stationary 3D air-to-air channels using the theory of algebraic curves. *IEEE Trans. Wireless Commun.*, 18(8):3767–3780, Aug. 2019.

[43] Hao Jiang, Zaichen Zhang, Cheng-Xiang Wang, Jiangfan Zhang, Jian Dang, Liang Wu, and Hongming Zhang. A novel 3D UAV channel model for A2G communication environments using AoD and AoA estimation algorithms. *IEEE Trans. Commun.*, 68(11):7232–7246, Nov. 2020.

[44] Zhangfeng Ma, Bo Ai, Ruisi He, Gongpu Wang, Yong Niu, Mi Yang, Junhong Wang, Yujian Li, and Zhangdui Zhong. Impact of UAV rotation on MIMO channel characterization for air-to-ground communication systems. *IEEE Trans. Veh. Technol.*, 69(11):12418–12431, Nov. 2020.

[45] Qixun Zhang, Huiqing Sun, Zhiyong Feng, Hui Gao, and Wei Li. Data-aided Doppler frequency shift estimation and compensation for UAVs. *IEEE Internet Thing J.*, 7(1):400–415, Jan. 2020.

[46] Quansheng Yuan, Yongjiang Hu, Changlong Wang, and Yongke Li. Joint 3D beamforming and trajectory design for UAV-enabled mobile relaying system. *IEEE Access*, 7:26488–26496, 2019.

[47] Lipeng Zhu, Jun Zhang, Zhenyu Xiao, Xianbin Cao, Xiang-Gen Xia, and Robert Schober. Millimeter-wave full-duplex UAV relay: Joint positioning, beamforming, and power control. *IEEE J. Select. Areas Commun.*, 38(9):2057–2073, Sep. 2020.

[48] Zhenyu Xiao, Hang Dong, Lin Bai, Dapeng Oliver Wu, and Xiang-Gen Xia. Unmanned aerial vehicle base station (UAV-BS) deployment with millimeter-wave beamforming. *IEEE Internet Thing J.*, 7(2):1336–1349, Feb. 2020.

[49] Zhiqiang Wei, Lou Zhao, Jiajia Guo, Derrick Wing Kwan Ng, and Jinhong Yuan. Multi-beam NOMA for hybrid mmWave systems. *IEEE Trans. Commun.*, 67(2):1705–1719, Feb. 2019.

[50] Xiaopeng Yuan, Yulin Hu, and Anke Schmeink. Joint design of UAV trajectory and directional antenna orientation in UAV-enabled wireless power transfer networks. *IEEE J. Select. Areas Commun.*, 2021 (Early access).

[51] Dongfang Xu, Yan Sun, Derrick Wing Kwan Ng, and Robert Schober. Multiuser MISO UAV communications in uncertain environments with no-fly zones: Robust trajectory and resource allocation design. *IEEE Trans. Commun.*, 68(5):3153–3172, May 2020.

[52] Wajiya Zafar and Bilal Muhammad Khan. Flying ad-hoc networks: Technological and social implications. *IEEE Technol. Soc. Mag.*, 35(2):67–74, Jun. 2016.

[53] Ram Ramanathan, Jason Redi, Cesar A. Santivanez, David P. Wiggins, and Stephen H. Polit. Ad hoc networking with directional antennas: a complete system solution. *IEEE J. Select. Areas Commun.*, 23(3):496–506, Mar. 2005.

[54] Hao Cai, Bo Liu, Lin Gui, and Min-You Wu. Neighbor discovery algorithms in wireless networks using directional antennas. In *Proc. IEEE Int. Conf. Commun.*, Ottawa, Canada, Jun. 2012.

[55] Bowen Zeng, Tian Song, and Jianping An. A dual-antenna collaborative communication strategy for flying ad hoc networks. *IEEE Commun. Lett.*, 23(5):913–917, May 2019.

[56] Kumari Sneha and Shivraj Singh. Free space dual antenna communication in flying ad hoc networks routing. In *Proc. Int. Conf. on Smart Syst. and Inventive Technol.*, Tirunelveli, India, Aug. 2020.

[57] Gabriel Astudillo and Michel Kadoch. Neighbor discovery and routing schemes for mobile ad-hoc networks with beamwidth adaptive smart antennas. *Telecommun. Systems*, 66(1):17–27, Jan. 2017.

[58] Wen Yang, Yue Wang, and Jian Yuan. Network construction in tactical UAV swarms with FSOC array antennas. In *Proc. IEEE Inf. Technol., Netw., Electron. and Automation Control Conf.*, Chengdu, China, Mar. 2019.

[59] Lin Chen, Yong Li, and Athanasios V. Vasilakos. On oblivious neighbor discovery in distributed wireless networks with directional antennas: Theoretical foundation and algorithm design. *IEEE/ACM Trans. Netw.*, 25(4):1982–1993, Aug. 2017.

[60] Ming Xu, Jian Xie, Yu Xia, Wei Liu, Rong Luo, Shunren Hu, and Daqing Huang. Improving traditional routing protocols for flying ad hoc networks: A survey. In *Proc. IEEE Int. Conf. on Comput. and Commun.*, Virtual, Dec. 2020.

[61] Ganbayar Gankhuyag, Anish Prasad Shrestha, and Sang-Jo Yoo. Robust and reliable predictive routing strategy for flying ad-hoc networks. *IEEE Access*, 5:643–654, 2017.

[62] Abdul Waheed, Abdul Wahid, and Munam Ali Shah. Laod: Link aware on demand routing in flying ad-hoc networks. In *Proc. IEEE Int. Conf. Commun. Workshops*, Shanghai, China, May 2019.

[63] Wissam Fawaz, Ribal Atallah, Chadi Assi, and Maurice Khabbaz. Unmanned aerial vehicles as store-carry-forward nodes for vehicular networks. *IEEE Access*, 5:23710–23718, 2017.

[64] Demeke Shumeye Lakew, Umar Sa'ad, Nhu-Ngoc Dao, Woongsoo Na, and Sungrae Cho. Routing in flying ad hoc networks: A comprehensive survey. *IEEE Commun. Surveys Tuts.*, 22(2):1071–1120, Second quarter 2020.

[65] Alexey V. Leonov. Modeling of bio-inspired algorithms anthocnet and beeadhoc for flying ad hoc networks (FANETs). In *Proc. Int. Sci.-Tech. Conf. Actual Probl. Electron. Instrum. Eng.*, volume 2, United States, Oct. 2016.

[66] Inam Ullah Khan, Ijaz Mansoor Qureshi, Muhammad Adnan Aziz, Tanweer Ahmad Cheema, and Syed Bilal Hussain Shah. Smart IoT control-based nature inspired energy efficient routing protocol for flying ad hoc network (FANET). *IEEE Access*, 8:56371–56378, 2020.

[67] Alexey V. Leonov. Application of bee colony algorithm for FANET routing. In *Proc. Int. Conf. of Young Specialists on Micro/Nanotechnologies and Electron Devices*, Erlagol, Altai, Jun. 2016.

[68] Zhenyu Xiao, Lipeng Zhu, Yanming Liu, Pengfei Yi, Rui Zhang, Xiang-Gen Xia, and Robert Schober. A survey on millimeter-wave beamforming enabled UAV communications and networking. *IEEE Commun. Surveys Tuts.*, 24(1):557–610, 1st Quart. 2022.

[69] Jiaxin Chen, Ping Chen, Qihui Wu, Yuhua Xu, Nan Qi, and Tao Fang. A game-theoretic perspective on resource management for large-scale UAV communication networks. *China Commun.*, 18(1):70–87, Jan. 2021.

[70] Haichao Wang, Jinlong Wang, Guoru Ding, Jin Chen, Yuzhou Li, and Zhu Han. Spectrum sharing planning for full-duplex UAV relaying systems with underlaid D2D communications. *IEEE J. Select. Areas Commun.*, 36(9):1986–1999, Sep. 2018.

[71] Zhiyong Feng, Lei Ji, Qixun Zhang, and Wei Li. Spectrum management for MmWave enabled UAV swarm networks: Challenges and opportunities. *IEEE Commun. Mag.*, 57(1):146–153, Jan. 2019.

[72] Samil Temel and Ilker Bekmezci. Scalability analysis of flying ad hoc networks (FANETs): A directional antenna approach. In *Proc. IEEE Int. Black Sea Conf. on Commun. and Networking*, Odessa, Ukraine, May 2014.

[73] Jaswinder Lota, Shu Sun, Theodore S. Rappaport, and Andreas Demosthenous. 5G uniform linear arrays with beamforming and spatial multiplexing at 28, 37, 64, and 71 GHz for outdoor urban communication: A two-level approach. *IEEE Trans. Veh. Technol.*, 66(11):9972–9985, Nov. 2017.

[74] Song Noh, Jiho Song, and Youngchul Sung. Fast beam search and refinement for millimeter-wave massive MIMO based on two-level phased arrays. *IEEE Trans. Wireless Commun.*, 19(10):6737–6751, Oct. 2020.

[75] Yun Hu, Jiang Zhan, Zhi Hao Jiang, Chao Yu, and Wei Hong. An orthogonal hybrid analog-digital multibeam antenna array for millimeter-wave massive MIMO systems. *IEEE Trans. Antennas Propag.*, 69(3):1393–1403, Mar. 2021.

[76] Jonathan Rodriguez. *Fundamentals of 5G Mobile Networks*. John Wiley & Sons, Ltd, United Kingdom, 2015.

[77] Kianoush Hosseini, Wei Yu, and Raviraj S. Adve. Large-scale MIMO versus network MIMO for multicell interference mitigation. *IEEE J. Sel. Topics Signal Process*, 8(5):930–941, Oct. 2014.

[78] Wonjae Shin, Mojtaba Vaezi, Byungju Lee, David J. Love, Jungwoo Lee, and H. Vincent Poor. Coordinated beamforming for multi-cell MIMO-NOMA. *IEEE Commun. Lett.*, 21(1):84–87, Jan. 2017.

[79] Qurrat-Ul-Ain Nadeem, Abla Kammoun, and Mohamed-Slim Alouini. Elevation beamforming with full dimension MIMO architectures in 5G systems: A tutorial. *IEEE Commun. Surveys Tuts.*, 21(4):3238–3273, Fourth quarter 2019.

[80] Younsun Kim, Hyoungju Ji, Juho Lee, Young-Han Nam, Boon Loong Ng, Ioannis Tzanidis, Yang Li, and Jianzhong Zhang. Full dimension MIMO (FD-MIMO): the next evolution of MIMO in LTE systems. *IEEE Wireless Commun.*, 21(2):26–33, Apr. 2014.

[81] Ping-Heng Kuo. A glance at FD-MIMO technologies for LTE. *IEEE Wireless Commun.*, 23(1):2–5, Feb. 2016.

[82] Xiao Li, Nana Qin, and Tingting Sun. Interference coordination for FD-MIMO cellular network with D2D communications underlaying. *China Commun.*, 15(12):75–88, Dec. 2018.

[83] Xiao Li, Zeyu Liu, Nana Qin, and Shi Jin. FFR based joint 3D beamforming interference coordination for multi-cell FD-MIMO downlink transmission systems. *IEEE Trans. Veh. Technol.*, 69(3):3105–3118, Mar. 2020.

[84] Mingbo Dai, Bruno Clerckx, David Gesbert, and Giuseppe Caire. A rate splitting strategy for massive MIMO with imperfect CSIT. *IEEE Trans. Wireless Commun.*, 15(7):4611–4624, Jul. 2016.

[85] Mingbo Dai and Bruno Clerckx. Multiuser millimeter wave beamforming strategies with quantized and statistical CSIT. *IEEE Trans. Wireless Commun.*, 16(11):7025–7038, Nov. 2017.

[86] Christo Kurisummoottil Thomas, Bruno Clerckx, Luca Sanguinetti, and Dirk Slock. A rate splitting strategy for mitigating intra-cell pilot contamination in massive MIMO. In *Proc. IEEE Int. Conf. Commun. Workshops*, Dublin, Ireland, Jun. 2020.

[87] Anastasios Papazafeiropoulos, Bruno Clerckx, and Tharmalingam Ratnarajah. Rate-splitting to mitigate residual transceiver hardware impairments in massive MIMO systems. *IEEE Trans. Veh. Technol.*, 66(9):8196–8211, Sep. 2017.

[88] Onur Dizdar, Yijie Mao, and Bruno Clerckx. Rate-splitting multiple access to mitigate the curse of mobility in (massive) MIMO networks. *IEEE Trans. Commun.*, 69(10):6765–6780, 2021.

[89] Michael Wu, Bei Yin, Aida Vosoughi, Christoph Studer, Joseph R. Cavallaro, and Chris Dick. Approximate matrix inversion for high-throughput data detection in the large-scale MIMO uplink. In *Proc. IEEE Int. Symp. Circuits Syst.*, Beijing, China, May 2013.

[90] Jung-Chieh Chen. A low complexity data detection algorithm for uplink multiuser massive MIMO systems. *IEEE J. Sel. Areas Commun.*, 35(8):1701–1714, Aug. 2017.

[91] Xinyu Gao, Linglong Dai, Yuting Hu, Yu Zhang, and Zhaocheng Wang. Low-complexity signal detection for large-scale MIMO in optical wireless communications. *IEEE J. Sel. Areas Commun.*, 33(9):1903–1912, Sep. 2015.

[92] Linglong Dai, Xinyu Gao, Xin Su, Shuangfeng Han, Chih-Lin I, and Zhaocheng Wang. Low-complexity soft-output signal detection based on Gauss-Seidel method for uplink multiuser large-scale MIMO systems. *IEEE Trans. Veh. Technol.*, 64(10):4839–4845, Oct. 2015.

[93] Qiufeng Liu, Hao Liu, Ying Yan, and Peng Wu. A distributed detection algorithm for uplink massive MIMO systems. In *Proc. IEEE Workshop Signal Process. Syst.*, Nanjing, China, Oct. 2019.

[94] Xueru Li, Emil Björnson, Erik G. Larsson, Shidong Zhou, and Jing Wang. Massive MIMO with multi-cell MMSE processing: exploiting all pilots for interference suppression. *EURASIP J. Wirel. Comm. Network*, 117(1), Jun. 2017.

[95] Anqi He, Lifeng Wang, Yue Chen, Kai-Kit Wong, and Maged Elkashlan. Uplink interference management in massive MIMO enabled heterogeneous cellular networks. *IEEE Wireless Commun. Lett.*, 5(5):560–563, Oct. 2016.

[96] Kazuki Maruta and Chang-Jun Ahn. Uplink interference suppression by semi-blind adaptive array with decision feedback channel estimation on multicell massive MIMO systems. *IEEE Trans. Commun.*, 66(12):6123–6134, Dec. 2018.

[97] Giovanni Interdonato, Emil Björnson, Hien Quoc Ngo, Pål Frenger, and Erik G Larsson. Ubiquitous cell-free massive MIMO communications. *EURASIP Journal on Wireless Commun. and Networking*, 2019(1):1–13, 2019.

[98] Hien Quoc Ngo, Alexei Ashikhmin, Hong Yang, Erik G. Larsson, and Thomas L. Marzetta. Cell-free massive MIMO versus small cells. *IEEE Trans. Wireless Commun.*, 16(3):1834–1850, Mar. 2017.

[99] Jiayi Zhang, Emil Bjornson, Michail Matthaiou, Derrick Wing Kwan Ng, Hong Yang, and David J. Love. Prospective multiple antenna technologies for beyond 5G. *IEEE J. Select. Areas Commun.*, 38(8):1637–1660, Aug. 2020.

[100] Hien Quoc Ngo, Alexei Ashikhmin, Hong Yang, Erik G. Larsson, and Thomas L. Marzetta. Cell-free massive MIMO: Uniformly great service for everyone. In *Proc. IEEE Int. Workshop Signal Process. Adv. Wireless Commun.*, Stockholm, Sweden, Jun. 2015.

[101] Ali Arshad Nasir, Hoang Duong Tuan, Hien Quoc Ngo, Trung Q Duong, and H Vincent Poor. Cell-free massive mimo in the short blocklength regime for urllc. *IEEE Trans. Wireless Commun.*, 20(9):5861–5871, 2021.

[102] Trang C. Mai, Hien Quoc Ngo, and Trung Q. Duong. Cell-free massive MIMO systems with multi-antenna users. In *Proc. IEEE Global Conf. Signal and Inf. Process.*, Anaheim, CA, Nov. 2018.

[103] Emil Björnson and Luca Sanguinetti. Scalable cell-free massive MIMO systems. *IEEE Trans. Commun.*, 68(7):4247–4261, Jul. 2020.

[104] Elina Nayebi, Alexei Ashikhmin, Thomas L. Marzetta, Hong Yang, and Bhaskar D. Rao. Precoding and power optimization in cell-free massive MIMO systems. *IEEE Trans. Wireless Commun.*, 16(7):4445–4459, Jul. 2017.

[105] Giovanni Interdonato, Pal Frenger, and Erik G. Larsson. Scalability aspects of cell-free massive MIMO. In *Proc. IEEE Int. Conf. Commun.*, Shanghai, China, May 2019.

[106] Sohan Gyawali, Shengjie Xu, Yi Qian, and Rose Qingyang Hu. Challenges and solutions for cellular based V2X communications. *IEEE Commun. Surveys Tuts.*, 23(1):222–255, First quarter 2021.

[107] Sherif Adeshina Busari, Muhammad Awais Khan, Kazi Mohammed Saidul Huq, Shahid Mumtaz, and Jonathan Rodriguez. Millimetre-wave massive MIMO for cellular vehicle-to-infrastructure communication. *IET Intelligent Transport Systems*, 13(7):983–990, Jun. 2019.

[108] Andrea Tassi, Malcolm Egan, Robert J. Piechocki, and Andrew Nix. Modeling and design of millimeter-wave networks for highway vehicular communication. *IEEE Trans. Veh. Technol.*, 66(12):10676–10691, Dec. 2017.

[109] Yijia Feng, Dazhi He, Yunfeng Guan, Yihang Huang, Yin Xu, and Zhiyong Chen. Beamwidth optimization for millimeter-wave V2V communication between neighbor vehicles in highway scenarios. *IEEE Access*, 9:4335–4350, 2021.

[110] Meilin Gao, Bo Ai, Yong Niu, Zhangdui Zhong, Yiru Liu, Guoyu Ma, Zhewei Zhang, and Dapeng Li. Dynamic mmWave beam tracking for high speed railway communications. In *Proc. IEEE Wireless Commun. Netw. Conf. Workshops*, Barcelona, Spain, Apr. 2018.

[111] Kui Xu, Zhexian Shen, Yurong Wang, and Xiaochen Xia. Location-aided mMIMO channel tracking and hybrid beamforming for high-speed railway communications: An angle-domain approach. *IEEE Syst. J.*, 14(1):93–104, Mar. 2020.

Index

Note: Locators in *italics* represent figures and **bold** indicate tables in the text.

For Product Safety Concerns and Information please contact our EU
representative GPSR@taylorandfrancis.com
Taylor & Francis Verlag GmbH, Kaufingerstraße 24, 80331 München, Germany

www.ingramcontent.com/pod-product-compliance
Lightning Source LLC
Chambersburg PA
CBHW080131220326
41598CB00032B/5032